Computational Methods and GIS Applications in Social Science

This textbook integrates GIS, spatial analysis, and computational methods for solving real-world problems in various policy-relevant social science applications. Thoroughly updated, the third edition showcases the best practices of computational spatial social science and includes numerous case studies with step-by-step instructions in ArcGIS Pro and open-source platform KNIME. Readers sharpen their GIS skills by applying GIS techniques in detecting crime hotspots, measuring accessibility of primary care physicians, forecasting the impact of hospital closures on local community, or siting the best locations for business.

FEATURES

- Fully updated using the latest version of ArcGIS Pro and open-source platform KNIME
- Features two brand-new chapters on agent-based modeling and big data analytics
- Provides newly automated tools for regionalization, functional region delineation, accessibility measures, planning for maximum equality in accessibility, and agent-based crime simulation
- Includes many compelling examples and real-world case studies related to social science, urban planning, and public policy
- Provides a website for downloading data and programs for implementing all case studies included in the book and the KNIME lab manual

Intended for students taking upper-level undergraduate and graduate-level courses in quantitative geography, spatial analysis, and GIS applications, as well as researchers and professionals in fields such as geography, city and regional planning, crime analysis, public health, and public administration.

Computational Methods and GIS Applications in Social Science

Third Edition

Fahui Wang and Lingbo Liu

CRC Press
Taylor & Francis Group
Boca Raton London New York

CRC Press is an imprint of the
Taylor & Francis Group, an **informa** business

Third edition published 2024
by CRC Press
2385 NW Executive Center Drive, Suite 320, Boca Raton FL 33431

and by CRC Press
4 Park Square, Milton Park, Abingdon, Oxon, OX14 4RN

CRC Press is an imprint of Taylor & Francis Group, LLC

© 2024 Fahui Wang and Lingbo Liu

First edition published by CRC Press 2006
Second edition published by CRC Press 2014

ISBN: 978-1-032-26681-7 (hbk)
ISBN: 978-1-032-27326-6 (pbk)
ISBN: 978-1-003-29230-2 (ebk)

DOI: 10.1201/9781003292302

Typeset in Times
by Apex CoVantage, LLC

Access the Support Material: https://www.routledge.com/9781032266817

Contents

PART I   GIS and Basic Spatial Analysis Tasks

PART II   Basic Computational Methods and Applications

PART III   Advanced Computational Methods and Applications

Foreword

In the last 20 years, geographic information systems have evolved to incorporate and relate to many spatial theories, models, and methods of how our cities and regions work that largely emerged during the second half of the last century. These ideas explained the spatial logic of cities and regions in terms of location theory as encapsulated in urban economics and regional science, articulating many tools and techniques that, from the 1990s on, came to be embedded in GIS. The essence of GIS, of course, is spatial representation in the first instance, but now the contemporary structure of the field is based on how representation can be consistently linked to these models and methods as well as being informed by key contemporary developments in computing. These involve big data, machine learning, parallel computing, simulation, as well as a host of new techniques and algorithms that make applications to spatial analysis ever more relevant.

Fahui Wang's book, now co-authored with Lingbo Liu, deals with how these methods and models might be clearly specified and then translated into a form that GIS can handle. The first edition in 2006 showed how desktop GIS could be extended, namely, ArcGIS, and then used to operationalize and implement many of these models. In a field which is as rapidly advancing as GIS, almost a decade later, the second edition in 2015 incorporated many new tools and techniques that had subsequently emerged. This third edition with Lingbo Liu contains all the models and methods that were subsequently dealt with as well as a series of new approaches of big data and agent-based modelling are key to the field.

However, the other feature of this text is the use of the open-source KNIME Analytics Platform that enables a user to easily implement all the case studies introduced within ArcGIS Pro. The companion lab manual to the book outlines the platform and also provides a bridge to many more advanced statistical analyses using R-based geoprocessing tools. This area of implementation is still under rapid change as new software opens up in more and more relevant ways for geospatial analysis. The KNIME workbook, in fact, allows GIS tasks to be automated in an open visual workflow-based platform. All GIS-related tasks are implemented in the ArcGIS Pro platform, and the lab manual implements the same case studies in the open-source KNIME Analytics Platform (mostly its Geospatial Analytics Extension). More advanced statistical analyses in ArcGIS Pro are conducted in R-based geoprocessing tools, with the support of R-ArcGIS Bridge. The KNIME workbook allows users to implement GIS automation tasks on the open, visual, workflow-based programming platform.

The other key feature of this third edition is the extension to several new case studies that deal with new applications, particularly to health and crime that have and continue to be major areas of research for the main author. The book is more comprehensive than previous editions, and thus it provides a very good complement to courses in regional science, urban economics, quantitative geography, even transportation planning and crime studies, and as well as, of course, to GIS, spatial data science, and geo-informatics. Part I has three chapters. The first chapter in the first part deals with data management and the basic spatial tools for representation. Spatial systems, of course, are dominated by distance, scale, size, diffusion, and decay, and in Chapter 2, all these ideas are presented as fundamental. In Chapter 3, the authors extend their treatment of space to interpolation, smoothing, and related programming techniques that are operationalized through ArcGIS. Part II has six chapters, and covers some basic computational methods that require little or no programming skills: functional region delineation, spatial accessibility, factor and cluster analysis, spatial statistics, and spatial clustering and regionalization. In Part III, five chapters cover more advanced topics: linear equations, linear and quadratic programming, Monte Carlo simulation, agent-based models, and big data analytics.

The focus on new case studies that also use various external open-source packages, as well as on new areas, such as criminology, widens the book's appeal, and it is now very much a book that is a "must-read" for any social scientist who wishes to get a rapid but thorough exposure to GIS and the desktop software that makes it work. In the pages that follow, Wang and Liu develop a very well-written operational guide to the most important GIS techniques available, and one of the great strengths of the book is that any potential user can pick it up and quickly adapt the techniques therein to their own problems. This is an important resource for computational social science, as well as for urban science itself and social physics.

Michael Batty

Centre for Advanced Spatial Analysis (CASA), University College London
London, UK

Preface

There are three notable trends in recent social science research, namely, turning *applied* (e.g., relevant to public policy), *computational* (e.g., data-driven), and *spatial* (e.g., precision public health).[1] *Geographic information systems (GIS)* have played an important role in this movement because of their capability of integrating and analyzing various datasets, in particular spatial data, and communicating the messages (e.g., via geo-visualization) intuitively, convincingly, and precisely. This book reflects the convergence of these forces.

Many of today's students in geography and other social science–related fields (e.g., sociology, anthropology, business management, city and regional planning, public administration, public health) all share the same excitement in GIS. But their interest in GIS may fade away quickly if its usage is limited to managing spatial data and mapping. In the meantime, a significant number of students complain that courses on statistics, quantitative methods, and spatial analysis are too dry and feel irrelevant to their interests in solving real-world problems. Over the years of teaching GIS, spatial analysis, and quantitative methods, I have learned the benefits of blending them together and practicing them in case studies. Students can sharpen up their GIS skills by applying GIS techniques in detecting crime hotspots, measuring accessibility of primary care physicians, forecasting the impact of hospital closures on local community, or siting the best locations for business. When students realize that they can use some of the computational methods and GIS techniques to solve a real-world problem in their own field, they become better motivated in class. In other words, technical skills in GIS or quantitative methods are best learned in the context of addressing substantive issues. Both are important for today's competitive job market.

This book is the result of my efforts of integrating *GIS* and *quantitative (computational) methods*, demonstrated in various policy-relevant *social science applications*. The applications are chosen with three objectives in mind. The first is to demonstrate the *diversity of issues* where GIS can be used to enhance the studies related to socioeconomic issues and public policy. Applications spread from popular themes in urban and regional analysis (e.g., functional areas, regional growth patterns, urban land use, and transportation) to issues related to crime and health analyses. The second is to illustrate *various computational methods*. Some methods become easy to use in automated GIS tools, and others rely on GIS to enhance the visualization of results. The third objective is to cover common tasks (e.g., distance and travel time estimation, spatial smoothing and interpolation, accessibility measures) and major issues (e.g., modifiable areal unit problem, rate estimate of rare events in small population, spatial autocorrelation) that are encountered often in *spatial analysis.*

One important feature of this book is that each chapter is *tasks driven*. Methods can be better learned in the context of solving real-world problems. While each method is illustrated in a specific area of application, it can be used to analyze a broad range of issues. Each chapter has one subject theme and introduces the method (or a group of related methods) most relevant to the theme. For example, spatial regression is used to examine the relationship between job access and homicide patterns, systems of linear equations are solved to predict urban land use patterns, linear programming is employed to solve the problem of wasteful commuting and allocate healthcare facilities, and Monte Carlo technique is illustrated in simulating urban population density patterns and commuting traffic.

Another important feature of this book is the emphasis on *implementation of methods*. All GIS-related tasks are implemented in the *ArcGIS Pro platform*, and a **companion lab manual** implements the same case studies in the *open-source KNIME Analytics Platform* (mostly its Geospatial Analytics Extension). More advanced statistical analyses in ArcGIS Pro are conducted in R-based geoprocessing tools with the support of R-ArcGIS Bridge. The KNIME workbook allows users to implement GIS automation tasks on the open, visual, and workflow-based programming platform. Most data used in the case studies are publicly accessible. Instructors and advanced readers may use the data sources and techniques discussed in the book to design their class projects or craft their own research projects. The **list of major GIS datasets and program files** summarizes all data and computer programs used in the case studies. Users can download them via the Harvard Dataverse link (https://doi.org/10.7910/DVN/4CM7V4) or my homepage (http://faculty.lsu.edu//fahui/), under "News and Blog."

The book has 14 chapters. Part I includes the first three chapters, covering some generic issues, such as an overview of data management in GIS and basic spatial analysis tools (Chapter 1), distance and travel time measurement and distance decay rule (Chapter 2), and spatial smoothing and interpolation (Chapter 3). Part II has six chapters covering some basic computational methods that require little or no programming skills: functional region delineation (Chapter 4), spatial accessibility measures (Chapter 5), function fittings (Chapter 6), factor and cluster analysis (Chapter 7), spatial statistics (Chapter 8), and spatial clustering or regionalization (Chapter 9). Part III uses five chapters to cover more advanced topics: a system of linear equations (Chapter 10), linear and quadratic programming (Chapter 11), Monte Carlo simulation (Chapter 12), agent-based model (Chapter 13), and big data analytics (Chapter 14).

Each chapter focuses on one computational method, except for the first chapter. In general, a chapter (1) begins with an introduction to the method, (2) discusses a theme to which the method is applied, and (3) uses a case study to implement the method using GIS. Some important issues, if not directly relevant to the main theme of a chapter, are discussed in appendices. Many important tasks are repeated in different projects to reinforce the learning experience. See the **quick reference for spatial analysis tasks**.

This revision includes at least 60% new materials. In addition to the *two new chapters* on agent-based crime simulation and spatiotemporal big data analytics, most noticeable *new topics* include estimating a transit time matrix (Subsection 2.2.3), estimating distance decay functions by the complementary cumulative distribution method (Subsection 2.4.2), estimating drive time and transit time matrices by Google Maps API (Appendix 2B), spatiotemporal kernel density estimation (STKDE) method (Section 3.6), refining the Dartmouth method (Section 4.3), spatialized network community detection methods (Section 4.4), inverted 2SFCA method

(Section 5.4), two-step virtual catchment area (2SVCA) method (Section 5.5), deriving the 2SFCA and i2SFCA methods (Appendix 5C), colocation analysis of two types of points (Section 8.4), GIS-automated regionalization methods (Section 9.2), the mixed-level regionalization (MLR) method (Section 9.3), quadratic programming and the maximal accessibility equality problem (MAEP) (Section 11.4), and improving measurement of wasteful commuting by Monte Carlo simulation (Appendix 12).

In addition to using more recent data or changing study areas in some case studies, 13 case studies are either significantly redesigned or new: Case Studies 2A, 2B, 3C, 4B, 8A, 8B, 9, 11B, 12A, 12B, 13, 14A, and 14B. The **list of automated tools** include:

1. *Google Maps API tool*, for estimating O–D travel time matrices via drive or transit (Chapter 2).
2. *Spatiotemporal KDE tool*, for predictive crime hotspot mapping and evaluation (Chapter 3).
3. *Generalized 2SFCA/I2SFCA Pro tool*, for implementing the two-step floating catchment area (2SFCA) and inverted-2SFCA (i2SFCA) methods (Chapter 5).
4. *2SVCA Pro tool*, for implementing the two-step virtual catchment area (2SVCA) method (Chapter 5).
5. *MLR tool*, for implementing the mixed-level regionalization method (Chapter 9).
6. *Garin–Lowry model tool*, for implementing the Garin–Lowry model (Chapter 10).
7. *WasteCommuteR tool*, for measuring wasteful commuting (Chapter 11).
8. *MinimaxR tool*, for solving the minimax location-allocation (p-center) problem (Chapter 11).
9. *MAEP tool*, for solving the maximal accessibility equality problem (MAEP) (Chapter 11).
10. *Agent-based crime simulation model* is an agent-based program for simulating spatiotemporal crime patterns (Chapter 13).
11. *XSTAR* is a free software for processing and analyzing the taxi trajectory dataset (Chapter 14).

In several case studies (e.g., 9, 12B, 13, 14A, and 14B), when computational time was reported, the *computing environment* was a laptop equipped with an i7 processor, 8 GB RAM, and 100 GB free hard disk storage space.

This book intends to mainly serve students in *geography, urban and regional planning, public policy*, and related fields. It can be used in courses such as (1) spatial analysis, (2) location analysis, (3) quantitative methods in geography, and (4) GIS applications (in business, social science, public health, crime analysis, or public administration). The book can also be useful for social scientists in general with research interests related to spatial issues. Some in *urban economics* may find the studies on urban structures and wasteful commuting relevant, and others in *business* may think the chapters on trade area analysis and accessibility measures useful. Case studies on crime patterns and crime simulation may interest *criminologists*, and those on hospital service areas and cancer data analysis may find the audience among *epidemiologists.* Some combinations of Chapters 1–9 may serve one or two upper-level undergraduate courses. Chapters 10–14 may be used for a graduate course. It is assumed that readers have some basic GIS and statistical knowledge equivalent to one introductory GIS course and one elementary statistical class. Some basic GIS knowledge (e.g., one introductory GIS course) will help readers navigate through the case studies smoothly but is not required.

It has been eight years since the release of the second version (*Quantitative Methods and Socio-Economic Applications in GIS*, 2015). I feel so fortunate to convince Dr. Lingbo Liu to join the force in preparing this edition before I could commit the undertaking. He has taken on the challenge of overhauling all case studies in ArcGIS Pro and automating many popular spatial analysis tasks. I hope that these efforts have truly lowered the technical barrier and will be appreciated by readers. He has also spearheaded completing all case studies in KNIME and drafting the companion lab manual. His diligence and dedication to the work has made the entire experience enjoyable and rewarding.

Lingbo and I are most grateful for the generous support of Drs. Haojie Zhu and Xiang Li for drafting Chapters 13 and 14 on two advanced and promising topics. Several case studies are built upon the thesis and dissertation works of my former students (in alphabetical order), including Yujie Hu, Peng Jia, Xuan Kuai, Cuiling Liu, Lingling Tian, Changzhen Wang, Guanxiong Wang, Shuai Wang, Yanqing Xu, Ling Zhang, and Haojie Zhu. We also appreciate the generous help from many of our friends and students in test-running the case studies and providing valuable feedback for this edition. They are (in alphabetical order) Weichuan Dong, Yuying Han, Yujie Hu, Hanqi Li, Shixuan Li, Chenxi Liu, Cehong Luo, Lan Mu, Tian Tian, Changzhen Wang, Yutian Zeng, and Zeliu Zheng. We also thank the joint team of Spatial Data Lab supported by Harvard's Center of Geographic Analysis and Geocomputation Center for Social Science for helping build and test KNIME workflows. They are (in alphabetical order) Weina Ding, Yan Ding, Lingjie Hu, Jiahao Wang, Yi Wang, and Yuxiang Zhong. It has been truly a rich and mutual learning experience between us and you all.

I thank Michael Batty for graciously writing the new Foreword on short notice.

Finally, I would like to thank an all-star editorial team at the CRC Press, especially acquisition editor Irma Britton, production editor Chelsea Reeves, and project manager Jayachandran Rajendiran. Thank you for guiding us through this arduous process.

NOTE

1 www.nature.com/articles/d41586-021-03819-2.

Authors

Fahui Wang is Associate Dean of the Pinkie Gordon Lane Graduate School and Cyril and Tutta Vetter Alumni Professor in the Department of Geography and Anthropology, Louisiana State University. He earned a BS in geography at Peking University, China, and an MA in economics and a PhD in city and regional planning at the Ohio State University. His research has revolved around the broad theme of spatially integrated computational social sciences, public policy, and planning in geographic information systems. He is among the top 1% most-cited researchers in geography in the world.

Lingbo Liu is Postdoctoral Fellow at the Center for Geographic Analysis, Harvard University, leading the development of Geospatial Analytics Extension for KNIME. He was a lecturer at the Department of Urban Planning, School of Urban Design, Wuhan University, from 2005 to 2022, and earned a PhD in digital urban administration and planning at Wuhan University in 2018. His research uses multi-source data and quantitative models to capture the spatiotemporal features of urban systems and provides decision support for public policy, sustainable urban planning, and design.

Major GIS Datasets and Program Files*

Case Study	Data Folder	GIS Dataset (Subfolder or Root File)	Program File	Study Area
1	BatonRouge	Census.gdb BR.gdb		East Baton Rouge Parish
2A	BatonRouge	BR_Road.gdb Hosp_Address.csv TransitNetwork Template.xml BR_GTFS	Google API Pro.tbx	East Baton Rouge Parish
2B	Florida	FL_HSA.gdb	R_ArcGIS_Tools.tbx (RegressionR)	Florida
3A	China_GX	GX.gdb		Guangxi, China
3B	BatonRouge	BR.gdb		East Baton Rouge Parish
3C	BatonRouge	BRcrime	R_ArcGIS_Tools.tbx (STKDE)	City of Baton Rouge
4A	BatonRouge	BRRoad.gdb		East Baton Rouge Parish
4B	Florida	FL_HSA.gdb FLplgnAdjAppend. csv	HSA Delineation Pro.tbx Huff Model Pro.tbx	Florida
5	BRMSA	BRMSA.gdb	Accessibility Pro.tbx	Baton Rouge MSA
6	Chicago	ChiUrArea.gdb	R_ArcGIS_Tools.tbx (RegressionR)	Chicago Urban Area
7	Beijing	BJSA.gdb bjattr.csv	R_ArcGIS_Tools.tbx (PCAandFA, BasicClustering)	Beijing, China
8A	Yunnan	YN.gdb	R_ArcGIS_Tools.tbx (SaTScanR)	Yunnan, China
8B	Jiangsu	JS.gdb		A city in Jiangsu, China
8C	Chicago	ChiCity.gdb cityattr.csv	R_ArcGIS_Tools.tbx (PCAandFA, SpatialRegressionModel)	Chicago City

(Continued)

---

* Data and program files are available for download at http://faculty.lsu.edu/fahui and https://doi. org/10.7910/DVN/4CM7V4.

(*Continued*)

Case Study	Data Folder	GIS Dataset (Subfolder or Root File)	Program File	Study Area
9	Louisiana	Louisiana.gdb	MLR Tools Pro.tbx R_ArcGIS_Tools.tbx (RegionalizationR)	Louisiana
10	SimuCity	SimuCity.gdb	Garin-Lowry.tbx	Hypothetical City
11A	Columbus	Columbus.gdb	R_ArcGIS_Tools.tbx (WasteCommuteR)	Columbus, Ohio
11B	Xiantao	XT.gdb	R_ArcGIS_Tools.tbx (MiniMaxR, MAEP)	Xiantao, Hubei, China
12A	Chicago	ZoneEffect.gdb		Chicago CMSA in Illinois
12B	BRMSA	BRMSAmc.gdb	MCSimulation.tbx	Baton Rouge MSA
13	ABMSIM	Data	ABM Crime Simulator for ArcGIS Pro.tbx	East Baton Rouge Parish
14	Shanghai	Shanghai.gdb SHtaxi.csv raw	XSTAR	Shanghai, China

Quick References for Spatial Analysis Tasks

Task	Section First Introduced (Step)	Figure Illustration	Sections (Steps) Repeated
Attribute query	Section 1.3.1 (1)	Figure 1.1a	Appendix 1, Section 2.4.1 (1), Section 2.4.2 (6), Section 4.2.1 (4), Section 4.2.2 (8,10), Section 5.6.1 (3), Section 5.6.3 (14), Section 6.5.1 (1), Section 8.3.1 (1), Section 8.3.2 (1), Section 8.8.1 (2), Section 11.2.2 (1), Section 11.5.2 (7,9), Section 12.2 (1), Section 14.2 (3)
Spatial query	Section 1.3.1 (2)	Figure 1.1b	Section 12.2 (3)
Project	Section 1.3.1 (3)	Figure 1.2	Section 2.2.1 (1)
Calculate geometry	Section 1.3.1 (4)	Figure 1.3a	Section 1.3.2 (9), Section 14.2 (1)
Calculate field	Section 1.3.1 (5)	Figure 1.3b	Section 1.3.2 (9,10), Section 1.3.2 (10), Section 2.2.2 (4,7,13), Section 2.4.1 (2), Section 2.4.2 (4)
Symbology visualization	Section 1.3.1 (6)	Figure 1.4	
Multiple ring buffer	Section 1.3.2 (7)	Figure 1.6a	
Intersect	Section 1.3.2 (8)	Figure 1.6b	Section 3.5.1 (1), Section 6.5.3 (14)
Dissolve	Section 1.3.2 (10)	Figure 1.7	Section 9.4.1 (4)
Create chart	Section 1.3.2 (11)	Figure 1.8	
Feature to point	Section 1.3.2 (12)		Section 2.2.1 (2), Section 12.2 (5)
Add spatial join	Section 1.3.2 (13)	Figure 1.9a	
Spatial join	Section 1.3.2 (13)	Figure 1.9b	Section 3.7 (1), Section 4.2.1 (1)
XY table to point	Section 2.2.1 (1)	Figure 2.2	Section 3.7 (1)
Geocode addresses	Section 2.2.1 (1)	Figure 2.3a	
Create locator	Section 2.2.1 (1)	Figure 2.3b	

(Continued)

xxiii

(Continued)

Task	Section First Introduced (Step)	Figure Illustration	Sections (Steps) Repeated
Generate near table	Section 2.2.1 (3)		Section 3.3.1 (2), Section5.6.3 (1)
Create network dataset	Section 2.2.2 (5)		Section 10.3 (1), Section 11.2.2 (3)
Build network	Section 2.2.2 (5)		Section 10.3 (1), Section 11.2.2 (3), Section 11.5.1 (1)
Define cost in network dataset	Section 2.2.2 (5)	Figure 2.4	Section 10.3 (1), Section 11.2.2 (3)
Add locations for OD cost matrix	Section 2.2.2 (6)	Figure 2.5	Section 2.2.3 (12)
GTFS to public transit data model	Section 2.2.3 (9)		
Connect public transit stops to streets	Section 2.2.3 (10)	Figure 2.7	
Build properties of a transit network dataset	Section 2.2.3 (11)	Figure 2.8	
Create network dataset from template	Section 2.2.3 (11)		
Summary statistics	Section 2.2.3 (13)	Figure 2.9	Section 3.3.1 (5), Section 3.5.1 (3), Section 4.2.1 (4)
Linear regression in RegressionR with single-variable model	Section 2.4.1 (3)	Figure 2.11	
Add Python toolbox	Appendix 2B	Figure 2A.2	
Google Maps API tool	Appendix 2B	Figure 2A.4	
Configure system variables for R path	Appendix 2C	Figure 2A.6	
Attribute join (add join)	Section 3.3.1 (3)	Figure 3.4	Section 6.5.3 (14)
Kernel density	Section 3.3.2 (9)		Section 13.4.1 (3)
IDW	Section 3.3.2 (10)		Section 11.5.1(5)
Areal weighting interpolation	Section 3.5.1 (1–3)		
Generate tessellation	Section 3.7 (1)		Section 12.2 (3), Section 14.2 (4)
Convert time field	Section 3.7 (2)		
Create space–time cube by aggregating points	Section 3.7 (3)		
Visualize space–time cube in 2D	Section 3.7 (3)		
Visualize space–time cube in 3D	Section 3.7 (4)		
Time series clustering	Section 3.7 (5)		

Task	Section First Introduced (Step)	Figure Illustration	Sections (Steps) Repeated
Emerging Hot Spot Analysis (EHSA)	Section 3.7 (6)		
STKDE tool	Section 3.7 (7)	Figure 3.14	
Create Thiessen polygons	Section 4.2.1 (1)		
Near	Section 4.2.1 (2)		Section 6.5.1 (3), Section 8.3.1 (1), Section 12.3 (5)
Build a spatial adjacency matrix	Section 4.5.1 (1)	Figure 4.9	
Automated Dartmouth method	Section 4.5.1 (2)	Figure 4.10	
Automated Huff model	Section 4.5.2 (3)	Figure 4.11	
Network community detection	Section 4.5.3 (5)	Figure 4.13	
Clone Python environment in ArcGIS Pro	Appendix 4B	Figure 4A.1	
Attribute join (join field)	Section 5.6.1 (7)		
Generalized 2SFCA/ I2SFCA Pro tool	Appendix 5B	Figure A5.1	
2SVCA Pro tool	Appendix 5B	Figure 5A.2	
Ordinary least squares (OLS) regression	Section 6.5.1 (4)		Section 8.8.2 (5)
Nonlinear regression in RegressionR	Section 6.5.1 (7)	Figure 6.4	
Pivot table	Section 6.5.2 (11)	Figure 11.6	Section 11.5.2 (10)
Principal component analysis by PCAandFA tool	Section 7.5 (1)	Figure 7.6	
Factor analysis by PCAandFA tool	Section 7.5 (2)		
Multivariate clustering	Section 7.5 (3)		
Multi-distance spatial cluster analysis (Ripley's K-function)	Section 8.2		
Mean center	Section 8.3.1 (2)		
Standard distance tool	Section 8.3.1 (3)		
SaTScanR tool	Section 8.3.2 (2)	Figure 8.3	
Colocation analysis tool	Section 8.5 (1)	Figure 8.5	
Generate spatial weights matrix	Section 8.5 (2)		
Generate network spatial weights	Section 8.5 (3)		
High–low clustering (Getis-Ord General G)	Section 8.8.1 (3)		

(Continued)

(Continued)

Task	Section First Introduced (Step)	Figure Illustration	Sections (Steps) Repeated
Spatial autocorrelation (Global Moran's I)	Section 8.8.1 (3)		
Cluster and outlier analysis (Anselin Local Moran's I)	Section 8.8.1 (4)		
Hotspot analysis (Getis-Ord Gi^*)	Section 8.8.1 (4)		
Spatial Regression Model	Section 8.8.2 (6)		
Geographically weighted regression (GWR)	Section 8.8.2 (7)	Figure 8.8	
Spatially constrained multivariate clustering	Section 9.4.1 (1)		
RegionalizationR tool	Section 9.4.1 (2)	Figure 9.4	
One-step regionalization in MLR	Section 9.4.2 (5)	Figure 9.6	
All-in-one MLR	Section 9.4.2 (7)	Figure 9.10	
Garin–Lowry model tool	Section 10.3 (4)	Figure 10.4	
WasteCommuteR	Section 11.2.3	Figure 11.2	
Define candidate facilities in location-allocation analysis	Section 11.5.2 (6)	Figure 11.5	
MiniMaxR tool	Section 11.5.2 (11)	Figure 11.7	
Append	Section 11.5.3 (12)		
MAEP tool	Section 11.5.3 (14)	Figure 11.8	
Pairwise intersect	Section 12.2 (1)		
Create random points	Section 12.2 (2)		
ProbSampling tool	Section 12.4 (2)	Figure 12.7 (a)	
RepeatRow	Section 12.4 (5)		
Agent-based crime simulation model	Section 13.4.1 (2)	Figure 13.6 and Figure 13.7	
Reclassify	Section 13.4.1 (4)		
Raster to polygon	Section 13.4.1 (5)		
Sort	Section 14.2 (2)		
Calculate geometry attribute	Section 14.2 (4)		
XSATR preprocessing data	Section 14.4.1 (1–4)	Figure 14.8	
XSATR OD analysis	Section 14.4.2 (5–7)	Figure 14.9	
XSATR destination analysis	Section 14.4.3 (8–9)	Figure 14.10	

Part I

GIS and Basic Spatial Analysis Tasks

1 Getting Started with ArcGIS

Data Management and Basic Spatial Analysis Tools

A *geographic information system (GIS)* is a computer system that captures, stores, manipulates, queries, analyzes, and displays geographically referenced data. Among the diverse set of tasks a GIS can do, mapping remains the primary function. The first objective of this chapter is to demonstrate how GIS is used as a computerized mapping tool. The key skill involved in this task is the management of spatial and aspatial (attribute) data and the linkage between them. However, GIS goes beyond mapping and has been increasingly used in various spatial analysis tasks, as GIS software has become more capable and also friendlier to use for these tasks. The second objective of this chapter is to introduce some basic spatial analysis tools in GIS.

Given its wide usage in education, business, and governmental agencies, ArcGIS has been chosen as the major software platform to implement GIS tasks in this book. Unless otherwise noted, studies in this book are based on ArcGIS Pro 3.0. All chapters are structured similarly, beginning with conceptual discussions to lay out the foundation for methods, followed by case studies to acquaint readers with related techniques.

This chapter serves as a quick warm-up to prepare readers for more advanced spatial analysis and computational methods in later chapters. Section 1.1 offers a quick tour of spatial and attribute data management in ArcGIS. Section 1.2 surveys basic spatial analysis tools in ArcGIS, including spatial queries, spatial joins, and map overlays. Section 1.3 uses a case study to illustrate the typical process of GIS-based mapping and practice some of the basic spatial analysis tools in ArcGIS. Section 1.4 concludes with a brief summary. In addition, Appendix 1 demonstrates how to use GIS-based spatial analysis tools to identify spatial relationships, such as polygon adjacency. *Polygon adjacency* defines spatial weights often needed in spatial statistical studies, such as spatial cluster and spatial regression analyses in Chapter 8 and regionalization in Chapters 4 and 9.

1.1 SPATIAL AND ATTRIBUTE DATA MANAGEMENT IN ArcGIS

As ArcGIS Pro is the primary software platform for this book, it is helpful to have a brief overview of its history and major modules and functions. *ArcGIS* was released in 2001 by the Environmental Systems Research Institute Inc. (ESRI) with a full-featured graphic user interface (GUI) to replace the older versions of *ArcInfo* that were

DOI: 10.1201/9781003292302-2

based on a command line interface. Prior to the release of ArcGIS, ESRI had an entry-level GIS package with a GUI called *ArcView*, but ArcView was not as powerful as ArcInfo. Both earlier versions of the book (Wang, 2006, 2015) used ArcGIS. *ArcGIS Pro*, first released in 2015, is the latest desktop GIS software from the ESRI.

Work in ArcGIS Pro is organized in a *Project*. Users create a project for work in a specific geographic area with related tasks. When a user creates a new project or opens an existing project, a *Contents* pane displays various map layers, and a *Catalog* pane shows items available for the project such as Maps, Toolboxes, Databases, Styles, Folders, and Locators. *Maps* are views in either 2D (regular map) or 3D (called "scene"). *Toolboxes* contain various task-specific tools for data analysis, such as data conversion, data management, spatial analysis, and spatial statistics. *Databases* include a home (default) geodatabase used for the project and others specified by directing to particular *Folders*. *Styles* organize map symbols for point markers, line styles, polygon shades and colors, etc. *Locators* are used for geocoding addresses.

1.1.1 MAP PROJECTIONS AND SPATIAL DATA MODELS

GIS differs from other information systems because of its unique capability of managing geographically referenced or spatial (location) data. Understanding the management of spatial data in GIS requires knowledge of the data's *geographic coordinate system* with longitude and latitude values and its representations on various *plane coordinate systems* with x- and y-coordinates (map layers). Transforming the Earth's spherical surface to a plane surface is a process referred to as *projection*. The two map projections commonly used in the United States are *Universal Transverse Mercator (UTM)* and *State Plane Coordinate System (SPCS)*. Strictly speaking, the SPCS is not a projection; instead, it is a set of more than 100 geographic zones or coordinate systems for specific regions of the United States. To achieve minimal distortion, north–south oriented states or regions use a Transverse Mercator projection, and east–west oriented states or regions use a *Lambert Conformal Conic projection*. Some states (e.g., Alaska, New York) use more than one projection. For more details, one may refer to an ArcGIS online documentation, "Understanding Map Projections."

For a new project, ArcGIS Pro displays two map layers by default: World Topographic Map and World Hillshade. Users can turn them on, off, or remove. In the Contents pane, right-click a map layer > Properties > Source > Spatial Reference; one can check the projected coordinate system for a layer (e.g., the World Topographic Map uses the WGS 1984 Web Mercator). ArcGIS automatically converts layers of different coordinate systems to the coordinate system of the first layer in map displays. It is good practice to use the same projection for all data layers in one project.

Traditionally, a GIS uses either the vector or the raster data model to manage spatial data. A *vector GIS* uses geographically referenced points to construct spatial features of points, lines, and areas, and a *raster GIS* uses grid cells in rows and columns to represent spatial features. The vector data structure is used in most socioeconomic applications, and here in most case studies in this book. The raster data structure is simple and widely used in environmental studies, physical sciences, and natural resource management. Most commercial GIS software can convert from vector to raster data, or vice versa. In ArcGIS Pro, the tools are available under Toolboxes > Conversion Tools.

Projection-related tasks are conducted under Toolboxes > Data Management Tools > Projections and Transformations. If the spatial data are vector-based, proceed to choose Project to transform the coordinate systems. The tool provides the option to define a new coordinate system or import the coordinate system from an existing geodatabase. If the spatial data are raster-based, choose Raster > Project Raster. This tool provides more options to select the resampling technique, output cell size, and geographic transformation. Step 3 in Section 1.3 illustrates the use of transforming a spatial dataset in a geographic coordinate system to a projected coordinate system.

Earlier versions of ESRI's GIS software used the *coverage* data model. Later, *shapefile*s were developed for the ArcView package. Since the release of ArcGIS, the *geodatabase* model has become available and represents the current trend of object-oriented data models. The object-oriented data model stores the geometries of objects (spatial data) with the attribute data, whereas the traditional coverage or shapefile model stores spatial and attribute data separately. Spatial data used for case studies in this book are provided entirely in geodatabases. Spatial and attribute data in socioeconomic analysis often come from different sources, and a typical task is to join them together in GIS for mapping and analysis. This involves attribute data management as discussed in the following.

1.1.2 ATTRIBUTE DATA MANAGEMENT AND ATTRIBUTE JOIN

A GIS includes both spatial and attribute data. Spatial data capture the geometry of map features, and attribute data describe the characteristics of map features. Attribute data are usually stored as a tabular file or table. ArcGIS reads several types of table formats. Shapefile attribute tables use the *dBase* format, and geodatabase tables use relational database format. ArcGIS can also read several text formats, including comma-delimited and tab-delimited text files.

Some of the basic tasks of data management are done in the Catalog pane. For example, to create a new table, under Databases, right-click a geodatabase where a new table will be placed > New > Table. Follow the steps to name the new table, define its fields, and save it. Similarly, an existing table can be exported, deleted, copied, or renamed.

Functions associated with modifying a table (e.g., adding a new field, deleting or updating an existing field) are performed in the Contents pane. For a stand-alone table, right-click it > Open; for a map layer, right-click it > Attribute Table to open it. For example, to add a field, in the window with the open table, click Add next to Field (the upper-left corner) > at the very bottom in the list of fields, input the new field name, define its data type and format, and confirm the changes. To delete a field, on the open table, right-click the field name and choose Delete. For updating values for a field in a table, on the open table, right-click the field and choose Calculate Field to access the tool. In addition, basic statistics for a field can be obtained by right-clicking the field and choosing Statistics. Steps 4 and 5 in Section 1.3 demonstrate how to create a new field and update its values.

In GIS, an *attribute join* is often used to link information in two tables based on a common field (key). The table can be an attribute table associated with a particular spatial layer or a stand-alone table. In an attribute join, the names of the fields used to link the two tables do not need to be identical in the two tables to be combined, but the data types must match. There are various relationships between tables when linking them: one-to-one, many-to-one, one-to-many, and many-to-many. For either one-to-one or many-to-one relationship, a *join* is used to combine two tables in ArcGIS. However, when the relationship is one-to-many or many-to-many, a join cannot be performed. ArcGIS uses a *relate* to link two tables while keeping the two tables separated. In a relate, one or more records are selected in one table, and the associated records are selected in another table. The case studies in this book mostly rely on *join*.

A join or relate is also accessed under the Contents pane. Right-click the spatial dataset or the table that is to become the *target table*, and choose Joins and Relates > Add Join (or Add Relate). For Join, in the Add Join dialog window, define (1) Input Table (the chosen target table is shown), (2) Input Join Field in the target table, (3) Join Table (the *source table*), and (4) Join Table Field in the source table. It is important to understand the different roles of the target and source tables in the process. As a result, the target table is expanded by adding the fields from the source table, and the source table remains intact.

A join is *temporary* in the sense that no new data are created, and the join relationship is lost once the project is exited without being saved. To preserve the resulting layer (or the combined table), right-click it under the Contents pane > Data > Export Features (or Export Table) to save it in a new feature (table). This can be termed "*permanent attribute join.*" Alternatively, select Analysis from the top menu > Tools > in the Geoprocessing window, search and choose the tool "Join Field" to implement a permanent attribute join.

Once the attribute information is joined to a spatial layer, mapping the data becomes straightforward (see step 6 in Subsection 1.3.1).

1.2 SPATIAL ANALYSIS TOOLS IN ARCGIS: QUERIES, SPATIAL JOINS, AND MAP OVERLAYS

Many spatial analysis tasks utilize information on how spatial features are related to each other in terms of location. Spatial operations such as "queries," "spatial joins," and "map overlays" provide basic tools in conducting such tasks.

Queries include attribute (aspatial) queries and spatial queries. An *attribute query* uses information in an attribute table to find attribute information (in the same table) or spatial information (features in a spatial data layer). In ArcGIS Pro, attribute queries are accessed either (1) under Map from the main menu bar > choose Select by Attributes, or (2) in an open table, click Select by Attributes. Both allow users to select spatial features based on a SQL query expression using attributes (or simply clicking attribute records from a table).

Compared to other information systems, a unique feature of GIS is its capacity to perform *spatial queries*, which find information based on locational relationships between features from multiple layers. Similarly, under Map from the main menu bar

> choose Select by Location. Use the dialog window to define the search for features in one layer based on their spatial relationships with features in another layer. Spatial relationships are defined by various operators, including "intersect," "within a distance," "contain," "within," etc.

In Section 1.3, step 1 demonstrates the use of attribute query, and step 2 is an example of spatial query. The selected features in a spatial dataset (or selected records in a table) either by an attribute query or a spatial query can be exported to a new dataset (or a new table) by following the steps explained in Subsection 1.1.2.

While an attribute join utilizes a common field between two tables, a *spatial join* uses the locations of spatial features, such as overlap or proximity between two layers. Similar to the attribute join, the *source layer* and *target layer* play different roles in a spatial join. Based on the join, information from the source layer is processed and transferred to the features in the target layer, and the result is saved in a new spatial dataset. If one object in the source layer corresponds to one or multiple objects in the target layer, it is a *simple join*. For example, by spatially joining a polygon layer of counties (source layer) with a point layer of school locations (target layer), attributes of each county (e.g., FIPS code, name, administrator) are assigned (populated) to schools that fall within the county boundary. If multiple objects in the source layer correspond to one object in the target layer, two operations may be performed: "summarized join" and "distance join." A *summarized join* summarizes the numeric attributes of features in the source layer (e.g., average, sum, minimum, maximum, standard derivation, or variance) and adds the information to the target layer. A *distance join* identifies the closest object (out of many) in the source layer from the matching object in the target layer and transfers all attributes of this nearest object (plus a distance field showing how far they are apart) to the target layer. For example, one may spatially join a point layer of geocoded crime incidents (source layer) with a polygon layer of census tracts (target layer) and generate aggregated crime counts in those census tracts, and one may also use a spatial join to join a point layer of transit stations (source layer) on a network to a point layer of student residences (target layer) and identify the nearest station for each student. The former uses a summarized join, and the latter a distance join.

There are a variety of spatial joins between different spatial features (Price, 2020, pp. 309–314). Table 1.1 summarizes all types of spatial joins in ArcGIS. Spatial joins are accessed in ArcGIS Pro similar to attribute joins: in the Contents pane, right-click the source layer > Joins and Relates > Add Spatial Join. Alternatively, select Analysis from the top menu > Tools, then search and choose the tool "Spatial Join" or "Add Spatial Join" to activate the tool dialog. In Subsection 1.3.2, step 13 illustrates the use of a spatial join, in particular a case of a summarized join.

Map overlays may be broadly defined as any spatial analysis involving modifying features from different layers. The following reviews some of the most used map overlay tools: Clip, Erase, Intersect, Union, Buffer, and Multiple Ring Buffer. A *Clip* truncates the features from one layer using the outline of another. An *Erase* trims away the features from the input layer that fall inside the erase layer. An *Intersect* overlays two layers and keeps only the areas that are common to both. A *Union* also overlays two layers but keeps all the areas from both layers. A *Buffer* creates areas by extending outward from point, line, or polygon features over a

TABLE 1.1

Types of Spatial Joins in ArcGIS

Source Layer (S)	Target Layer (D)	Simple Join	Distance Join	Summarized Join
Point	Point		For each point in T, find its closest point in S, and transfer attributes of that closest point to T.	For each point in T, find all the points in S closer to this point than to any other point in T, and transfer the points' summarized attributes to T.
Line	Point		For each point in T, find the closest line in S, and transfer attributes of that line to T.	For each point in T, find all lines in S that intersect it, and transfer the lines' summarized attributes to T.
Polygon	Point	For each point in T, find the polygon in S containing the point, and transfer the polygon's attributes to T.	For each point in T, find its closest polygon in S, and transfer attributes of that polygon to T.	
Point	Line		For each line in T, find its closest point in S, and transfer the point's attributes to T.	For each line in T, find all the points in S that either intersect or lie closest to it, and transfer the points' summarized attributes to T.
Line	Line	For each line in T, find the line in S that it is part of, and transfer the attributes of source line to T.		For each line in T, find all the lines in S that intersect it, and transfer the summarized attributes of intersected lines to T.
Polygon	Line	For each line in T, find the polygon in S that it falls completely inside of, and transfer the polygon's attributes to T.	For each line in T, find the polygon in S that it is closest to, and transfer the polygon's attributes to T.	For each line in T, find all the polygons in S crossed by it, and transfer the polygons' summarized attributes to T.

TABLE 1.1 (*Continued*)
Types of Spatial Joins in ArcGIS

Source Layer (S)	Target Layer (D)	Simple Join	Distance Join	Summarized Join
Point	Polygon		For each polygon in *T*, find its closest point in *S*, and transfer the point's attributes to *T*.	For each polygon in *T*, find all the points in *S* that fall inside it, and transfer the points' summarized attributes to *T*.
Line	Polygon		For each polygon in *T*, find its closest line in *S*, and transfer the line's attributes to *T*.	For each polygon in *T*, find all the lines in *S* that intersect it, and transfer the lines' summarized attributes to *T*.
Polygon	Polygon	For each polygon in *T*, find the polygon in *S* that it falls completely inside of, and transfer the attributes of source polygon to *T*.		For each polygon in *T*, find all the polygons in *S* that intersect it, and transfer the summarized attributes of intersected polygons to *T*.

specified distance. A *Multiple Ring Buffer* generates buffer features based on a set of distances. Similarly, in ArcGIS Pro, select Analysis from the top menu > Tools > search and choose any of the above map overlay tools in the Geoprocessing window. In Subsection 1.3.2, step 7 uses the Multiple Ring Buffer tool, and step 8 uses the Intersect tool.

One may notice the similarity among spatial queries, spatial joins, and map overlays. Indeed, many spatial analysis tasks may be accomplished by any one of the three. Table 1.2 summarizes the differences between them. A spatial query only finds and displays the information on screen and does not create new datasets (unless one chooses to export the selected records or features). A spatial join always saves the result in a new layer. There is an important difference between spatial joins and map overlays. A spatial join merely identifies the location relationship between spatial features of input layers and does not change existing spatial features or create new features. In the process of a map overlay, some of the input features are split, merged, or dropped in generating a new layer. In general, a map overlay operation takes more computation time than a spatial join, and a spatial join takes more time than a spatial query.

TABLE 1.2

Comparison of Spatial Query, Spatial Join, and Map Overlay

Basic Spatial Analysis Tools	Function	Whether a New Layer Is Created	Whether New Features Are Created	Computation Time
Spatial query	Finds information based on location relationship between features from different layers and displays on-screen	No (unless the selected features are exported to a new dataset)	No	Least
Spatial join	Identifies location relationship between features from different layers and transfers the attributes to target layer	Yes	No	Between
Map overlay	Overlays layers to create new features and saves the result in a new layer	Yes	Yes (splitting, merging, or deleting features to create new)	Most

1.3   CASE STUDY 1: MAPPING AND ANALYZING POPULATION DENSITY PATTERN IN BATON ROUGE, LOUISIANA

For readers with little previous exposure to GIS, nothing demonstrates the value of GIS and eases the fear of GIS complexity better than the experience of mapping data from the public domain with just a few clicks. As mentioned earlier, GIS goes far beyond mapping, and the main benefits lie in its spatial analysis capacity. The first case study illustrates how spatial and attribute information are managed and linked in a GIS, and how the information is used for mapping and analysis. Part 1 in Subsection 1.3.1 uses the 2020 US Census data to map the population density pattern across census tracts in a county, and Part 2 in Subsection 1.3.2 utilizes some basic spatial analysis tools to analyze the distance decay pattern of population density across concentric rings from the city center. The procedures are designed to incorporate most functions introduced in Sections 1.1 and 1.2.

1.3.1   MAPPING THE POPULATION DENSITY PATTERN

A GIS project begins with data acquisition and often uses existing data. For socioeconomic applications in the United States, the *Topologically Integrated Geographical Encoding and Referencing (TIGER)* files from the Census Bureau are the major source for spatial data, and the 2020 decennial census data from the same agency are the major source for attribute data. This case study uses the 2020 Census P.L 94–171 Redistricting File for population data and the corresponding TIGER/Line file for spatial data.[1] For historical census data in GIS format, visit the National Historical Geographic Information System (NHGIS) website at www.nhgis.org.

For this case study, the study area is East Baton Rouge Parish of Louisiana, where the state capital city of Baton Rouge resides. *Parish* is a county-equivalent unit in Louisiana. The 2020 Census data for the study area is processed, and the following datasets are provided under the data folder `BatonRouge`:

1. Features `State`, `County`, and `Tract` in geodatabase `Census.gdb` cover the state of Louisiana at the state, county, and census tract level, respectively.
2. Feature `BRBLK` in geodatabase `BR.gdb` includes census blocks in Baton Rouge, with fields `Popu2020` for population and `Area` for area in square meters.
3. Feature `BRCenter` in geodatabase `BR.gdb` represents the city center of Baton Rouge.

Throughout this book, all computer file names, field names, and computer commands are in the `Courier New` font. The following is a step-by-step guide for the project.

Step 0. Getting Started with ArcGIS Pro: Activate ArcGIS Pro and create a new project by clicking on ▨ Map under Blank Template.[2] Input `Case1` as the Project Name, and click the folder button ▨ to set its location, uncheck the box "Create a new folder for this project," and press OK to create the new project. All subsequent outputs in the project will be saved in geodatabase `Case1.gdb`. Note that the World Topographic Map and World Hillshade are automatically added to the Contents pane. One may click the Map tab and use ▨ Basemap to change the default Basemap to other world maps, such as satellite imagery.

Under the Map tab (alternatively, right-click Map in the Contents pane), click ▨ Add Data to add all three features in geodatabase `Census.gdb` and the feature `BRCenter` in geodatabase `BR.gdb` to the project.

In the Contents pane, selecting a feature will bring out the corresponding context tab. Choose one of the files, click the Feature Layer tab, use the slider next to ▨ Transparency to adjust its transparency, or click ▨ Symbology button to bring up the symbology sidebar to change the symbols. Users can also revoke the Symbology pane by right-clicking on the feature and choosing *symbology*.

In many cases, GIS analysts need to delineate a study area contained in a larger region. The task begins with extracting the census tracts in East Baton Rouge Parish.

Step 1. Using an Attribute Query to Extract the Parish Boundary: From the Map tab, click ▨ Select By Attributes. In the dialog window shown in Figure 1.1a, make

(a)

Select By Attributes   ?   ×

Input Rows
County   ∨   📁

Selection Type
New selection   ∨

Expression
📁 Load   💾 Save   ✕ Remove

🔢 ✓   SQL ⬤ ⚙

Where   NAMELSAD   ▾   is equal ▾   East Baton Rou ▾   ✕

+ Add Clause

☐ Invert Where Clause

Apply   OK

(b)

Select By Location   ?   ×

Input Features ⊙
Tract   ∨   📁

∨   📁

Relationship
Within Clementini   ∨

Selecting Features
EBRP   ∨   📁 ✏ ▾

Search Distance
Unknown   ∨

Selection Type
New selection   ∨

☐ Invert Spatial Relationship

Apply   OK

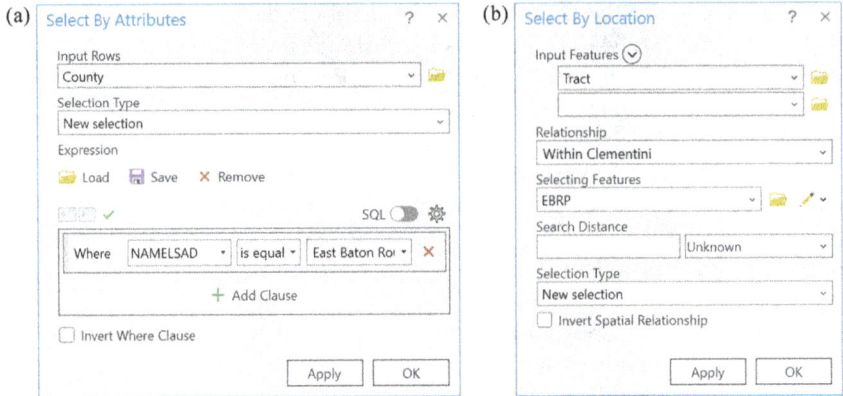

FIGURE 1.1 Dialog windows for (a) attribute query and (b) spatial query.

sure that the Input Rows is County and Selection Type is New selection; in the Expression box, click ✚ Add Clause and select values from the drop-down menu to set the expression as NAMELSAD is equal East Baton Rouge Parish, as shown, and click Apply (or OK). The parish is selected and highlighted on the map. The selection is based on the attributes of a GIS dataset, and thus it is an *attribute query*.

In the layers window under Contents, right-click the layer County > Data > Export Features. In the dialog, make sure County in Input Features and Case1. gdb in Output Location and input EBRP for Output Name, and click OK. The exported data is added as a new layer in the Map pane.

Step 2. Using a Spatial Query to Extract Census Tracts in the Parish: From the Map tab, click 🔲 Select By Location. In the dialog window shown in Figure 1.1b, choose Tract under Input Features, choose Within Clementini under Relationship, choose EBRP under Selecting Features, and click OK. All census tracts having their boundaries in the parish are selected.[3] The selection is based on the spatial relationship between two layers, and thus it is a *spatial query*.

Similar to step 1, return to the layers window, and right-click the layer Tract > Data > Export Features. Export the selected features to a new feature EBRPtrt in geodatabase Case1.gdb. EBRPtrt is added to the map.

Step 3. Transforming the Spatial Dataset to a Projected Coordinate System: In the Contents pane, right-click the layer EBRPtrt > Properties. In the Layer Properties window, click Source in the left window, and then scroll down in the right window to unfold the information in Spatial Reference. It shows the Extent of the feature is in decimal degrees (dd), and its Geographic Coordinate System is NAD 1983 (GCS_ North_American_1983). Repeat on other layers (Tract, County, and State) to confirm that they are all in the same geographic coordinate system, as are all TIGER data. Now, check the layer BRCenter. It has the Extent measured in meters and a Projected Coordinate System (NAD_1983_UTM_Zone_15N).[4]

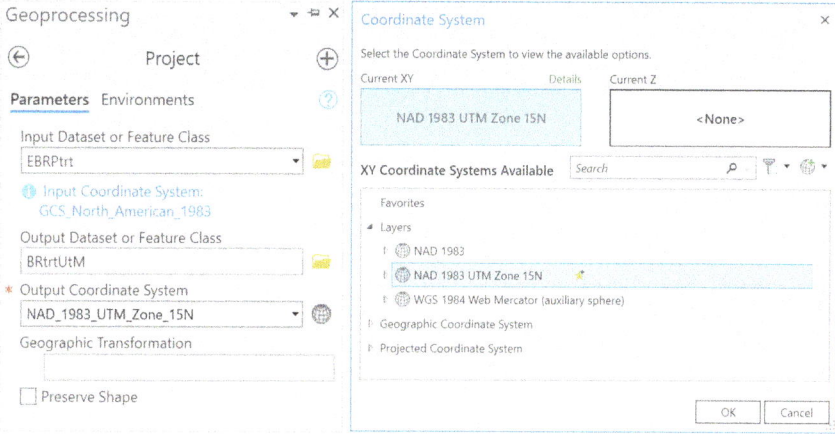

FIGURE 1.2 Dialog window for projecting a spatial dataset.

For reasons discussed previously, we need to transform the census tracts dataset EBRPTrt in a geographic coordinate system to one in a projected coordinate system. For convenience, we use the projection system that is predefined in the layer BRCenter.

Open Toolbox by clicking the icon 🧰 Tools in the Analysis tab, then the Geoprocessing sidebar comes up. In Toolboxes, go to Data Management Tools > Projections and Transformations > click the tool Project. In the Project dialog (as shown on the left of Figure 1.2), choose EBRPtrt for Input Dataset or Feature Class; name the Output Dataset or Feature Class as BRTrtUtm (make sure that it is in geodatabase Case1.gdb). Under Output Coordinate System, choose BRcenter, then the coordinate system of BRcenter is automatically chosen and shown in the input box as NAD _ 1983 _ UTM _ Zone _ 15N,[5] and click Run to execute the Project tool. Feature BRTrtUtm is added to the map. Check and verify its coordinate system (as shown on the right of Figure 1.2).

Step 4. Calculating Area in Census Tracts: In the Contents pane, right-click the layer BRTrtUtm > ▦ Attribute Table. In the table, click the graphic button ▦ Add Field. In the Fields tab, enter Area for Field Name, and choose Double for Data Type, set Numeric for Number Format, and close Fields tab. In the dialog of Unsaved Changes Found, click Yes to save all changes. A new field Area is now added to the table as the last column.

In the open table, scroll to the right side of the table, right-click the field Area > choose Calculate Geometry. In the Calculate Geometry dialog (Figure 1.3a), choose Area for Property and Square meters for Area Unit,[6] choose BRTrtUtm for Coordinates System, and click OK to update areas.

Step 5. Calculating Population Density in Census Tracts: Still on the open attribute table of BRTrtUtm, follow the previous procedure to add a new field PopuDen (choose Float for Data Type). Right-click the field PopuDen > choose Calculate Field. In the dialog (Figure 1.3b), under "PopuDen =," input the

(a) **Calculate Geometry** ? ×

ⓘ This tool modifies the input data. ×

Input Features
BRTrtUtm ▾ 📁

Geometry Attributes
Field (Existing or New) ⌄ | Property
Area ▾ | Area ▾
▾ | ▾

Area Unit
Square meters ▾

Coordinate System
NAD_1983_UTM_Zone_15N ▾ 🌐

Current Map [Map1]
Tract
State
County
BRTrtUtm
EBRPtrt
EBRP

OK

(b) **Calculate Field** ? ×

ⓘ This tool modifies the input data. ×

Input Table
BRTrtUtm ▾ 📁

Field Name (Existing or New)
PopuDen ▾

Expression Type
Python 3 ▾

ⓘ Expression

Fields ▾ | Helpers ▾
DP0230001 | .as_integer_ratio()
DP0230002 | .capitalize()
Area | .conjugate()

Insert Values ▾ | * / + - =

PopuDen =
!POP100! * 1000000 / !Area!

Code Block

Enable Undo ⬤ | Apply | OK

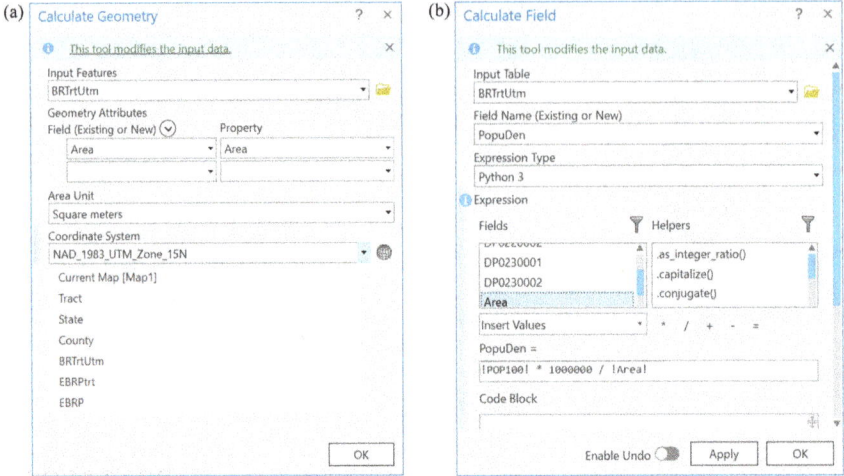

FIGURE 1.3 Dialog windows for (a) Calculate Geometry and (b) Calculate Field.

formula `!POP100! * 1000000/!Area!`, and click OK to calculate the field.[7] The area unit is square meters, and the formula converts area unit in square meters to square kilometers, and thus the unit of population density is persons per square kilometer. In subsequent case studies in this book, the exclamation points (!!) for each variable may be dropped in a formula for simplicity, for example, the previous formula may be written in instructions as `PopuDen= POP100*1000000/ Area`.

It is often desirable to obtain some basic statistics for a field. Still on the open table, right-click the field `PopuDen` > choose Statistics. The result includes count, minimum, maximum, sum, mean, standard deviation, and a frequency distribution. Close the table window.

Step 6. Mapping Population Density across Census Tracts: In the Contents pane, right-click the layer `BRTrtUtm` > Symbology. In the dialog (Figure 1.4a), under the Primary symbology, choose Graduated Colors; select `PopuDen` for Field, and keep other default settings. The map is automatically refreshed.

For showing the labels in integers, click the graphic icon 📧 Advanced symbology options on the top right of the dialog. In the dialog (Figure 1.4b), under Format labels, for Numeric, Rounding, set 0 in Decimal places, and close the Advanced symbology options sub-window. Experiment with different classification methods, number of classes, and color schemes, and close the Primary symbology window.

From the main ribbon, choose Insert tab > 🖼 New Layout > ISO-Portrait > A4 to create a new layout. Click Insert again > 🖼 Map Frame > choose the symbolized map, and drag your mouse to draw the map frame in the layout. To remove unrelated layers, uncheck them under the Map Frame of the Contents pane. Click Layout tab >

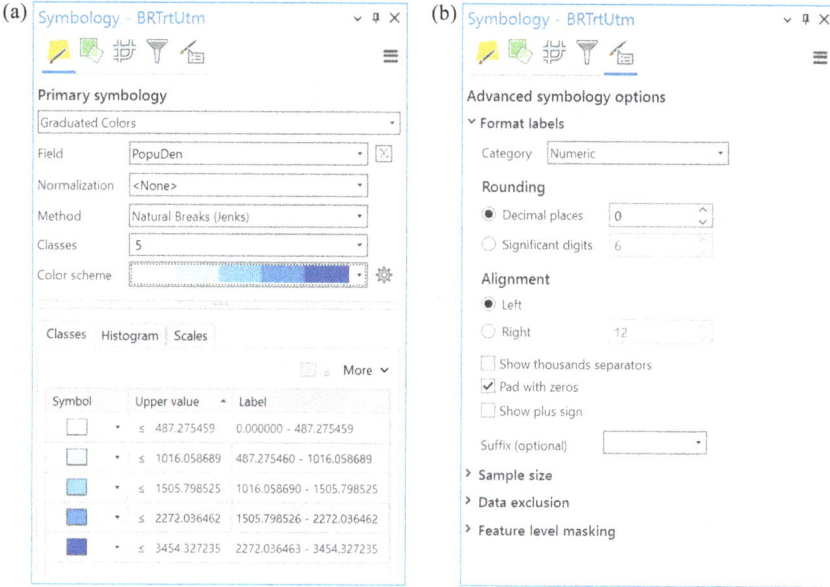

FIGURE 1.4 Dialog window for (a) Primary symbology and (b) Advanced symbology options.

Activate to adjust the map scale in the layout; if needed, manually revise the number under the bottom left of the layout to adjust the scale. Users can also drag the map to an appropriate location. Click Layout tab > Close Activation. Click the Insert tab again and choose ⬚ Legend, ▱ Scale Bar, and ↑N North Arrow to add map elements. You can also adjust the format and design of them by right-clicking the elements and choosing Properties. When you are satisfied with the map, go to the main ribbon > Share tab > ⬚ Export Layout to save the map product in a chosen format. The population density map shown in Figure 1.5 uses a Natural Breaks classification for density classes.

Save the project with the final map by going to the main ribbon > Project > Save.

1.3.2 ANALYZING THE POPULATION DENSITY PATTERN ACROSS CONCENTRIC RINGS

A general observation from the population density map in Figure 1.5 is that densities near the city center are higher than in more remote areas. This is commonly referred to as the distance decay pattern for urban population densities. The second part of the case study analyzes such a pattern while illustrating several spatial analysis tools, mainly, spatial join and map overlay, discussed in Section 1.2.

Step 7. Generating Concentric Rings around the City Center: In the main ribbon, choose Analysis tab > ⬚ Tools. In the Geoprocessing pane, choose Toolboxes tab > Analysis Tools > Proximity > Multiple Ring Buffer. You may also search this tool in

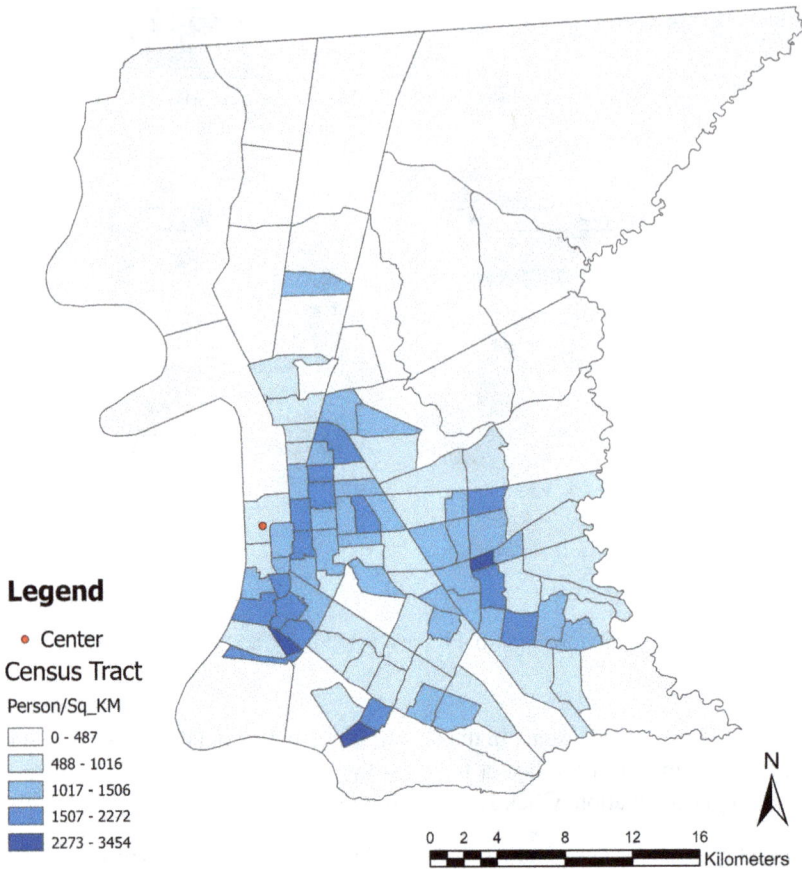

FIGURE 1.5 Population density pattern in Baton Rouge 2020.

the Geoprocessing pane and click it. In the dialog (Figure 1.6a), choose BRCenter for Input Features and name Output Feature Class as Rings under Case1.gdb; input values 2,000, 4,000, and so on till 26,000 with an increment of 2,000 under

Distances (using the ⊕ sign to add one by one); choose Meters for Distance Unit; keep distance for Buffer Distance Field Name, Non-overlapping(rings) for Dissolve Option, and Planar for Method; and click Run. The output feature has 13 concentric rings with an increment of 2 kilometers. While the traditional Buffer tool generates a buffer area around a feature (point, line, or polygon), the Multiple Ring Buffer tool creates multiple buffers at defined distances. Both are popular map overlay tools.

Step 8. Overlaying the Census Tracts Layer and the Rings Layer: Add the census block group layer BRBKG to the map. Again, in the Geoprocessing sidebar, search and activate the tool Intersect. In the dialog (Figure 1.6b), select Rings and

(a)

Geoprocessing   ▾ ⊟ ✕

◉   Multiple Ring Buffer   ⊕

Parameters Environments   ⑦

Input Features
| BRcenter   ▾ | 📁 | ✎ ▾ |

Output Feature class
| Rings   📁 |

ⓘ Distances
	2000
	4000
	6000
	8000
	10000
	12000
	14000
	16000
	18000
	20000
	22000
	24000
	26000

⊕ Add another

Distance Unit
| Meters   ▾ |

Buffer Distance Field Name
| distance |

Dissolve Option
| Non-overlapping (rings)   ▾ |

▶ Run ▾

(b)

Geoprocessing   ▾ ⊟ ✕

◉   Intersect   ⊕

ⓘ The Pairwise Intersect tool provides   ✕
   enhanced functionality or performance.

Parameters Environments   ⑦

Input Features ⌄   Ranks
Rings ▾	📁		
BRTrtUtm ▾	📁		
▾	📁		

Output Feature Class
| TrtRings   📁 |

Attributes To Join
| All attributes   ▾ |

Output Type
| Same as input   ▾ |

▶ Run ▾

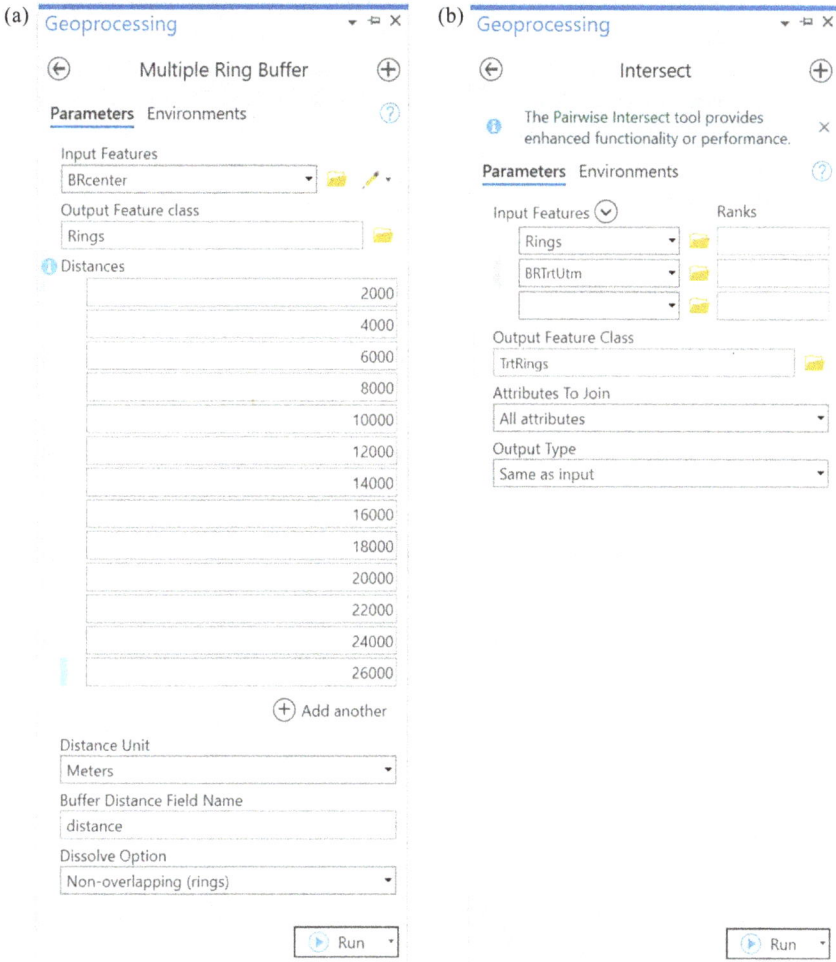

FIGURE 1.6 Dialog windows for (a) Multiple Ring Buffer and (b) Intersect.

BRTrtUtm as Input Features, name the Output Feature Class as TrtRings, keep other default settings, and click Run. Only the overlapping areas of the two input features are preserved in the output feature, where any census tracts crossing multiple rings is split. This step uses another popular map overlay tool Intersect.

In the Contents pane, right-click the layer TrtRings > Attribute Table. Note that it inherits the attributes from both input features. Among the fields in the table, values for the fields such as POP100 (population) and Area are those for the whole tracts and thus no longer valid for the new polygons, but values for the fields such as PopuDen and distance remain valid if assuming that population is uniformly distributed within a tract and areas in a ring are at the same distance from the city center.

Step 9. Calculating Area and Estimating Population in the Intersected Areas:
Still on the open attribute table of `TrtRings`, follow the procedure illustrated in
steps 4 and 5 to add two new fields `InterArea` (choose `Float` for Data Type) and
`EstPopu` (choose `Float` for Data Type), and use Calculate Geometry to update
the area size for `InterArea` (i.e., select `Area` for Property and `Square meters`
for Area Unit), and use the formula "`!PopuDen!*!InterArea!/1000000`" in
the Calculate Field to update `EstPopu`. Here, population in each new polygon is
estimated by its population density (inherited from the tract layer) multiplied by
its area.

Step 10. Computing Population and Population Density in Concentric Rings:
Another popular map overlay tool, Dissolve, is introduced here for data aggre-
gation. In the Geoprocessing pane, click Toolboxes > Data Management Tools >
Generalization > Dissolve. In the dialog (Figure 1.7), choose `TrtRings` for Input
Features, name the Output Feature Class `DissRings`, and select `distance`
for Dissolve Field(s); for Statistics Field(s), choose a Field `InterArea` and its

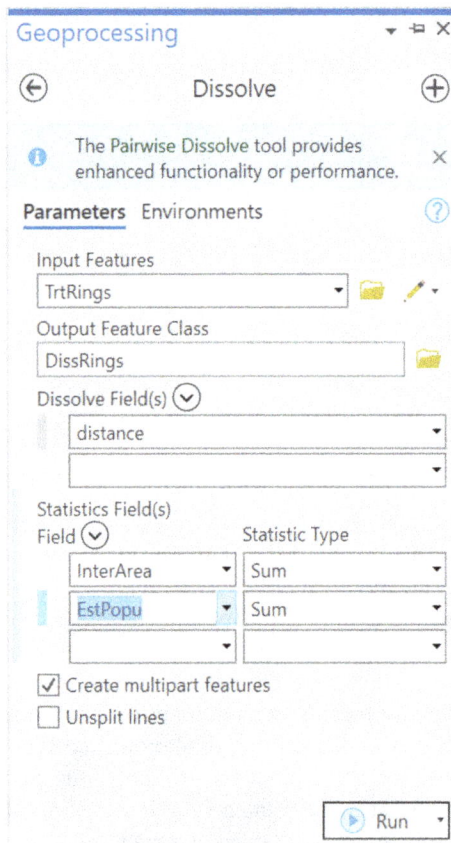

FIGURE 1.7 Dialog window for Dissolve.

Statistic Type Sum, and choose another Field EstPopu and its Statistic Type Sum. Keep other options as default, and click Run. In essence, the Dissolve tool merges polygons sharing the same values of a Dissolve Field (here distance) and aggregates attributes according to the defined fields and their corresponding statistics.

Right-click the layer DissRings in the layers window > Attribute Table. Similar to step 5 in Subsection 1.3.1, add a new field PopuDen (choose Float for Type) and calculate it as !SUM _ EstPopu!*1000000/!SUM _ InterArea! Note that the fields SUM _ EstPopu and SUM _ InterArea are the total population and total area for each ring, and the unit for population density of each ring is persons per square kilometer.

Step 11. Graphing and Mapping Population Density across Concentric Rings:

Again, right-click the DissRings layer > Create Chart > Line Chart. In the dialog of Chart Properties (Figure 1.8, right), click Data > choose distance for Date or Number, leave the default <none> for Aggregation, and click under Numeric field(s) to select PopuDen and Apply. The line chart of the population density shows below the main map. In the Chart Properties pane, explore more styles by adjusting other parameters, such as Format, General, etc. Click Export on the top menu of the resulting graph to save it in a chosen format, such as JPEG, SVG, PNG, etc.

Similar to step 6, map the population density pattern across the concentric rings.

The following illustrates the process to analyze the concentric density pattern based on the block-level data. One key step uses a *spatial join* tool.

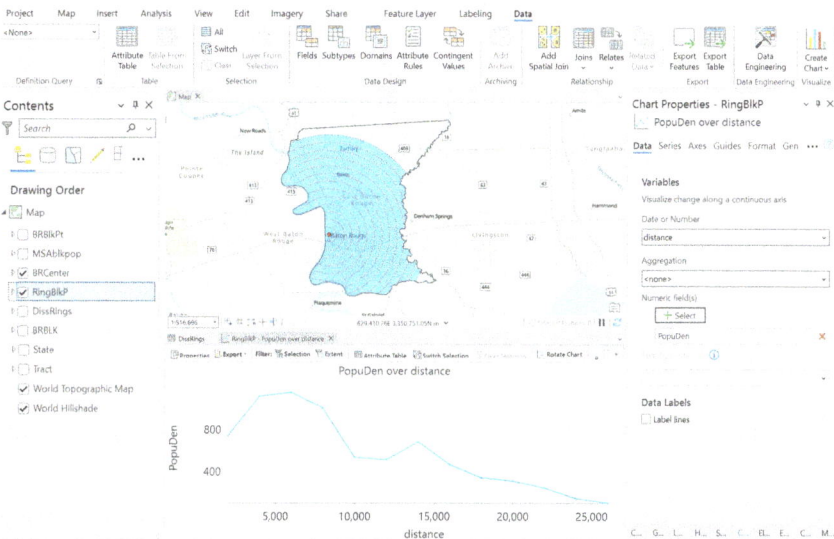

FIGURE 1.8 Dialog window for creating a graph in ArcGIS.

Step 12. Generating the Block Centroids: In the Geoprocessing pane, search and activate the tool Feature to Point. In the dialog, choose BRBLK for Input Features, enter BRBlkPt for Output Feature Class, check the "Inside (optional)" box, and click Run.

Step 13. Using a Spatial Join to Aggregate Population and Area from Blocks to Rings: In the Layers window, right-click the layer DissRings > Joins and Relates > Add Spatial Join. In the dialog (Figure 1.9a), choose BRBlkPt for Join Features, Contains for Match Option; unfold the menu of Fields to choose Popu2020 on the left side and then set Sum for Merge Rule on the right side. Repeat on another field Area and Sum for Merge Rule, remove other unnecessary fields, keep other options as default, and click Run to perform the spatial join.

The Spatial Join tool sums up attributes of the joined layer BRBlkPt and attaches the derived attributes to the target layer DissRings based on the spatial relationship between the two layers. Export the target feature DissRings to a new layer RingBlkP, whose attribute table includes fields Popu2020 and Area for total population and total area for each ring, respectively.

User can also implement this step by using another Spatial Join tool in the Geoprocessing Toolboxes > Analysis Tool > Overlay > Spatial Join. As shown in the dialog (Figure 1.9b), users need to set Join one to one for Join Operation, input

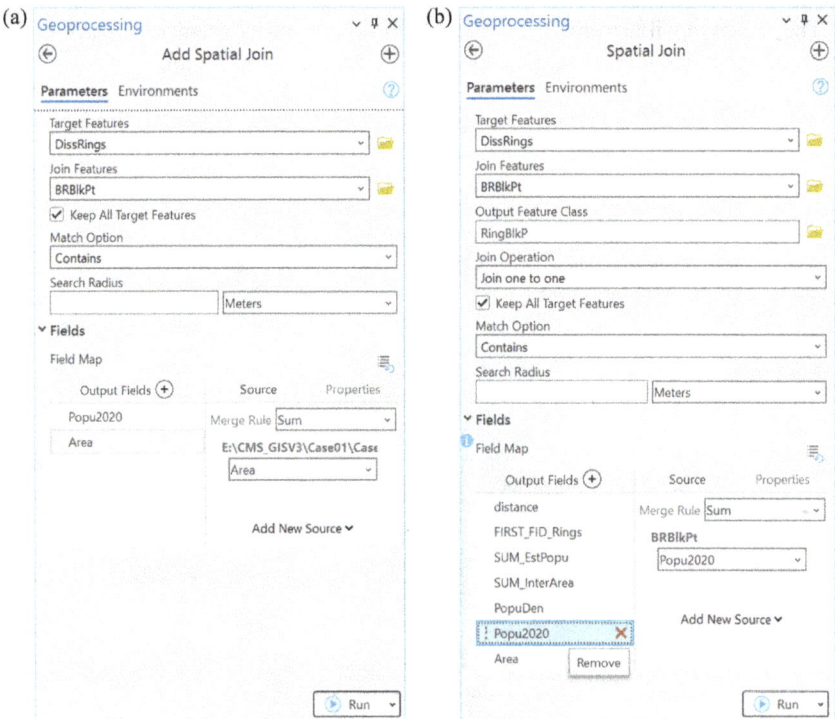

FIGURE 1.9 Dialog windows for (a) Add Spatial Join and (b) Spatial Join.

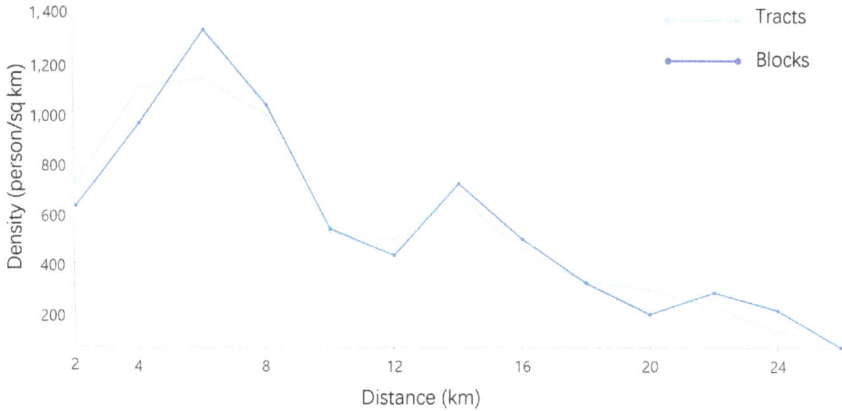

FIGURE 1.10 Population density patterns at the census tract and block levels.

`RingBlkP` as Output Feature Class, keep the key variables of both target and join features, and click Run.[8]

Step 14. Analyzing and Comparing Population and Population Density Estimates in Rings: In the attribute table of `RingBlkP`, similar to step 10, add a new field `PopuDenBlk` with `Double` as the Data Type, and calculate it as `!Popu2020!* 1000000/! Area!` One may map the population density based on the field `PopuDenBlk` and compare it to that from step 10.

The attribute table of `RingBlkP` also inherits a field `PopuDen` from step 10. Figure 1.10 highlights the general consistency and minor discrepancy for the variations of population density across the concentric rings based on `PopuDen` and `PopuDenblk`, respectively.

Figure 1.11 summarizes the steps for Part 1 and Part 2 of the case study.

At the end of the project, one may use the Catalog pane to organize the data (e.g., deleting unneeded data to save disk space, renaming some layers, etc.).

In sum, steps 8–11 use a simple areal interpolation (more details in Section 3.4 of Chapter 3) to transform population at one area unit (census tract) to another area unit (rings), and steps 12–14 use a spatial join to aggregate point-based (block) population to an area unit (ring). By doing so, we are able to examine whether a research finding is stable when data of different areal units are used. This is commonly referred to as the *"modifiable areal unit problem"* (*MAUP*) (Fotheringham and Wong, 1991).

1.4 SUMMARY

Major GIS and spatial analysis skills learned in this chapter include the following:

1. Map projections and transformation between them.
2. Spatial data management (copying, renaming, and deleting a spatial dataset).

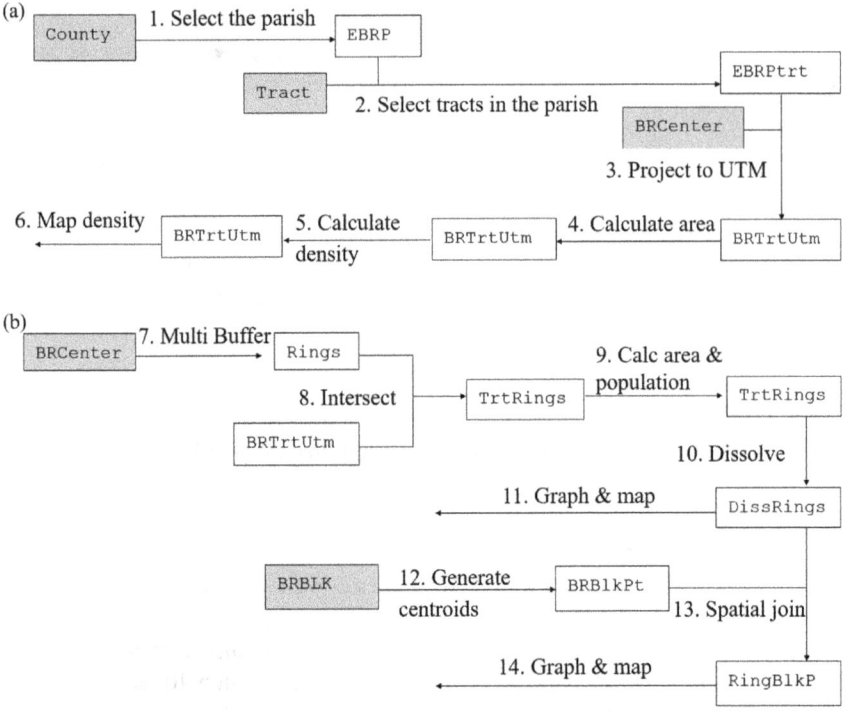

FIGURE 1.11 Flowchart for Case Study 1: (a) steps 1–6, (b) steps 7–14 (numbers index steps, boxes represent datasets).

3. Attribute data management (adding, deleting, and updating a field, calculating polygon areas, and obtaining basic statistics of a field in a table).
4. Mapping attributes (symbolizing a feature layer).
5. Attribute query vs. Spatial query.
6. Attribute join vs. Spatial join.
7. Map overlay operations (multiple ring buffer, intersect and dissolve; buffer, clip, and erase to be learned in Appendix 1).

Spatial query finds information based on location relationship between features, and the result may be saved to generate a new layer. Spatial join uses location relationship between source and target layers to process the feature from the source layer and attaches the processed data as added attributes to the target layer. Map overlay usually creates a new feature from one or multiple input layers and saves the result in a new layer. Among the three spatial analysis tools, the computational time increases from spatial query to spatial join, and to map overlay.

Other tasks introduced include finding spatial and attribute data in the public domain, generating centroids from a polygon layer to create a point layer, and graphing the relationship between variables. Projects in other chapters will also utilize these skills.

For the additional practice of GIS-based mapping, readers can download the census data and TIGER files for a familiar area and map some demographic (population, race, etc.) and socioeconomic variables (income, poverty, educational attainment, family structure, housing characteristics, etc.) in the area.

APPENDIX 1: IDENTIFYING CONTIGUOUS POLYGONS BY SPATIAL ANALYSIS TOOLS

Deriving a polygon adjacency matrix is an important task in spatial analysis. For example, in Chapter 8, both area-based spatial cluster analysis and spatial regression utilize a spatial weight matrix in order to account for spatial autocorrelation. Polygon adjacency is one of several ways to define spatial weights that capture spatial relationships between spatial features (e.g., neighboring features). Adjacency between polygons may be defined in two ways: (1) *rook contiguity* defines adjacent polygons as those sharing edges, and (2) *queen contiguity* defines adjacent polygons as those sharing edges or nodes (Cliff and Ord, 1973). This appendix illustrates how the queen contiguity can be defined by using some basic spatial analysis tools (Shen, 1994), as discussed in Section 1.2.

Here we use a census tract from the same study area BRTrtUtm in Case Study 1 to illustrate how to identify the contiguous polygons for one specific tract. Take a tract with its TRACTCE = 000602 as an example. As shown in Figure A1.1, based on the rook contiguity labeled "R," it has six neighboring tracts (000500, 000900, 001000, 001103, 000601, and 000702). Using instead the queen contiguity labeled "Q," tract 000701 is added as neighboring tract in addition to the six rook-contiguous tracts.

The following steps (Figure A1.2) illustrate how to implement the process of identifying queen-contiguous polygons. Each step uses a spatial analysis tool, highlighted in *italics*.

0. Use an *attribute query* to select the tract with "TRACTCE = 000602" on the layer BRTrtUtm and export it to Zone (N=1); and select the tract and its surrounding tracts and export it to AppendA1 (N=8).
1. *Buffer* a small distance (e.g., 100 meters) around the tract Zone to create a layer Zone_Buff.
2. *Use* Zone_Buff (as Clip Feature) to *clip* from the study area AppendA1 (as Input Features or Dataset) to generate a layer Zone_Clip (be sure to set the XY Tolerance as 10 meters with high accuracy under the Environments tab).
3. *Erase* Zone (as Erase Features) from Zone _ Clip (as Input Features) to yield another new layer Zone _ Neighbor (also set the XY Tolerance as 10 meters).

The resulting layer Zone _ Neighbor contains all eight neighboring tracts (i.e., their IDs) of Zone based on the queen contiguity.

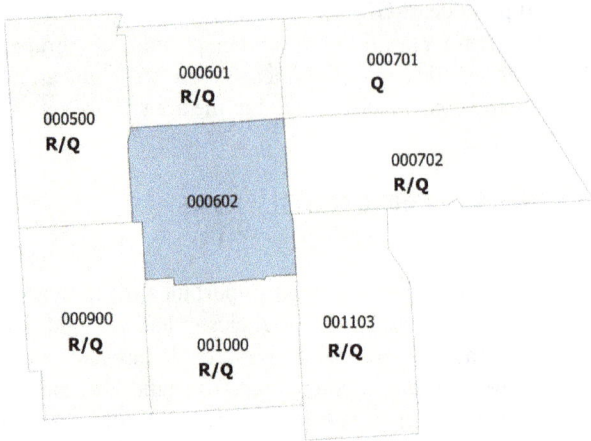

FIGURE A1.1 Rook vs. queen contiguity.

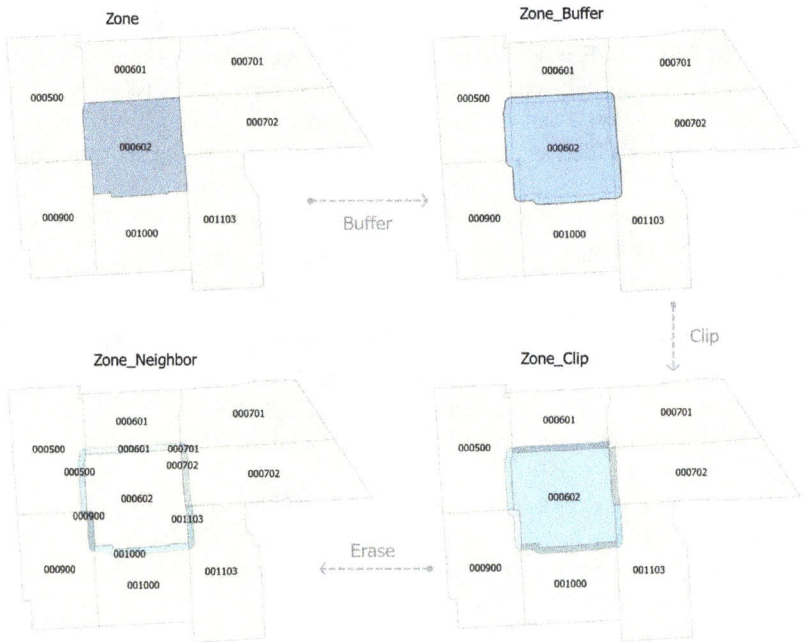

FIGURE A1.2 Workflow for defining queen contiguity.

NOTES

1 Advanced users may download (1) the TIGER data at www.census.gov/geographies/ mapping-files/time-series/geo/tiger-line-file.2021.html and (2) the 2020 Census Data Products at www.census.gov/programs-surveys/decennial-census/about/rdo/summary-files.html and use Microsoft Access, SAS, or R statistical software to process the census data and then ArcGIS to link them together. For downloading census data of a large size, use www2.census.gov.

2 Alternatively, start a new project by choosing "Start without a template" (you can save it later). Click the Insert tab on the top ribbon, and then choose New Map to automatically insert base world maps.

3 If one chooses "Have their center in" for Relationship, it would leave out one census tract whose centroid falls right on the border.

4 As we move the cursor in a map, the bottom right corner displays the corresponding coordinates in the current map units, in this case Decimal Degrees. In other words, maps in the data (or layout) view are in geographic coordinates, which is set by the coordinate system of the first layer added to the project.

5 Alternatively, click the graphic icon 🌐 to activate the Coordinates System dialog window (on the right of Figure 1.2).

6 For a polygon feature in geodatabase, its attribute table is embedded with a field Shape_Area (implying a unit of square meters, given the projection definition). Here we introduce this step of adding area so that an area field is explicitly defined with a chosen unit (e.g., a user may choose "square kilometers").

7 Field names such as POP100 and Area in the formula are entered by double-clicking the field names from the top box to save time and minimize typos.

8 The Add Spatial Join tool accessed from the Contents pane and the Spatial Join tool accessed from the Geoprocessing pane are similar. The difference is, the former does not save the result itself (here the result is saved by exporting), and the latter saves the result in a new layer.

2 Measuring Distance and Travel Time and Analyzing Distance Decay Behavior

This chapter discusses two tasks encountered very often in spatial analysis: measuring distance or time and modeling distance decay behaviors. After all, spatial analysis is about how the physical environment and human activities vary across space—in other words, how these activities change with distances from reference locations or objects of interest. The advancement and wide availability of GIS have made the task much easier than it was in the past. Once the distance or travel time measure is obtained, we can examine whether spatial interactions in movement of people, commodity, financial transaction, or information decline with distance or travel time between places. It is a spatial behavior commonly referred to as the *distance decay rule*, also referred to as the "*first law of geography*" (Tobler, 1970).

The tasks of estimation of distance or travel time and derivation of the best-fitting distance decay function are found throughout this book. For example, spatial smoothing and spatial interpolation in Chapter 3 utilize distance to determine which objects are considered within a catchment area and how much influence the objects have upon each other. In functional region delineations by the Huff model in Chapter 4, distances or travel time between facilities and residents affect how frequently facilities are visited. In the discussion of accessibility measures in Chapter 5, spatial impedance in distance or travel time and distance decay functions serve as the building blocks of either the floating catchment area method or its generalized version. Chapter 6 examines how population density or land use intensity declines with distance from a city or regional center. Measurements of distance and time can also be found in other chapters.

This chapter is structured as follows. Section 2.1 provides an overview of various distance and travel time measures. Section 2.2 uses a case study to illustrate the implementation of those measures in ArcGIS Pro, especially on computing a drive time matrix and a transit time matrix. Section 2.3 discusses the distance decay rule and presents two approaches, namely, the spatial interaction model and the complementary cumulative distribution curve, to deriving the best-fitting distance decay function. A case study demonstrates how these two methods are implemented in Section 2.4. The chapter concludes with a brief summary in Section 2.5.

DOI: 10.1201/9781003292302-3

2.1 MEASURES OF DISTANCE AND TRAVEL TIME

Distance measures include Euclidean distance, geodesic distance, Manhattan distance, network distance, topological distance, and others.

Euclidean distance, also referred to as straight-line distance or "as the crow flies," is simply the distance between two points connected by a straight line on a flat (planar) surface. Prior to the widespread use of GIS, researchers needed to use mathematical formulas to compute distance, and the accuracy was limited depending on the information available and tolerance of computational complexity. If a study area is small in terms of its geographic territory (e.g., a city or a county), Euclidean distance between two points (x_1, y_1) and (x_2, y_2) in Cartesian coordinates is approximated as:

$$d_{12} = [(x_1 - x_2)^2 + (y_1 - y_2)^2]^{\frac{1}{2}}. \tag{2.1}$$

If the study area covers a large territory (e.g., a state or a nation), the geodesic distance is a more accurate measure. *Geodesic distance* between two points is the distance through a great circle assuming the Earth as a sphere. Given the geographic coordinates of two points as (a, b) and (c, d) in radians (*rad*) that may be converted from decimal degrees (*dd*) by using $rad = \pi * dd / 180$, the geodesic distance between them is:

$$d_{12} = r * arccos\left[\sin b * \sin d + \cos b * \cos d * \cos(c - a)\right], \tag{2.2}$$

where r is the radius of the Earth (approximately 6,367.4 kilometers).

As the name suggests, *Manhattan distance* describes a rather-restrictive movement in rectangular blocks, as in the Manhattan borough of New York City. Manhattan distance is the length of the change in the x direction plus the change in the y direction. For instance, the Manhattan distance between two nodes (x_1, y_1) and (x_2, y_2) in Cartesian coordinates is simply computed as:

$$d_{12} = |x_1 - x_2| + |y_1 - y_2|. \tag{2.3}$$

Like Euclidean distance, Manhattan distance is only meaningful in a small study area, such as a city or county. Manhattan distance is often used as an approximation for network distance if the street network is close to a grid pattern. It has become obsolete in the era of GIS, with the wide availability of street (road) network data.

Network distance is the shortest-path (or least-cost) distance through a transportation network. A transportation network consists of a set of nodes (or vertices) and a set of arcs (or edges or links) that connect the nodes and is usually a directed network, as some of the arcs are one-way streets and most of the arcs have two permissible directions. Finding the shortest path from a specified origin to a specified destination is the *shortest-route problem*, which records the shortest distance or the least travel time, depending on whether the impedance value is defined as distance or time on each arc. Most algorithms for solving the problem are built upon the *label setting algorithm* (Dijkstra, 1959). Appendix 2A introduces the valued-graph (or L-matrix) method that is easy to follow but not nearly as efficient as the Dijkstra algorithm.

The popular label-setting method assigns "labels" to nodes, and each label is the shortest distance from a specified origin. To simplify the notation, the origin is assumed to be node 1. The method takes four steps:

1. Assign the *permanent* label $y_1 = 0$ to the origin (node 1), and a *temporary* label $y_j = M$ (a very large number) to every other node. Set $i = 1$.
2. From node i, recompute the temporary labels $y_j = \min (y_j, y_i+d_{ij})$, where node j is temporarily labeled and $d_{ij} < M$ (d_{ij} is the distance from i to j).
3. Find the minimum of the temporary labels, say, y_i. Node i is now permanently labeled with value y_i.
4. Stop if all nodes are permanently labeled; go to step 2 otherwise.

The following example illustrates the method. Figure 2.1a shows the network layout with nodes and links. The number next to a link is the impedance value for the link.

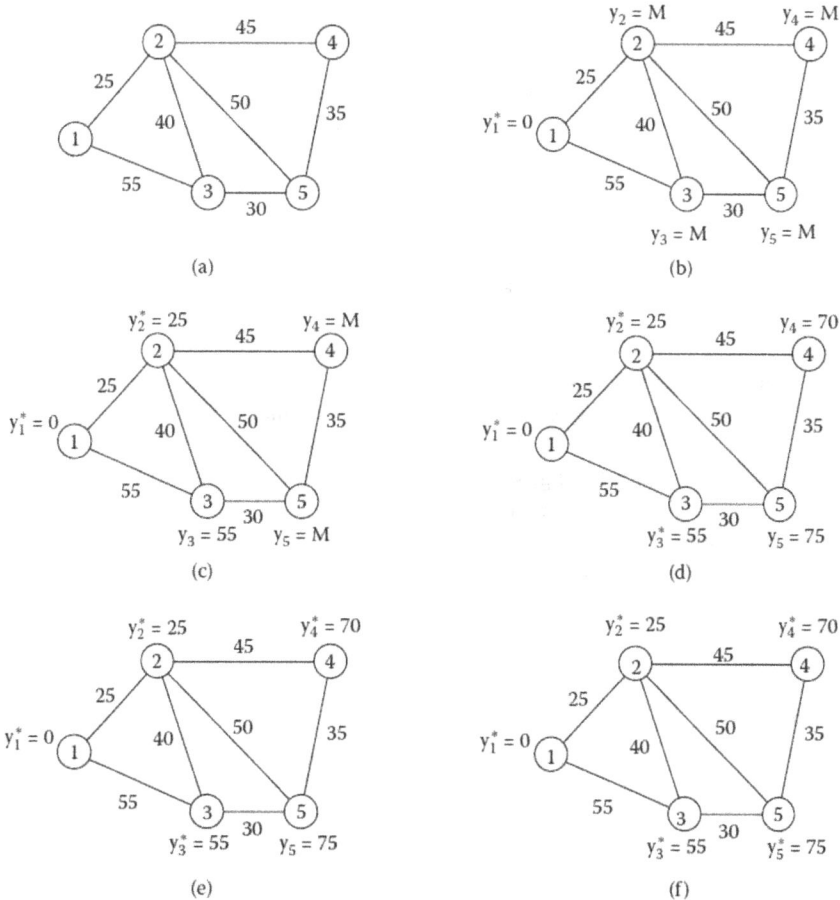

FIGURE 2.1 An example for the label-setting algorithm.

Following step 1, permanently label node 1 and set $y_1 = 0$; temporarily label $y_2 = y_3 = y_4 = y_5 = M$. Set $i = 1$. A permanent label is marked with an asterisk (*). See Figure 2.1b.

In step 2, from node 1 we can reach nodes 2 and 3 that are temporarily labeled. $y_2 = \min (y_2, y_1 + d_{12}) = \min (M, 0 + 25) = 25$, and similarly, $y_3 = \min (y_3, y_1 + d_{13}) = \min (M, 0 + 55) = 55$.

In step 3, the smallest temporary label is min $(25, 55, M, M) = 25 = y_2$. Permanently label node 2, and set $i = 2$. See Figure 2.1c.

Back to step 2 as nodes 3, 4, and 5 are still temporarily labeled. From node 2, we can reach temporarily labeled nodes 3, 4, and 5. $y_3 = \min (y_3, y_2 + d_{23}) = \min (55, 25 + 40) = 55$, $y_4 = \min (y_4, y_2 + d_{24}) = \min (M, 25 + 45) = 70$, $y_5 = \min (y_5, y_2 + d_{25}) = \min (M, 25 + 50) = 75$.

Following step 3 again, the smallest temporary label is min $(55, 70, 75) = 55 = y_3$. Permanently label node 3, and set $i = 3$. See Figure 2.1d.

Back to step 2 as nodes 4 and 5 are still temporarily labeled. From node 3 we can reach only node 5 (still temporarily labeled). $y_5 = \min (y_5, y_3 + d_{35}) = \min (75, 55 + 30) = 75$.

Following step 3, the smallest temporary label is min $(70, 75) = 70 = y_4$. Permanently label node 4, and set $i = 4$. See Figure 2.1e.

Back to step 2 as node 5 is still temporarily labeled. From node 4 we can reach node 5. $y_5 = \min (y_5, y_4 + d_{45}) = \min (75, 70 + 35) = 75$.

Node 5 is the only temporarily labeled node, so we permanently label node 5. By now all nodes are permanently labeled, and the problem is solved. See Figure 2.1f.

The permanent labels y_i give the shortest distance from node 1 to node i. Once a node is permanently labeled, we examine arcs "scanning" from it only once. The shortest paths are stored by noting the "scanning node" each time a label is changed (Wu and Coppins, 1981, p. 319). The solution to the previous example can be summarized in Table 2.1.

A transportation network has many network elements, such as link impedances, turn impedances, one-way streets, overpasses, and underpasses, that need to be defined (Chang, 2004, p. 351). Putting together a road network requires extensive data collection and processing, which can be very expensive or infeasible for many applications. For example, a road layer extracted from the TIGER/Line files does not contain nodes on the roads, turning parameters, or speed information. When such information is not available, one has to make various assumptions to prepare

TABLE 2.1
Solution to the Shortest-Route Problem

Origin–Destination Nodes	Arcs on the Route	Shortest Distance
1, 2	(1, 2)	25
1, 3	(1, 3)	55
1, 4	(1, 2), (2, 4)	70
1, 5	(1, 2), (2, 5)	75

the network dataset. For example, in Luo and Wang (2003), speeds are assigned to different roads according to the CFCC codes (census feature class codes) used by the US Census Bureau in its TIGER/Line files, and whether the roads are in urban, suburban, or rural areas. Wang (2003) uses regression models to predict travel speeds by land use intensity (business and residential densities) and other factors. Section 2.2 uses a case study to illustrate the technical details on how to use ArcGIS Pro for estimating Euclidean distance, geodesic distance, drive time, and transit time.

All the aforementioned measures of distance and travel time are metric. In contrast, *topological distance* emphasizes the number of connections (edges) between objects instead of their total lengths. For example, the topological distance between two locations on a transportation network is the fewest number of edges that take to connect them. Topological distances in a transportation network can be calibrated similarly by using the preceding label-setting algorithm. However, in a topological network, the edge length between two nodes is coded as 1 if they are directly connected, and 0 otherwise. Topological distances between polygons can be computed by constructing a topological network representing the polygon connectivity based on rook adjacency (see Appendix 1) and then applying the shortest-route algorithm. For example, the topological distance between two neighboring polygons is 1, and 2 if they are connected through a common neighbor, and so on. Topological distance measure is used in applications such as urban centrality index (Hillier, 1996) or social network analysis (Freeman, 1979). Spatial interaction may also conform to the distance decay rule when distance is measured in topological distance. For example, as reported in Liu et al. (2014), relatedness between provincial units in China declines with topological distance between them.

Other distance terms, such as *behavioral distance*, psychological distance (e.g., Liberman et al., 2007), mental distance, and social distance, emphasize that the aforementioned physical distance measures may not be the best approach to model spatial decision-making that varies by individual attributes, environmental considerations, or the interaction between them. For example, one may choose the safest but not necessarily the shortest route to return home, and the choice may be different at various times of the day. One traveling with people with health concerns may choose the route with the least exposure to air pollution. The routes from the same origin to the same destination may vary between a resident with local knowledge and a new visitor to the area.

2.2 CASE STUDY 2A: ESTIMATING DRIVE AND TRANSIT TIMES IN BATON ROUGE, LOUISIANA

In many spatial analysis applications, a distance or travel time matrix between a set of origins and a set of destinations is needed. This section uses a case study to illustrate how to estimate a Euclidean or geodesic distance matrix, a drive time matrix via a road network, and a transit time matrix via a transit network in ArcGIS Pro. Appendix 2B shows how to use an online routing app such as Google Maps API to bypass the data preparation and estimate a drive or transit time matrix that accounts for real traffic conditions. It enables us to tap into the dynamically updated transportation network data and the routing rules maintained by Google and possibly obtain a more current and reliable estimate of travel times.

This case study illustrates how to measure distance and travel time between residents at the census block group level and acute hospitals in Baton Rouge, Louisiana. The census block group data in East Baton Rouge Parish (EBRP) is based on the 2020 Census, as explained in Case Study 1 in Chapter 1. The hospitals data is extracted from the membership directory of Louisiana Hospital Association (2021). As noted earlier, measuring distance and time is a fundamental task in spatial analysis. For instance, results from this project will be used in Case Study 4A of Chapter 4 that delineates hospital service areas by the Huff model, and in Case Study 5 of Chapter 5 to measure spatial accessibility.

The following datasets are provided under the data folder `BatonRouge`:

1. A comma-separated text file `Hosp _ Address.csv` contains addresses, geographic coordinates, and numbers of staffed beds for the five acute hospitals in EBRP.[1]
2. A geodatabase `BR_Road.gdb`[2] contains a feature class `BR_Bkg` for the census block group data, a feature dataset `BR_MainRd` that includes a single feature class `BR_MainRd`[3] for main roads, and a street address TIGER file `tl_2021_22033_addrfeat` for geocoding, all in EBRP.
3. A Network Dataset Template file `TransitNetworkTemplate.xml` will be used for creating a network dataset for transit time estimation.
4. A folder `BR _ GTFS` contains nine files for transit time estimation.

2.2.1 Estimating a Euclidean or Geodesic Distance Matrix

Step 1. Geocoding Hospitals: In GIS, *geocoding* is a process to transform locations in coordinates, addresses, or other descriptions to a set of points. The case study begins with introducing two geocoding approaches: one based on geographic coordinates, and another (optional) based on street addresses.

Geocoding based on geographic coordinates. Activate ArcGIS Pro and add an unprojected layer by navigating to the data folder of Case Study 1 in Chapter 1 and loading `Census.gdb\State` to the project.[4] From the main ribbon, choose the Map tab > Add Data > XY Point Data > XY Table to Point. In the dialog (Figure 2.2),[5] choose `Hosp _ Address.csv` as Input Table; set the Output Feature Class as `Hosp _ geo`; keep the default X Field and Y Field (LONGITUDE and LATITUDE); click ⊕ to open the dialog window of Coordinate system, click ⊕ and select Import Coordinate System to open its dialog window; navigate to geodatabase `Census.gdb`, select `State`,[6] and click OK; and click Run to execute the tool and also OK in the warning message window. The resulting layer `Hosp _ geo` has the same geographic coordinate system as `State`.

In the Geoprocessing pane, search and activate the tool "Project." In the dialog, choose `Hosp _ geo` for Input Dataset or Feature Class and name the Output Dataset or Feature Class as `Hosp` under the geodatabase `BR _ Road.gdb`; under Output

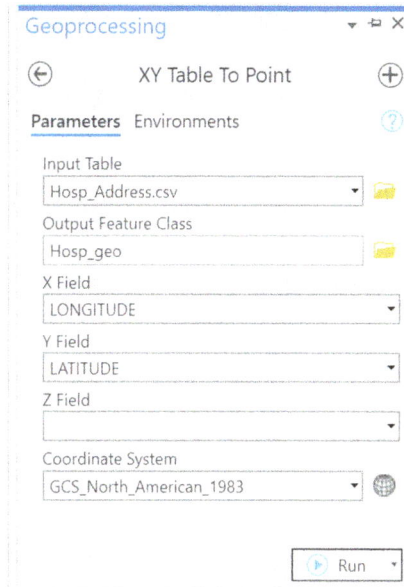

FIGURE 2.2 Dialog window for XY Table to Point.

Coordinate System, choose BR _ MainRd (also under the geodatabase BR _ Road. gdb) to import the coordinate system NAD _ 1983 _ UTM _ Zone _ 15N, and click Run. This transforms the feature Hosp _ geo in geographic coordinates to a feature Hosp in a UTM projection. If needed, refer to step 3 and Figure 1.2 in Subsection 1.3.1 of Chapter 1.

Optional: Geocoding based on street addresses. This step can be done by the toolbox Geocode Addresses[7] under the Geocoding Tools. There are two Address Locators: the built-in ArcGIS World Geocoding Service, and a user-customized locator.

For the built-in locator, click Analysis tab > ▨ Tools > Geoprocessing pane > Toolboxes tab > Geocoding Tools > Geocode Addresses. In the dialog (Figure 2.3a), choose Hosp _ Address.csv as Input Table and select ArcGIS World Geocoding Service as Input Address Locator; choose Multiple Field for Input Address Fields, and then set corresponding fields for Address or Place, City, State, and ZIP. Name Hosp1 for Output Feature Class, check United States for Country, select Address location for Preferred Location Type, and check Address for Category; keep other options as default, and click Run. The layer Hosp1 is added.

A user-defined locator requires a street layer, and here tl _ 2021 _ 22033 _ addrfeat is used. Similarly, click Toolboxes tab > Geocoding Tools > Create Locator. In the dialog (Figure 2.3b), select United States for Country or Region; choose the street layer tl _ 2021 _ 22033 _ addrfeat as Primary Table(s) and Street Address for Role; set LFROMHN, LTOHN, RFROMHN, RTOHN, and

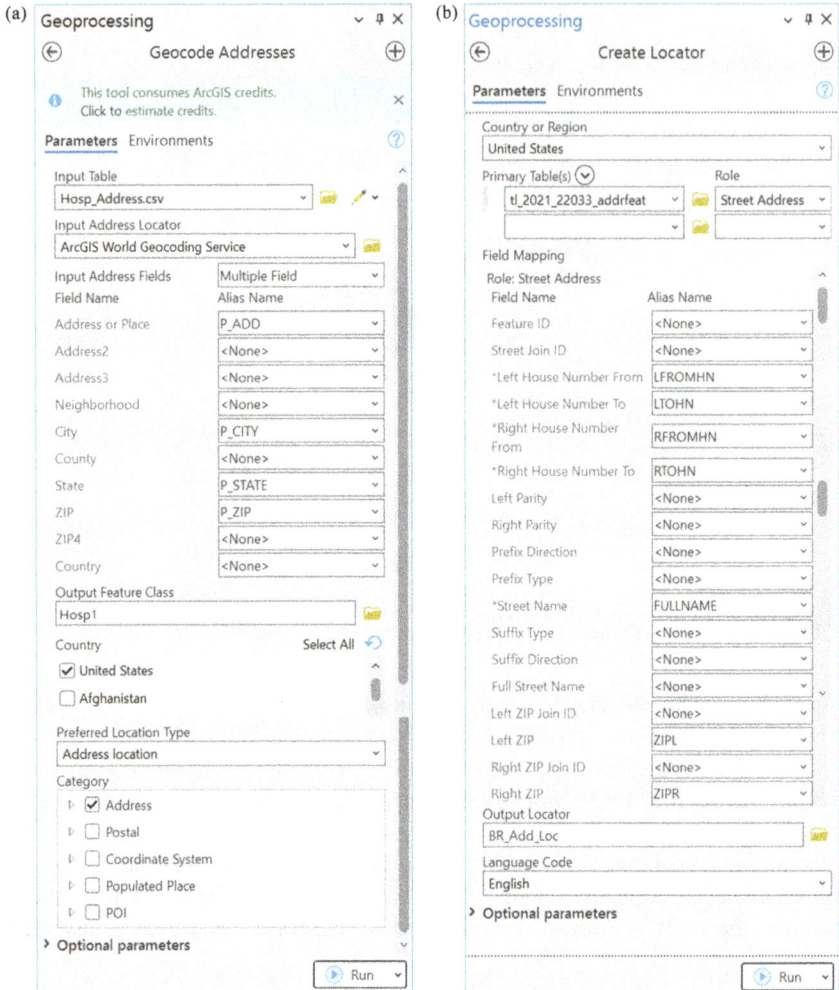

FIGURE 2.3 Dialog windows: (a) ArcGIS World Geocoding Service, (b) Create Locator.

FULLNAME for all five required inputs, such as Left House Number From, Left House Number To, Right House Number From, Right House Number To, and Street Name, respectively; choose ZIPL and ZIPR for Left ZIP and Right ZIP; and for Output Locator, navigate to the project folder and name it BR _ Add _ Loc, select English for Language Code, keep other options as default, and click Run.

Again, activate the tool Geocode Addresses. In the dialog similar to Figure 2.3a, select Hosp _ Address.csv as Input Table, choose BR _ Add _ Loc as Input Address Locator, set the Address Fields as the displayed values in Figure 2.3a, name Hosp2 for Output Feature Class, check Address for Category, keep other options as default, and click Run.

One may observe that the geocoding results (Hosp, Hosp1, Hosp2) are largely consistent. Note that the layer Hosp2 misses one hospital (Our Lady of the Lake Regional Medical Center) because the TIGER street file used in geocoding does not contain its address. Therefore, the built-in ArcGIS World Geocoding Service is recommended when the number of geocoded addresses is not large.

Step 2. Generating Census Block Group Centroids: In the Geoprocessing pane, click Toolboxes > Data Management Tools > Features > Feature to Point. In the dialog, choose BR _ Road.gdb\BR _ Bkg for Input Features; for Output Feature Class, navigate to the geodatabase BR _ Road.gdb and name the feature class BRPt, and check the option Inside. Click Run.

Step 3. Computing Euclidean and Geodesic Distances: In the Geoprocessing pane, click Toolboxes > Analysis Tools > Proximity > Generate Near Table. In the dialog, choose BRPt as Input Features and Hosp as Near Features, and name the Output Table Dist (under the geodatabase BR _ Road.gdb); the default unit is Meters; uncheck "Find only closest feature," keep other options default, and click Run. There is no need to define a search radius, as all distances are needed. Note that there are 324 (block groups) × 5 (hospitals) = 1,620 records in the distance table, where the field IN _ FID inherits the values of OBJECTID _ 1 in the attribute table of BRPt, the field NEAR _ FID inherits the values of OBJECTID in the attribute table of Hosp, and the field NEAR _ DIST is Euclidean distance between them in meters. For Method, choose Planar or Geodesic for computing Euclidean or geodesic distances, respectively.

2.2.2 Estimating a Drive Time Matrix

Network analysis often takes significant computational power and time. For saving the computational time, this project uses only the main roads (BR _ MainRd) instead of all streets as the road network for estimating travel time between census block groups and hospitals. The main roads are extracted from the streets layer, which include the interstate, US, and state highways with Feature Class Codes (FCC) A11-A38. The spatial extent of roads in BR _ MainRd is slightly larger than the EBRP to permit some routes through the roads beyond the parish (county unit) borders.

Step 4. Preparing the Road Network Source Layer: The source feature dataset must have fields representing network impedance values, such as distance and travel time. ArcGIS recommends naming these fields as the units of impedance such as Meters and Minutes so that they can be detected automatically.

In ArcGIS Pro, open the attribute table of the feature class BR _ MainRd and note that the field Shape _ Length is automatically generated by the system with the unit in meters. Add fields Minutes (Double as Data Type). Right-click the field Minutes, and calculate its values using "60*(!Shape _ Length!/1609)/!SPEED!" when the Expression Type is Python 3. Note that the unit of speed is miles per hour, and the formula returns the travel time in minutes.

Step 5. Building the Network Dataset: On the Geoprocessing pane, click Toolboxes > Network Analyst Tools > Network Dataset > Create Network Dataset. In the dialog, under Parameters, choose BR _ MainRd for Target Feature Dataset, type BR _ MainRd _ ND for Network Dataset Name, check BR _ MainRd as Source Feature Classes, leave the default setting "Elevation fields" for Elevation Model, and click Run to execute the tool. Now, under the feature dataset BR _ MainRd, the newly created network dataset BR _ MainRd _ ND and feature class BR _ MainRd _ ND _ Junctions are added.

Remove the network dataset layer BR _ MainRd _ ND from the map so that it can be edited. In the Catalog pane, right-click the network dataset BR _ MainRd _ ND > Properties > Travel Attributes > Costs tab > Distance (Figure 2.4a). Under Properties, change the Name to Dist and assign Meters for Units. Under Evaluators, for BR _ MainRd(Along), choose Field Script for Type, click

on the right side of the Value to open the Field Script dialog window, type [Shape _ Length] for Result, and click OK. Still under Costs tab, click the Menu

button and choose " New." Define a new cost field for Time (Figure 2.4b): update the name to Time _ min, and assign Minutes for Units. Similarly, under Evaluators, for BR _ MainRd(Along), choose Field Script for Type and type [Minutes] in the Field Script dialog window for Result, and click OK. All others are left as default. Click OK to commit the changes.

Now we are ready to build the network. Similarly, in the Geoprocessing pane, click Toolboxes > Network Analyst Tools > Network Dataset > Build Network. In the dialog, under Parameters, choose the network dataset BR_MainRd_ND for Input Network Dataset and click Run. It takes 1–2 minutes to build the network dataset.

Step 6. Creating the OD Cost Matrix: On the Analysis tab, click Network Analysis drop-down menu; under Network Data Source, make sure that the network data source is BR _ MainRd _ ND. Click Network Analysis drop-down menu again and

select   Origin–Destination Cost Matrix. The OD Cost Matrix layer with several sublayers is added to the Contents pane.

In the Contents pane, select the layer OD Cost Matrix. The OD Cost Matrix tab appears in the Network Analysis group at the top. Click the OD Cost Matrix tab to see the controls, including Import Origins, Import Destinations, Travel Settings, and Run, to be defined as follows:

a. Click   Import Origins to activate the Add Locations dialog window (Figure 2.5a). Layers OD Cost Matrix and Origins are automatically set as Input Network Analysis Layer and Sub Layer, respectively. Select BRPt for Input Locations; under Advanced, input 10000 for Search Tolerance,[8] and select Meters as its units. Leave other parameters as default settings. Uncheck Append to Existing Locations to avoid generating duplicated records when running this tool for multiple times. Click OK to load the 324 census block group points as origins.

b. Similarly, click   Import Destinations to activate a new Add Locations dialog.[9] Layers OD Cost Matrix and Destinations are set as Input

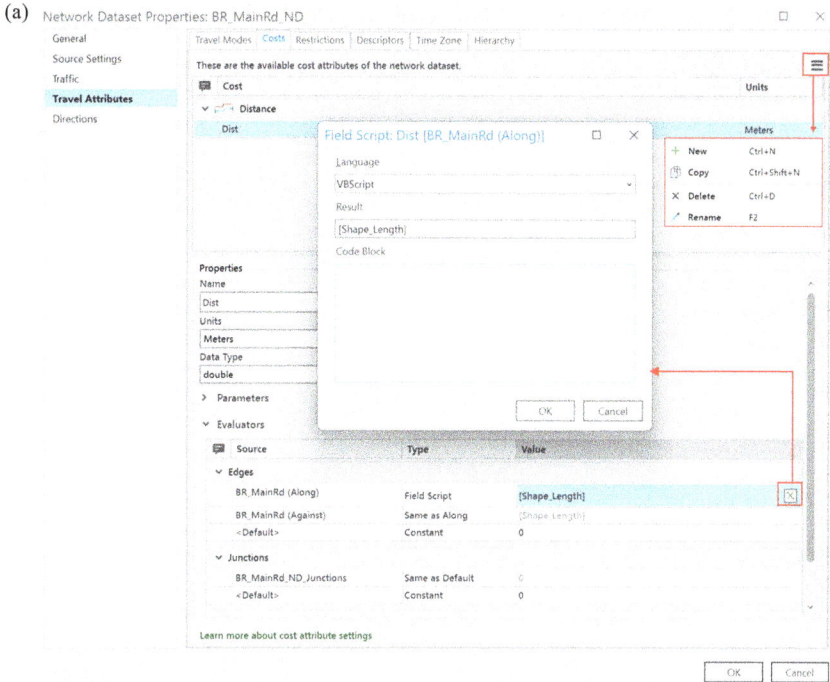

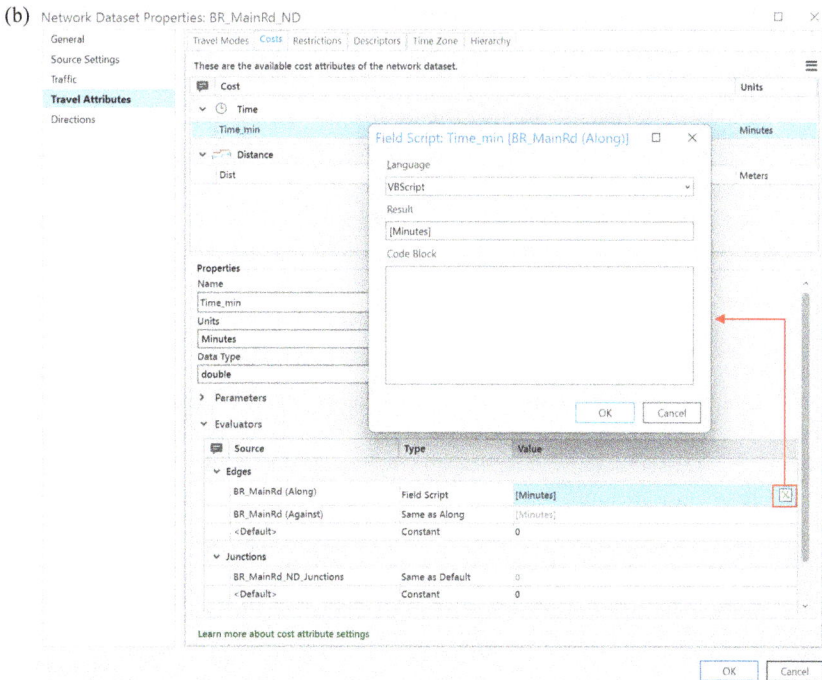

FIGURE 2.4 Dialog windows for defining cost in network dataset by (a) distance and (b) time.

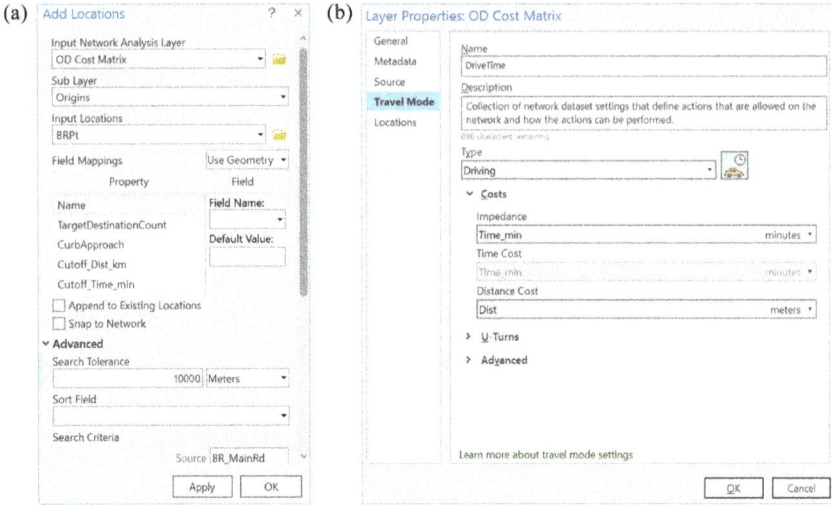

FIGURE 2.5 Dialog windows for OD cost matrix: (a) add locations and (b) travel settings.

Network Analysis Layer and Sub Layer, respectively. Select Hosp for Input Locations, use the default 5000 Meters for Search Tolerance, and leave other parameters as default settings. Uncheck Append to Existing Locations when running this tool for multiple times. Click OK to load the five hospitals as destinations.

c. Click the icon ⬚ beside Travel Settings to open the dialog (Figure 2.5b). For Travel Mode, input DriveTime as Name, choose Driving for Type, define items under Costs (e.g., Time_min for Impedance, Time_min for Time Cost, and Dist for Distance Cost), and leave the default settings for others. Click OK to accept the settings.

d. Click ▶ Run to execute the tool.

In the Contents pane, under the OD Cost Matrix group layer, right-click the sublayer Lines, and select Attribute Table to view the result. It contains OriginID and DestinationID that are identical to the field OBJECTID _ 1 in BRPt and the field OBJECTID in Hosp, respectively. Right-click Lines > Data > Export Table to save the result as a table OD _ Drive _ Time. The number of records is 1,620 (= 324×5).

Step 7. Merging the OD Distance and Drive Time Tables by a Concatenate Field: The table Dist generated in step 3 includes a field NEAR _ DIST for Euclidean (geodesic) distance, and the table OD _ Drive _ Time has a field DriveTime representing drive time for the same 1,620 OD pairs. This step joins them together based on a field that represents the unique OD pairs.

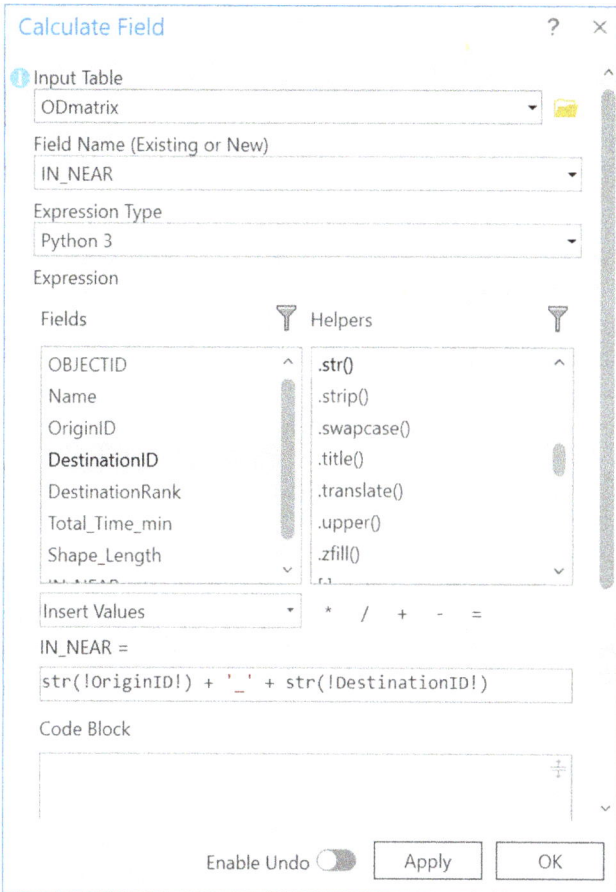

FIGURE 2.6 Dialog window for calculating a concatenate field.

Right-click on the table `OD_Drive_Time` in the Contents pane, add a new field and name it as `IN_NEAR`, set Data Type as `TEXT`, and calculate it as "`str(!-OriginID!)+'_'+ str(!DestinationID!)`" (Figure 2.6). The unique identifier `IN_NEAR` for each OD pair is generated. Similarly, on the table `Dist`, add a same field `IN_NEAR` and update it as "`str(!IN_FID!)+'_'+ str(!NEAR_FID!)`." Join the table `OD_Drive_Time` to `Dist` based on their common filed `IN_NEAR`.

On the open table `Dist` > click the icon ☰ > Export. In the dialog, name the Output Table as `DistDrive` (under `BR _ Road.gdb`); under Fields, for Output Fields, select two fields `IN _ NEAR` and `Total _ Time _ min` to keep, and click OK to preserve the joined table `DistDrive`.

Save the project as `Case02A` for future reference.

2.2.3   ESTIMATING A TRANSIT TIME MATRIX

Computing a transit time matrix is a special scenario of network analysis that utilizes a transit network dataset. The General Transit Feed Specification (GTFS) data model developed by Google is used to describe the fixed-route transit system (Antrim and Barbeau, 2013). It includes a series of tables in the form of comma-delimited text files (https://developers.google.com/transit/gtfs/ reference). These tables use data with predefined field names to describe multiple components of a transit system, such as agency information (`agency.txt`), transit stops (`stops.txt`), routes (`routes.txt`), trips belonging to each route (`trips.txt`), arrival and departure time of each stop (`stop _ times.txt`), and operation calendar types of weekly schedules or holidays (`calendar.txt` or `calendar _ dates.txt` or both), etc.

The GTFS data for the study area, EBRP, extracted from the open transit data website Transitland (www.transit.land/feeds/f-baton~rouge~cats), is saved under the data folder `BR _ GTFS`. The following steps implement estimation of the transit time matrix from census block groups to hospitals in EBRP.

Step 8. Viewing and Preparing the Transit and Road Network Datasets: Under the data folder `BR _ GTFS`, there are nine files, including the aforementioned six required text tables. Open `stop _ times.txt` and examine the two fields `arrival _ time` and `departure _ time` to ensure no blank values for either column. Open `calendar.txt`, examine the two fields `start _ date` and `end _ date` for date ranges and ensure no overlapping, and check the weekday fields Monday through Sunday with values 1 or 0. Close all txt files.

Create a new project, add the feature class `BR _ MainRd` to the Map pane, and open the attribute table of `BR _ MainRd` to review two required fields `RestrictPedestrians` and `ROAD _ CLASS`.[10]

Step 9. Creating Transit Stops, Lines, and Schedules from the GTFS Data: In the Catalog pane, right-click `BR _ Road.gdb` > New > Feature Dataset. In the dialog, define the Feature Dataset Name as `BR _ Transit`; for Coordinate System, import the spatial reference of feature dataset `BR _ MainRd`, and click Run to create it.

Under the Analysis tab, click Tools > Geoprocessing pane > Toolboxes > Public Transit Tools > Conversion > GTFS to Public Transit Data Model. In the dialog, select `BR _ GTFS` for the Input GTFS Folders and `BR _ Transit` for the Target Feature Dataset, leave other parameters unchecked, and click Run.

Under the feature dataset `BR _ Transit`, load the two newly generated feature classes `Stops` and `LineVariantElements` to the Map pane. The layer `Stops` represents the spatial locations of the transit stops, located on the transit line segments `LineVariantElements`.

Step 10. Connecting the Transit Stops to the Streets: In the Geoprocessing pane, click Toolboxes > Conversion Tools > To Geodatabase > Feature Class to Feature Class. In the dialog, select `BR _ MainRd` for the Input Features and `BR _ Transit` for Output Location; define the Output Name as `Streets`, and click Run.

Under the Analysis tab, click Tools > Geoprocessing pane > Toolboxes > Public Transit Tools > Conversion > Connect Public Transit Data Model to Streets.

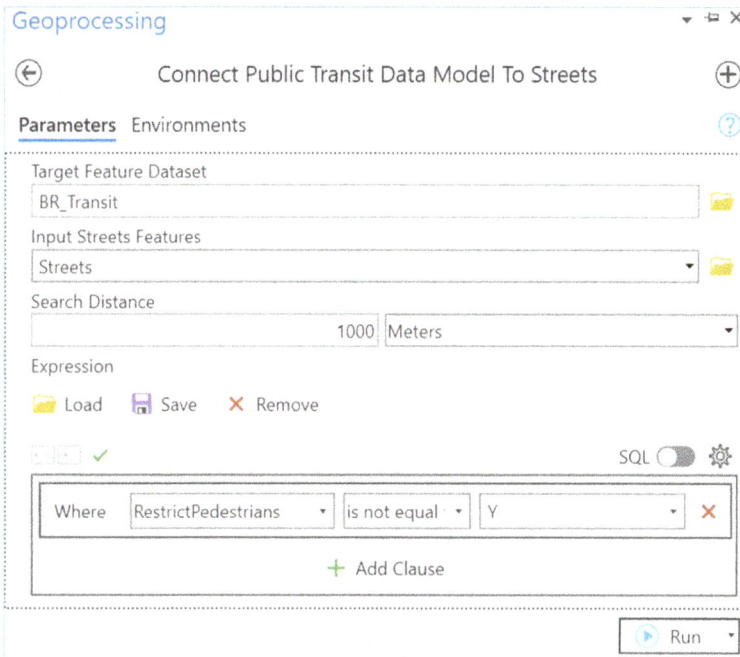

FIGURE 2.7 Dialog window for connecting public transit stops to streets.

In the dialog (Figure 2.7), select BR _ Transit for the Target Feature Dataset and Streets for the Input Streets Features, and 1000 Meters for the Search Distance. Under Expression, choose the field RestrictPedestrians, and set where clause to "is not equal to" and the value to "Y." Click Run. Two newly generated feature classes StopsOnStreets and StopConnectors are added to the Contents pane.

Zoom in to any transit stop. A short line in StopConnectors connects points in Stops and the intersections in StopsOnStreets.

Step 11. Creating and Building the Transit Network Dataset: Under the Analysis tab, click Tools > Geoprocessing pane > Toolboxes > Network Analyst Tools > Network Dataset > Create Network Dataset from Template. In the dialog, select the provided XML file TransitNetworkTemplate.xml for Network Dataset Template and feature dataset BR _ Transit for Output Feature Dataset, and click Run. It creates a new network dataset TransitNetwork _ ND. Remove it from the Contents pane so that it can be edited.

In the Catalog pane, right-click network dataset TransitNetwork _ ND > Properties. In the dialog (Figure 2.8), click Travel Attributes > Costs tab. There are three Cost attributes: PublicTransitTime, WalkTime, and Length. Select PublicTransitTime; under Evaluators, select Constant for the Type of LineVariantElements (Against), and set the Value as -1 to indicate that the traversal is not allowed in the backward direction of the road network. For StopConnectors

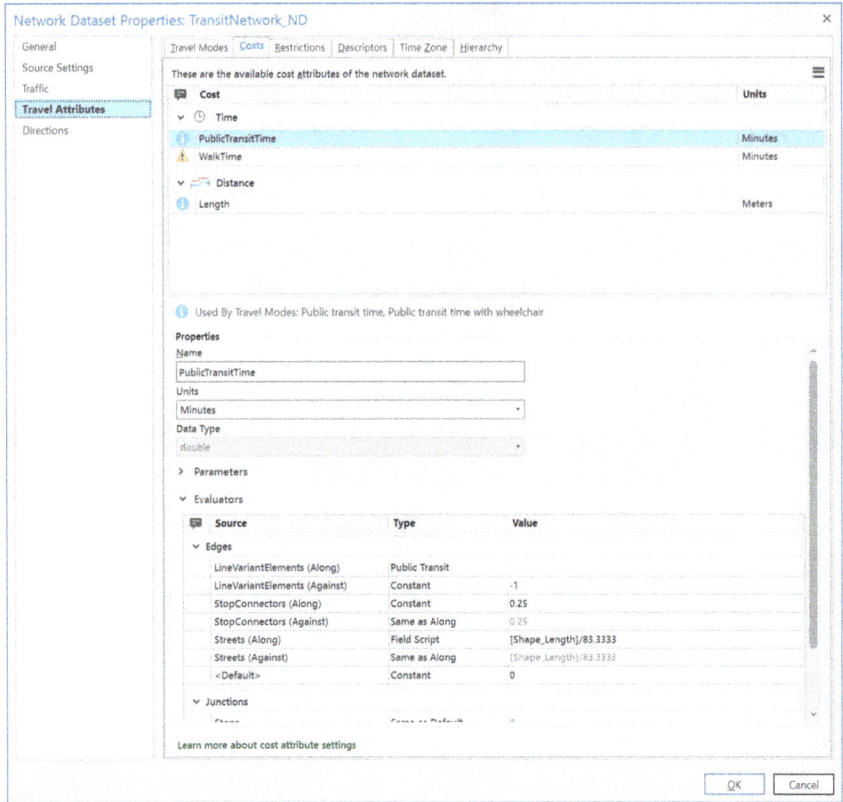

FIGURE 2.8 Building Properties of a Transit Network Dataset

(Along) and StopConnectors (Against), select Constant for the Type and set the Value as 0.25 to assume that the boarding and exiting time is around 0.25 minutes. For Streets (Along) and Streets (Against), select Field Script for the Type and [Shape _ Length]/83.3333 for the Value to assume that 83.3333 meters per minute is the default walking speed defined in the Travel Modes tab. Since we are only interested in PublicTransitTime, keep the default settings for the other two cost attributes (Walktime and Length).

In the Geoprocessing pane, click Toolboxes > Network Analyst Tools > Network Dataset > Build Network. In the dialog, choose the network dataset TransitNetwork _ ND for Input Network Dataset, and click Run.[11]

Step 12. Creating the OD Cost Matrix: Similar to step 6 in Section 2.2.2 for estimating a drive time matrix in a regular road network, we use BRPt and Hosp to define Origins and Destinations, and TransitNetwork _ ND as network dataset. On the Analysis tab, click Network Analysis drop-down menu; under Network Data Source, make sure that the network data source is TransitNetwork _ ND.

Click Network Analysis drop-down menu again and select Origin–Destination Cost Matrix. The OD Cost Matrix layer with several sublayers is added to the

Contents pane. Click the layer OD Cost Matrix to select it. The OD Cost Matrix tab appears in the Network Analysis group at the top. Define parameters as follows:

a. Click Import Origins. In the Add Locations dialog, OD Cost Matrix is Input Network Analysis Layer, and Origins is Sub Layer. Select BRPt for Input Locations, uncheck Append to Existing Locations, input 10000 for Search Tolerance, select Meters as its units, and leave other parameters as default settings. Click Apply and OK to load the 324 census block group points as origins.

b. Similarly, click Import Destinations. In a new Add Locations dialog, OD Cost Matrix is Input Network Analysis Layer, and Destinations is Sub Layer. Select Hosp for Input Locations, uncheck Append to Existing Locations, and leave the default 5000 Meters for the Search Distance and other parameters as default settings. Click Apply and OK to load the five hospitals as destinations.

c. Click the icon [icon] beside Travel Settings to open the Travel Settings dialog. For Travel Mode, keep default Name setting (Public transit time), keep Other for Type, keep default name under Costs (e.g., PublicTransitTime for Impedance and Time Cost), and take the default settings for others. Click OK to accept the settings.

d. In the Date and Time tab, select Day of Week, 8AM, and Monday, just for example. Click Run to execute the tool.[12] In the Contents pane, under the OD Cost Matrix group layer, right-click the sublayer Lines > Attribute Table to view the result. It contains OriginID and DestinationID that are identical to the field OBJECTID _ 1 in BRPt and the field OBJECTID in Hosp, respectively. Export the table to save the result as a table OD _ Transit.

Step 13. Preparing Data for Comparing Drive Times and Transit Times: The travel time ratios between the two modes play an important role in affecting transit ridership (Kuai and Wang, 2020). Here we add an extra step to compare the drive times estimated from Section 2.2.2 and transit times calibrated earlier.

Similar to step 7 in Section 2.2.2, we need to join data based on a field that combines OriginID and DestinationID. In the Contents pane, right-click the table OD _ Transit, add a new field IN _ NEAR, set its data type as TEXT, and update it as "str(!OriginID!)+' _ '+ str(!DestinationID!)." Join table OD _ Transit to DistDrive based on the common key filed IN _ NEAR.

Export the expanded table DistDrive to a new table DistDrive _ Join, which includes two fields, Total _ PublicTransitTime for transit time and Total _ Time _ min for drive time.

In the Contents pane, right-click the table DistDrive _ Join > Open. On the open table, right-click the field OriginID > Summarize. In the Summary Statistics dialog (Figure 2.9), name Output Table as DistDrive _ Mean; under Statistics Field(s), choose the fields Total _ Time _ min and Total _ PublicTransitTime, and set Mean for Statistic Type for both. Click OK to run it.[13] In the new table DistDrive _ Mean, fields MEAN _ Total _ Time _ min

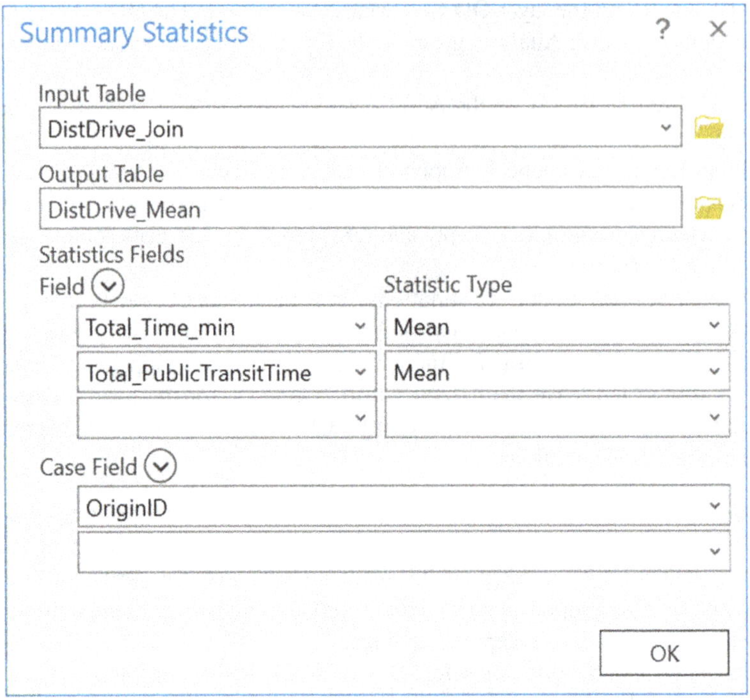

FIGURE 2.9 Dialog window for Summary Statistics.

and MEAN _ Total _ PublicTransitTime are the average travel times from each census block group to hospitals by driving and transit, respectively. Add a new field TransitToDrive (Double as Data Type), and calculate it as !MEAN _ Total _ PublicTransitTime!/!MEAN _ Total _ Time _ min!.

Figure 2.10 maps the transit-to-driving time ratios across block groups in EBRP. Driving a private vehicle is much faster than taking a public transit. The latter involves time spent on walking, waiting, boarding, and deboarding and also runs with a typically slower speed. The disadvantage is most pronounced with higher ratios in the southeast of EBRP. In the central and north EBRP, the travel time ratios are lower, as those areas have busier traffic for car driving but more bus routes and stops on designated lanes, and the public transit is a reasonable alternative for residents in those neighborhoods.

2.3 ESTIMATING DISTANCE DECAY FUNCTIONS

The *distance decay* rule demonstrates that the interaction between physical or socioeconomic objects declines with distance between them. As a result, the density, intensity, frequency, or other measures of a feature decline with distance from a reference point. It is also referred to as Waldo R. Tobler's (1970) *first law of geography* that "all things are related, but near things are more related than far things."

There are many examples of the distance decay rule. For example, the volume of air passengers between cities or the number of commuters in a city tends to decline

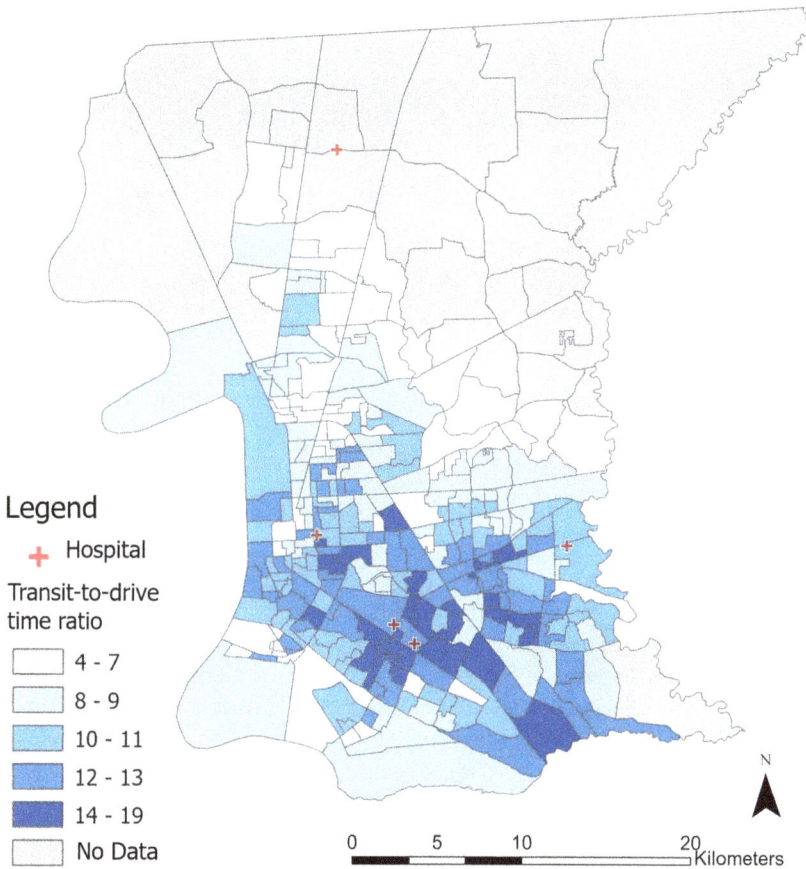

FIGURE 2.10 Transit-to-drive time ratios for census block groups in EBRP.

with trip distances (Jin et al., 2004; De Vries et al., 2009). Analysis of the journey-to-crime data suggests that the number of crime incidents drops with increasing distances between offenders' residences and crime locations (Rengert et al., 1999). The frequency of hospitalization also declines with distance from patients' residences (Delamater et al., 2013).

The distance decay rule is the foundation for many spatial analysis methods. In Chapter 3, both the kernel density estimation and the inverse distance weighted methods use a distance decay function to interpolate the value at any location based on known values at observed locations. In Chapter 4, the delineation of service areas by the Huff model assumes that the influence of a facility (or probability of customers or clients visiting a facility) declines with distance from the facility until its influence (or probability of being visited) is equal to that of other facilities. In Chapter 5, the measure of accessibility to a service facility is also modeled as a distance decay function. Chapter 6 focuses on identifying functions that capture the spatial pattern of population density decay with distance from the city center.

Some continuous distance decay functions commonly found in the literature include:

1. The *power function* $f(x) = x^{-\beta}$ (e.g., Hansen, 1959).
2. The *exponential function* $f(x) = e^{-\beta x}$ (e.g., Wilson, 1969).
3. The *square-root exponential function* $f(d) = e^{-\beta\sqrt{d}}$ (e.g., Taylor, 1983).
4. The *Gaussian function* $f(d) = \dfrac{1}{\sigma\sqrt{2\pi}} e^{-\frac{1}{2}\left(\frac{d-\mu}{\sigma}\right)^2}$ (e.g., Shi et al., 2012), and

   when $\mu = 0$ and simplifying the constant terms, it becomes the *normal function* $f(d) = e^{-\beta d^2}$.
5. The *log-normal function* $f(d) = e^{-\beta(lnd)^2}$ (e.g., Taylor, 1983).
6. The *log-logistic function* $f(d) = 1/\left(1+(d/\alpha)^{\beta}\right)$ (e.g., Delamater et al., 2013).
7. The *compound power-exponential function* $f(d) = e^{-\alpha d^{\beta}}$ (Halás et al., 2014).

The first five functions (1–5) contain one parameter β, often referred to as the *distance friction coefficient*, and functions 6–7 are more complex, with two parameters, α and β. Therefore, in terms of fitness of power such as R^2, it is not a fair comparison between them. The distance decay rule can also be modeled as a discrete set of choices or a hybrid form, such as a kernel function (i.e., a continuous function within a certain distance range and fixed values beyond), to be discussed in Subsection 5.3.2 in Chapter 5.

How do we choose a distance decay function and set its parameter values? The question can only be answered by analyzing the actual trip behavior to find the best-fitting function. Take the power function as an example. It is critical to find an appropriate value for the distance decay parameter (or distance friction coefficient) β. The power function is based on an earlier version of the *gravity model* or *spatial interaction model*, such as:

$$T_{ij} = aO_i D_j d_{ij}^{-\beta},\qquad(2.4)$$

where T_{ij} is the number of trips between areas i and j, O_i is the size of an origin i (e.g., population in a residential area), D_j is the size of a destination j (e.g., a store size), a is a scalar (constant), d_{ij} is the distance or travel time between them, and β is the distance friction coefficient.

Rearranging equation 2.4 and taking logarithms on both sides yield:

$$ln\left[T_{ij}/(O_i D_j)\right] = lna - \beta ln d_{ij}.$$

A more general gravity model is written as:

$$T_{ij} = aO_i^{\alpha_1} D_j^{\alpha_2} d_{ij}^{-\beta},\qquad(2.5)$$

TABLE 2.2

Distance Decay Functions and Log-Transforms for Regression

Distance Decay Function	Function Form	Log-Transform Used in Linear Regression	Restrictions
Power	$f(d) = d^{-\beta}$	$\ln I = A - \beta \ln d$	$I \neq 0$ and $d \neq 0$
Exponential	$f(d) = e^{-\beta d}$	$\ln I = A - \beta d$	$I \neq 0$
Square-root exponential	$f(d) = e^{-\beta \sqrt{d}}$	$\ln I = A - \beta \sqrt{d}$	$I \neq 0$
Normal	$f(d) = e^{-\beta d^2}$	$\ln I = A - \beta d^2$	$I \neq 0$
Log-normal	$f(d) = e^{-\beta(\ln d)^2}$	$\ln I = A - \beta(\ln d)^2$	$I \neq 0$

where α_1 and α_2 are the added exponents for origin O_i and destination D_j. The logarithmic transformation of equation 2.5 is:

$$\ln T_{ij} = \ln a + \alpha_1 \ln O_i + \alpha_2 \ln D_j - \beta \ln d_{ij}. \tag{2.6}$$

A similar logarithmic transformation can be applied to the other four 1-parameter distance decay functions, as summarized in Table 2.2. After their logarithmic transformations, all functions can be estimated by a *linear ordinary least square* (LOLS) *regression*.

In the conventional gravity model, the task is to estimate flows T_{ij} given nodal attractions O_i and D_j. The notion of "*reverse gravity model*" seeks to reconstruct the gravitational attractions O_i and D_j from network flow data T_{ij} (O'Kelly et al., 1995). In both the conventional and reverse gravity models, the distance friction coefficient β is a critical parameter capturing the distance decay effect. It can be estimated by the regression method as illustrated earlier, or other methods, such as linear programming (O'Kelly et al., 1995), the Particle Swarm Optimization (PSO) (Xiao et al., 2013), and the Monte Carlo simulation (Liu et al., 2012).

In some studies, there are no data on measures of the attractions (sizes) of origins and destinations (O_i and D_j), and only the interaction intensity between them (T_{ij}) is available. In that case, it is not feasible to derive the distance decay function by the gravity (or spatial interaction) model. The *complementary cumulative distribution method* provides a solution. One may imagine that all origins are located at the center and the connected destinations are distributed in different rings by distance. The interaction intensity in the center is not measurable and often set as 1 when $d_{ij} = 0$. The outward interaction intensity gradually decreases and approaches zero with increasing distance from the center.

For example, the result from Case Study 2B in the next section shows that the hospitalization volume for a corresponding band of estimated travel time (e.g., a 1-minute interval) (or its percentage out of total hospitalizations) increases up to 15 minutes and declines to 0 when the travel time approaches 180 minutes (Figure 2.13a). A complementary cumulative distribution is calibrated inversely from 180 to 0 minutes, where the cumulative percentage is 0% at 180 minutes and increases to 100%

at 0 minutes (Figure 2.13b). In other words, the complementary cumulative percentage at any travel time is the percentage of patients who travel beyond that time. The functions outlined previously can be used to capture the distribution curve, and the best-fitting one is identified by regression analysis. As the travel time interval (band) varies, the complementary cumulative distribution pattern is largely consistent, and the parameters in the fitted corresponding function change slightly.

2.4   CASE STUDY 2B: ANALYZING DISTANCE DECAY BEHAVIOR FOR HOSPITALIZATION IN FLORIDA

This case study is developed from a project reported in Wang and Wang (2022). The main data source is the State Inpatient Database (SID) from the Healthcare Cost and Utilization Project (HCUP) sponsored by the Agency for Healthcare Research and Quality (AHRQ, 2011). Specifically, the SID in Florida in 2011 is used, and it includes individual inpatient discharge records from community hospitals. The patient flow volumes are aggregated to OD flows between ZIP code areas.

The case study uses an OD flow table OD_All_Flows with 209,379 flows in geodatabase FL_HSA.gdb under the folder Florida, with the following fields:

1. Hosp _ ZoneID, Hosp _ X _ MC, and Hosp _ Y _ MC for each hospital's ID and XY-coordinates, and NUMBEDS for its staffed bed size.
2. PatientZipZoneID, XCoord, and YCoord for each ZIP code area's ID and XY-coordinates, and POPU for its population.
3. AllFlows and Total _ Time _ Min for patient service flow volume and drive time (minutes) on each OD pair.

2.4.1   Estimating Distance Decay Functions by the Spatial Interaction Model

Step 1. Extracting Nonzero OD Flows: In ArcGIS Pro, create a new project. Add OD _ All _ Flows to the map and open it. On the opened table, choose Select by Attributes. Enter the expression AllFlows > 0 to select 37,180 records with $d_{ij} > 0$. Export it to another table OD _ All _ Flows _ Reg in FL _ HSA.gdb.

Step 2. Logarithm Transformation for Variables: Open the table OD _ All _ Flows _ Reg, add five new fields (all have Double as Data Type): LnI, LnD, SqRoot _ D, Sq _ D, and Sq _ lnD. They represent lnI, lnd, $\sqrt{d}$, d^2, and $(lnd)^2$ in Table 2.2. Calculate them as follows:

```
LnI= math.log(!AllFlows!/(!NUMBEDS!*!POPU!))

LnD = math.log(!TOTAL_TIME_MIN!)

SqRoot_D = math.sqrt(!TOTAL_TIME_MIN!)

Sq_D = math.pow(!TOTAL_TIME_MIN!, 2)

Sq_lnD = math.pow(!LnD!, 2)
```

Step 3. Implementing OLS Regression of the Spatial Interaction Model in RegressionR Tool: The three regression tools in ArcGIS Pro, Ordinary Least Square (OLS), Generalized Linear Regression, and Explanatory Regression, are designed to be applied on spatial features. Here we introduce a prepared geoprocessing tool based on R programming language[14] to implement the OLS regression on a table. The same tool will be used for implementing more advanced (e.g., weighted and nonlinear) regressions in Section 6.5 of Chapter 6. See Appendix 2C for details on installing the R-based tools.

In the Catalog pane, Folders > Add Folder Connection. In the dialog, choose a disk or folder which contains CMGIS-V3-Toolbox, and click OK. In the newly added folder, browse to CMGIS-V3-Toolbox > R _ ArcGIS _ Tools > unfold the toolbox R _ ArcGIS _ Tools.tbx; double-click RegressionR to activate the tool.

In the dialog (Figure 2.11a), select OD _ All _ Flows _ Reg as Input Table, choose LnI as Dependent Variable; under Explanatory Variables, check all the five

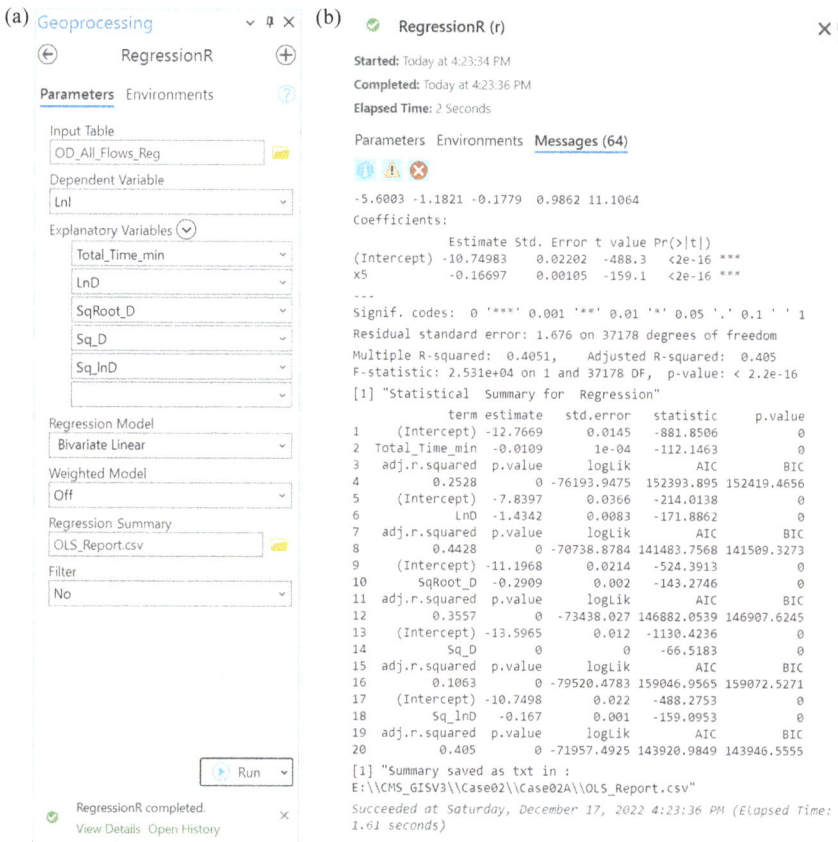

FIGURE 2.11 RegressionR with linear for single-variable model: (a) Dialog window, and (b) View Details.

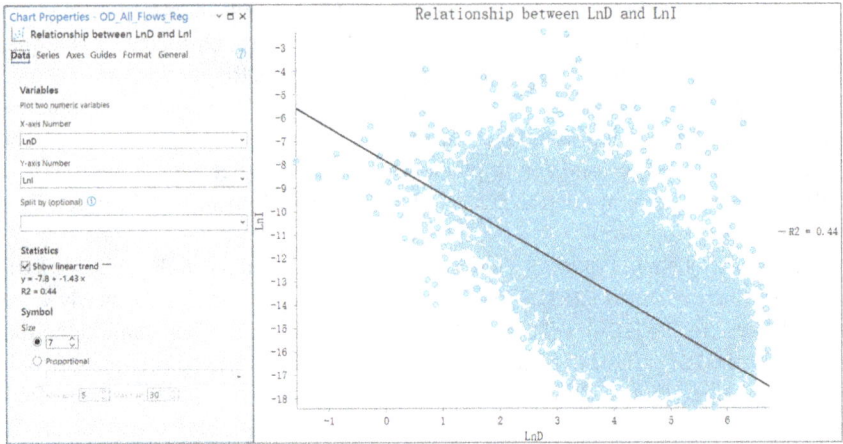

FIGURE 2.12 Scatter plot for regression on the spatial interaction model.

variables for regression: `Total _ Time _ min, LnD, SqRoot _ D, Sq _ D, Sq _ lnD`. Choose `Bivariate Linear` for Regression Model, input `OLS _ Report. csv` for Regression Summary, keep `Off` for Weighted Model and `No` for Filter, and click Run.

In View Details (Figure 2.11b), it reports all the variables, their alias names in regression (e.g., x1, x2, . . .), coefficients, standard error, AIC, BIC, Adjusted R^2, and other statistic indices.

Based on the regression results, the best-fitting distance decay function with the highest $R^2 = 0.443$ is the power function, $lnI = -7.8397 - 1.4342lnd$, where its friction coefficient $\beta = 1.434$. Figure 2.12 shows the scatter plot of lnI_{ij} and lnd_{ij} overlaid with the fitted power function and demonstrates the power function as a good fit for the observations.

2.4.2 ESTIMATING DISTANCE DECAY FUNCTIONS BY THE COMPLEMENTARY CUMULATIVE DISTRIBUTION METHOD

Step 4. Calculating Flow Volumes by 1-Minute Interval: Open the table `OD _ ALL _ Flows _ Reg`, add a field `TimeBin` (data type `Double`), and calculate it as "`math.ceil(!Total _ Time _ min!)`," which rounds a number up to the next largest integer.

Still on the table `OD _ ALL _ Flows _ Reg`, right-click the field `TimeBin` and choose Summarize. Input `OD _ Statistics` for Output Table, choose `AllFlows` for the field to be summarized, select `Sum` for its corresponding Statistic Type, and click OK. The field `SUM _ AllFlows` is summed-up flows by an increment of 1 minute. The result is Figure 2.13a, as previously discussed.

Step 5. Calculating Percentage and Complementary Cumulative Percentage of Flows: Open the table `OD _ Statistics`, right-click the field `SUM _ AllFlows`, and choose Statistics to obtain total flow volumes (SUM = 2,392,066). Add two fields `TotalFlows` and `ccdf` (both set `Double` as data type) to table `OD _ Statistics`.

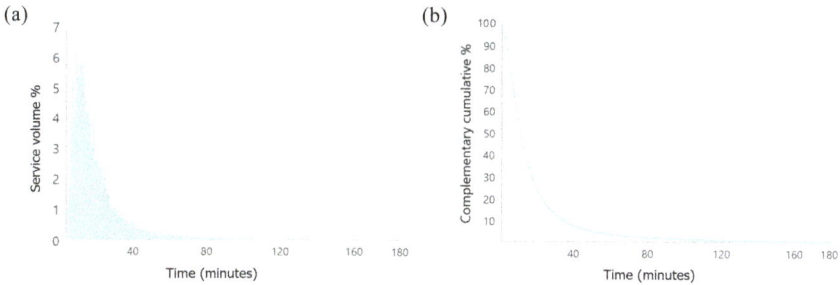

FIGURE 2.13 (a) Service volume % vs. travel time, and (b) complementary cumulative % vs. travel time.

Calculate the field `TotalFlows` as "!SUM _ AllFlows!/2392066*100" as percentage of flows.

For calculating the cumulative percentage `ccdf`, it needs traversing rows, and thus advanced coding. In the dialog of Calculate Field for `ccdf` (Figure 2.14), input "accumulate(!TotalFlows!)" in the expression box; under Code Block, input:

```
total = 0
def accumulate(increment):
    global total
    if total:
      total += increment
    else:
      total = increment
    ccdf= 100- total
return ccdf
```

Click ✔ to verify the expression in the code block box. Click Apply and OK to calculate the value for `ccdf`. The result is Figure 2.13b, as previously discussed.

Step 6. Estimating Distance Decay Function by the Complementary Cumulative Distribution: In preparing for the regression analysis, we need to cut off the long tail of distribution of `ccdf` vs. `TimeBin`. On the table OD _ Statistics > Select By Attributes > use the expression ("TimeBin","is less than","181") to select 180 records within 180 minutes, and export it to a new table OD _ Stat180.

Similar to step 2 in Section 2.4.1, add five new fields: `LnI`, `LnD`, `SqRoot _ D`, `Sq _ D`, and `Sq _ lnD` to table OD _ Stat180. Update them such as:

```
LnI= math.log(!ccdf!)

LnD = math.log(!TimeBin!)

SqRoot_D = math.sqrt(!TimeBin!)

Sq_D = math.pow(!TimeBin!,2)

Sq_lnD = math.pow(!LnD!,2)
```

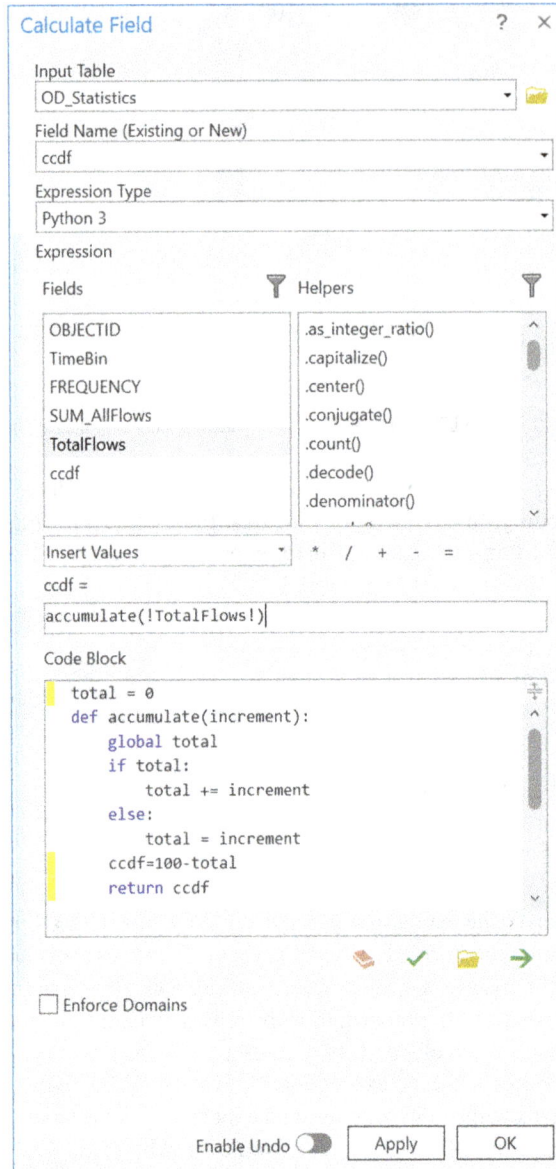

Calculate Field   ?   ✕

Input Table
OD_Statistics ▾ 📁

Field Name (Existing or New)
ccdf ▾

Expression Type
Python 3 ▾

Expression

Fields   ▽   Helpers   ▽

Fields	Helpers
OBJECTID	.as_integer_ratio()
TimeBin	.capitalize()
FREQUENCY	.center()
SUM_AllFlows	.conjugate()
TotalFlows	.count()
ccdf	.decode()
	.denominator()

Insert Values ▾   *   /   +   -   =

ccdf =

```
accumulate(!TotalFlows!)
```

Code Block

```
total = 0
def accumulate(increment):
    global total
    if total:
        total += increment
    else:
        total = increment
    ccdf=100-total
    return ccdf
```

📘   ✓   📁   →

☐ Enforce Domains

Enable Undo ◉   Apply   OK

FIGURE 2.14 Calculate complementary cumulative distribution percentages.

Similar to step 3 in Section 2.4.1, run the OLS regressions by the RegressionR tool. In the dialog similar to Figure 2.11(a), replace Total _ Time _ min with TimeBin, input OLS _ Report _ ccdf.csv for Regression Summary, and all other parameters are the same. Based on the results summarized in Table 2.3, the best-fitting distance decay function is the log-normal function $ln I = 4.9563 - 0.2096 \, ln \, Sq\_lnd$ with $R^2 = 0.997$.

Figure 2.15a–b summarizes the steps for case studies 2A and 2B.

TABLE 2.3

Distance Decay Function Regression Results

Distance Decay Function in Log-Transform	By the Spatial Interaction Model (n = 37,180)			By the Complementary Cumulative Distribution Method (n = 180)		
	A	β	Adjust R²	A	β	Adjust R²
$\ln I = A - \beta d$	−12.7669	0.0109	0.253	3.3574	0.0255	0.900
$\ln I = A - \beta \ln d$	−7.8397	1.4342	0.443	7.1015	1.4368	0.951
$\ln I = A - \beta \sqrt{d}$	−11.1968	0.2909	0.356	5.0099	0.4411	0.982
$\ln I = A - \beta d^2$	−13.5965	0.0000*	0.106	2.3764	0.0001	0.716
$\ln I = A - \beta (\ln d)^2$	−10.7498	0.1670	0.405	4.9563	0.2096	0.997

Note: *A very small value, precisely 2.498e-07, as reported in the View Details.

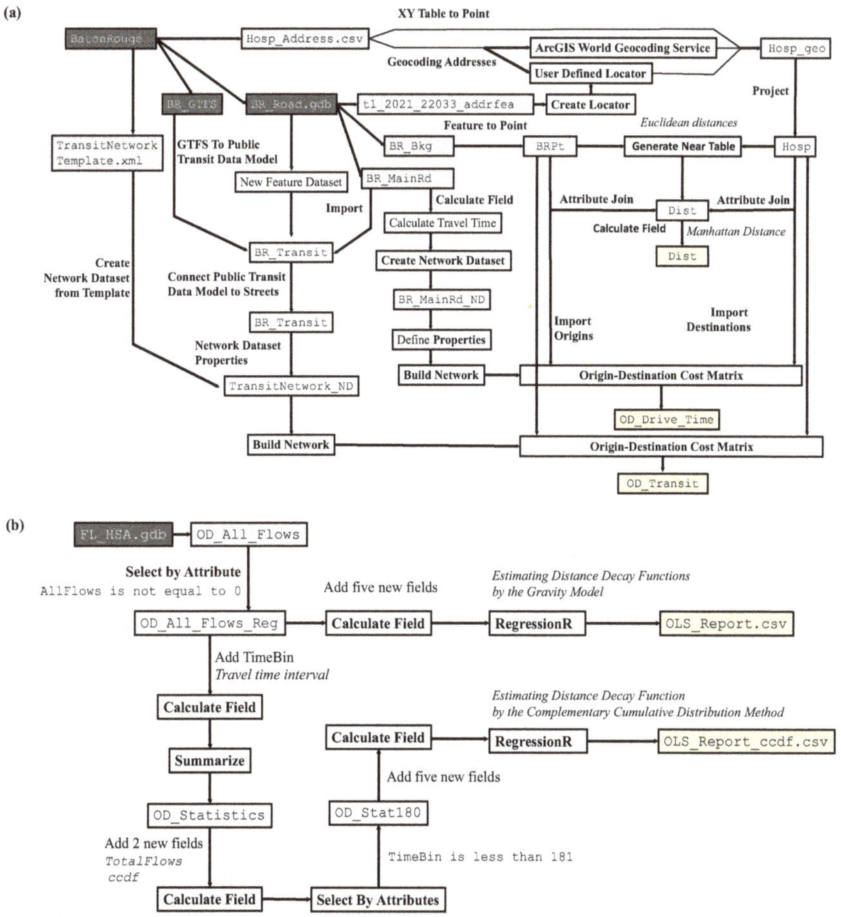

FIGURE 2.15 Flowcharts for (a) Case Study 2A and (b) Case Study 2B.

2.5　SUMMARY

This chapter covers several basic spatial analysis tasks:

1. Geocoding from geographic coordinates and street addresses
2. Measuring Euclidean and geodesic distances
3. Measuring drive distance or time via a road network
4. Measuring transit travel time via a transit network
5. Estimating distance decay functions by the gravity (spatial interaction) model
6. Estimating distance decay functions by the complementary cumulative distribution method

Calibrating Euclidean and geodesic distances is straightforward in GIS. Computing network travel time via driving or public transit requires the corresponding network data and takes more effort to implement. Most GIS data of transit systems (e.g., in GTFS format) are publicly accessible and can be converted and integrated into ArcGIS with reasonable effort. Calibrating a drive or transit time matrix can also be achieved by using online routing platforms such as the Google Maps API (Appendix 2B). While the use of public transit is limited in geographic coverage and ridership, especially in the USA, it is an important transportation mode to be considered in many studies (e.g., disparity in spatial accessibility of health care, mobility, risk in exposure to infectious diseases). Several case studies in other chapters need to compute Euclidean or geodesic distance, network distance or travel time, and thus provide additional practice for developing these basic skills in spatial analysis.

A distance decay function is an analytical way to capture the spatial behavior of population. A flatter gradient in a distance decay function usually corresponds to a longer range of travel on average. For example, in seeking health care, a longer travel range by a demographic group may indicate more travel burden and thus a disadvantaged location for the group; in the meantime, it may also indicate a group with higher mobility and being more selective in health cares received (e.g., bypassing closer providers for more distant and perhaps higher-quality services) (Jia et al., 2019). It is similar to the paradox of residential choice as city dwellers balance proximity to jobs for saving in commuting versus more spacious housing, as explained in Section 6.1.1 of Chapter 6. As a result, average commuting range tends to increase with income, but the highest income earners retreat to shorter commutes (Wang, 2003). A distance decay function serves as an important building block in many advanced spatial models, such as the Garin–Lowry model in Chapter 10, urban traffic simulation in Chapter 12, and the agent-based crime simulation model in Chapter 13.

APPENDIX 2A: THE VALUED GRAPH APPROACH TO THE SHORTEST-ROUTE PROBLEM

The *valued graph*, or *L-matrix*, provides another way to solve the shortest-route problem (Taaffe et al., 1996, pp. 272–275).

For example, a network is shown as in Figure A2.1. The network resembles the highway network in north Ohio, with node 1 for Toledo, 2 for Cleveland, 3 for

Toledo ①——— 116 ———② Cleveland

155   142   113

77   ④   76

Dayton ⑤————————③ Cambridge
        Columbus

Two-step connection 1–3
(1,1) + (1,3) = 0 + M = M
(1,2) + (2,3) = 116 + 113 = 229
(1,3) + (3,3) = M + 0 = M
(1,4) + (4,3) = M + 76 = M
(1,5) = (5,3) = 155 + M = M

Nodes	1	2	3	4	5
1	0	116	M	M	155
2	116	0	113	142	M
3	M	113	0	76	M
4	M	142	76	0	77
5	155	M	M	77	0

Nodes	1	2	3	4	5
1	0	116	229	232	155
2		0	113	142	219
3			0	76	153
4				0	77
5					0

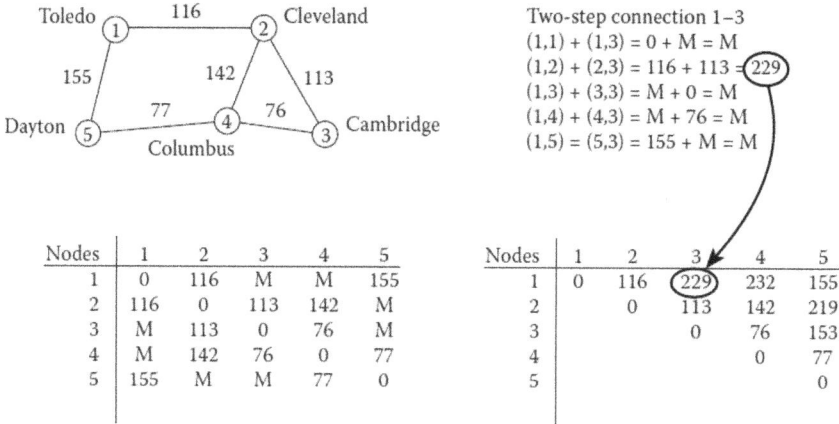

FIGURE A2.1 A valued-graph example.

Cambridge, 4 for Columbus, and 5 for Dayton. We use a matrix L^1 to represent the network, where each cell is the distance on a direct link (one-step link). If there is no direct link between two nodes, the entry is M (a very large number). We enter 0 for all diagonal cells $L^1(i, i)$ because the distance is 0 to connect a node to itself.

The next matrix L^2 represents two-step connections. All cells in L^1 with values other than M remain unchanged because no distances by two-step connections can be shorter than a one-step (direct) link. We only need to update the cells with the value M. For example, $L^1(1, 3) = M$ needs to be updated. All possible "two-step" links are examined:

$$L^1(1, 1) + L^1(1, 3) = 0 + M = M$$
$$L^1(1, 2) + L^1(2, 3) = 116 + 113 = 229$$
$$L^1(1, 3) + L^1(3, 3) = M + 0 = M$$
$$L^1(1, 4) + L^1(4, 3) = M + 76 = M$$
$$L^1(1, 5) + L^1(5, 3) = 155 + M = M$$

The cell value $L^2(1, 3)$ is the minimum of all the preceding links, which is $L^1(1, 2) + L^1(2, 3) = 229$. Note that it records not only the shortest distance from 1 to 3 but also the route (through node 2).

Similarly, other cells are updated, such as $L^2(1, 4) = L^1(1, 5) + L^1(5, 4) = 155 + 77 = 232$, $L^2(2, 5) = L^1(2, 4) + L^1(4, 5) = 142 + 77 = 219$, $L^2(3, 5) = L^1(3, 4) + L^1(4, 5) = 76 + 77 = 153$, and so on. The final matrix L^2 is shown in Figure A2.1.

By now, all cells in L^2 have values other than M, and the shortest-route problem is solved. Otherwise, the process continues until all cells have values other than M. For example, L^3 would be computed as:

$$L^3(i,j) = \left\{ L^1(i,k) + L^2(k,j), \forall k \right\}$$

APPENDIX 2B: ESTIMATING DRIVE TIME AND TRANSIT TIME MATRICES BY GOOGLE MAPS API

Google launched the Google Maps API, a JavaScript API, to allow customization of online maps in 2005. The Google Maps API enables one to estimate the travel time without reloading the web page or displaying portions of the map. An earlier version of the Python program was developed to use the Google Maps API for computing an OD travel time matrix (Wang and Xu, 2011). An improved version `GoogleODTravelTimePro.py` by Wang and Wang (2022), available under the subfolder Scripts, is used here to illustrate its usage.

The program reads the origin (O) and destination (D) layers of point or polygon features in a *projected coordinate system* and automatically calculates the latitudes and longitudes of all features by their point locations or geographic centroids of the polygons. The data are fed into a tool in Python that automates the process of estimating the travel time and distance matrix between a set of origins and a set of destinations at a time by calling the Google Maps Distance Matrix API. The iterations stop when the program reaches the last OD combination. The result is saved in an ASCII file (with a file extension. txt) listing each origin, each destination, travel time (in *minutes*), and distance (in *miles*) between them. The origins and destinations are represented by the OBJECTIDs of the input features, which can be used to link the travel time result.

Before using this service, an API key is needed to authenticate each request associated with this program.[15] The program can be added to ArcGIS Pro as a tool, and the following steps illustrate its implementation (Figure A2.2).

1. Create a new project. In the Catalog sidebar, right-click Toolboxes > Add Toolbox. In the dialog "Add Toolbox," navigate to the folder CMGIS-V3-Toolbox > ArcPy-tools, select "Google API Pro.tbx," and click OK to add it.

2. Click the newly added Toolbox "Google API Pro.tbx." Under the Catalog pane, right-click the script named "Google OD Travel Time" and select Properties to activate the Tool Properties dialog. Under Execution, click Browse 📁 on the right side of the Script File, navigate to the "Scripts" subfolder in the "ArcPy-tools" folder, and select GoogleODTravelTimePro.py.

3. Under the Tool Properties > Parameters, there are five predefined parameters (API Key, Input Features as Origins, Input Features as Destinations, Travel Mode, and Output Table in Data Types of String, Feature Class, Feature Class, String, and Text File, respectively), as shown in Figure A2.3. All parameters are required, and the former four parameters have a Direction of Input, and the last one has a Direction of Output. For the Travel Mode, Value List is selected for Filter. In the Value List Filter dialog, there are two inputs: Driving and Transit. Click OK to close the window. Driving is set for Default. Click OK to close the Tool Properties dialog.

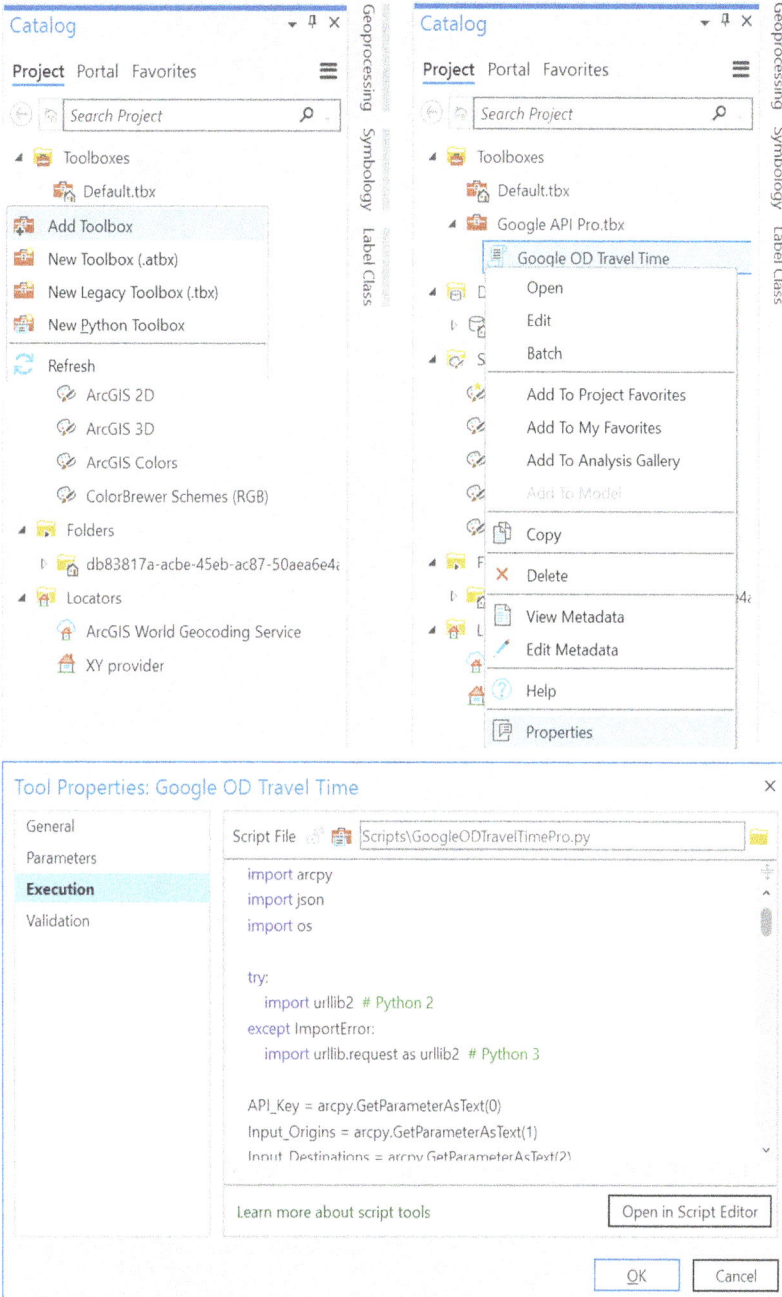

FIGURE A2.2 Add Google API Pro toolbox.

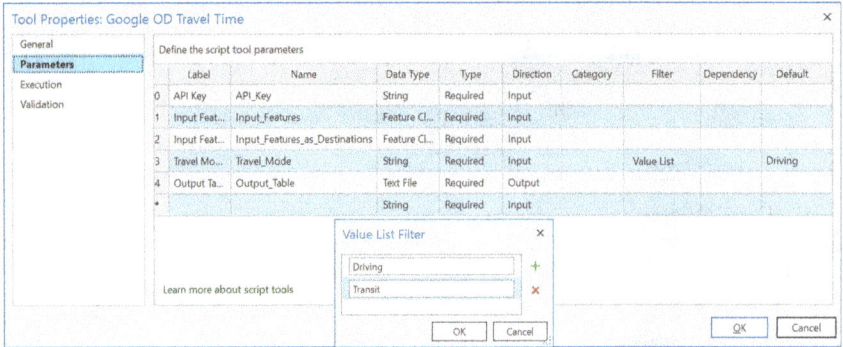

FIGURE A2.3 Tool property for Google API Pro toolbox.

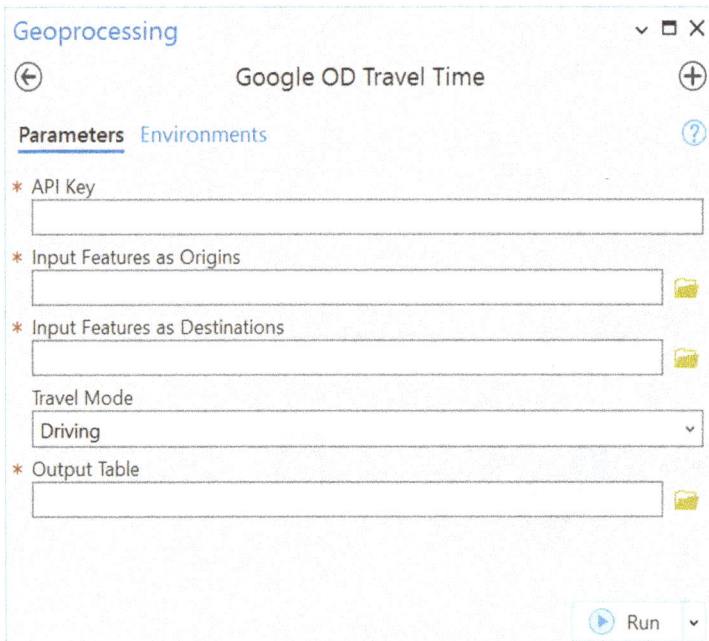

FIGURE A2.4 Google Maps API tool for computing OD travel time matrix.

4. Double-click the script "Google OD Travel Time" under the "Google API Pro.tbx"[16] to activate the program in the Geoprocessing pane.

As shown in Figure A2.4, a user needs to input the Google API key, two point or polygon features in a projected coordinate system as origins and destinations, and a text file name as the result. For example, as in Section 2.2, one may select census block group layer (BRPt) and hospital layer (Hosp) for the Input Features as Origins and Input Features as Destinations, respectively, and name the Output Table as (1) `DriveTime.txt` when Driving is chosen for the Travel Mode, or (2) `TransitTime.txt` when Transit is chosen for the Travel Mode. The total computational time depends on the size of OD travel time matrix, the performance of the running computer, the Internet connection speed, and the request limit.[17]

Note that some OD trips derived from the transit mode may have blank values for travel time and distance. When neither the `departure _ time` nor the `arrival _ time` is specified in the request, the `departure _ time` defaults to the current time running the program, which may happen to have no public transit in operation for the routes. In addition, the Google Maps Distance Matrix API provides multiple transit modes, such as bus, subway, train, rail, and tram, which can be specified in the `transit _ mode` parameter as one or more composite modes of transit. The preferences, such as `less _ walking` and `fewer _ transfers`, can also be specified in the `transit _ routing _ preference` parameter for the transit requests. Other travel modes, such as walking and bicycling, are also supported in the Distance Matrix API when the pedestrian paths, sidewalks, bicycle paths, and preferred streets are available on the road network. For simplicity, this program only defines driving or transit in the mode parameter for the OD trip requests and does not specify any detailed transit modes and preference.[18]

Wang and Xu (2011) computed the travel time between the city center and each census tract centroid in Baton Rouge and found that travel time by the Google Maps API approach was consistently longer than that by the ArcGIS Network Analyst approach (Figure A2.5). The estimated travel time by either method correlates well with the Euclidean distance from the city center (with a $R^2 = 0.91$ for both methods). However, the regression model of travel time against corresponding distances by

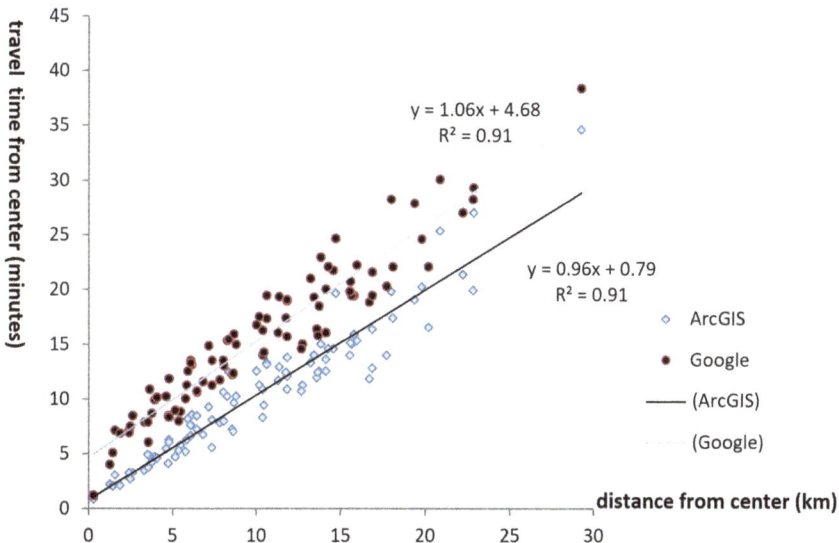

FIGURE A2.5 Estimated travel times by ArcGIS and Google.

Google has a significant intercept of 4.68 minutes (vs. a negligible 0.79-minute intercept in the model of travel time by ArcGIS). The 4.68-minute intercept by Google
probably reflects the elements of starting and ending time on a trip (getting on and
off a street on the road network). This is consistent with our daily travel experience
and empirical data (Wang, 2003, p. 258). The regression model for the travel time
by Google also has a slightly steeper slope (1.06) than the slope in the model for the
travel time by ArcGIS, but this difference is minor.

Some advantages of using the Google Maps API approach include the convenience of not preparing a road network dataset and the ability of tapping into Google's
updated road data and algorithm accounting for the impact of traffic. There are also
some major limitations. The most concerning one is that Google's query limit for
geolocation requests (currently 2,500 per day) may not meet the demand of many
spatial analysis tasks. The computational time also seems long. Another drawback
for many advanced researchers is that a user has neither control over the data quality
nor any editing permissions. Furthermore, the tool can only generate the current
travel time and thus is not suitable for work that needs the travel time in the past.

APPENDIX 2C: INSTALLING GEOPROCESSING TOOLS BASED ON R

R-based geoprocessing tools refer to the tools built with the support of R-ArcGIS Bridge
and related R programming packages. The R package *arcgisbinding* acts as a bridge to
connect the statistical capabilities of the R language with the spatial science of ArcGIS.

For detailed information, users can go the R-ArcGIS Bridge website (www.esri.
com/en-us/arcgis/products/r-arcgis-bridge/overview) to check the new updates, and
go to GitHub repository (https://r.esri.com/) for more information.

To use the R tools for ArcGIS Pro, users need to install R, core package *arcgisbinding*, and other packages wrapped in the tools.

1 Download and Install R

Go to the official website (https://cran.r-project.org/) to download R. To match
ArcGIS Pro 3.0, download R 4.2.x (x64). For example, the link for R 4.2.2 is: https://
cloud.r-project.org/bin/windows/base/R-4.2.2-win.exe.

It is strongly suggested not to install R in the C drive, which may cause the default
library unwritable.

2 Install Arcgisbinding Package

There are three ways for installing the core package.

1. *Install it in R (recommended).* Input and run the following code:

```
install.packages("arcgisbinding", repos="https://r.esri.
com", type="win.binary")
```

2. *Install it with local zip in R.* Download the zip file from https://r.esri.com/
   bin/, and install it as a local package in R.
3. *Install with r-bridge-install Python tools.* Go to the GitHub website (https://
   github.com/R-ArcGIS/r-bridge-install) > click the green button Code

   < > Code ▼ , and choose 🔽 Download Zip to download the zip file. Unzip
   it in a folder > access the folder in ArcGIS Pro Catalog > R Integration >

Install R bindings. It also contains the other three tools: Print R Version, R installation Details, and Update R bindings.

3 Check Installation and Update It via ArcGIS Pro

Open ArcGIS Pro > ⚙ Settings Settings > Options.[19] In the Options dialog, go to Application > Geoprocessing > R-ArcGIS Support > click 📂 Browse icon to choose R home directories (or use the one auto-detected by ArcGIS) > click Check for Updates > Download Latest Version.

4 Configure the System Variable for R Path

This step specifies the R path for the *system variable*. In Windows, Search > Input Edit the system environment variables. In the pop-up dialog of System Properties (Figure A2.6), click Environment Variables. In the dialog of Environment Variables, under User Variables, click New > for New User Variable, and enter Variable

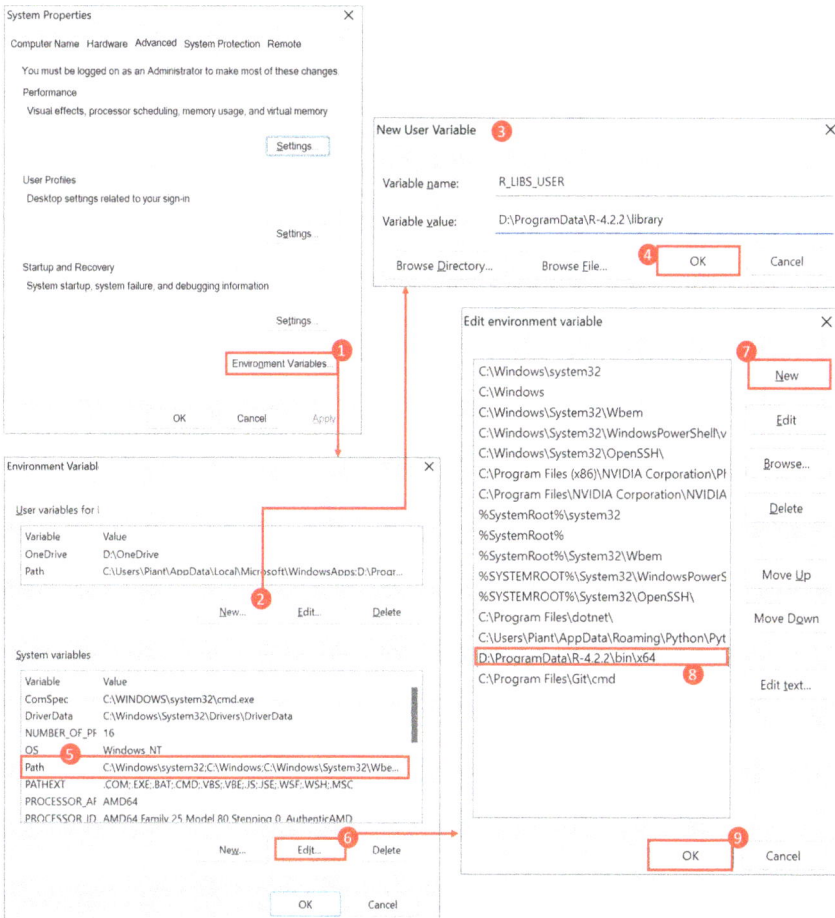

FIGURE A2.6 Configure system variable for R path.

Name R _ LIBS _ USER, and Variable Value D:\ProgramData\R-4.2.1\ library for the location of R; under System Variables, in Path > click Edit > for Edit environment variable. Click New and add the path D:\ProgramData\R-4.2.1\ bin\x64 for the corresponding location of R, and click OK to close the dialogs.

5 Install Related R Packages

This step installs packages in R. Similar to step 2, one may input and run the command:

```
install.packages("packagename")
```

Where packagename can be any package. It installs a package one at a time. One can also use a combination list to install multiple packages at once:

```
packagelist <- c("smacpod","sf")
```

```
install.packages(packagelist)
```

Here we use a combination list to install all packages used in the R-ArcGIS-Tools:

```
packagelist<-c(

"dplyr","broom","tidyr",#basic data manipulation

"minpack.lm",# linear and nonlinear regression

"psych","GPArotation", # PCA and FA

"quadprog", "Rglpk","lpSolve", # Linear Programming

"sf","sp","raster","rgdal","maptools", #Basic spatial data
manipulation

"smacpod","spatstat", # Spatial statistics

"rgeoda","spatialreg","spdep", #Spatial modelling

"np","spatialEco","rstudioapi" #STKDE)
```

```
install.packages(packagelist)
```

The script is also saved in the file R _ tool _ package.txt under the folder CMGIS-V3-Toolbox > R _ ArcGIS _ Tools. Users can copy and paste to run it in R.

To add all R-based toolboxes in this book, go to the Catalog pane, right-click Toolboxes > Add Toolbox to open its dialog window, add R _ ArcGIS _ Tools. tbx under the folder R _ ArcGIS _ tools, and click OK.

NOTES

1   Two hospitals, Baton Rouge General Medical Center (Mid-City Campus) and the General, share the same location (street address) and are thus combined in one record.
2   Another geodatabase BRRoad.gdb includes additional features Hosp and BRPt and network dataset BR_MainRd_ND, which are generated from steps 1, 2, and 8. It is provided so that readers can work on Case Study 4A without completing this case study.

3   The network source dataset needs to be a feature dataset so that other feature classes and data created in the next step can be saved there. If the initial source is a road network as a feature class (or in other formats), create a new feature dataset and import that feature class to the feature dataset.

4   The coordinates recorded in the hospitals file Hosp_Address.csv are in geographic coordinates, so the project environment also needs to be unprojected.

5   This tool could also be accessed in the Geoprocessing pane: under Toolboxes > Data Management Tools > Features > XY Table to Point, or by searching "XY Table to Point."

6   The feature class State may appear under Layers > NAD 1983, which can be clicked directly.

7   Note that the Geocode Addresses service consumes a user's ArcGIS Pro credits.

8   As some of BRPt is more than 5,000 meters away from the nearest street in the road network BR_MainRd, we revise the default search tolerance of 5,000 meters to 10,000 meters to ensure all census block group points will be snapped to the road network.

9   Close a warning message on the length of field "Name."

10 We added the two fields for estimating transit time. Based on the corresponding FCC codes, the field ROAD_CLASS is used to configure the walking directions, and its values are equal to the field ROADCLASS. The field RestrictPedestrians is assigned a value "N" to prohibit pedestrians for highways and other roads passable only by vehicles, and "Y" otherwise. More information about CFCC code lookup table refers to https://support.esri.com/en/technical-article/000001496.

11 Close a warning sign for errors to proceed.

12 There are warnings under the Solve message to show that no destinations were found for some origins. Therefore, fewer than $324 \times 5 = 1,620$ OD pairs are retained.

13 A warning indicates that input fields contain null values because some OD pairs do not have public transit times. The summarize tool automatically skips the null values and identifies the minimal transit time from those valid records.

14 For more information about R programming, refer to https://cran.r-project.org/.

24 Readers must go to the Google Cloud Platform Console to create a project with a billing account to enable the service of Distance Matrix API and then create an API key. The Google Maps Platform uses a pay-as-you-go pricing model. It provides a $200 free monthly credit that allows 40,000–50,000 OD records per month, enough to support this case study. Visit https://developers.google.com/maps/documentation/distance-matrix/overview for more detail.

16 You can click "Google API Pro.tbx" to use it directly. If it is desirable to have it loaded by default every time you open ArcGIS Pro, right-click "Google API Pro.tbx" and select "Add to Favorites,"

17 See details on the billing via https://developers.google.com/maps/documentation/distance-matrix/usage-and-billing. For the case study, it took us approximately 15 and 21 seconds for the two travel time matrices.

18 More details about transportation mode in Distance Matrix API can be found via https://developers.google.com/maps/documentation/distance-matrix/overview.

19 In an existing ArcGIS Project, click Project in the top ribbon > Options.

3 Spatial Smoothing and Spatial Interpolation

This chapter covers two generic tasks in GIS-based spatial analysis: spatial smoothing and spatial interpolation. Spatial smoothing and spatial interpolation are closely related and are both useful to visualize spatial patterns and highlight spatial trends. Some methods (e.g., kernel density estimation, or KDE) can be used in either spatial smoothing or interpolation. There are varieties of spatial smoothing and spatial interpolation methods. This chapter covers those most commonly used.

Conceptually similar to moving averages (e.g., smoothing over a longer time interval), *spatial smoothing* computes the averages using a larger spatial window. Section 3.1 discusses the concepts and methods for spatial smoothing. *Spatial interpolation* uses known values at some locations to estimate unknown values at other locations. Section 3.2 presents several popular *point-based spatial interpolation* methods. Section 3.3 uses a case study of place-names in southern China to illustrate some basic spatial smoothing and interpolation methods. Section 3.4 discusses *area-based spatial interpolation*, which estimates data for one set of areal units with data for a different set of areal units. Area-based interpolation is useful for the aggregation and integration of data based on different areal units. Section 3.5 presents a case study on spatial interpolation between different census areal units. Section 3.6 extends the popular KDE analysis to spatiotemporal KDE (STKDE) by adding a temporal dimension. A set of spatiotemporal visualization tools in ArcGIS Pro enables us to visualize temporal trends, spatial patterns, and the intersection between them with a space–time cube in 3D. A case study demonstrates how the STKDE is implemented to detect spatiotemporal crime hotspots in Section 3.7. Section 3.8 concludes the chapter with a brief summary.

3.1 SPATIAL SMOOTHING

Like moving averages that are calculated over a longer time interval (e.g., five-day moving average temperatures), *spatial smoothing* computes the value at a location as the average of its nearby locations (defined in a spatial window) to reduce spatial variability. Spatial smoothing is a useful method for many applications. One is to address the *small population problem*, which will be explored in detail in Chapter 9. The problem occurs in areas with small populations, where the rates of rare events, such as cancer or homicide, are unreliable due to random errors associated with small numbers. The occurrence of one case can give rise to unusually high rates in some areas, whereas the absence of cases leads to a zero rate in many areas. Another application is for examining spatial patterns of point data by converting discrete point data to a continuous density map, as illustrated in Case Study 3A (Part 1). This section discusses two common spatial smoothing methods (floating catchment area method and kernel density estimation), and Appendix 3A introduces the *empirical Bayes estimation*.

DOI: 10.1201/9781003292302-4

3.1.1 FLOATING CATCHMENT AREA (FCA) METHOD

The *floating catchment area (FCA) method* draws a circle or square around a location to define a filtering window and uses the average value (or density of events) within the window to represent the value at the location. The window moves across the study area until averages at all locations are obtained. The average values have less variability and are thus spatially smoothed values. The FCA method may also be used for other purposes, such as accessibility measures (see Section 5.2).

Figure 3.1 shows part of a study area with 72 grid-shaped tracts. The circle around tract 53 defines the window containing 33 tracts (a tract is included if its centroid falls within the circle), and therefore the average value of these 33 tracts represents the spatially smoothed value for tract 53. A circle of the same size around tract 56 includes another set of 33 tracts that defines a new window for tract 56. The circle centers on each tract centroid and moves across the whole study area until smoothed values for all tracts are obtained. Note that windows near the borders of a study area do not include as many tracts and cause a less degree of smoothing. Such an effect is referred to as *edge effect*.

The choice of window size has a significant effect on the result. A larger window leads to stronger spatial smoothing and thus better reveals regional than local patterns, and a smaller window generates reverse effects. One needs to experiment with different sizes and choose one with balanced effects. Furthermore, the window size may vary to maintain a certain degree of smoothing across a study area. For example, for mapping a reliable disease rate, it is desirable to have the rate estimate in

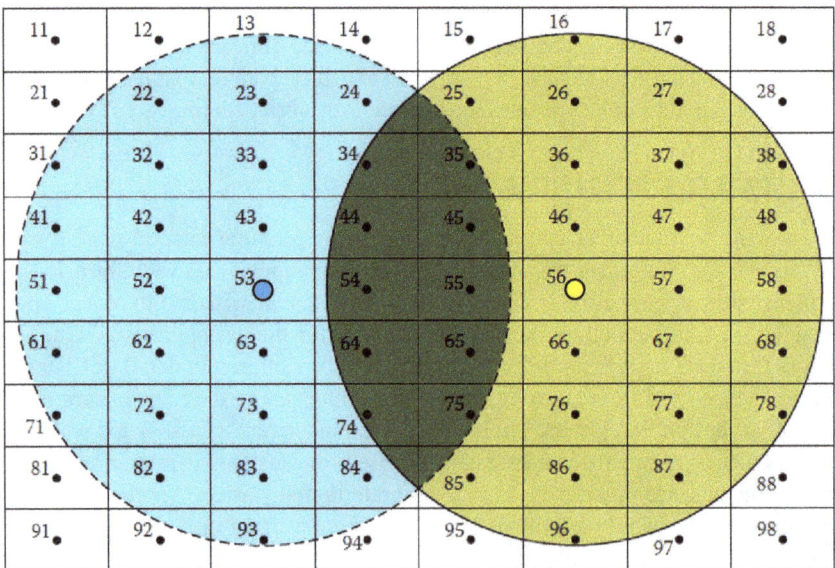

FIGURE 3.1 The floating catchment area (FCA) method.

a spatial window, including population above a certain threshold (or of a similar size). In order to do so, one has to vary the spatial window size to obtain a comparable base population. This is a strategy adopted in *adaptive spatial filtering* (Tiwari and Rushton, 2004; Beyer and Rushton, 2009).

Implementing the FCA in ArcGIS Pro is demonstrated in Section 3.3.1 in detail. We first compute the distances between all objects, and then distances less than or equal to the threshold distance are extracted.[1] In ArcGIS Pro, we then summarize the extracted distance table by computing average values of attributes by origins. Since the table only contains distances within the threshold, only those objects (destinations) within the window are included and form the catchment area in the summarization operation. This eliminates the need for programming that implements iterations of drawing a circle and searching for objects within the circle.

3.1.2 KERNEL DENSITY ESTIMATION (KDE) METHOD

The *kernel density estimation* (KDE) bears some resemblance to the FCA method. Both use a filtering window to define neighboring objects. Within the window, the FCA method does not differentiate between far and nearby objects, whereas the kernel density estimation weights nearby objects more than far ones. The method is particularly useful for analyzing and displaying point data. The occurrences of events are shown as a map of scattered (discrete) points, which may be difficult to interpret. The kernel density estimation generates a density of the events as a continuous field and thus highlights areas of concentrated events as peaks and areas of lower densities as valleys.

A kernel function looks like a bump centered at each point x_i and tapering off to 0 over a bandwidth or window. See Figure 3.2 for the illustration. The kernel density at point x at the center of a grid cell is estimated to be the sum of bumps within the bandwidth:

$$\hat{f}(x) = \frac{1}{nh^d} \sum_{i=1}^{n} K\left(\frac{x - x_i}{h}\right),$$

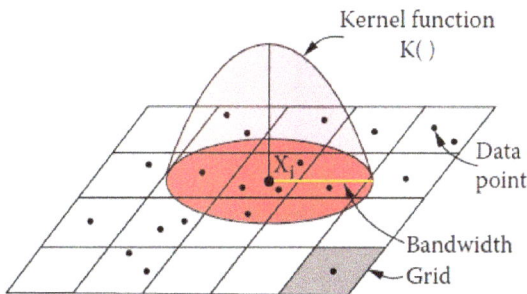

FIGURE 3.2 Kernel density estimation.

where $K()$ is the kernel function, h is the bandwidth, n is the number of points within the bandwidth, and d is the data dimensionality. Silverman (1986, p. 43) provides some common kernel functions. For example, when $d = 2$, a commonly used kernel function is defined as:

$$\hat{f}(x) = \frac{1}{nh^2\pi} \sum\nolimits_{i=1}^{n} [1 - \frac{(x-x_i)^2 + (y-y_i)^2}{h^2}]^2, \qquad (3.1)$$

where $(x-x_i)^2 + (y-y_i)^2$ measures the deviation in x- and y-coordinates between points (x_i, y_i) and (x, y).

Similar to the effect of window size in the FCA method, larger bandwidths tend to highlight regional patterns, and smaller bandwidths emphasize local patterns (Fotheringham et al., 2000, p. 46). The KDE method discussed in this context derives spatially smoothed values based on neighboring observations around each point from the input layer. It may also be used for spatial interpolation, where KDE-derived values are estimated at any given locations, e.g., raster cells. ArcGIS Pro has a built-in tool to implement KDE for spatial interpolation. The KDE tool can be accessed in the Geoprocessing pane > Toolboxes > Spatial Analyst Tools > Density > Kernel Density.

3.2 POINT-BASED SPATIAL INTERPOLATION

Point-based spatial interpolation includes global and local methods. *Global interpolation* utilizes all points with known values (control points) to estimate an unknown value. *Local interpolation* uses a sample of control points (e.g., points within a certain distance) to estimate an unknown value. As Tobler's (1970) *first law of geography* states, "everything is related to everything else, but near things are more related than distant things." The choice of global vs. local interpolation depends on whether faraway control points are believed to have an influence on the unknown values to be estimated. There are no clear-cut rules for choosing one over the other. One may consider the scale from global to local as a continuum. A local method may be chosen if the values are most influenced by control points in a neighborhood. A local interpolation also requires less computation than a global interpolation (Chang, 2004, p. 277). One may use *validation* techniques to compare different models. For example, the control points can be divided into two samples: one sample is used for developing the models, and the other sample is used for testing the accuracy of the models. This section surveys two global interpolation methods briefly and focuses on three local interpolation methods.

3.2.1 GLOBAL INTERPOLATION METHODS

Global interpolation methods include trend surface analysis and regression model. *Trend surface analysis* uses a polynomial equation of x- and y-coordinates to approximate points with known values, such as:

$$z = f(x, y),$$

where the attribute value z is considered as a function of x- and y-coordinates (Bailey and Gatrell, 1995). For example, a cubic trend surface model is written as:

$$z(x,y) = b_0 + b_1 x + b_2 y + b_3 x^2 + b_4 xy + b_5 y^2 + b_6 x^3 + b_7 x^2 y + b_8 xy^2 + b_9 y^3.$$

The equation is usually estimated by an ordinary least square regression. The estimated equation is then used to project unknown values at other points.

Higher-order models are needed to capture more complex surfaces and yield higher R-square values (goodness of fit) or lower RMS in general.[2] However, a better fit for the control points is not necessarily a better model for estimating unknown values. Validation is needed to compare different models. If the dependent variable (i.e., the attribute to be estimated) is binary (i.e., 0 and 1), the model is a *logistic trend surface model* that generates a probability surface. A local version of trend surface analysis uses a sample of control points to estimate the unknown value at a location and is referred to as *local polynomial interpolation*.

ArcGIS offers up to 12th-order trend surface model. To access the method, in the Geoprocessing pane > Toolboxes > Spatial Analyst Tools > Interpolation > Trend.

A *regression model* uses a linear regression to find the equation that models a dependent variable based on several independent variables and then uses the equation to estimate unknown points (Flowerdew and Green, 1992). Regression models can incorporate both spatial (not limited to x- and y-coordinates) and attribute variables, whereas trend surface analysis only uses x- and y-coordinates as predictors.

3.2.2 Local Interpolation Methods

The following discusses three popular local interpolators: inverse distance weighted, thin-plate splines, and Kriging.

The *inverse distance weighted* (IDW) method estimates an unknown value as the weighted average of its surrounding points, in which the weight is the inverse of distance raised to a power (Chang, 2004, p. 282). Clearly, the IDW utilizes Tobler's first law of geography. The IDW is expressed as:

$$z_u = \left(\sum_{i=1}^{s} z_i d_{iu}^{-k} \right) / \left(\sum_{i=1}^{s} d_{iu}^{-k} \right),$$

where z_u is the unknown value to be estimated at u, z_i is the attribute value at control point i, d_{iu} is the distance between points i and u, s is the number of control points used in estimation, and k is the power. The higher the power, the stronger (faster) the effect of distance decay is (i.e., nearby points are weighted much higher than remote ones). In other words, distance raised to a higher power implies stronger localized effects.

The *thin-plate splines* method creates a surface that predicts the values exactly at all control points and has the least change in slope at all points (Franke, 1982). The surface is expressed as:

$$z(x,y) = \sum_{i=1}^{n} A_i d_i^2 \ln d_i + a + bx + cy,$$

where x and y are the coordinates of the point to be interpolated, $d_i = \sqrt{(x-x_i)^2+(y-y_i)^2}$ is the distance from the control point (x_i, y_i), and A_i, a, b, and c are the $n + 3$ parameters to be estimated. These parameters are estimated by solving a system of $n + 3$ linear equations (see Chapter 10), such as:

$$\sum_{i=1}^{n} A_i d_i^2 \ln d_i + a + bx_i + cy_i = z_i,$$

$$\sum_{i=1}^{n} A_i = 0,$$

$$\sum_{i=1}^{n} A_i x_i = 0,$$

$$\sum_{i=1}^{n} A_i y_i = 0.$$

Note that the first equation represents n equations for $i = 1, 2, \ldots, n$, and z_i is the known attribute value at point i.

The thin-plate splines method tends to generate steep gradients ("overshoots") in data-poor areas. Other methods, such as *spline with tension, completely regularized spline, multiquadric function,* and *inverse multiquadric function,* are proposed to mitigate the problem (see Chang, 2004, p. 285). These advanced interpolation methods are grouped as *radial basis functions (RBF).*

Kriging (Krige, 1966) models the spatial variation as three components: a spatially correlated component, representing the regionalized variable; a "drift" or structure, representing the trend; and a random error. To measure spatial autocorrelation, kriging uses the measure of *semivariance* (1/2 of variance):

$$\gamma(h) = \frac{1}{2n} \sum_{i=1}^{n} \left[z(x_i) - z(x_i + h) \right]^2,$$

where n is the number of pairs of the control points that are distance (or spatial lag) h apart, and z is the attribute value. In the presence of spatial dependence, $\gamma(h)$ increases as h increases. A *semivariogram* is a plot that shows the values of $\gamma(h)$ along the y-axis and the distances h along the x-axis.

Kriging fits the semivariogram with a mathematical function or model and uses it to estimate the semivariance at any given distance, which is then used to compute a set of spatial weights. For instance, if the spatial weight for each control point i and a point s (to be interpolated) is W_{is}, the interpolated value at s is:

$$z_s = \sum_{i=1}^{n_s} W_{is} z_i,$$

where n_s is the number of sampled points around the point s, and z_s and z_i are the attribute values at s and i, respectively. Similar to the kernel density estimation, kriging can be used to generate a continuous field from point data.

In ArcGIS Pro, all three basic local interpolation methods are available in Geoprocessing Toolbox > Spatial Analyst Tools > Interpolation > IDW (Kriging or Spline). The other four advanced radial basis function methods can be accessed in Geoprocessing Toolbox > Geostatistical Analyst Tools > Interpolation > Radial Basis Functions.

3.3 CASE STUDY 3A: MAPPING PLACE-NAMES IN GUANGXI, CHINA

This case study examines the distribution pattern of contemporary Zhuang place-names (toponyms) in Guangxi, China, based on a study reported in Wang et al. (2012). Zhuang, part of the Tai language family, are the largest ethnic minority in China and mostly live in the Guangxi Zhuang Autonomous Region (a provincial unit simply referred to as "Guangxi" here). The Sinification of ethnic minorities, such as the Zhuang, has been a long and ongoing historical process in China. The impact has been uneven in the preservation of Zhuang place-names. The case study is chosen to demonstrate the benefit of using GIS in historical-linguistic-cultural studies. Mapping is a fundamental function of GIS. However, direct mapping Zhuang place-names as in Figure 3.3 has limited value. Spatial analysis techniques such as spatial smoothing and spatial interpolation methods can help enhance the visualization of the spatial pattern of Zhuang place-names.

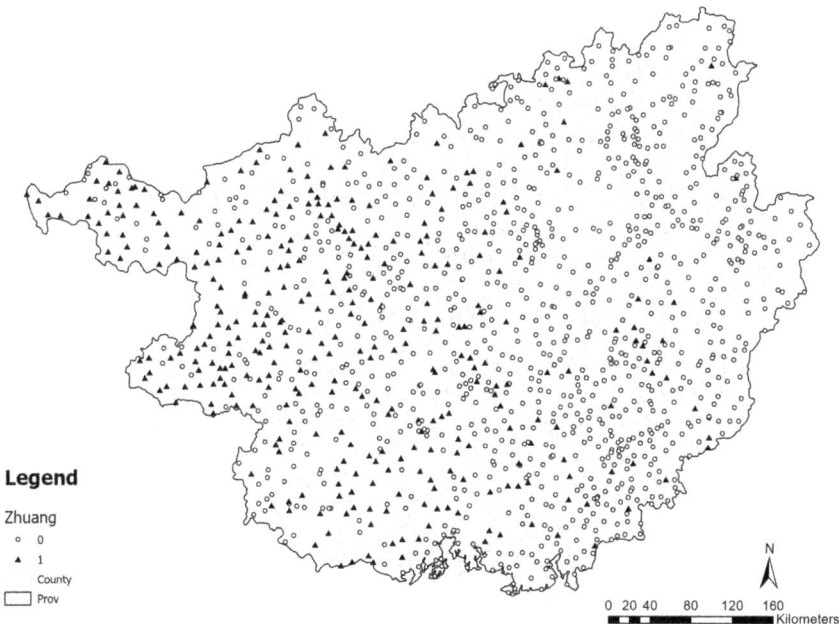

FIGURE 3.3 Zhuang and non-Zhuang place-names in Guangxi, China.

A geodatabase `GX.gdb` is provided under the data folder `China _ GX` and includes the following features:

1. Point feature `Twnshp` for all townships in the region, with the field `Zhuang` identifying whether a place-name is Zhuang (= 1) or non-Zhuang (= 0) (mostly Han).
2. Two polygon features `County` and `Prov` for the boundaries of counties and the provincial unit, respectively.

3.3.1   IMPLEMENTING THE FLOATING CATCHMENT AREA METHOD

We first implement the floating catchment area method. Different window sizes are used to help identify an appropriate window size for an adequate degree of smoothing to balance the overall trends and local variability. Within the window around each place, the ratio of Zhuang place-names among all place-names is computed to represent the concentration of Zhuang place-names around that place. In implementation, the key step is to utilize a distance matrix and extract the places that are within a specified search radius from each place.

Step 1. Direct Mapping of Zhuang Place-Names: In ArcMap Pro, add all features in `GX.gdb` to a new project > right-click the layer `Twnshp` > Symbology > under Primary Symbology, choose Unique Values; choose `Zhuang` for Field 1, then click the icon 📑 Add All Value. Click the drop-down menu in More and uncheck "Show all other values" to map the field `Zhuang` (as shown in Figure 3.3).

Step 2.   Computing Distance Matrix between Places: In the Contents pane, right-click the layer `Twnshp` > Data > Export Feature, then save it as `Twnshp1`.[3] In the Geoprocessing Toolboxes pane, choose Analysis Tools > Proximity > Generate Near Table > Choose `Twnshp` for the Input Features and `Twnshp1` for the Near Features. Name the output table `Dist _ 50km`, type `50000` for Search Radius, keep default unit Meters, uncheck "Find only closest feature," and click Run. If needed, refer to step 3 in Subsection 2.2.1 of Chapter 2 for a detailed illustration of computing the Euclidean distance matrix. By defining a wide search radius of 50 kilometers, the distance table allows us to experiment with various window sizes ≤ 50 kilometers. In the table `Dist _ 50km`, the field `IN _ FID` identifies the "from" (origin) place, and the `NEAR _ FID` identifies the "to" (destination) place.

Step 3. Attaching Attributes of Zhuang Place-Names to Distance Matrix: In the window of Contents, right-click the table `Dist _ 50km` > choose Joins and Relates > Add join. In the dialog (Figure 3.4), for Input Join Field, choose the field `NEAR _ FID`; for Join Table, choose the layer `Twnshp4`; for Join Table Field, choose the field `OBJECTID`; and click OK.

This step uses an *attribute join* to transfer or attach fields in `Twnshp` to the distance table `Dist _ 50km`. Hereafter in the book, such an attribute join operation is simply stated as: "join the attribute table of `Twnshp` (based on the field `OBJECTID`) to the target table `Dist _ 50km` (based on the field `NEAR _ FID`)." The expanded table `Dist _ 50km` now contains the field `Zhuang` that identifies each destination place-name as either Zhuang or non-Zhuang.

Add Join ? ✕

Input Table

Dist_50km ▾ 📁

Input Join Field

NEAR_FID ▾

Join Table

Twnshp ▾ 📁

Join Table Field

OBJECTID ▾

☑ Keep All Target Features

☐ Index Joined Fields

Validate Join

OK

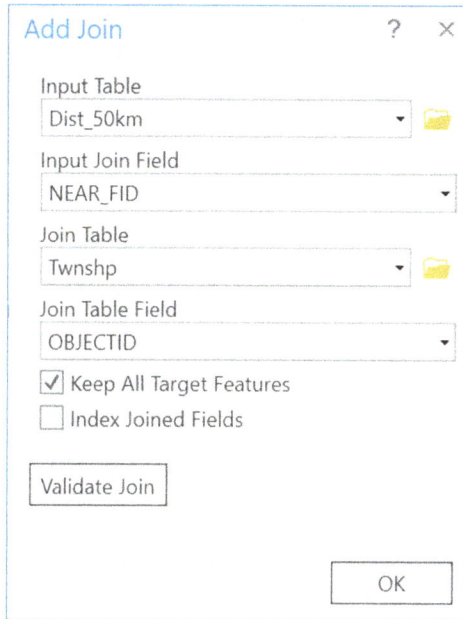

FIGURE 3.4 Dialog window for attribute join.

Step 4. Selecting Places within a Window around Each Place: For example, we define the window size with a radius of 10 kilometers. Right-click the table Dist _ 50km > Open 📋. On the opened table, click 🗒 Select By Attributes. In the dialog, enter the Expression "NEAR _ DIST," "is less than or equal to," "10000," and click OK. For each origin place, only those destination places within 10 kilometers are selected.

Step 5. Summarizing Places within a Window around Each Place: On the open table Dist _ 50km, with selected records being highlighted, right-click the field IN _ FID and choose Summarize. In the dialog, name the output table Sum _ 10km; under Statistics Fields, choose Twnshp.Zhuang for Field and choose Sum as the Statistics Type, keep the default IN _ FID under Case field, and click OK. Alternatively, one may access the Summarize tool in the Geoprocessing Toolboxes pane > Analysis Tools > Statistics > Summary Statistics. If needed, refer to step 13 of Subsection 2.2.3 of Chapter 2 for details of the Summary Statistics tool.

The *Summarize* operation on a table is similar to the spatial analysis tool Dissolve on a spatial dataset as used in step 10 in Subsection 1.3.2 of Chapter 1. It aggregates the values of a field or fields (here only Zhuang) to obtain its basic statistics, such as minimum, maximum, average, sum, standard deviation, and variance (here only "sum") by the unique values of the field to summarize (here IN _ FID). In the resulting table Sum _ 10km, the first field IN _ FID is from the source table; FREQUENCY is the number of distance records within 10 kilometers from each unique INPUT _ FID and thus the total number of places around each place; and SUM _ Twnshp.Zhuang indicates the number of Zhuang place-names within the range (because a Zhuang place-name has a value 1 and a non-Zhuang place-name has a value 0).

Step 6. Calculating Zhuang Place Ratios around Each Place: Add a new field ZhuangR with Double as Data Type to the table Sum_10km, and calculate it as "!SUM_Twnshp_Zhuang! /!FREQUENCY!" This ratio measures the portion of Zhuang place-names among all places within the window that is centered at each place.

Step 7. Mapping Zhuang Place-Name Ratios: Join the table Sum _ 10km (common key IN _ FID) to the layer Twnshp (common key OBJECTID). Right-click the layer Twnshp > Symbology > under Primary Symbology, choose Proportional Symbols to map the field ZhuangR, which represents the Zhuang place-name ratios within a 10-kilometer radius around a place across the study area, as shown in Figure 3.5.

This completes the FCA method for spatial smoothing, which converts a binary variable Zhuang to a continuous numerical variable ZhuangR.

Step 8. Sensitivity Analysis: Experiment with other window sizes, such as 5 kilometers and 15 kilometers, and repeat steps 4–7. Compare the results with Figure 3.5 to examine the impact of window size. A larger window size leads to stronger spatial smoothing.

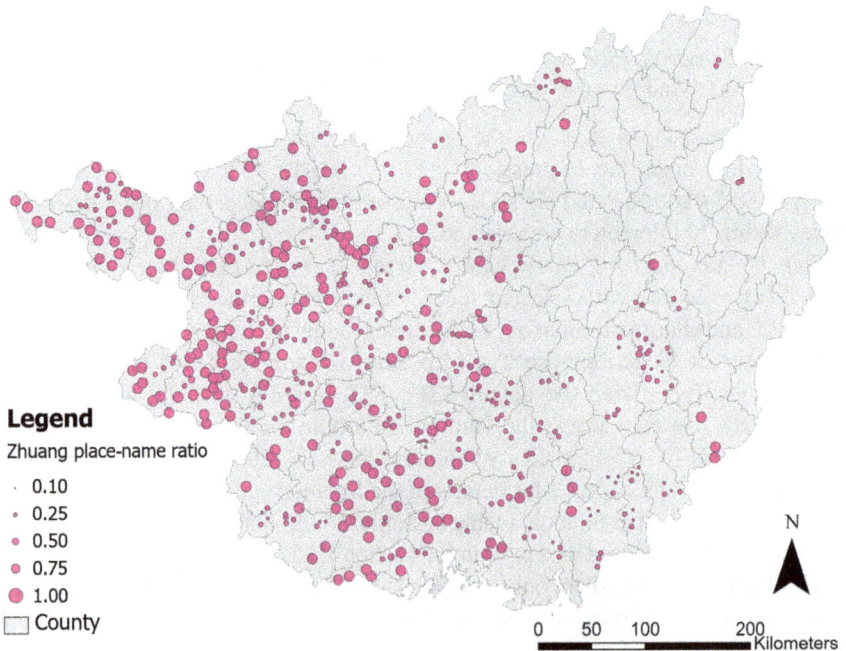

FIGURE 3.5 Zhuang place-name ratios in Guangxi by the floating catchment area method.

3.3.2 Implementing Spatial Interpolation Methods

Step 9. Implementing the Kernel Density Estimation Method: In the Geoprocessing Toolboxes pane > Spatial Analyst Tools > Density > Kernel Density. In the dialog, choose Twnshp for Input point or polyline features, choose Zhuang for Population field, name the Output Raster KDE_20km, input 20000 for Search Radius (which defines the bandwidth in KDE as 20km), keep default value for Output Cell Size and other default choices; click Environments > under Processing Extent, in the drop-down box, choose County below Same as Layer for Extent (which uses the spatial extent of the layer County to define the rectangular extent for the raster); under Raster Analysis, also choose County as the Mask (which filters out cells outside the boundary of the layer County); and click Run to execute the tool. A raster layer KDE_20km within the study area is generated.

The kernel function is as previously discussed in equation 3.1. By default, estimated kernel densities are categorized into nine classes, displayed as different hues. Figure 3.6 is based on five classes and natural breaks (Jenks) classification with the county layer as the background. The kernel density map shows the distribution of Zhuang place-names as a continuous surface so that areas of concentrated Zhuang place-names can be easily spotted. However, the density values simply indicate relative degrees of concentration and cannot be interpreted as a meaningful ratio like ZhuangR in the FCA method.

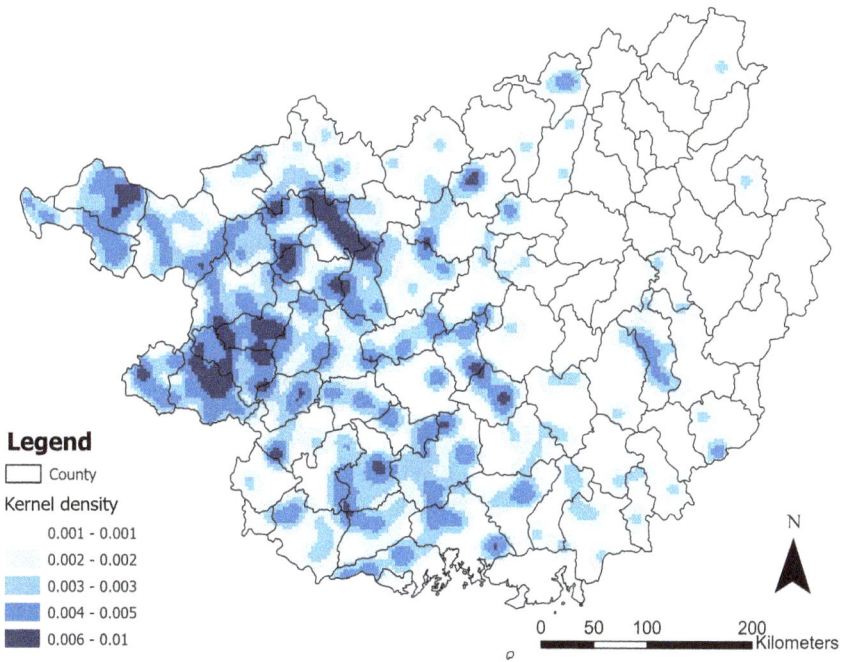

Legend

County

Kernel density

0.001 - 0.001
0.002 - 0.002
0.003 - 0.003
0.004 - 0.005
0.006 - 0.01

0   50   100   200
                        Kilometers

N

FIGURE 3.6 Kernel density estimation of Zhuang place-names in Guangxi.

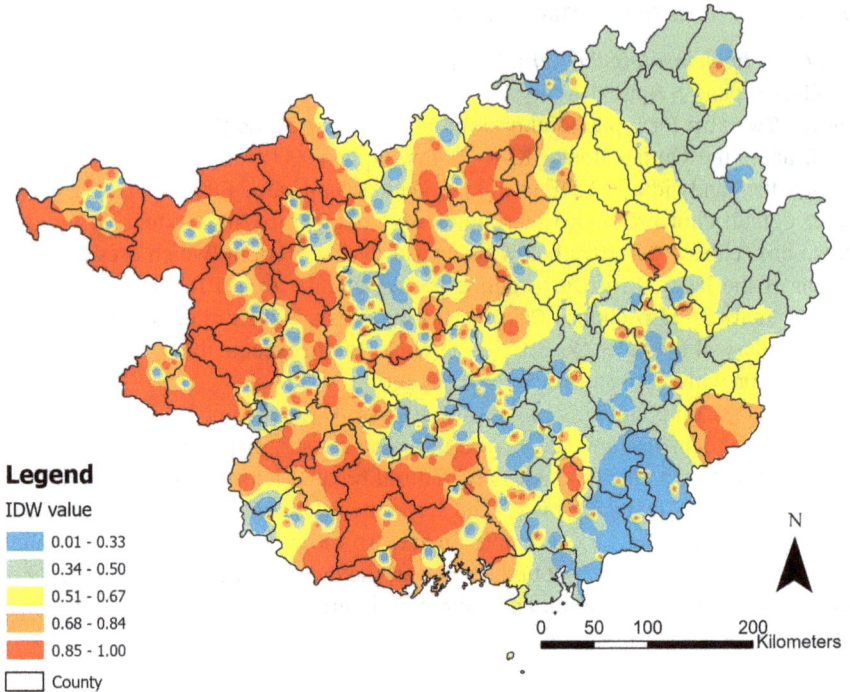

FIGURE 3.7 IDW interpolation of Zhuang place-names in Guangxi.

Step 10. Implementing the Inverse Distance Weighting Method: In the Geoprocessing Toolboxes pane > Geostatistical Analyst Tools > Interpolation > choose IDW. In the dialog, under Parameters, choose Twnshp for Input features, and Zhuang for Z value field; input IDW for Output geostatistical layer, IDWraster for Output raster; use the default power of 2, Standard neighborhood type, 15 maximal and 10 minimal neighboring points, and other default settings; click on Environments > similar to step 9, use the layer County to define both the Extent and Mask; and click Run.

A surface raster IDWraster and a geostatistical layer IDW for the study area are generated (shown in Figure 3.7). Note that all interpolated values are within the same range as the original, that is, between 0 and 1.

Experiment with other methods (global/local polynomial interpolation, kriging, empirical Bayesian kriging, etc.) available in the Geostatistical Analyst Tools.

3.4 AREA-BASED SPATIAL INTERPOLATION

Area-based (Areal) interpolation is also referred to as *cross-area aggregation*, which transforms data from one system of areal units (source zones) to another (target zones). A point-based method such as kriging or polynomial trend surface analysis can also be used as a bridge to accomplish the transformation from one area unit to another.

Specifically, the source polygon dataset may be represented by their centroids and is first interpolated to a grid surface by a point-based interpolation method; then the values of grid cells are aggregated to predict the attributes of the target polygon dataset. In ArcGIS Pro, use the Areal Interpolation Layer to Polygons tool under the Geostatistical Analyst Tools to implement it. This approach is essentially point-based interpolation. Another point-based interpolation is to simulate individual points from a polygon feature by the Monte Carlo simulation (see Section 12.2 of Chapter 12) and then aggregate the point feature back to any areal units (Wang F et al., 2019). This section focuses on area-based interpolation methods (Goodchild et al., 1993).

The simplest and most widely used is *areal weighting interpolator* (Goodchild and Lam, 1980). The method apportions the attribute value from each source zone to target zones according to the areal proportion. The method assumes that the attribute value is evenly distributed within each of the source zones. In fact, steps 8–10 in Subsection 1.3.2 of Chapter 1 already used this assumption to interpolate population from one area unit (census tracts) to another (concentric rings), and the interpolation was done by passing the "population density" variable through various polygon features.

The following discusses an interpolation method, termed *target-density weighting (TDW)* by Schroeder (2007). It is particularly useful for interpolating census data across different years.

Using s and t denoting the source and target zones, respectively, the task is to interpolate the variable of interest y from s to t. The analysis begins with the spatial intersect operation between the source and target zones that produces the segments (atoms) denoted by st. Assume that additional information such as the ancillary variable z in the target zones t is also available. The method distributes the ancillary variable z in the target zones t to the atoms st proportionally to area sizes. Unlike the simple areal weighting interpolation that assigns weights proportional to areas of atoms, the TDW interpolation assigns weights according to the value of ancillary feature z to honor its pattern. In other words, the TDW assumes that y is more likely to be consistent with the pattern of z than simply proportional to areas. Finally, it aggregates the interpolated y in atoms st to the target zone t.

The TDW can be expressed as:

$$\widehat{y}_t = \sum_s \widehat{y}_{st} = \sum_s \frac{\widehat{z}_{st}}{\widehat{z}_s} y_s = \sum_s \frac{\left(A_{st}/A_t\right)Z_t}{\sum_t \left(A_{st}/A_t\right)Z_t} y_s, \qquad (3.2)$$

where A denotes the area, and other notations are as previously stated.

The TDW method is shown to be more accurate than the simple areal weighting interpolation in the temporal analysis of census data (Schroeder, 2007). It can be easily implemented in ArcGIS by using a series of "join" and "summarize" operations.

Another area-based interpolation method is very similar to the concept of the aforementioned TDW method by utilizing an ancillary variable such as road network (Xie, 1995). Specifically, the *network hierarchical weighting (NHW) method* (1) intersects the road network layer with source and target area layers, (2) assigns weights for different road categories in the US Census Bureau's TIGER files as population or business densities vary along various road classes, (3) allocates population in source area units to road segments, and (4) aggregates the allocated population on road segments back to target areas.

3.5   CASE STUDY 3B: AREA-BASED INTERPOLATIONS OF POPULATION IN BATON ROUGE, LOUISIANA

Transforming data from one area unit to another is a common task in spatial analysis for integrating data of different scales or resolutions. If the area units are strictly hierarchal (e.g., a state is composed of multiple whole counties, and a county is composed of multiple census tracts), the transformation does not require any interpolation but simply aggregates data from one areal unit to another. By using a table linking the correspondence between the units, the task can be accomplished without GIS. There are also cases that one area unit is not completely contained in the other but assumed so for simplicity when the approximation is reasonable with negligible errors. For example, step 13 in Subsection 1.3.2 assumes that census blocks are entirely contained in the concentric rings and uses a spatial join to aggregate the population data. This case study uses areal interpolation methods for better accuracy.

Data needed for the project are provided in the geodatabase BR.gdb under the folder BatonRouge. It contains:

1. Features BRTrt2010 and BRTrtUtm for census tracts in the study area in 2010 and 2020, respectively.
2. Feature BRUnsd for unified school districts in the study area in 2020.

Note that the feature BRTrtUtm is the output from step 3 in Subsection 1.3.1.

3.5.1   Implementing Areal Weighting Interpolation

This subsection illustrates how to use the areal weighting interpolation method to transform data from census tracts to school districts in 2020.

Step 1. Overlaying the Census Tract and the School District Layers: In the Geoprocessing Toolboxes pane, choose Analysis Tools > Overlay > Intersect. In the dialog, select BRTrtUtm and BRUnsd as Input Features; for the Output Feature, input TrtSD, and click OK. In the attribute table of TrtSD, the field Area is the area size of each census tract inherited from BRTrtUtm, and the field Shape _ Area is the area size of each newly intersected polygon ("atom").

Step 2. Apportioning Attribute to Area Size in Intersected Polygons: On the open attribute table of TrtSD, add a field EstPopu and calculate it as !POP100!*!Shape _ Area!/!Area! This is the interpolated population for each polygon in the intersected layer (atom) by the areal weighting method.[5] In Subsection 1.3.1, the population in the rings is reconstructed from population density, which is passed on from the tracts to the atoms. It makes the same assumption of uniform population distribution.

Step 3. Aggregating Data to School Districts: On the attribute table of TrtSD, right-click the field DISTRICT_N (school district codes) and choose Summarize.

In the dialog, name the output table SD_Pop; under Statistics Field(s), select EstPopu for Field and Sum for its corresponding Statistics Type, and click OK. The interpolated population for school districts is in the field Sum_EstPopu in the table SD_Pop, which may be joined to the layer BRUnsd for mapping or other purposes.

3.5.2 IMPLEMENTING TARGET-DENSITY WEIGHTING INTERPOLATION

It is common that census tract boundaries in a study area change from one decennial year (say, 2010) to another (say, 2020). In order to map and analyze the population change over time, one has to convert the population in two different years to the same areal unit. The task here is to interpolate the 2020 population in the 2020 census tracts to the 2010 census tracts by the TDW method so that population change rates can be assessed in the 2010 census tracts. The 2010 population pattern is used as an ancillary variable to improve the interpolation.

According to the TDW, the layers of 2010 and 2020 census tracts are intersected to create a layer of atoms, and the ancillary variable (z or 2010 population) in the target zone (t or 2010 census tracts) is firstly distributed (disaggregated) across the atoms (st), assuming uniformity, and then aggregated in the source zone (s, that is, 2020 census tracts), producing the proportion $\frac{\hat{z}_{st}}{\hat{z}_s}$. The variable of interest (y or 2020 population) in the source zone is subsequently disaggregated to atoms according to this proportion to honor the spatial distribution pattern of the ancillary variable (z or 2010 population). The disaggregated y (2020 population) is then aggregated in the target zone (2010 census tracts) to complete the interpolation process. In summary, the TDW re-distributes the population of 2020 to match the 2010 census tract boundary in a manner to honor the historical distribution pattern of the population of 2010.

Step 4. Overlaying the Two Census Tract Layers: Similar to step 1, create a new project and use the Intersect tool to intersect BRTrtUtm and BRTrt2010 to obtain InterTrt.

Step 5. Disaggregating 2010 Population to the Atoms: On the open attribute table of InterTrt, add a field EstP2010 and calculate it as !DP0010001!*!Shape_Area!/!Area_1!, where Shape_Area is the area size of atoms and Area_1 is the area size of 2010 tracts. Similar to step 2, this proportions the 2010 population to the areas of atoms. The field EstP2010 is $\hat{z}_{st}$.

Step 6. Aggregating Estimated 2010 Population to 2020 Tracts: On the open attribute table of InterTrt, right-click the field GEOID (2020 tract codes) and choose Summarize. In the dialog, name the output table Trt2020_P10, select EstP2010 for Field and Sum for its corresponding Statistics Type, and click OK. The field Sum_EstP2010 is $\hat{z}_s$.

Step 7. Computing the Ratio of 2010 Population in Atoms Out of 2020 Tracts: In the Contents pane, right-click the layer InterTrt > Joins and Relates > Join. Use an attribute join to join the table Trt2020_P10 to the attribute table of InterTrt by their common key GEOID. On the open attribute table of InterTrt, add a field P10_rat and calculate it as !InterTrt.EstP2010!/!Trt2020_P10.

Sum_EstP2010!, which is the proportion $\frac{\hat{z}_{st}}{\hat{z}_s}$.

Step 8. Disaggregating 2020 Population to Atoms according to the Ratio: On the open attribute table of `InterTrt`, add another field `EstP2020` and calculate it as `!InterTrt.POP100!*!InterTrt.P10 _ rat!`

In other words, the IDW method interpolates 2020 population in atoms proportionally to 2010 population instead of area sizes as the areal interpolator does.

Step 9. Aggregating Estimated 2020 Population to 2010 Tracts: On the open attribute table of `InterTrt`, right-click the field `GEOID10` (2010 tract codes) and choose Summarize > select field `EstP2020` and corresponding statistics Sum, and name the output table `Trt2010 _ P20`, where the field `Sum _ EstP2020` is the interpolated 2020 population in the 2010 tracts.

Step 10. Mapping Population Change Rates from 2010 to 2020 in 2010 Tracts: Use an attribute join to join the table `Trt2010 _ P20` to the layer `BRTrt2010` based on their common key `GEOID10`. Add a field `ChngRate` to the attribute table of `BRTrt2010` and calculate it as `(!Trt2010 _ P20.SUM _ EstP2020! -!BRTrt2010.DP0010001!)/!BRTrt2010.DP0010001!`

Map the field `ChngRate` (population change rate) as shown in Figure 3.8. Areas near the city center lost significant population, and suburban areas experienced much growth, particularly in the northwest and south.

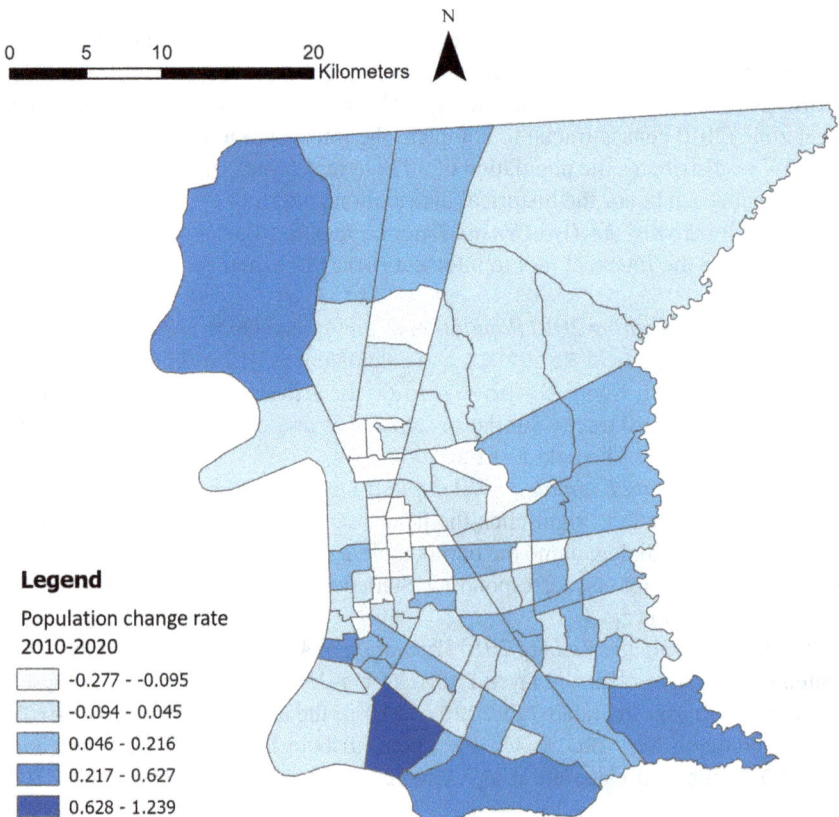

FIGURE 3.8 Population change rates across census tracts in Baton Rouge, 2010–2020.

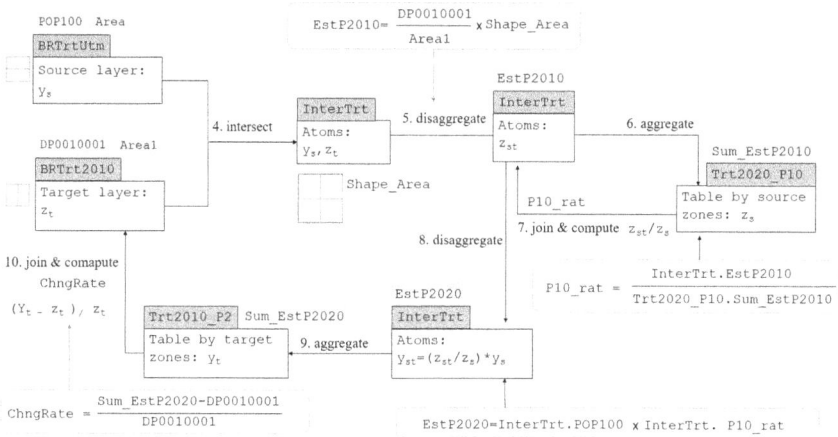

FIGURE 3.9 Flowchart for implementing the TDW method.

The process is illustrated in a flowchart as shown in Figure 3.9.

The preceding project design uses the 2010 population pattern as an ancillary variable to facilitate transforming the 2020 population in the 2020 tracts to the 2010 tracts, and the changes are assessed in the 2010 tracts. Similarly, one may transform the 2010 population data to the 2020 tracts and assess the changes in the 2020 tracts. In that case, the ancillary variable will be the 2020 population.

3.6 SPATIOTEMPORAL KERNEL DENSITY ESTIMATION (STKDE) METHOD

Similar to the notion of distance decay in the regular (spatial) kernel density estimation (KDE) method as discussed in Subsection 3.1.2, temporal decay can be introduced to model that an event is more likely to occur around one in more recent than distant past. For example, Brunsdon et al. (2007) proposed a spatiotemporal kernel density estimation (STKDE) by multiplying the spatial KDE by a temporal kernel function. The model is written as:

$$\hat{f}(x,y,t) = \frac{1}{nh_s^2 h_t} \sum_{i=1}^{n} K_s\left(\frac{x - x_i}{h_s}, \frac{y - y_i}{h_s}\right) K_t\left(\frac{t - t_i}{h_t}\right), \quad (3.3)$$

where (x_i, y_i, t_i) represents each event $i \in (1, \ldots, n)$; (x, y, t) is the location in the space–time domain, where the density $\hat{f}$ is being estimated; $K_s(\cdot)$ is a bivariate kernel function for the spatial domain; $K_t(\cdot)$ is a univariate kernel for the temporal domain; and h_s and h_t are the spatial and temporal bandwidths, respectively. In comparison to the spatial KDE model in equation 3.1, the temporal kernel K_t is newly added.

Zhang Z et al. (2011) designed a slightly different STKDE for disease risk estimation. The difference lies in the way of modeling the spatial component of the STKDE. Instead of using a bivariate kernel K_s in equation 3.3, Zhang Z et al. (2011)

utilizes two univariate kernel functions to capture possibly different distribution patterns along x and y dimensions and defines the STKDE as:

$$\hat{f}(x,y,t) = \frac{1}{nh_x h_y h_t} \sum_{i=1}^{n} k\left(\frac{x-x_i}{h_x}\right) \times k\left(\frac{y-y_i}{h_y}\right) \times k\left(\frac{t-t_i}{h_i}\right) \tag{3.4}$$

Our case study on predicting crime hotspots in the next section is based on Hu et al. (2018), which adopted the formulation of STKDE in equation 3.4. Since crimes may have different distribution patterns in x and y dimensions due to the layout, transportation network, and land use pattern of a study area, an identical search bandwidth h_s for both x and y dimensions may give rise to biased estimations. Figure 3.10 shows a conceptual illustration of STKDE. The kernel density at point (x, y, t) at the center of a grid cell is estimated to be the sum of bumps along three dimensions within their respective bandwidths.

Similar to the regular KDE for examining spatial concentrations or hotspots of events, the STKDE can be used for detecting hotspots not only spatially but also temporally. For example, crime tends to cluster temporally in addition to spatially. In other words, crime is concentrated in certain parts of a study area during certain times of the day. Ignoring the temporal component of crime deprives researchers and practitioners of the opportunity to target specific time periods with elevated crime risks. Temporal dimension is also integral to the study of crime displacement and diffusion of benefits, as crimes may displace to different areas, time, or even types when law enforcements adopt hotspot patrolling.

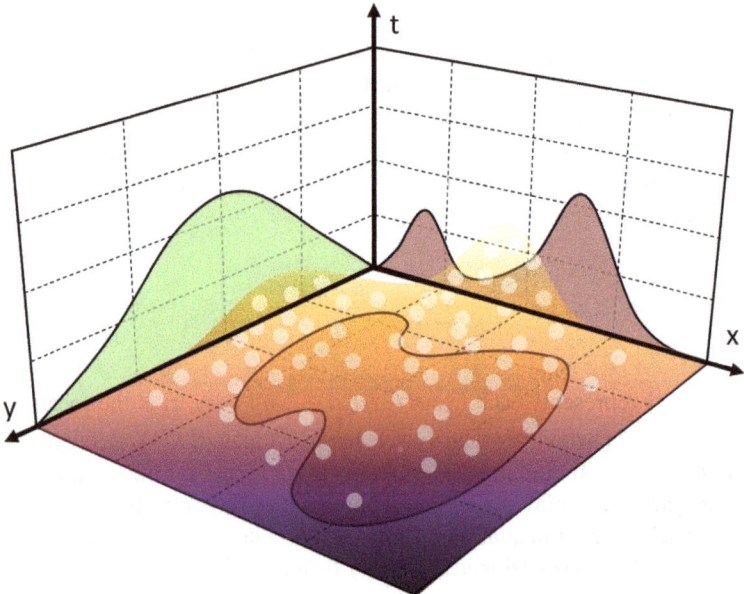

FIGURE 3.10 A conceptual illustration of STKDE.

ArcGIS Pro has a set of tools for analyzing spatial and temporal clusters separately and their integration, that is, spatiotemporal clusters. For example, the tool "Emerging Hotspot Analysis (EHSA)" examines the data spatially and temporally simultaneously and classifies patterns in a space–time cube into 1 of 17 possible categories. It determines whether a space–time cube bin's neighbors contain a number of check-ins that are significantly higher (hotspot) or lower (cold spot) than the global average. Once every location in the space–time cube has been designated a hotspot, a cold spot, or neither, EHSA examines variations in each location's z-score over time to determine whether the location is a consecutive, intensifying, diminishing, or sporadic hot- or cold spot. It accounts for both spatial and temporal variations in the data.

3.7 CASE STUDY 3C: DETECTING SPATIOTEMPORAL CRIME HOTSPOTS IN BATON ROUGE, LOUISIANA

This case study explores the spatiotemporal patterns of residential burglary crimes in Baton Rouge, Louisiana, in 2011, as reported in Hu et al. (2018).

Data needed for the project are provided in a sub-folder BRcrime under the folder BatonRouge. It contains:

1. Raster feature Brraster.tif for a background land use pattern, including residential areas
2. Table BRcrime.csv for residential burglary crimes records with XY coordinates and time label
3. R-based tool STKDE in the toolbox R _ ArcGIS _ Tools.tbx under the folder CMGIS-V3-Toolbox > R-ArcGIS-Tools

The following steps illustrate how to detect spatial and temporal clusters, create a space–time cube, and analyze emerging hotspots.

Step 1. Aggregating Crime Data in Hexagon Grids: In ArcGIS Pro, add BRraster. tif and BRcrime.csv in a new project map. Click No on the prompt window of Calculate Statistic for BRraster.tif. In the Geoprocessing pane, search for and activate the tool "XY Table to Point." In the dialog, select BRcrime.csv for Input Table; name BRcrime for Output Feature Class; choose X for X field, Y for Y field, and BRraster.tif for Coordinate System; and click Run. It generates a point layer for crime incidents. Refer to step 1 in Subsection 2.2.1 of Chapter 2 if needed.

Search for and activate the tool *Generate Tessellation*. Use the name Hexagon _ Tessellation for Output Feature Class, choose BRcrime for Extent, define 1 for Size and its unit Square Kilometers, and click Run. It creates a polygon layer of uniform hexagons.

Search for and open the Spatial Join tool. Choose Hexagon _ Tessellation for Target Features and BRcrime for Join Features, name BRcrimeCts for Output Feature Class, and click Run. Field Join _ count in the layer BRcrimeCts is crime counts in hexagons (Figure 3.11). One may use the tools introduced in Subsection 8.6.2 of Chapter 8 to identify spatial clusters of high-crime areas.

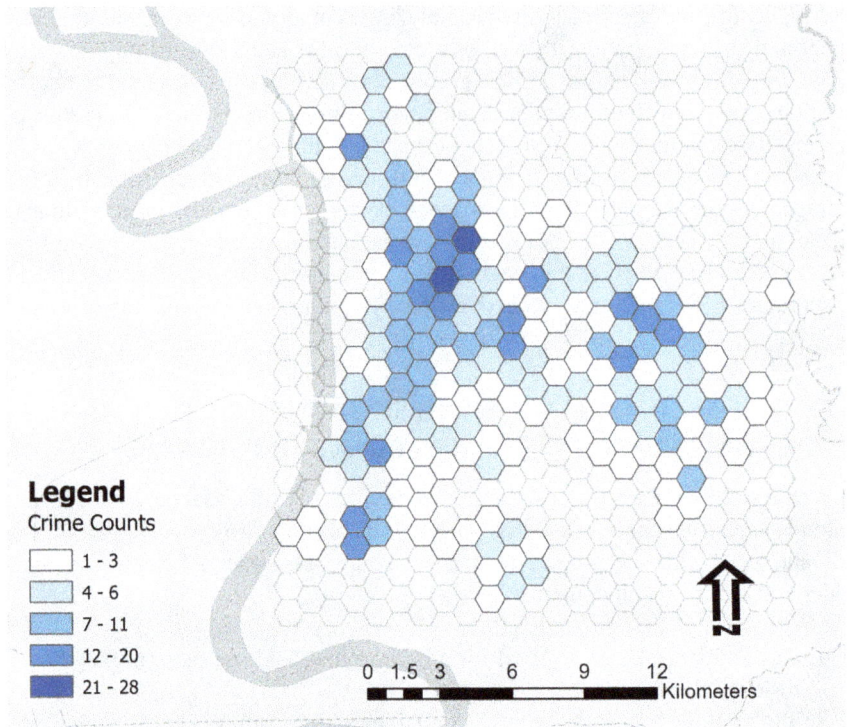

FIGURE 3.11 Aggregating crime data in hexagons.

Step 2. Analyzing Temporal Trend of Crime with Data Clock: Search for and open the tool *Convert Time Field*. Choose BRcrime for Input Table, choose OFFENSE _ DATE for Input Time Field, input MM/dd/yy for Input Time Format, input time for Output Time Field, make sure Date as Output Time Type, and click Run. This standardizes the time field to a format in ArcGIS Pro, where the old field OFFENSE _ DATE is changed to a new Date field time (e.g., a value of 10/1/11 will be identified as 10/1/2011).

In the Contents pane, right-click BRcrime > Create Chart > Data Clock. In the Chart Properties pane, choose time for Date, choose Weeks for Rings, Days of the week for Wedges, and set Count for Aggregation, and keep other default settings. Figure 3.12a shows fewer residential burglaries on Sunday or Saturday. Users may change the parameter settings to explore other temporal trends, and the map is updated simultaneously.

Step 3. Visualizing Spatiotemporal Trends with a Space–Time Cube in 2D: Search for and activate the tool Create Space Time Cube by Aggregating Points in the Geoprocessing pane. In the dialog, choose BRcrime for Input Features, type BRcrimeCube.nc for Output Space Time Cube,[6] choose time for Time Field, enter 1 and unit Weeks for Time Step Interval, choose Hexagon grid for Aggregation

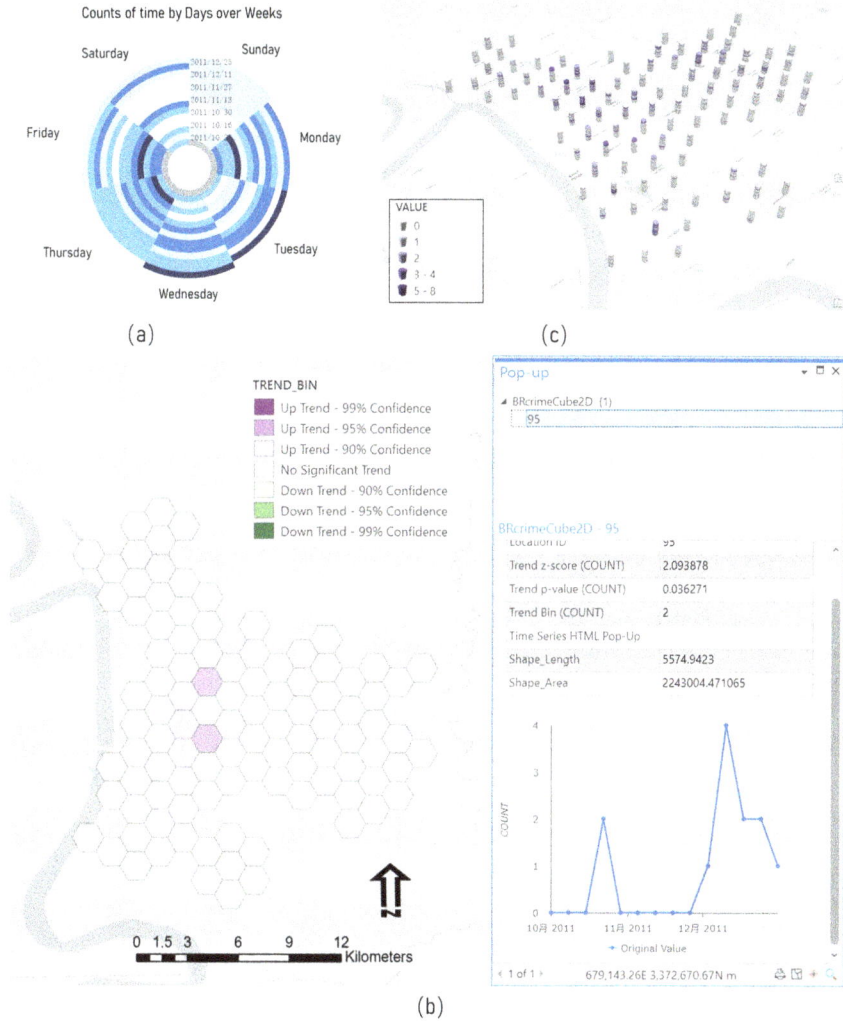

(a)

(b)

(c)

FIGURE 3.12 Visualization in (a) data clock, (b) space–time cube in 2D, and (c) space–time cube in 3D.

Shape Type, enter 1 and unit US Survey Miles for Distance Interval, keep other default settings, and click Run.

Search for and open the tool Visualize Space Time Cube in 2D. In the dialog, choose BRcrimeCube.nc for Input Space Time Cube, COUNT for Cube Variable, and Trends for Display Theme; check Enable Time Series Pop-ups; input BRcrimeCube2D for Output Features; and click Run. On the map shown in Figure 3.12b, click a hexagon bin in the layer BRcrimeCube2D to bring up the pop-up chart showing the temporal trend of crime at that location. Note that the hexagon color represents increasing or decreasing crimes over time.

Step 4. Visualizing Spatiotemporal Trends with a Space–Time Cube in 3D: On the ribbon, click Insert tab > Click the New Map drop-down arrow, and choose New Local Scene to add a Scene view.

In the Geoprocessing pane, search for and open the tool Visualize Space Time Cube in 3D. In the dialog, choose BRcrimeCube.nc for Input Space Time Cube, COUNT for Cube Variable, and Value for Display Theme; type BRcrimeCube3D for Output Features; and click Run.[7] On the 3D scene, as shown in Figure 3.12c, explore the scene by holding down the left mouse button to pan, scrolling the mouse wheel to zoom, and holding the mouse wheel to tilt. Each hexagon bin has a height composed of segments, with each segment corresponding to a different month and its color indicating the crime count in that area during that week.

Step 5. Detecting Time Series Clusters: Search for and open the tool *Time Series Clustering*. In the dialog, choose BRcrimeCube.nc for Input Space Time Cube and COUNT for Analysis Variable, input BRCubeTimeCl for Output Features, choose Value for Characteristic of Interest, type 3 for Number of Clusters, check Enable Time Series Pop-ups, enter BRcCluster for Output Table for Charts, and click Run.[8]

As shown in Figure 3.13a, the layer BRCubeTimeCl visualizes the clustered hexagon bins in three groups (1 for low, 2 for medium, and 3 for high crime risk). Users can also open the table BRcCluster and double-click Average Time Series per Cluster shown in Figure 3.13c. The chart in Figure 3.13b shows the hexagons with historically elevated or lower burglary records. Temporal clusters are grouped by temporal proximity, whereas spatial clusters are grouped by spatial proximity.

Step 6. Detecting Spatiotemporal Hotspots: Switch back to Map view from Scene view. Search for and open the tool *Emerging Hot Spot Analysis (EHSA)*. In the dialog, choose BRcrimeCube.nc for Input Space Time Cube, choose COUNT for Analysis Variable, input BRcrimeHotSpot for Output Features, enter 1 and unit US Survey Miles for Neighborhood Distance, keep other default settings, and click Run.

Figure 3.13d shows three types of hotspots: (1) New Hotspots identify locations that are statistically significant hotspot for the first time, (2) Consecutive Hotspots identify locations with a single uninterrupted run of statistically significant hotspot bins, and (3) Sporadic Hotspots are on-again-then-off-again ones.

Step 7. Applying R-Based Tool STKDE for Predictive Hotspot Evaluation: This step uses an R-based tool "STKDE" for predictive crime hotspot mapping and evaluation, as discussed in Section 3.6.

Go to the Catalog > under the folder R-ArcGIS-Tools, double-click the tool STKDE in R-ArcGIS-Tools.tbx to activate it. In the dialog shown in Figure 3.14a, select the point feature BRcrime generated in the step 2; choose the field X for Longitude_X, Y for Latitude_Y, time for Time Offend, and 2011-11-01 for Test Data Starting From; set the parameters 7 for Grid Size for Time, 100 for Grid Size for XY, 359.6488 for Bandwidth X, 702.1862 for Bandwidth Y, and 230 for Bandwidth T; input StkdeMean.tif for the resulting STKDE Raster and BRcrime _ STKDE for Significance Test Result; and click Run.

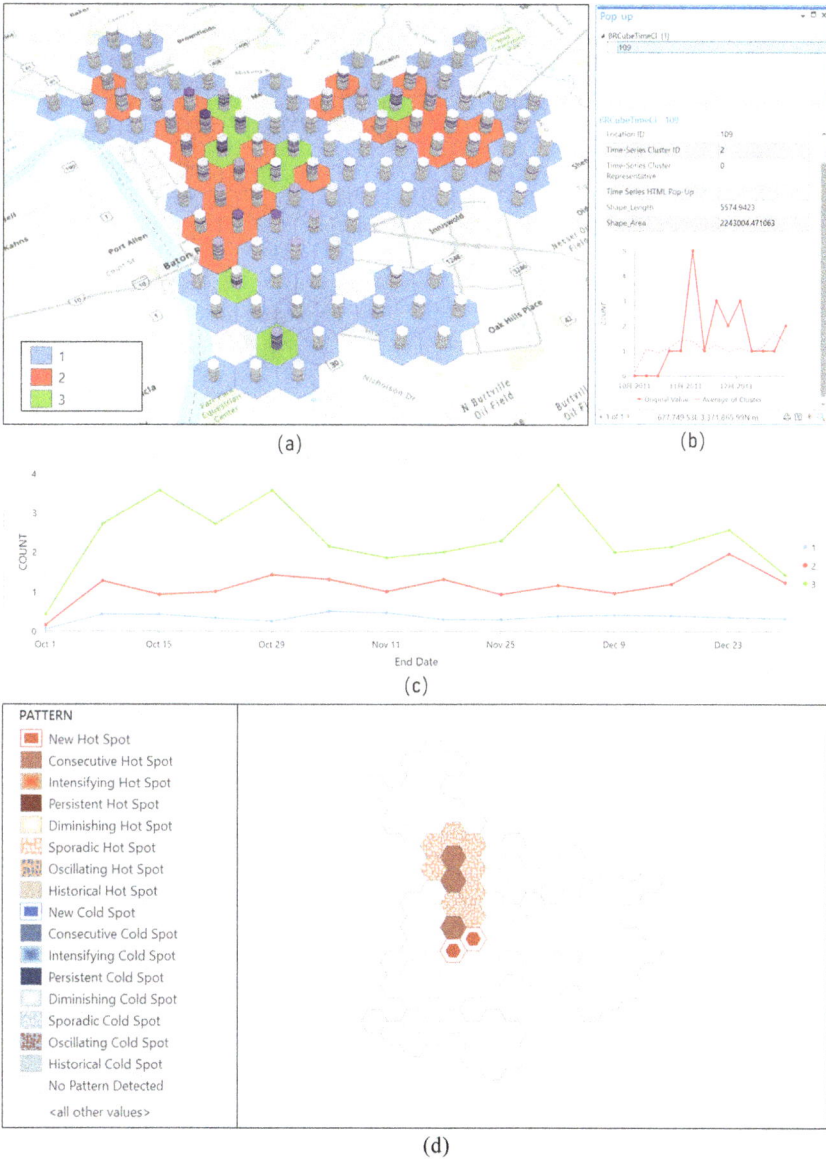

(a)

(b)

(c)

(d)

FIGURE 3.13 (a) Time series clusters, (b) pop-up window, (c) average time series per cluster, and (d) spatiotemporal hotspots.

The previous settings build a 3D grid based on 100 × 100 meters and 7 days, and the three bandwidths, X, Y, and T, correspond to h_x, h_y, and h_i in equation 3.4. StkdeMean.tif indicates the mean value of estimated residential burglary risk after 2011-11-01, as shown in Figure 3.14b. For a similar visualization, users can

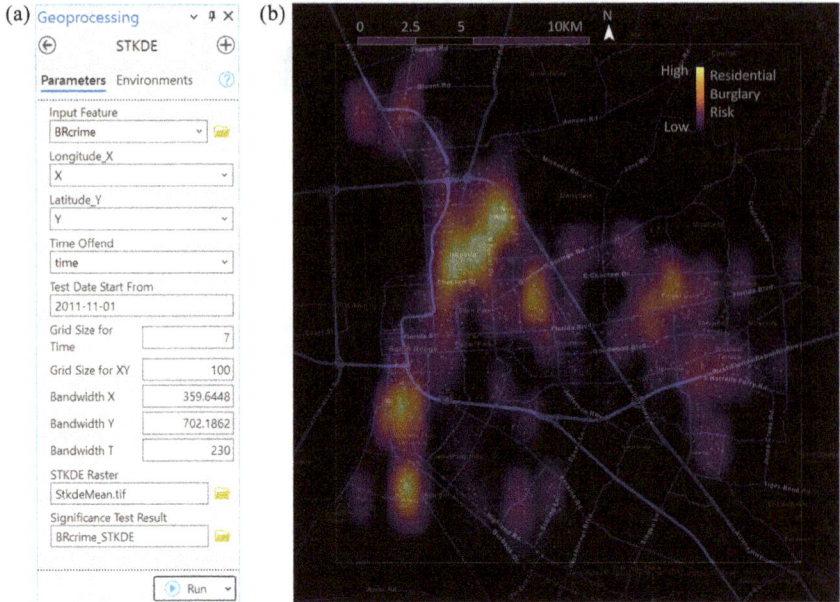

FIGURE 3.14 STKDE tool for predictive hotspots: (a) dialog window and (b) mean residential burglary risk.

change the base map to Streets (Night), revise the color style in the Symbology pane (say, Magma) and the transparence degree in the Raster Layer tab on the top ribbon.

`BRcrime _ STKDE` records the result of the statistical significance test based on the records from 2011-11-01 to 2011-11-07.

Several metrics have been commonly used to evaluate predictive accuracy of the identified crime hotspots. A popular index is hit rate (H), such as:

$$H = n / N,$$

where n is the number of robberies inside predicted hotspots, and N is the total number of reported robberies. Another metric is *prediction accuracy index (PAI)* or *forecast precision index (FPI)*:

$$PAI = \left(\frac{n}{N}\right) / \left(\frac{a}{A}\right)$$

That is, the ratio of the hit rate (n/N) to the area percentage (a/A, where a is area of identified hotspots from Time 1 and A is the study area size).

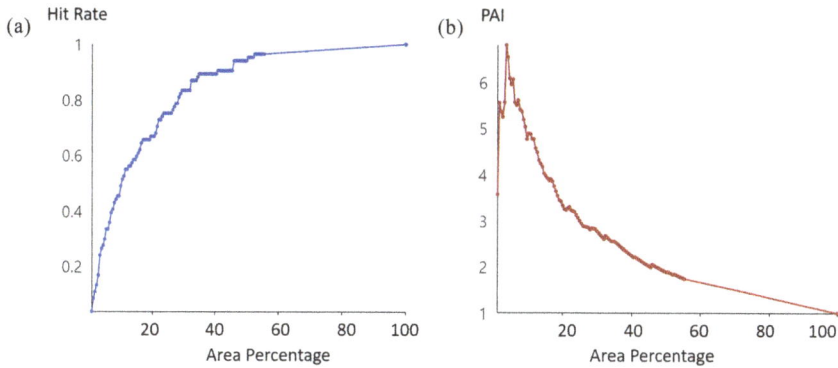

FIGURE 3.15 Line plots: (a) hit rate vs. area percentage, and (b) PAI vs. area percentage.

Obviously, as more areas are identified as hotspots (e.g., lowering the criterion), chances for the crimes being captured by the hotspots increase, but it does not necessarily indicate a better prediction model. In practice, more police force would need to be called upon to enforce hotspot policing too, which may not be realistic and defeat the very purpose of focusing limited resources on small areas. As shown in Figure 3.15a (based on the table BRcrime _ STKDE), hit rate increases with area percentage. Therefore, PAI standardizes hit rate by dividing it by area percentage. Figure 3.15b shows that the PAI peaks when area percentage reaches 3.5.

3.8 SUMMARY

A rich set of new spatial analysis skills is introduced in this chapter:

1. Implementing the FCA method for spatial smoothing
2. Kernel density estimation (KDE) for mapping point data
3. Trend surface analysis (including logistic trend surface analysis)
4. Local interpolation methods, such as inverse distance weighting, thin-plate splines, and kriging
5. Areal weighting interpolation
6. Target-density weighting interpolation
7. Generating tessellation (e.g., uniform hexagons)
8. A set of tools for spatiotemporal visualization (e.g., data clock; 2D and 3D space–time cube)
9. Spatiotemporal kernel density estimation (STKDE) and hotspot analysis

Spatial smoothing and spatial interpolation are often used for mapping spatial patterns, like in Case Study 3A on mapping Zhuang place-names in southern China. However, surface mapping is merely descriptive. Identified spatial patterns, such as concentrations or lack of concentrations, can be arbitrary. Where are concentrations statistically significant instead of random? The answer relies on rigorous statistical analysis, for example, spatial cluster analysis—a topic to be covered in Chapter 8.

Area-based spatial interpolation is often used to convert data from different sources to one areal unit for an integrated analysis. It is also used to convert data across different spatial resolutions (area units) for examining the modifiable areal unit problem (MAUP). It is a popular spatial analysis task that has become increasingly routine in GIS. For example, in Case Study 6 on urban density functions, the technique is used to aggregate data from census tracts to townships so that functions based on different areal units can be compared.

This chapter also extends the popular KDE analysis to spatiotemporal KDE (STKDE) by adding a temporal dimension. A set of spatiotemporal visualization tools enables us to visualize temporal trends, spatial patterns, and the intersection between them with a space–time cube in 3D. Once again, a visual clustering pattern or hotspots in 3D may not be statistically significant. Answering this question calls for simulation of a large number of random scenarios, which will be discussed in Chapter 12.

APPENDIX 3: EMPIRICAL BAYES ESTIMATION FOR SPATIAL SMOOTHING

Empirical Bayes estimation is another commonly used method for adjusting or smoothing variables (particularly rates) across areas (e.g., Clayton and Kaldor, 1987; Cressie, 1992). Recall that the joint probability of two events is the product of one event and the probability of the second event conditionally upon the first event. Bayesian inference may be expressed as being the inclusion of prior information or belief about a dataset in estimating the probability distribution of the data (Langford, 1994, p. 143), that is:

$$likelihood\ function \times prior\ belief = posterior\ belief.$$

Using a disease risk as an example, the likelihood function can be said to be the Poisson distributed numbers of observed cases across a study area. The prior belief is on the distribution of relative risks (rates) conditional on the distribution of observed cases: for example, relative risks in areas of larger population size are likely to be more reliable than those in areas of smaller population size. In summary, (1) the mean rate in the study area is assumed to be reliable and unbiased, (2) rates for large population are adjusted less than rates for small population, and (3) rates follow a known probability distribution.

Assume that a common distribution, gamma, is used to describe the prior distribution of rates. The gamma distribution has two parameters, namely, the shape parameter α and the scale parameter v, with the mean $= v/\alpha$ and the variance $= v/\alpha^2$. The two parameters α and v can be estimated by a mixed maximum likelihood and moments procedure discussed in Marshall (1991). For an area i with population P_i and k_i cases of disease, the crude incidence rate for the area is k_i/P_i. It can be shown that the posterior expected rate or empirical Bayes estimate is:

$$E_i = \frac{k_i + v}{P_i + \alpha}$$

If area i has a small population size, the values of k_i and P_i are small relative to v and α, and the empirical Bayes estimate E_i will be "shrunken" toward the overall mean v/α. Conversely, if area i has a large population size, the values of k_i and P_i are large relative to v and α, and the empirical Bayes estimate E_i will be very close to the crude rate k_i/P_i. Compared to the crude rate k_i/P_i, the empirical Bayes estimate E_i is smoothed by the inclusion of v and α.

Empirical Bayes (EB) estimation can be applied to a whole study area where all rates are smoothed toward the overall rate. This is referred to as *global empirical Bayes smoothing*. It can also be applied locally by defining a neighborhood around each area and smoothing the rate toward its neighborhood rate. The process is referred to as *regionalized empirical Bayes smoothing*. A neighborhood for an area can be defined as all its contiguous areas plus itself. Contiguity may be defined as rook contiguity or queen contiguity (see Appendix 1), first-order or second-order contiguity, and so on.

GeoDa, a free package developed by Luc Anselin and his colleagues (https://spatial.uchicago.edu/geoda), can be used to implement the EB estimation for spatial smoothing (Anselin, 2018). The tool is available in GeoDa by choosing Map > Rates-Calculated Map > Empirical Bayes.

NOTES

1 One may use the threshold distance to set the search radius in distance computation and directly obtain the distances within the threshold. However, starting with a table for distances between all objects gives us the flexibility of experimenting with various window sizes.

2 RMS (root mean square) is measured as $RMS = \sqrt{\sum \left(z_{i,obs} - z_{i,est} \right)^2 / n}$.

3 This prepares an identical layer of Twnshp with a different name so that the result from "Generate Near Table" includes distance records from a place-name to itself (NEAR_DIST=0). When the input feature and the near feature are the same, the result would not include those zero-distance records.

4 NEAR_FID is identical to OBJECTID of Twnshp1, as Twnshp1 and Twnshp are identical.

5 For validation, add a field popu_valid, and calculate it as popu_valid = popu-den*Shape_Area /1000000, which should be identical to EstPopu. The areal weighting method assumes that population is distributed uniformly within each census tract, and thus a polygon in the intersected layer resumes the population density of the tract, of which the polygon is a component.

6 Notice that the file BRcrimeCube.nc will not be saved in the geodatabase but in an upper-level folder.

7 It shows a warning message. The time it takes to render the cube in three dimensions may vary considerably based on the number of features and the graphics card associated with your CPU.

8 If the Value or Profile (Fourier) options of the Characteristic of Interest parameter are chosen, the locations of the space–time cube are clustered using the k-means algorithm. The k-means algorithm can provide different clustering results depending on the initial random cluster representatives.

Part II

Basic Computational Methods and Applications

4 Delineating Functional Regions and Application in Health Geography

A *functional region* is an area around a node, facility, or hub connected by a certain function (e.g., retail distribution, healthcare service, advertisement for a media company, or telecom coverage). Delineation of functional regions occurs by defining regions that are coherent in terms of connections between supply and demand for a service. By defining a functional region, a distinctive market area has a geographic boundary that encompasses many smaller areas more closely connected with the central node(s) than beyond.

The literature in various fields uses different terms that possess the same or similar properties as a functional region. For example, *trade area* is "the geographic area from which a store draws most of its customers and within which market penetration is highest" (Ghosh and McLafferty, 1987, p. 62). *Catchment area (CA)* refers to an area around a facility where most of its clients, patients, or patrons reside (Paskett and Hiatt, 2018). A *hospital service area* (*HSA*) is an area within which patients receive most of their hospital care (www.dartmouthatlas.org). In urban and regional studies, *hinterland* (Wang, 2001a) or *urban sphere of influence* (Berry and Lamb, 1974) includes the rural area that maintains the highest volume of commerce, service, and other connections with and around a city.

Defining functional regions is a classic task in geography. It can be as straightforward as the proximal area method that assigns areas to their nearest facility or by the Huff model that accounts for the joint effects of facility sizes and their distances from residents. They are used in the absence of data on actual interactions between them (e.g., customer or patient flow volumes), as discussed in Section 4.1. Section 4.2 illustrates both methods in a case study of estimating the service areas of acute hospitals in Baton Rouge. Section 4.3 introduces the Dartmouth method that pioneered the delineation of HSAs by a simple plurality rule, and Section 4.4 introduces the network community detection approach to defining functional regions. Both rely on data of observed service volumes between supply and demand locations. Their implementations are demonstrated in a case study of defining hospital service areas in Florida in Section 4.5. The chapter is concluded with some remarks in Section 4.6.

4.1 BASIC GEOGRAPHIC METHODS FOR DEFINING FUNCTIONAL REGIONS

Much of the early application of functional region has been in trade area analysis, a common and important task in the site selection of a retail store. Defining trade

DOI: 10.1201/9781003292302-6

areas for the existing or proposed stores helps project market potentials, determines the focus areas for promotional activities, and highlights geographic weak-
ness in their customer bases versus their competitors'. The proximal area method
and the Huff model are two popular approaches widely used in retail and marketing analysis. The proximal area method simply assumes that residents only choose
the nearest facility for services, and thus a trade area is composed of residential
units sharing the same nearest facility. In addition to the distance decay effect (see
Section 2.3 in Chapter 2), the Huff model also considers that a more attractive
facility (e.g., larger and of higher quality) may outdraw other competitors closer to
residents. Understanding the proximal area method helps us build the foundation
for the Huff model.

4.1.1 Proximal Area Method

The *proximal area method* assumes that residents choose the nearest facility among
alternatives providing the same service. The proximal area method implies that
people only consider spatial impedance (distance, travel time, or by extension, travel
cost) in their choice of service, and thus a functional region is simply made of demand
areas that are closer to the anchoring facility than any others. How to define proximal
areas in GIS depends on the measurement of spatial impedance (see Section 2.1 in
Chapter 2).

When proximity is simply measured in Euclidean or geodesic distance, we may
define proximal areas by starting with either the supply locations or the demand locations. The *supply-based approach* constructs *Thiessen (or Voronoi) polygons* from
supply locations, overlays the polygons with demand locations (e.g., a census tract or
ZIP code area layer with population information) to identify resident areas within
each polygon, and aggregates the data by the polygon IDs. For example, Roos (1993)
used the Thiessen polygons to define hospital service areas (HSAs) in Manitoba,
Canada. Figures 4.1a–c show how the Thiessen polygons are constructed from five
points. First, five points are scattered in the study area, as shown in Figure 4.1a.
Secondly, in Figure 4.1b, lines are drawn to connect points that are near each other,
and lines are drawn perpendicular to the connection lines at their midpoints. Finally,
in Figure 4.1c, the Thiessen polygons are formed by the perpendicular lines. The
demand-based approach utilizes a tool "Near" to identify the nearest facility for
each demand area and then integrates demand areas sharing identical facilities to
form the functional regions. Both are illustrated in Section 4.2 (steps 1–2) and yield
similar results, and the latter is recommended for its simplicity.

When the spatial impedance is defined as the travel distance or time through a
transportation network, we can use the tool "Closest Facility" in the ArcGIS Pro
Network Analysis module to derive the proximal areas. Step 3 in Section 4.2 illustrates the implementation. If a predefined OD travel time matrix between demand and
supply locations is available, we can utilize the matrix to identify what areas share
the same nearest facilities and aggregate them by the facility IDs accordingly. In
ArcGIS, use the Summarize tool ("Minimum" for Statistic Type) to identify nearest
facilities and then use the Dissolve tool to aggregate demand areas sharing the identical nearest facilities (similar to step 4 in Section 4.2).

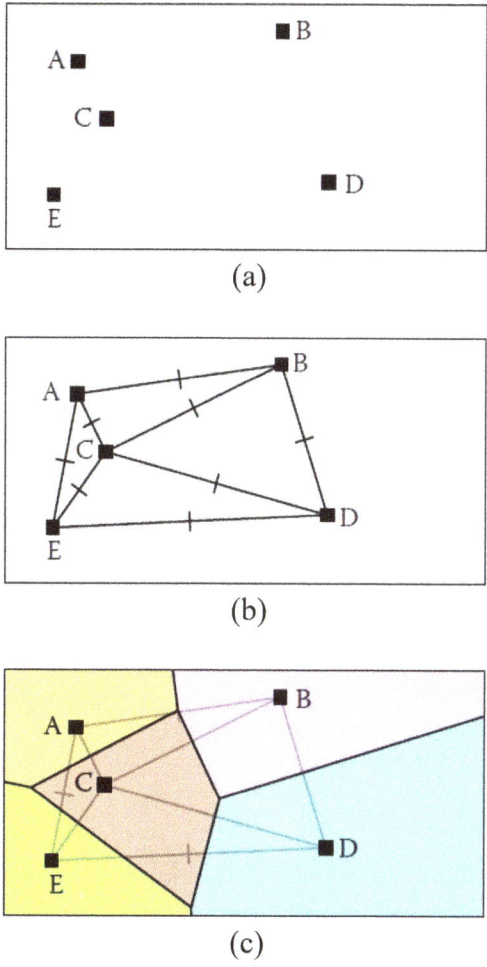

FIGURE 4.1 Constructing Thiessen polygons.

4.1.2 HUFF MODEL

The proximal area method only considers distance (or time) in defining trade areas. However, residents may bypass the closest facility to patronize others with better prices, better services, larger assortments, or a better image. Here we start with Reilly's law of retail gravitation for delineating trade areas between two stores (Reilly, 1931).

Consider two stores 1 and 2 that are at a distance of d_{12} from each other (Figure 4.2). Assume that the attractions for stores 1 and 2 are measured as S_1 and S_2 (e.g., in square footage of a store), respectively. The question is to identify the *breaking point* (BP) that separates trade areas of the two stores. The BP is d_{1x} from store 1, and d_{2x} from store 2, that is:

$$d_{1x} + d_{2x} = d_{12}. \tag{4.1}$$

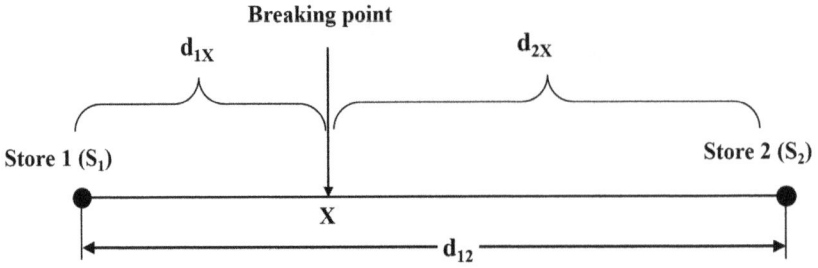

FIGURE 4.2 Illustration of Reilly's law.

By the notion of the *gravity model*, the retail gravitation by a store is in direct proportion to its attraction and in reverse proportion to the square of distance. Consumers at the BP are indifferent in choosing either store, and thus the gravitation by store 1 is equal to that by store 2, such as:

$$S_1 / d_{1x}^{\ 2} = S_2 / d_{2x}^{\ 2}. \tag{4.2}$$

Using equation 4.1, we obtain $d_{1x} = d_{12} - d_{2x}$. Substituting it into equation 4.2 and solving for d_{1x} yield:

$$d_{1x} = d_{12} / \left(1 + \sqrt{S_2 / S_1}\right). \tag{4.3}$$

Similarly:

$$d_{2x} = d_{12} / \left(1 + \sqrt{S_1 / S_2}\right). \tag{4.4}$$

Equations 4.3 and 4.4 define the boundary between two stores' trading areas and are commonly referred to as Reilly's law.

A more general gravity model method is the *Huff model* that defines trade areas of multiple stores (Huff, 1963). The model's widespread use and longevity "can be attributed to its comprehensibility, relative ease of use, and its applicability to a wide range of problems" (Huff, 2003, p. 34). The behavioral foundation for the Huff model may be similar to that of the multi-choice logistic model. The probability that some-one chooses a particular store among a set of alternatives is proportional to the perceived utility of each alternative. That is:

$$P_{ij} = U_j / \sum_{k=1}^{n} U_k, \tag{4.5}$$

where P_{ij} is the probability of an individual i selecting a store j, U_j and U_k are the utility choosing the stores j and k respectively, and k are the alternatives available ($k = 1, 2, \ldots, n$).

In practice, the utility of a store is measured as a *gravity kernel*. Like in equation 4.2, the gravity kernel is positively related to a store's attraction (e.g., its size in square footage) and inversely related to the distance between the store and a consumer's residence. That is:

$$P_{ij} = S_j d_{ij}^{-\beta} \Big/ \sum_{k=1}^{n} \left(S_k d_{ik}^{-\beta} \right), \qquad (4.6)$$

where S is a store's size, d is the distance, $\beta > 0$ is the *distance friction coefficient*, and other notations are the same as in equation 4.5.

Note that the gravity kernel in equation 4.6 is a more general form than in equation 4.2, where the distance friction coefficient β is assumed to be 2. The term $S_j d_{ij}^{-\beta}$ is also referred to as *potential*, measuring the impact of a store j on a demand location at i. Reilly's law may be considered as a special case of Huff model. In other words, when the distance friction coefficient β is assumed to be 2 and the choices are limited to two stores ($k = 1, 2$), $P_{ij} = 0.5$ defines the breaking point (BP), and the Huff model in equation 4.6 is regressed to Reilly's law in equation 4.2.

To illustrate the property of the Huff model, it is useful to stay with the simple Reilly's law but assume a general friction coefficient β. Similar to equation 4.2, we have:

$$S_1 d_{1x}^{-\beta} = S_2 d_{2x}^{-\beta}.$$

The BP is solved as:

$$d_{1x} = d_{12} \Big/ \left[1 + \left(S_2 / S_1 \right)^{1/\beta} \right],$$

$$d_{2x} = d_{12} \Big/ \left[1 + \left(S_1 / S_2 \right)^{1/\beta} \right].$$

Therefore, if store 1 grows faster than store 2 (i.e., $S_1 > S_2$ increases), d_{1x} increases and d_{2x} decreases, indicating that the market area for store 1 expands at the cost of store 2. That is straightforward.

However, how does the change in β affect the market areas? When β decreases, if $S_1 > S_2$, that is, $S_2 / S_1 < 1$, $\left(S_2 / S_1 \right)^{1/\beta}$ decreases, and thus d_{1x} increases and d_{2x} decreases, indicating that a larger store is expanding its market area farther. That is to say, when the β value decreases over time due to improvements in transportation technologies or road network, travel impedance matters to a less degree, giving even a stronger edge to larger stores. This explains some of the challenges of small hospitals, especially those in rural areas, in the era of modern transportation and telecommunication technologies. According to the Sheps Center for Health Services Research (2021), 181 rural hospitals have closed since January 2005, and 138 closed since 2010 in the USA.

Using the gravity kernel to measure utility may be purely a choice of empirical convenience. However, the gravity models (also referred to as "*spatial interaction models*") can be derived from individual utility maximization (Niedercorn and Bechdolt, 1969; Colwell, 1982) and thus have its economic foundation (see Appendix 4A). Wilson (1967, 1975) also provided a theoretical base for the gravity model by an entropy-maximization approach. Wilson's work also led to the discovery of a family of gravity models: a production-constrained model, an attraction-constrained model, and a production-attraction-constrained, or doubly-constrained, model (Wilson, 1974; Fotheringham and O'Kelly, 1989).

Furthermore, the gravity kernel assumes a power function. It is only one of many choices in modeling distance decay, and the best function needs to be identified by regression analysis of real-world travel data, as discussed in Section 2.3. Jia et al. (2017) used a general distance decay function $f(d)$ to refine the Huff model as:

$$P_{ij} = S_j f\left(d_{ij}\right)\Big/\sum_{k=1}^{n}\left(S_k f\left(d_{ik}\right)\right),\tag{4.7}$$

where a unitary exponent for facility size S is adopted to preserve the model's simplicity.

Based on equation 4.7, consumers, clients, or patients in an area visit facilities with various probabilities, and an area is assigned to the market area of a facility that is visited with the highest probability. In practice, given a demand location i, the denominator in equation 4.7 is identical for various facilities j, and thus the highest value of numerator identifies the facility with the highest probability. The numerator $S_j f\left(d_{ij}\right)$ is also known as "potential" for facility j at distance d_{ij}. In other words, one only needs to identify the facility with the highest potential for defining the market area. Implementation in ArcGIS can take full advantage of this property. However, if one desires to show a continuous surface of probabilities of individual facilities, equation 4.7 needs to be fully calibrated. In fact, one major contribution of the Huff model is the suggestion that market areas are continuous, complex, and overlapping, unlike the non-overlapping geometric areas of central place theory (Berry, 1967).

Implementing Huff model in ArcGIS utilizes a distance matrix between facilities and demand locations, and probabilities are computed by using equation 4.7. The result is not simple functional regions with clear boundaries but a continuous probability surface. Based on the probabilities, the traditional functional regions can be defined as areas where residents choose a particular facility with the highest probability.

4.2   CASE STUDY 4A: DEFINING SERVICE AREAS OF ACUTE HOSPITALS IN BATON ROUGE, LOUISIANA

This case study uses the ArcGIS network analyst module to define the service areas of five acute hospitals in Baton Rouge based on travel time through the road network. In the Huff model, we assume the traditional power function for the distance decay effect, and thus equation 4.6, where $\beta = 3.0$ and the capacity of each hospital S is number of beds.

Datasets needed for the project are the same as in Case Study 2 under the data folder BatonRouge. Several datasets generated from Case Study 2 (especially Subsection 2.2.2) are also used. The geodatabase BRRoad.gdb includes:

1. A projected census block group layer BR _ Bkg
2. A hospitals layer Hosp generated from step 1 in Case Study 2A

3. A census block group centroids layer BRPt from step 2 in Case Study 2A, whose OBJECTID field is identical to the field OBJECTID in BR_Bkg
4. A major road network dataset BR_MainRd_ND from step 5 in Case Study 2A
5. An OD drive time matrix file OD _ Drive _ Time (its field Total _ Time _ min representing drive time) from step 6 of Case Study 2A

4.2.1 DEFINING HSAs BY THE PROXIMAL AREA METHOD

In this subsection, steps 1 and 2 use Euclidean distance to define proximal areas by the supply-based and demand-based approach, respectively. There are also two ways to define proximal areas by travel time. Step 3 uses an ArcGIS network analysis tool to identify the nearest facility for each demand location via transport network, and step 4 takes advantage of a pre-calibrated travel time matrix to implement the same task.

Step 1. Defining Hospital Service Areas with Thiessen Polygons: In ArcGIS Pro, add the aforementioned data to the project. In the Geoprocessing pane > Toolboxes, select Analysis Tools > Proximity > Create Thiessen Polygons. In the dialog, under Parameters, choose Hosp for Input Features, enter HospThiessen for Output Feature Class, choose All fields for Output Fields, and under Environments, Processing Extent, choose "As Specified Below" and then BR_ Bkg, and click Run.

In the Geoprocessing pane, click Toolboxes > Analysis Tools > Overlay > Spatial Join. In the dialog, choose BR _ Bkg for Target Features and HospThiessen for Join Features, input BRBkg _ Thiessen for Output Feature Class, check Keep All Target Features, choose Have their center in for Match Option, and click Run. The field OID _ in BRBkg _ Thiessen, inherited from HospThiessen, represents the unique Thiessen polygons or service areas around hospitals and can be mapped in Unique Values in Symbology, as shown in Figure 4.3a.

Step 2. Defining Hospital Service Areas with the Near Tool: In the Geoprocessing pane, select Toolbox > Analysis Tools > Proximity > Near. In the dialog, choose BRPt for Input Features and Hosp for Near Features, keep other default settings, and click Run. The new field Near _ FID in BRPt indicates the nearest hospital for every census block group area. The attribute table of BRPt can be joined to BR _ Bkg based on the common ID OBJECTID _ 1, which represents the unique IDs for hospital service areas similar to Figure 4.3a.

Step 3. Finding the Nearest Hospitals by the Closest Facility Tool: Similar to step 6 in Case Study 2A (see Subsection 2.2.2 of Chapter 2), click the Analysis tab > Network Analysis. Make sure BR _ MainRd _ ND is the Network Data Source, select the tool "Closest Facility" to add the composite network analysis layer Closest Facility to the Contents pane, and click it. The Closest Facility Layer tab appears in the Main menu at the top. Its controls include the elements to be defined, such as: (1) for Import Facilities, use Hosp as Input Locations and 5000 Meters for Search Tolerance; (2) for Import Incidents, use BRPt as Input Locations and 15000 Meters for Search Tolerance; (3) for Travel Settings, under Travel Mode, choose Driving for Type, and under Costs, define Time _ min for Impedance. Click Run to execute the tool. The solution is saved in the layer Routes.

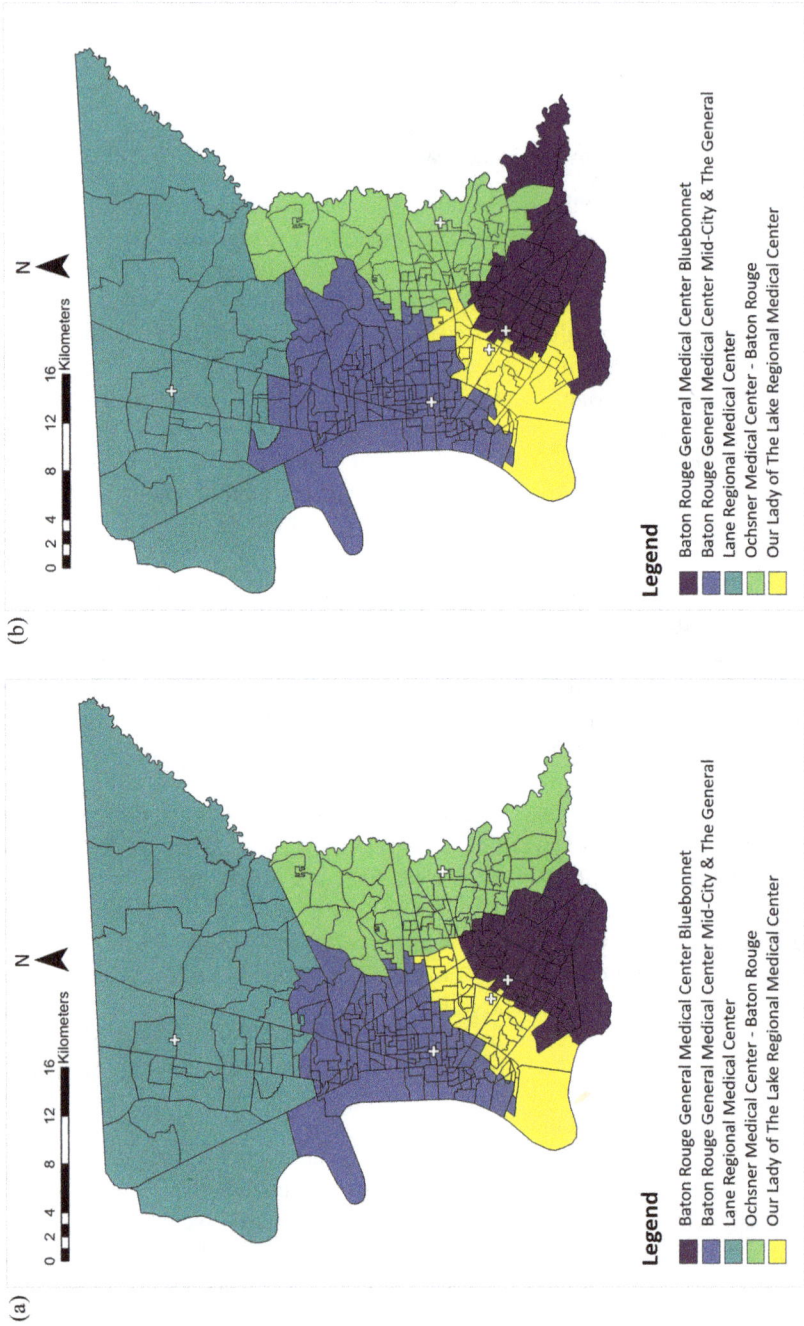

FIGURE 4.3 Proximal areas for the hospitals in EBRP based on (a) Euclidean distance and (b) drive time.

Join the attribute table of `Routes` (common key `IncidentID`) to the layer `BR _ Bkg` (common key `OBJECTID`, remove all joins prior to this step). Similar to step 10 in Case Study 1 in Section 1.3 of Chapter 1, use the tool "Dissolve" to aggregate block groups (`BR _ Bkg` as Input Features) to form proximal areas (`ProxArea` as Output Feature Class) based on the unique hospitals (`FacilityID` as Dissolve Fields). One may obtain total population within each proximal area by choosing `Popu2020` for Statistics Field and `Sum` for corresponding Statistics Type. See the result visualized in Figure 4.3b.

Step 4. Identifying the Nearest Hospitals from an OD Travel Time Matrix: This step uses the OD travel time table `OD _ Drive _ Time` derived from Case Study 2A. Its field names `OriginID` and `DestinationID` correspond to field `OBJECTID` of layer `BRPt` and field `OBJECTID _ 1` of layer `Hosp`, respectively.

Open the table `OD _ Drive _ Time`, right-click its field `OriginID`, and select Summarize. In the dialog, set `OD _ Drive _ Time` as the Input Table and name the Output Table as `HospMinTime`; under Statistics Fields, select `Total _ Time _ min` for Field and `Minimum` for its corresponding Statistic Type and make sure that the Case Field is `OriginID`. Click OK. It identifies the minimum travel time from each unique block group and all hospitals.

Join the table `HospMinTime` back to `OD _ Drive _ Time` based on the common field `OriginID`. On the expanded table `OD _ Drive _ Time`, apply Select by Attribute with the condition "`Total _ Time _ min`," "`is equal to`," "`MIN _ Total _ Time _ min`,"[1] and export the result to a new table `ProxTime`.

Join the table `ProxTime` (common field `OriginID`) to the feature class `BR _ Bkg` (common field `OBJECTID`, remove all joins prior to Join), and use the Dissolve tool to aggregate `BR _ Bkg` by the field `DestinationID` to generate the proximal areas. The result is the same as Figure 4.3b.

4.2.2 DEFINING HSAs BY THE HUFF MODEL

Step 5. Measuring Hospital Potentials: Join the attribute table of `Hosp` (common field `OID _`) to the table `OD _ Drive _ Time` (common field `DestinationID`, remove all joins and selections before doing this). Add a new field `Potent` (`Double` as Data Type) to `OD _ Drive _ Time`, and calculate it as `!Hosp.Bed _ size!*` `math.pow(!OD _ Drive _ Time.Total _ Time _ min!,-3)` when the Expression Type is "Python 3." This step computes the potential term $S_j d_{ij}^{-\beta}$ in equation 4.6.

Step 6. Calibrating Total Hospital Potential for Each Census Block Group: Apply Removes All Joins on `OD _ Drive _ Time` and open the table > right-click the field `OriginID` > Summarize. In the dialog, name Output Table as `SumPotent`; under Statistics Fields, choose `Potent` for Field and `Sum` for corresponding Statistics Type, make sure `OriginID` is selected for Case field, and click OK. In the resulting table `SumPotent`, the field `SUM _ Potent` is the total hospital potential for each block group. This step computes the denominator $\sum_{k=1}^{n} \left(S_k d_{ik}^{-\beta} \right)$ in equation 4.6. For each block group area indexed by *i*, it has a unique value.

Step 7. Calibrating the Probability of Each Hospital Being Visited by a Block Group: Join the table SumPotent back to OD _ Drive _ Time (based on the common field OriginID for both) to transfer the field SUM _ Potent. Join the attribute table of BRPt (common field OBJECTID) to OD _ Drive _ Time (common field OD _ Drive _ Time.OriginID) to attach the field Popu2020.

Add two new fields Prob and EstFlow (Data Type defined as Double for both) in OD _ Drive _ Time, and calculate them as Prob=!Potent!/!SUM _ Potent!² and EstFlow =!Prob!*!Popu2020! This step completes the calibration of probability $P_{ij} = S_j d_{ij}^{-\beta} \Big/ \sum_{k=1}^{n} \left(S_k d_{ij}^{-\beta} \right)$ in equation 4.6, and multiplying it by the corresponding population in a block group area yields the predicted patient flow from a block group area *i* to a hospital *j*.

Step 8. Identifying Hospitals Being Visited with the Highest Probability: On the table OD _ Drive _ Time, right-click the field OriginID, then choose Summarize. In the Summary Statistics dialog, make sure that Input Table is OD _ Drive _ Time and name Output Table as MaxProb; under Statistics Fields, choose OD _ Drive _ Time.Prob for Field and Maximum for corresponding Statistics Type, and make sure OriginID is selected for Case field. Click OK. In the resulting table MaxProb, the field MAX _ OD _ Drive _ Time _ Prob is the maximum probability for each block group area to visit a particular hospital.

Join the table MaxProb back to OD _ Drive _ Time (again, based on the common field OriginID, remove all joins before doing this) to transfer the field MAX _ Prob. On the table OD _ Drive _ Time, select records with the criterion "Prob," "is equal to," "MAX _ OD _ Drive _ Time _ Prob" (324 records selected), and export the result to a new table named Hosp _ MaxProb, which identifies the hospitals being visited by each block group area with the highest probability. If one replaces the field OD _ Drive _ Time.Prob by the field EstFlow and repeats this step, it yields the same result, that is, identifying the same hospital experiencing the highest flow volume from a block group.

Step 9. Defining the Service Areas by the Huff Model: Apply Removes All Joins on BR _ Bkg. Join the table Hosp _ MaxProb to the feature class BR _ Bkg (recall that the values in field OriginID of Hosp _ MaxProb correspond to those in field OBJECTID of BR _ Bkg) to transfer the field DestinationID, which indicates the hospitals being visited with the highest probability by each block group area.

Similarly, use the tool Dissolve to aggregate BR _ Bkg by the field DestinationID to derive the hospital service areas (HSAs), named HSA _ Huff. Here, block group areas visiting the same hospital with the highest probability are merged to form an HSA for that hospital (as shown in Figure 4.4).

Step 10. Mapping the Probability Surface for a Hospital's Visitation: As explained in step 9, the service area of a hospital in Figure 4.4 includes census block groups whose residents visit the hospital with the highest probability. However, one major lesson from the Huff model is that residents are likely to visit all hospitals with different probabilities. This step uses the largest hospital, that is, Our Lady of the Lake Regional Medical Center (OLTLRMC), to illustrate the implementation of calibrating and mapping the probabilities of hospital visits.

On the table OD_Drive_Time from step 7, select the records for the hospital (DestinationID = 5) and export to a new table Prob_OLTLRMC. Similar to

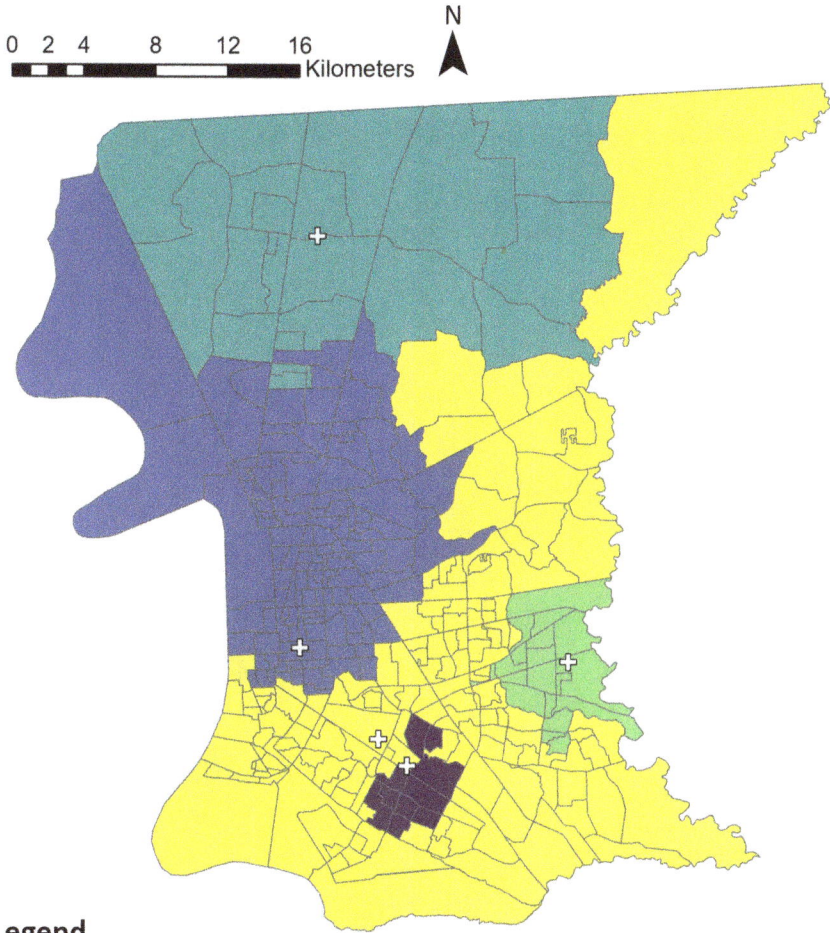

FIGURE 4.4 Hospital Service Areas in EBRP by Huff Model ($\beta = 3.0$).

step 9, join the table `Prob_OLTLRMC` to `BR_Bkg` based on their common fields (`OriginID` and `OBJECTID`, respectively), and map the field `Prob`, that is, probabilities of residents visiting the OLTLRMC among the five hospitals, as shown in Figure 4.5.

A toolkit "Huff Model Pro.tbx" is developed in ArcGIS Pro to automate the implementation of the Huff model. See Section 4.5 for more details.[3]

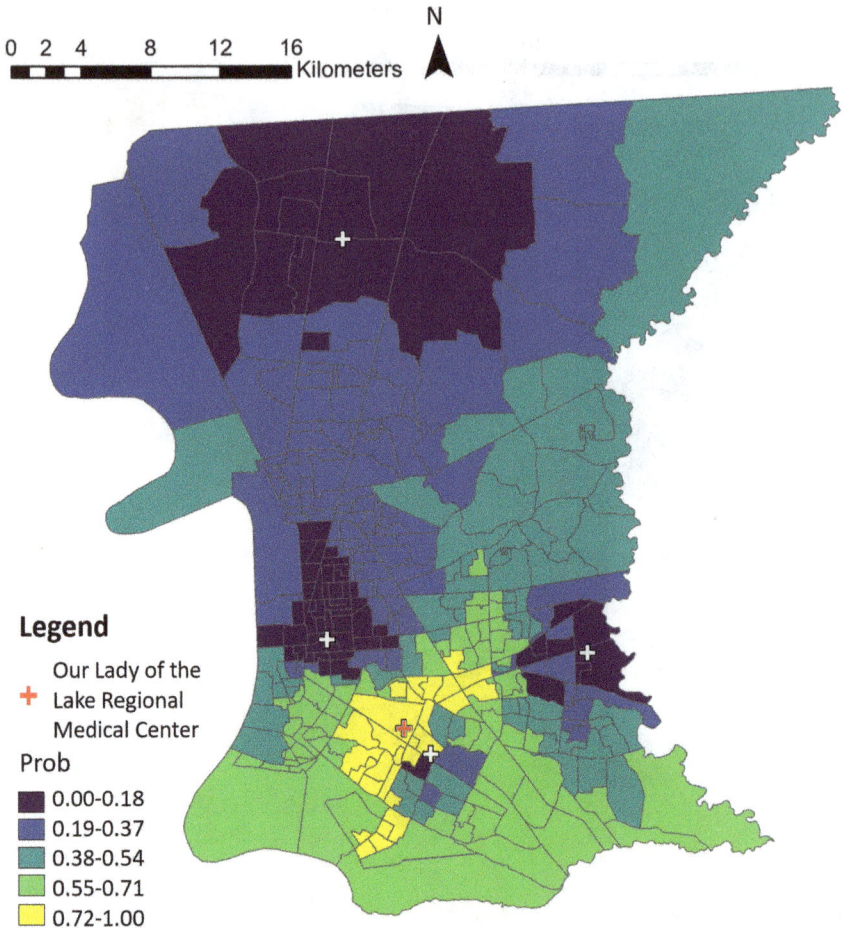

FIGURE 4.5 Probabilities of choosing one hospital by the Huff model ($\beta = 3.0$).

4.3 REFINING THE DARTMOUTH METHOD

The *Dartmouth Atlas of Health Care Project* began in 1993 to study the geographic variations of healthcare resources and their utilization in the USA (www.dartmouthatlas.org). For small area analysis and mapping, administrative units below the state level (e.g., county or city) may not capture healthcare markets, so the project piloted the delineations of *hospital service areas (HSAs)* and *hospital referral regions (HRRs)* as analysis units for local hospital market areas and regional tertiary areas, respectively. Their method, commonly referred to as the *Dartmouth method*, is primarily used for defining various healthcare markets, such as *pediatric surgical areas (PSAs)*, *primary care service areas (PCSAs)*, and *cancer care service areas (CSAs)*, in addition to HSAs and HRRs (Wang and Wang, 2022, pp. 1–2). The concept and methodology are equally applicable for defining functional regions in general. For convenience, we use HSAs as an illustrative example in this section.

The Dartmouth method defines HSAs in three steps:

1. A hospital is assigned to the town or city in which it is located to form a candidate HSA.
2. All hospitalization records are aggregated from each ZIP code area of patients to each hospital. If a town or city has multiple hospitals, the records are combined. A *plurality rule* is applied to assign each ZIP code area to the HSA, with which the hospitals are most often used.
3. The ZIP codes being assigned to the hospitals in the same town form an HSA. The HSAs are examined visually and adjusted to ensure all HSAs are geographically contiguous.

Wang and Wang (2022, pp. 63–67) clarified some uncertainties in the steps and refined the Dartmouth method so that the HSAs can be defined consistently and reliably. An example is used to illustrate its step-by-step implementation.

As shown in Figure 4.6, the example has nine nodes (say, ZIP code areas) indexed as 1–9 in circles. The solid lines labeled with numbers refer to the service flow volumes, with arrows pointing from the ZIP code of patients to the ZIP code of

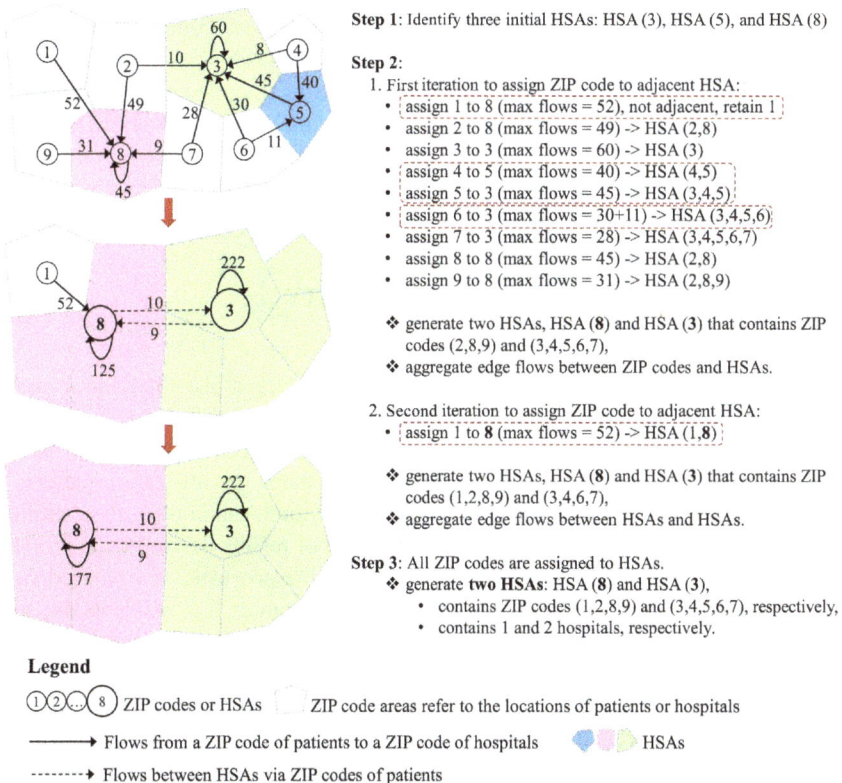

Step 1: Identify three initial HSAs: HSA (3), HSA (5), and HSA (8)

Step 2:
1. First iteration to assign ZIP code to adjacent HSA:
   - assign 1 to 8 (max flows = 52), not adjacent, retain 1
   - assign 2 to 8 (max flows = 49) -> HSA (2,8)
   - assign 3 to 3 (max flows = 60) -> HSA (3)
   - assign 4 to 5 (max flows = 40) -> HSA (4,5)
   - assign 5 to 3 (max flows = 45) -> HSA (3,4,5)
   - assign 6 to 3 (max flows = 30+11) -> HSA (3,4,5,6)
   - assign 7 to 3 (max flows = 28) -> HSA (3,4,5,6,7)
   - assign 8 to 8 (max flows = 45) -> HSA (2,8)
   - assign 9 to 8 (max flows = 31) -> HSA (2,8,9)

   ❖ generate two HSAs, HSA (**8**) and HSA (**3**) that contains ZIP codes (2,8,9) and (3,4,5,6,7),
   ❖ aggregate edge flows between ZIP codes and HSAs.

2. Second iteration to assign ZIP code to adjacent HSA:
   - assign 1 to **8** (max flows = 52) -> HSA (1,**8**)

   ❖ generate two HSAs, HSA (**8**) and HSA (**3**) that contains ZIP codes (1,2,8,9) and (3,4,6,7),
   ❖ aggregate edge flows between HSAs and HSAs.

Step 3: All ZIP codes are assigned to HSAs.
   ❖ generate **two HSAs**: HSA (**8**) and HSA (**3**),
      - contains ZIP codes (1,2,8,9) and (3,4,5,6,7), respectively,
      - contains 1 and 2 hospitals, respectively.

Legend

①②…⑧ ZIP codes or HSAs   ZIP code areas refer to the locations of patients or hospitals

——→ Flows from a ZIP code of patients to a ZIP code of hospitals   ◆◆◆ HSAs

------→ Flows between HSAs via ZIP codes of patients

FIGURE 4.6 Schematic workflow of the refined Dartmouth method.

hospitals, including two self-loops (nodes 3 and 8) with patients visiting hospitals within the same ZIP code areas.

1. Step 1 identifies three destination ZIP codes with hospitals as three initial HSAs: HSA(3), HSA(5), and HSA(8).
2. Step 2 has two iterations to assign each ZIP code to the adjacent HSA connected by the highest service volume. The first iteration initially assigns nodes 2, 8, and 9 to HSA(8); nodes 3, 5, 6, and 7 to HSA(3); and node 4 to HSA(5). As node 1 goes to a non-adjacent HSA(8) with the highest volume, it is unassigned at this stage. Node 4 goes to HSA(5), which goes to HSA(3), and according to the rule for consolidating the "non-independent HSAs," HSA(5) is merged to HSA(3), which now contains five nodes (3, 4, 5, 6, and 7). The second iteration finds node 1 unassigned and is then assigned to the expanded HSA(8) composed of three ZIP code areas (2, 8, and 9), which become adjacent to node 1.
3. Step 3 ends by verifying that all nodes have been assigned, forming two final HSAs: HSA(8) contains four nodes (1, 2, 8, and 9), and HSA(3) contains five nodes (3, 4, 5, 6, and 7).

Subsection 4.5.2 uses an automated tool for the Dartmouth method to define HSAs in Florida. The tool also provides two optional rules that are commonly considered in defining HSAs or functional regions in general: (1) each HSA with a minimum population, such as 1,000 used in defining the PCSAs (Goodman et al., 2003) and 120,000 in defining the CSAs (Wang et al., 2020b), and (2) each HSA with a minimum *localization index (LI)*, such as 0.5 used in defining the HSAs (Haynes et al., 2020). LI refers to the proportion of patients that receive services in hospitals within an HSA out of total patients from the same HSA. One may impose either rule or both rules.

4.4   SPATIALIZED NETWORK COMMUNITY DETECTION METHODS

Both the Huff model and the Dartmouth method employ a plurality rule to assign demand areas to facilities being utilized most often, and then areas assigned to the same facilities form unique functional regions. While the Dartmouth method uses actual utilization data, the Huff model interpolates the utilization pattern. In fact, one may employ the automated tool for Dartmouth method on the estimated demand-to-facility service flows for post-treatment of results derived from the Huff model to ensure spatial contiguity and other desirable properties for derived functional regions. However, the plurality rule cannot guarantee that residents use most services within the defined regions, as each assignment only considers the greatest proportion of demands to one facility at a time. In other words, it lacks a systematic perspective of ensuring the maximal total utilization volume within derived regions. This section introduces the network community detection approach, considered the most promising technique for functional region delineation. It follows a simple principle of grouping what is linked while keeping apart what is not (Reichardt and Bornholdt, 2006). As a result, functional regions are formed by grouping the most densely interconnected areas in terms of service volumes. The maximal service volumes within the functional regions, while minimal between them, are achieved by the innovative network community detection methods.

Based on a network of service utilization patterns between nodes (e.g., census tracts or ZIP code areas), *community detection* groups nodes with dense connections into initial communities and continues to group dense connections between initial communities into higher-level communities until a single community for a whole study area is attained. The process is guided by maximizing a quality measure, "modularity," for detected communities and generates a series of functional regions where lower-level (smaller) regions are nested within higher-level (larger) regions. This section adopts two popular community detection methods, "*Louvain algorithm*" (Blondel et al., 2008) and "*Leiden algorithm*" (Traag et al., 2019), named after the University of Louvain and Leiden University, with which the research team is affiliated, respectively. The latter was proposed to address a deficiency of the former in its tendency to generate arbitrarily poorly connected or even internally disconnected communities. Most recently, C Wang et al. (2021) have built upon these two algorithms and developed their spatially constrained versions, termed "*spatially constrained Louvain (ScLouvain)*" and "*spatially constrained Leiden (ScLeiden)*" methods, for application in defining cancer service areas (CSAs).

Given a spatial network, both algorithms use modularity, one of the best-known quality measures in network science, to detect communities within which the interactions are maximal but minimal between different communities. *Modularity* is formulated by incorporating a tunable resolution parameter (Reichardt and Bornholdt, 2006):

$$Q = \sum_{c \in C} \left(\frac{l_{in}^c}{m} - \gamma \left(\frac{k_{tot}^c}{2m} \right)^2 \right), \tag{4.8}$$

where Q represents the modularity value that sums over each community $c \in C$, l_{in}^c is the total number of edge weights between all nodes within community c, k_{tot}^c is the sum of the edge weights between nodes in community c and nodes in other communities, and m is the total number of edge weights in the network. The constant $\gamma > 0$ is the resolution parameter, and a higher resolution γ leads to a larger number of communities. Such an important property makes the ScLouvain and ScLeiden methods scale flexible. In general, a higher modularity represents higher-quality communities, and thus a more stable and robust network structure. Here, *node* refers to analysis areal unit (e.g., census tract or ZIP code area), *edge weight* is the service volume between them, and *community* is functional region to be derived.

The ScLouvain method has three steps, namely, (1) local nodes moving, (2) network aggregation, and (3) spatial contiguity and minimal region size guarantees, while the ScLeiden method contains four phases: (1) local nodes moving, (2) refinement, (3) network aggregation based on the refined partition, and (4) spatial contiguity and minimal region size guarantees. We use a simple network of 19 ZIP codes to briefly illustrate their implementations in delineating three HSAs in Figures 4.7 and 4.8, respectively.

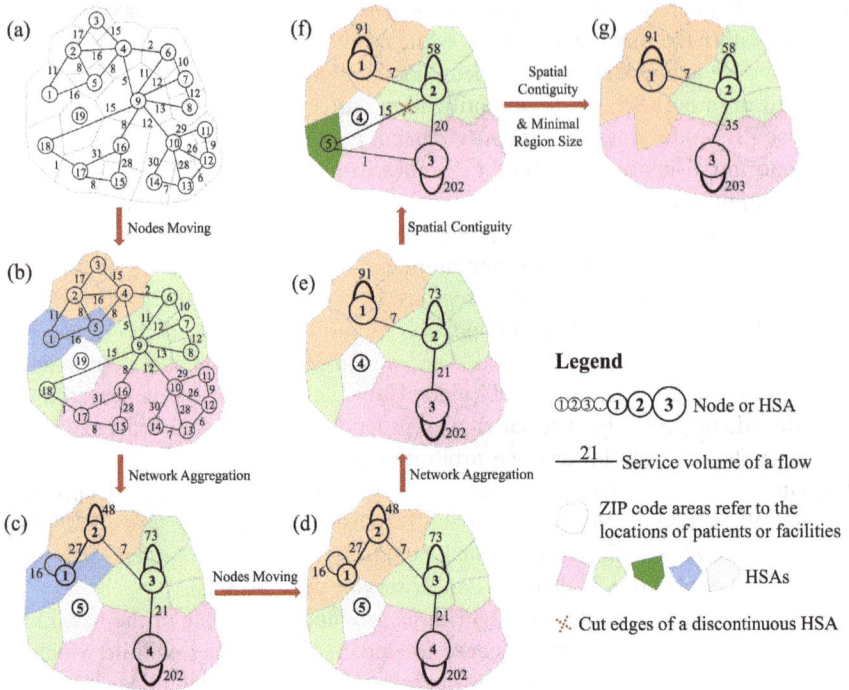

FIGURE 4.7 Schematic workflow of ScLouvain method.

In the nodes moving phase, both methods assign 19 ZIP codes to 19 individual HSAs in Figure 4.7a and Figure 4.8a. Then the ScLouvain method repeatedly moves each individual ZIP code to different HSAs to find the one with the greatest gain of modularity. Eventually, it detects 5 HSAs with different colors in Figure 4.7b. In contrast, the ScLeiden method uses the fast local move procedure to speed up the process, as not all ZIP codes are necessary to be moved, and it derives 5 HSAs with different colors in Figure 4.8b. The highlighted HSA has two separate sets of ZIP codes (15, 16, and 17 and 10, 11, 12, 13, and 14) at the bottom, forming two sub-HSAs in terms of the local subnetworks but as whole for the entire network. This is attributable to the role of ZIP code 9 acting as a bridge to connect two sub-HSAs. It is later moved to the upper HSA for gaining modularity, leaving two sub-HSAs internally disconnected. ZIP code 9 also has a strong connection with the nonadjacent ZIP code 18 at the lower left, which is assigned to the same HSA for the greatest modularity gain.

Next, the ScLouvain method creates a new network, where five HSAs become five new renumbered nodes and the service volumes between any new nodes are updated (see Figure 4.7c). However, the ScLeiden method adopts the smart local move and random neighbor move to refine the five HSAs by splitting the highlighted HSAs into two sub-HSAs in which two sets of nodes are well-connected, respectively, in Figure 4.8c. This is a significant difference between the two methods. The ScLeiden method continues to initialize the new network by five HSAs, but with

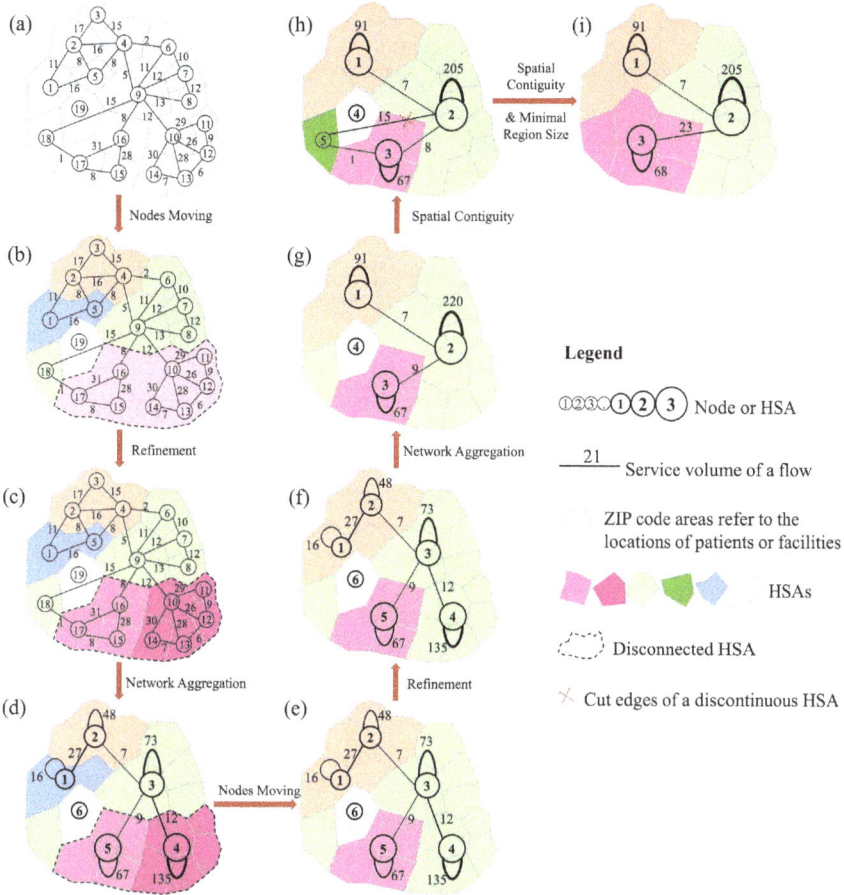

FIGURE 4.8 Schematic workflow of ScLeiden method.

the highlighted HSA locally separated into two sub-HSAs in Figure 4.8d for guaranteeing a monotonic increase of modularity. Although the poorly connected HSAs detected in this phase have small edge weights within each HSA, having one large HSA in the aggregated network will have modularity higher than that derived from multiple sub-HSAs detected in the refinement phase.

Then both methods start the second iteration. Since nodes 1 and 2 at the top left corner have strong connections with each other, they are moved into one HSA by both methods. The ScLeiden method even moves node 4 to node 3 to maximize the modularity gain (see Figure 4.8e). The ScLouvain method continues to aggregate the network into a new one in Figure 4.7e, while the ScLeiden method refines the partition and aggregates the network to a different one in Figure 4.8g. Both obtain the best aspatial community structure with the highest modularity.

The final phase is to enforce the spatial contiguity and minimal region size rules. Using a predefined spatial adjacency matrix, both methods split four non-spatially continuous HSAs in Figures 4.7e or 4.8g to five HSAs (nodes) in Figures 4.7f or 4.8h at a cost of decreasing modularity. As explained in Appendix 1 of Chapter 1, a spatial adjacency matrix may be traditionally defined in two ways: (1) rook contiguity defines adjacent areas as those sharing edges, and (2) queen contiguity defines adjacent areas as those sharing edges or nodes. We recommend using a predefined spatial adjacency matrix to account for some special scenarios. For example, some are separated by natural barriers, such as rivers, wildlife management areas, or parks, but are still connected via physical roads, bridges, or ferryboats. One may examine local geography (e.g., via Google Maps) to validate the connection and adjust the spatial adjacency matrix accordingly.

Similar to the Dartmouth method, we impose two constraints on the final derived regions: each has a minimal regional size and contains at least one facility. As shown in Figures 4.7g and 4.8i, both methods aggregate 19 ZIP code areas to three contiguous HSAs so that the total service volumes are maximal within each HSA and minimal between different HSAs. Note that the two methods delineate HSAs with different spatial configurations, as the ScLeiden method can identify poorly or disconnected HSAs and refine them, while such a merit is absent in the ScLouvain method. Subsection 4.5.3 illustrates how to implement the two methods in a case study of delineating HSAs in Florida.

4.5   CASE STUDY 4B: AUTOMATED DELINEATION OF HOSPITAL SERVICE AREAS (HSAS) IN FLORIDA

This case study uses toolkits developed by Python modules to delineate the hospital service areas in Florida. It uses three methods: Huff model, Dartmouth method, and network community detection methods. Huff model is implemented in a toolkit "Huff Model Pro.tbx," while the other two are integrated in a toolkit "HSA Delineation Pro.tbx." Both are saved in the folder CMGIS-V3-Toolbox > ArcPy-tools. The Florida folder contains all data for Case Study 4B, listed as follows:

1. Geodatabase FL_HSA.gdb contains a polygon feature class ZIP_Code_Area for 983 ZIP code areas in Florida, and a table for hospitalization volumes between these ZIP code areas OD_All_Flows with 37,180 nonzero flows and a total service volume of 2,392,066. Details about the descriptions of hospitalization table are referred to the Case Study 2B in Section 2.4 of Chapter 2.

2) A file FLplgnAdjAppend.csv is an adjusted spatial adjacency matrix.

4.5.1   DELINEATING HSAS BY THE REFINED DARTMOUTH METHOD

The following process illustrates how to use the toolkit of "HSA Delineation Pro.tbx" to delineate HSAs in Florida by the refined Dartmouth method.

Step 1. Building an Adjusted Spatial Adjacency Matrix: Activate the tool Build a Spatial Adjacency Matrix under the toolkit HSA Delineation Pro.tbx. Define the parameters as shown in Figure 4.9 (left): ZIP_Code_Area for Input

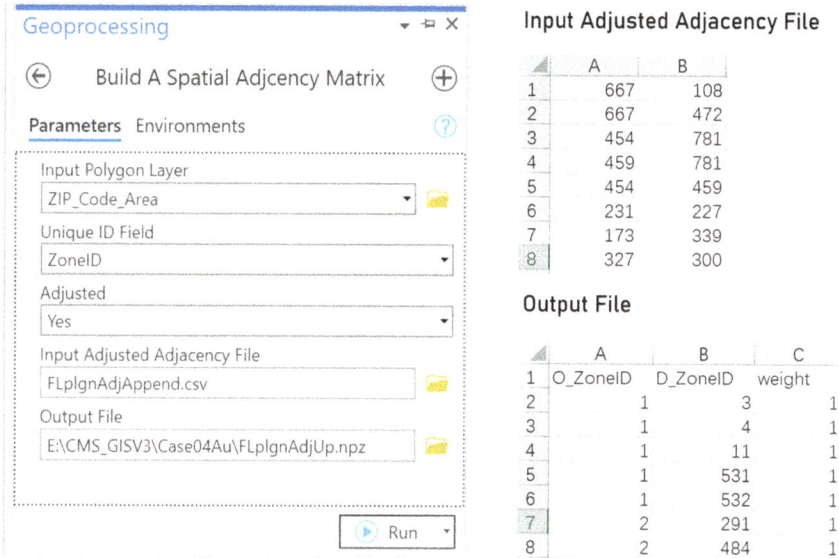

FIGURE 4.9 Interface for Building a Spatial Adjacency Matrix and sample input and output data.

Polygon Layer, its `ZoneID` for Unique ID Field starting from 1, `Yes` for Adjusted,[4] `FLplgnAdjAppend.csv` under the same folder of `FL _ HSA.gdb` for Input Adjusted Adjacency File, name `FLplgnAdjUp.npz` for Output File,[5] and click Run.

As shown in Figure 4.9 (top bottom), the input adjusted adjacency file `FLplgnAdjAppend.csv` contains two columns without header names, corresponding to origin `ZoneID` and destination `ZoneID` with amended connectivity. In addition to the output file `FLplgnAdjUp.npz` with an npz format for improving computational efficiency, the tool also outputs the same spatial adjacency matrix in a csv file format under the same path. The csv file only includes polygons that are spatially adjacent or linked by aforementioned adjustments and has three columns representing origin ID field, destination ID field, and weight (= 1) (see Figure 4.9, bottom right).[6]

Step 2. Using the Automated Dartmouth Method to Delineate HSAs: Activate the tool `Dartmouth Method (HSAs or HRRs)`. As shown in Figure 4.10, select `ZIP_Code_Area` for Input Polygon Layer, its `ZoneID` for Unique ID Field, and `POPU` for Population Field; select the table `OD_All_Flows` for Input Edge File, its `PatientZipZoneID` for Origin ID Field, `Hosp_ZoneID` for Destination ID Field, `AllFlows` for Service Flow Field, and `Total_Time_min` for Distance (or Travel Time) Field; select `Dartmouth Method` for Delineation Method and `FLplgnAdjUp.npz` for Input Spatial Adjacency Matrix File[7]; leave `Threshold Size` as the default value of 1,000 and Minimum Localization Index as blank; select `Dissolved` for Output Type; and save the result as `DHSAs`. Click Run.

This defines HSAs by the refined Dartmouth method. The input polygon feature should have a projected coordinate system to ensure the geographic compactness is

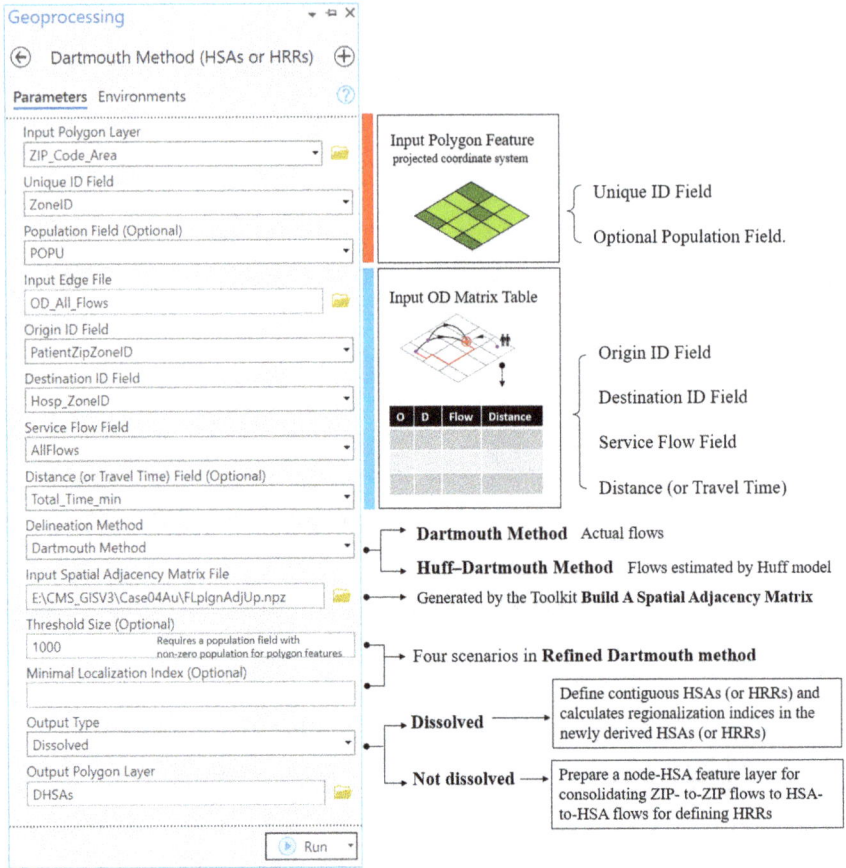

FIGURE 4.10 Interface for automated Dartmouth method

calibrated properly. The default values for threshold size and LI (Localization Index) are 1,000 and 0.5, respectively. The usage of threshold size requires a population field with nonzero value in the input polygon feature. The Output Type has two options: the option "`Dissolved`" generates contiguous HSAs and calculates several regionalization indices in the newly derived HSAs, and the option "`Not dissolved`" generates a layer by adding a field to the input feature identifying which ZIP code areas are contained in which HSAs.

The output feature class `DHSAs` has 136 spatially continuous HSAs. The fields `COUNT _ ZoneID` and `Num _ DZoneID` record the number of ZIP code areas and the number of destination ZIP code areas (with hospitals) within each HSA, respectively. Four additional fields, `LI`, `POPU`, `Compactness`, and `EstTime`, represent the localization index, population size, geographic compactness, and average travel time of each, respectively.[8]

Readers may input 0.5 for Minimum Localization Index or change other constraints and repeat the analysis to generate a new set of HSAs.

4.5.2 Delineating HSAs by Integrating the Huff Model and Dartmouth Method

An improved toolkit of Huff Model Pro.tbx is developed in ArcGIS Pro to automate the implementation of the Huff model. It includes two tools: Huff model, and Huff Model (w External Distance Table). The former calculates either Euclidean or geodesic distances between the supply and demand locations and then uses them to implement the Huff model, and the latter utilizes a pre-prepared external distance (or travel time) table. Here, we use the second tool.

Step 3. Defining Preliminary HSAs by the Huff Model with a Predefined External Distance Table: Activate the tool "Huff Model (w External Distance Table)," and enter the parameters as shown in Figure 4.11. Users can place the mouse on the red star next to an item to view its detailed description.

For the items pertaining to the external distance table, the ID fields for customer and facility locations must be consistent with the corresponding layers and ID fields. For the distance decay function *f(x)*, the choices include one-parameter continuous functions (power, exponential, square-foot exponential, Gaussian, and log-normal) or two-parameter continuous functions (log-logistic and compound power-exponential) and their associated coefficients.[9] The weight scale factor is designed to inflate the values for potentials, with a default value of 10,000 to avoid too small values in the process of calibration.

For the output type (associated table or feature class), choose the default option "Predicted OD flows." It is a table with fields of probability, estimated flows, and other fields related to the potentials and can be used as an input for the Dartmouth method. When the option "Dissolved preliminary HSAs" is chosen, it generates a feature class that contains the dissolved HSAs with predicted total patient volumes. Note that the dissolved HSAs from this option are preliminary, as some HSAs may not be spatially contiguous.

Step 4. Refining HSAs by Using the Huff–Dartmouth Method: As stated previously, both the Huff model and the Dartmouth method use the plurality rule by assigning a demand (patient residence) area to a hospital being visited most. However, the Dartmouth method is based on actual patient flows, and the Huff model is based on estimated flows in absence of such data. The Huff model is prone to derive spatially non-contiguous HSAs, and some HSAs do not even contain any hospitals. Here we use the refined Dartmouth method to further process the estimated flows from the Huff model in step 3 to enforce spatial contiguity and accommodate additional constraints, such as a threshold population size and/or a minimum LI value. In other words, the Dartmouth method is used here as a post treatment on the predicted OD flows by the Huff model.

Activate the tool "Dartmouth Method (HSAs or HRRs)." In the dialog (similar to Figure 4.10), choose ZIP _ Code _ Area for Input Polygon Layer, its ZoneID for Unique ID Field, and POPU for Population Field; choose OD _ ALL _ Flows for Input Edge File, its PatientZipZoneID for Origin ID Field, its Hosp _ ZoneID for Destination ID Field, EstFlow for Service Flow Field, and Total _ Time _ min for Distance (or Travel Time) Field; choose Huff-Dartmouth Method for Delineation Method; set FLplgnAdjUp.npz as Input

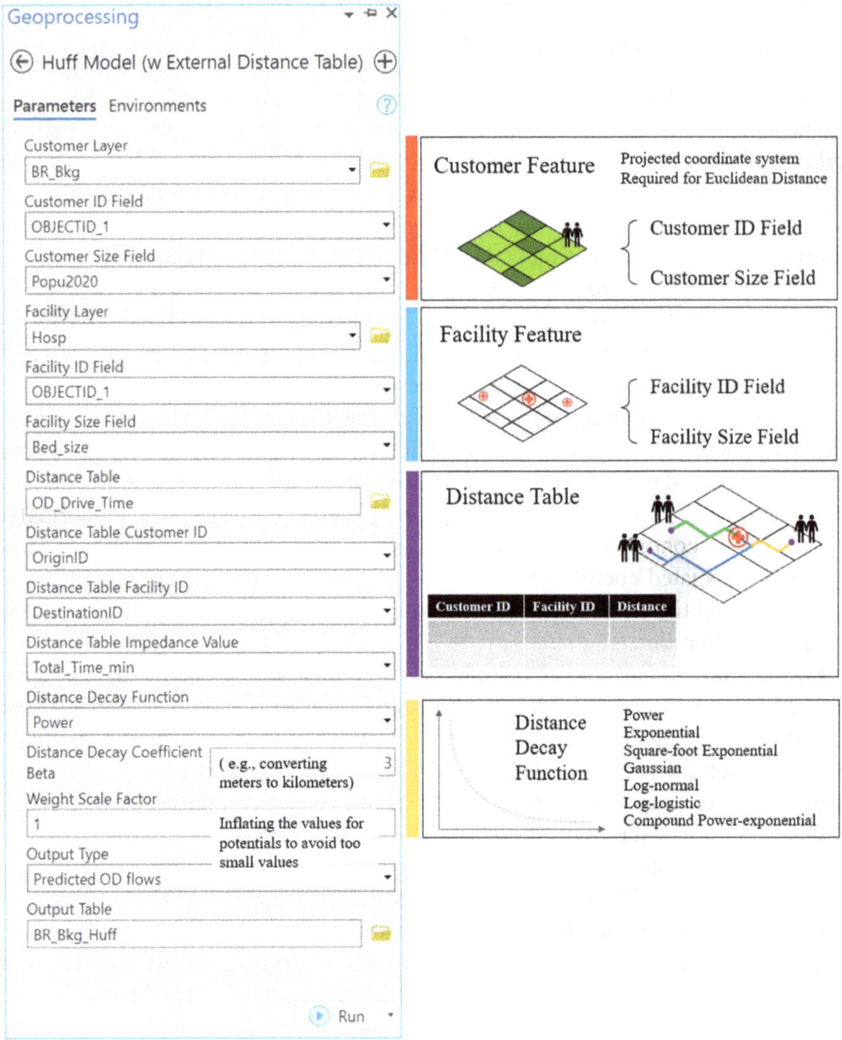

FIGURE 4.11 Interface of automated Huff model with a predefined external distance table.

Spatial Adjacency Matrix File, keep 1000 as the default Threshold Size, and leave Minimal Localization Index blank; choose Dissolved for Output Type and name Output Polygon Layer as HDHSAs; and click Run.

Here the option "Huff-Dartmouth Method" extracts the maximum estimated flows from each of the ZIP code areas to start the process, whereas the other option, "Dartmouth Method," uses all actual flows. This strategy is adopted for improving computational efficiency. For a given number of demand sites (*m*) and a number of supply sites (*n*), the matrix of simulated flows between them is a complete *m* × *n* and sizable matrix, whereas the actual flow matrix is often a small fraction of that size. However, all estimated flows participate in the calculation of indices, such as localization index (LI) and average travel time.

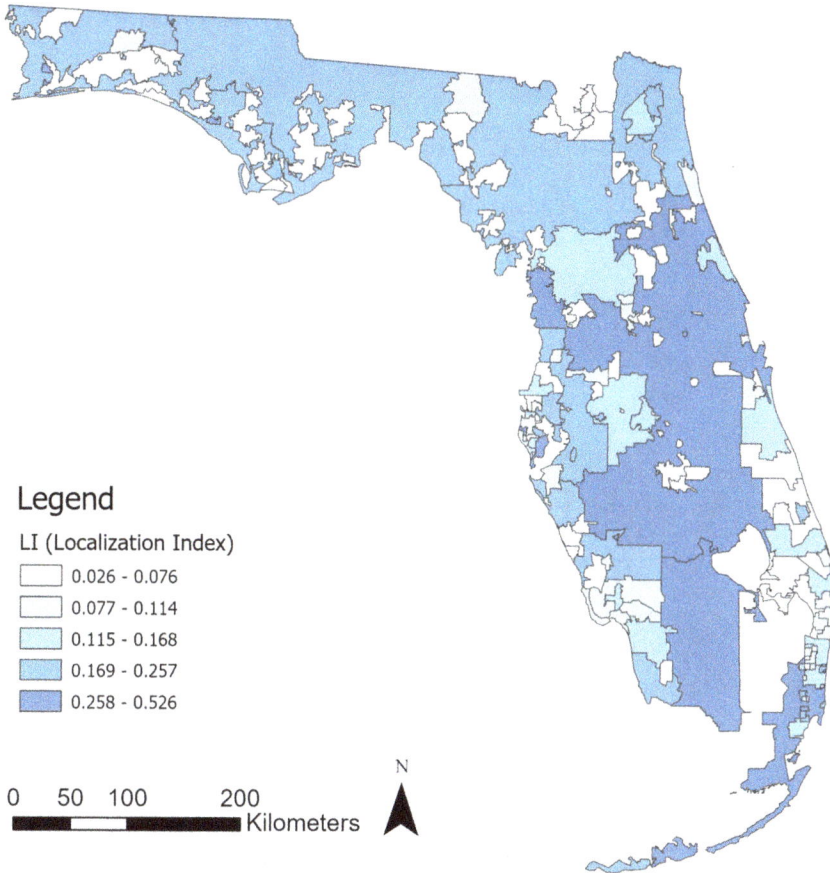

FIGURE 4.12 HSAs in Florida delineated by Huff–Dartmouth method.

The Huff–Dartmouth method generates 113 continuous HSAs. Figure 4.12 shows the variations of LI across 113 HSAs. Note that the so-called LIs are calculated from estimated hospitalization flows from the Huff model and have relatively low values.

4.5.3 Delineating HSAs by Spatialized Network Community Detection Methods

Step 5. Delineating HSAs by ScLeiden Method: In ArcGIS Pro, activate the tool "Network Community Detection Method (ScLeiden or ScLouvain)" under the toolkit "HSA Delineation Pro.tbx." Set the parameters as shown in Figure 4.13. The parameter "resolution," $\gamma > 0$ in equation 4.8, has a default value of 1 that usually corresponds to the scenario with the global maximal modularity. The method enables users to delineate different numbers of communities (here, HSAs) by changing this parameter. For options associated with imposing spatial contiguity, (1) if choosing "Yes," users need to input a predefined spatial adjacency matrix in the.

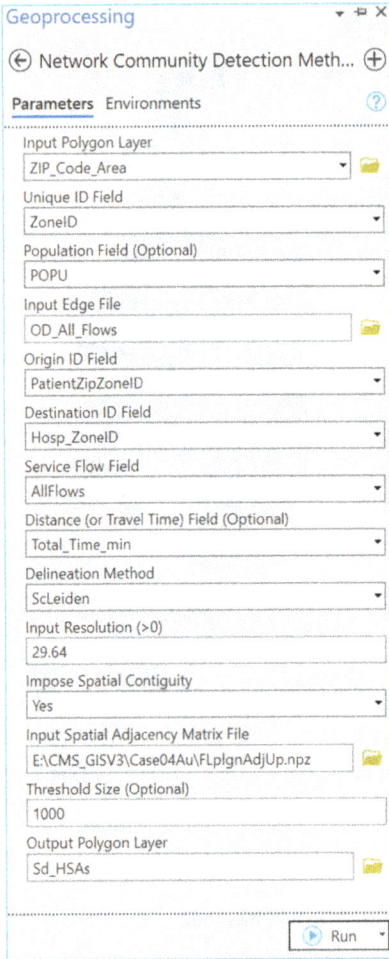

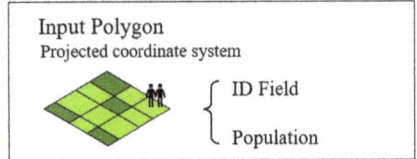

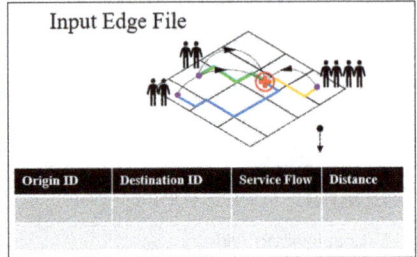

FIGURE 4.13 Interface for the ScLeiden method.

npz format and an option to define `Threshold Size (Optional)` with a default population value of 1,000 (note that the input layer must contain a population field), and (2) if choosing "No," it will generate aspatial communities without imposing any spatial constraints, and the option "`Threshold Size (Optional)`" will not be available.

The result is the Output Polygon Layer, here `Sd _ HSAs`, with 136 contiguous HSAs. It contains fields: `HSAID` for each HSA ID, `COUNT _ ZoneID` and `Num _ DZoneID` for numbers of (origin) ZIP code areas and destination ZIP code areas (or hospitals) within each HSA, respectively, and four additional fields, `LI`, `POPU`, `Compactness`, and `EstTime`.

Under the same path, there is a non-dissolved polygon layer with an extended name "`NotDis`" (i.e., `Sd _ HSAsNotDis`) to record the relationship between the

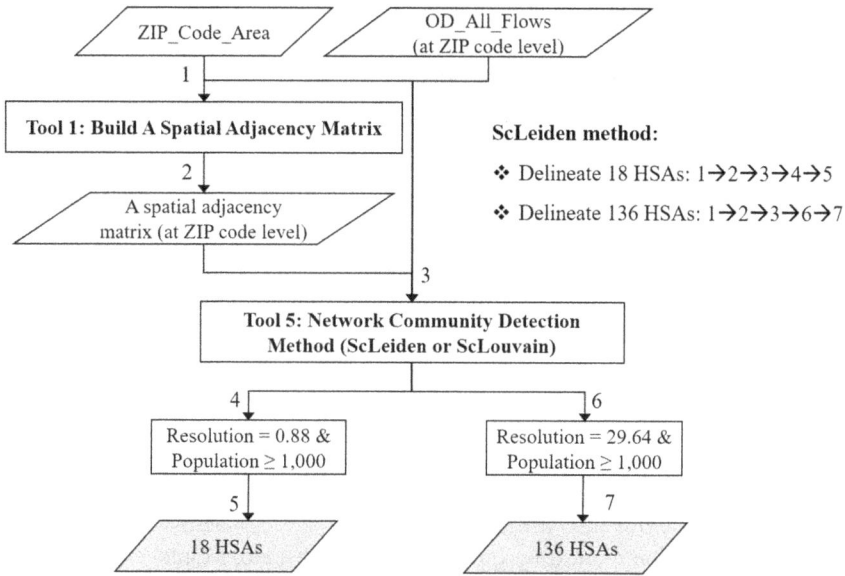

FIGURE 4.14 Workflow of using two tools to delineate HSAs by the ScLeiden method.

input polygon feature and the delineated communities (here, HSAs). That is to say, the tool can be used to generate aspatial (or spatially contiguous) communities without (or with) any spatial constraints. Click the two options, "View Details" and "Open History," at the bottom of the dialog to review additional information.

Change the resolution value for the ScLeiden method to 0.88 and 29.64 to derive 18 HSAs with a global optimal modularity and 136 HSAs, respectively.

Switch the Delineation Method to "ScLouvain" and repeat the same steps to delineate HSAs by the ScLouvain method. The suggested resolution values are 1.38 and 30.23 for deriving 22 and 136 HSAs, respectively.

A workflow is illustrated in Figure 4.14 to show the process of delineating HSAs by the ScLeiden method. A similar workflow applies to the ScLouvain method.

In the study of Wang and Wang (2022), upon which the case study is built, either spatialized network community detection method ScLouvain or ScLeiden outperforms the Dartmouth method in various regionalization assessment measures, such as the localization index (LI), compactness, and balance of region size.

4.6 SUMMARY

This chapter covers various methods for defining functional regions, within which encompassed areas are more closely connected than between. It begins with the proximal area method that assigns areas to their nearest facility, based on the commonly known "least-effect principle" in geography (Zipf, 1949). As a result, a function region is made of areas sharing the same nearest facility. Road network distance or travel time is generally a better measure than a straight-line (Euclidean) distance to capture spatial impedance. However, network distance or travel time may not be the best measure. Travel cost, convenience, comfort, or safety may also be important in influencing individual spatial behaviors, as discussed in Section 2.1 of Chapter 2. Research indicates that people of various socioeconomic or demographic characteristics perceive the same distance differently, that is, a difference between cognitive and physical distances (Cadwallader, 1975). Defining network distance or travel time also depends on a particular transportation mode. This makes distance or time measurement more than a routine task. Accounting for interactions by telecommunication, internet, and other modern technologies adds further complexity to the issue.

The Huff model estimates the probabilities of residents choosing facilities based on a gravity model, which accounts for the joint effects of facility attractions and their distances from residents. In addition to a distance measure (d), the Huff model in equation 4.6 has two more variables: facility attraction (S) and travel friction coefficient (β). Attraction can be measured as simply as a single variable (e.g., hospital bed size in the Case Study 4), or in more complex models to account for multiple factors (Nakanishi and Cooper, 1974; Weisbrod et al., 1984). The travel friction coefficient β is also difficult to define, as it varies across time and space, among transportation modes, by type of commodities, etc. Moreover, the distance decay function may not be best captured by the popular power function, as discussed in Section 2.3 of Chapter 2. This leads to the proposition of a general Huff model as defined in equation 4.7.

In short, the proximal area method or the Huff model is used in absence of data on actual connection strengths between supply and demand locations. They offer us an opportunity to develop a preliminary estimation of market areas or help us predict the impact of changes, such as closing, opening, expansion, downscaling, or relocation of facilities. Reliable and realistic delineation of functional regions relies on actual origin–destination (OD) flow data. The Dartmouth method, proposed for defining hospital service areas (HSAs), uses a simple plurality rule by assigning a demand area to a facility being used by its residents most often. This plurality rule is also implied by the Huff model. The difference is that the OD flows are estimated in the Huff model but observed in the Dartmouth method. Therefore, one can use the Huff model to predict the OD flows and then use the Dartmouth method as post treatment to define functional regions, a strategy adopted in Case Study 4B.

The plurality rule in the Dartmouth method emphasizes the largest flow originated from a demand area and lacks a systematic perspective. It does not ensure that the derived regions encompass the maximum flows. The network community detection approach explicitly maximizes connections within regions and minimizes connections between them. The spatialized network community detection methods, namely, ScLouvain and ScLeiden, introduced in Section 4.4 account for spatial constraints, such as area contiguity and threshold population size. Both methods are computationally efficient and automated in a user-friendly GIS tool. They also outperform the Dartmouth method in various regionalization assessment measures. Moreover, the methods delineate a series of functional regions of various sizes corresponding to user-defined resolution values and are thus termed *scale-flexible*. Such an important property enables an analyst to examine the modifiable area unit problem (MAUP) as discussed in Section 1.3 of Chapter 1 or the uncertain geographic context problem (UGCoP) (Kwan, 2012). Therefore, when supporting data are available, the ScLouvain and ScLeiden methods are recommended.

APPENDIX 4A: ECONOMIC FOUNDATION OF THE GRAVITY MODEL

The gravity model is often criticized, particularly by economists, for its lack of foundation in individual behavior. This appendix follows the work of Colwell (1982) to provide a theoretical base for the gravity model. For a review of other approaches to derive the gravity model, see Fotheringham et al. (2000, pp. 217–234).

Assume a trip utility function in a Cobb–Douglas form, such as:

$$u_i = ax^{\alpha}z^{\gamma}t_{ij}^{\tau_{ij}},\qquad (A4.1)$$

where u_i is the utility of an individual at location i, x is a composite commodity (i.e., all other goods), z is leisure time, t_{ij} is the number of trips taken by an individual at i to j, $\tau_{ij} = \beta P_j^{\phi} / P_i^{\xi}$ is the trip elasticity of utility that is directly related to the destination population P_j and reversely related to the origin population P_i, and a, α, β, γ, ϕ, and ξ are positive parameters. Colwell (1982, p. 543) justifies the particular way of defining trip elasticity of utility on the ground of central place theory: larger places serve many of the same functions as smaller places plus higher-order functions not found in smaller places, and thus the elasticity τ_{ij} is larger for trips from the smaller to the larger place than trips from the larger to the smaller place.

The budget constraint is written as:

$$px + rd_{ij}t_{ij} = wW,\qquad (A4.2)$$

where p is the price of a composite commodity x, r is the unit distance cost for travel, d_{ij} is the distance between point i and j, t_{ij} is the number of trips taken by an individual at i to j, w is the wage rate, and W is the time worked.

In addition, the time constraint is:

$$sx + hd_{ij}t_{ij} + z + W = H,\qquad (A4.3)$$

where s is the time required per unit of x consumed, h is the travel time per unit of distance, and H is total time.

Combining the two constraints in equations A4.2 and A4.3 yields:

$$(p + ws)x + (rd_{ij} + whd_{ij})t_{ij} + wz = wH.\qquad (A4.4)$$

Maximizing the utility in equation A4.1 subject to the constraint in equation A4.4 yields the following Lagrangian function:

$$L = ax^{\alpha}z^{\gamma}t_{ij}^{\tau_{ij}} - \lambda\left[(p + ws)x + (rd_{ij} + whd_{ij})t_{ij} + wz - wH\right]$$

Based on the four first-order conditions, that is, $\partial L / \partial x = \partial L / \partial z = \partial L / \partial t_{ij} = \partial L / \partial \lambda = 0$, we can solve for t_{ij} by eliminating λ, x, and z:

$$t_{ij} = \frac{wH\tau_{ij}}{(r+wh)d_{ij}(\alpha+\gamma+\tau_{ij})}. \tag{A4.5}$$

It is assumed that travel cost per unit of distance r is a function of distance d_{ij}, such as:

$$r = r_0 d_{ij}^{\sigma}, \tag{A4.6}$$

where $r_0 > 0$ and $\sigma > -1$ so that total travel costs are an increasing function of distance. Therefore, the travel time per unit of distance, h, has a similar function:

$$h = h_0 d_{ij}^{\sigma}, \tag{A4.7}$$

So that travel time is proportional to travel cost. For simplicity, assume that the utility function is homogeneous to degree 1, that is:

$$\alpha + \gamma + \tau_{ij} = 1. \tag{A4.8}$$

Substituting equations A4.6, A4.7, and A4.8 into A4.5 and using $\tau_{ij} = \beta P_j^{\phi} / P_i^{\xi}$, we obtain:

$$t_{ij} = \frac{wH\beta P_i^{-\xi} P_j^{\phi}}{(r_0 + wh_0)d_{ij}^{1+\sigma}}. \tag{A4.9}$$

Finally, multiplying equation A4.9 by the origin population yields the total number of trips from i to j:

$$T_{ij} = P_i t_{ij} = \frac{wH\beta P_i^{1-\xi} P_j^{\phi}}{(r_0 + wh_0)d_{ij}^{1+\sigma}}, \tag{A4.10}$$

which resembles the gravity model.

APPENDIX 4B: INSTALLING GEOPROCESSING TOOLS BASED ON PYTHON

Python (python.org) is a free cross-platform and open-source programming language. It has been introduced to the ArcGIS community since the release of ArcGIS 9.0 and acts as the scripting language for geoprocessing tools in ArcGIS Pro. Geoprocessing tools developed for ArcGIS products other than ArcGIS Pro might not work in ArcGIS Pro, as the latter uses Python 3 as the backend Python platform.

Users can try Python scripting in ArcGIS Pro by using the built-in Jupyter Notebook by clicking the icon 🗒️ New Notebook on the Insert tab. For learning Python in ArcGIS Pro, visit the official website (https://pro.arcgis.com/en/pro-app/latest/arcpy/get-started/installing-python-for-arcgis-pro.htm). For using Python functionality not provided by ArcGIS Pro or running a third-party geoprocessing tool, one needs to install related Python packages. A new Python environment needs to be cloned in ArcGIS Pro in preparing for installing and running the toolkits (Wang F and Wang C, 2022, p. 69).

In ArcGIS Pro, click the Project tab > Package Manager. The default Python environment is shown in Active Environment (Figure A4.1). Click the icon ⚙️ for Environment Manager to activate it. In the dialog, click on the default environment symbol "..." next to `argispro-py3` and then choose "Clone ...." In the dialog of Clone Environment, keep the default settings. A default clone Python environment will be loading (e.g., `arcgispro-py3-clone`). Note the file location under Destination is for the cloned environment. Click OK to close the dialog, and restart ArcGIS Pro.

To run all Python-based geoprocessing tools provided in this book, five additional packages need to be installed to the cloned environment: (1) SciPy (https://www.scipy.org/), (2) NetworkX (https://networkx.org/), (3) igraph (https://igraph.org/), (4) leidenalg (https://github.com/vtraag/leidenalg), and (5) Shapely (https://shapely.readthedocs.io/en/stable/manual.html). The first four packages are for the tools of Chapter 4, and the other package will be used in Chapter 13.

To check the installation status of the packages in ArcGIS Pro, also go to Package Manager. In the dialog, click Installed tab and input the package name in the Search box to check whether it is already included, or click Add packages tab to locate and install the package name by searching in the input box and pressing the Install button. Make sure all required packages are among the list of installed packages.

For a Windows system, right-click the Start Menu button ⊞ > Run to launch the Run command window > enter `cmd` in the Open box, and click OK to open the command prompt window. Use the command "`CD . . .`" to navigate the path of the cloned environment.[10] To install all packages, use the file `requirements.txt` under the data folder `Florida`. For example, in our case, type "`python -m pip install -r D:\CMGISV3\CMGIS-V3-Toolbox\ArcPy-tools\requirements.txt.`" Change your file path if needed to point to the location of "`requirements.txt.`"

Now we are ready to add the toolkits. In ArcGIS Pro, under the Catalog view, right-click Toolboxes > Add Toolbox. In the dialog, select and add the toolkits from the folder `ArcPy-tools`. Before using these tools, right-click each tool > Properties, and ensure that the Script File of each tool under the Execution tab points to the same Python script under the subfolder `Scripts` under `ArcPy-tools`. Refer to Appendix 2B in Chapter 2 for adding the Python toolkit. In addition, users need to save all layers and tables in a geodatabase to avoid truncated field names that may not be identified by the Python tools.

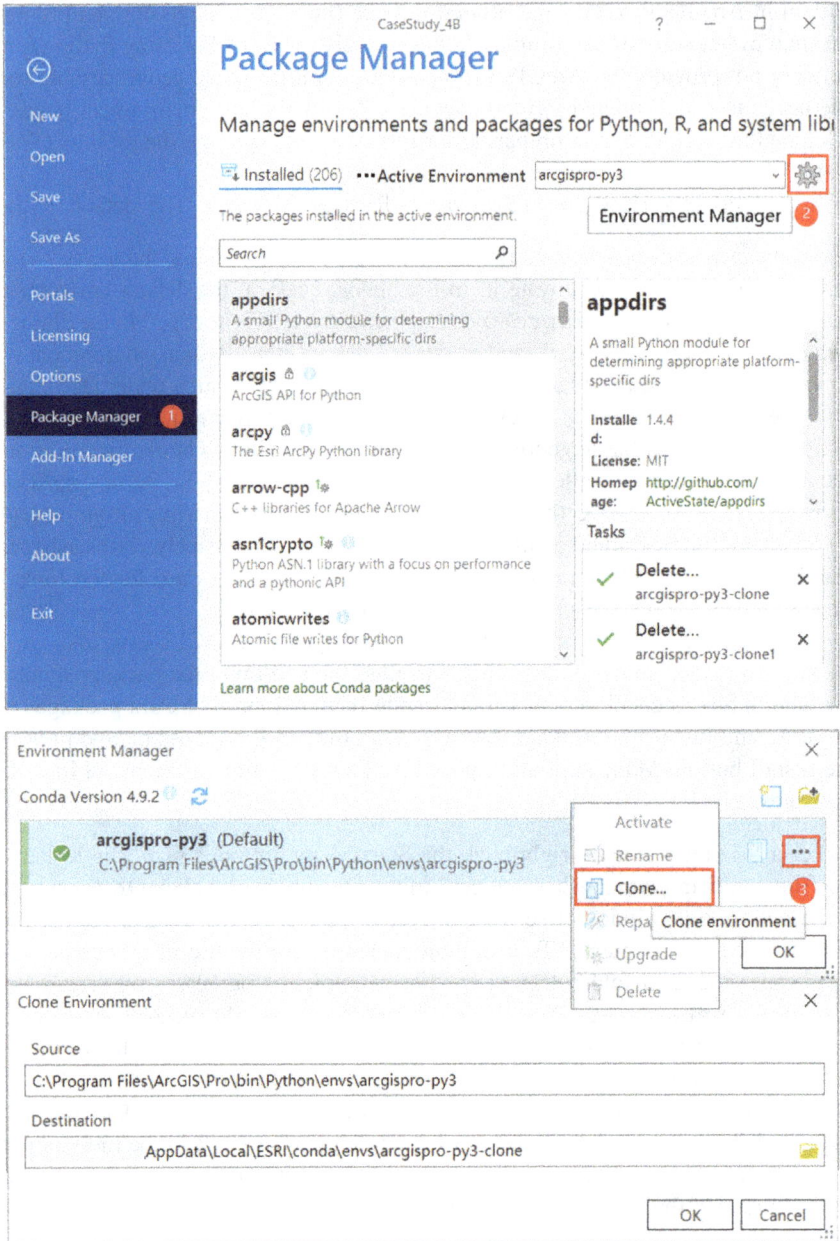

FIGURE A4.1 Clone Python environment in ArcGIS Pro.

NOTES

1 In the third box of Expression of Select by Attributes, scroll down in the drop-down menu to locate and check Fields in order to skip the specific values and find the field name "`MIN_Total_Time_min`."

2 This is an abbreviated form from a complete expression. The same applies to the calculation of `EstFlow`.

3 Users may also test the Huff Model tool in ArcGIS Pro by navigating to Geoprocessing Tools > Business Analyst toolbox > Modeling toolset > Huff Model.

4 In the study area, some ZIP code areas are separated by natural barriers (e.g., rivers, lakes) but bridged by some man-made infrastructures (e.g., roads, ferries). We select "Yes" for Adjusted so that these additional links identified by an Input Adjusted Adjacency File will be appended to those spatial adjacent ones. Otherwise, leave the default setting "No."

5 The ".npz" file format can improve computational efficiency when using a large spatial adjacency matrix.

6 Users can join the geographic coordinates of origin ID field and destination ID field and then use the "XY to Line" tool in ArcGIS Pro to generate a flow map and validate whether the adjacent polygons are linked by lines or not.

7 If the `.npz` file cannot be loaded, users need to copy the full path of the spatial adjacency matrix (e.g., `E:\CMS_GISV3\Case04Au\FLplgnAdjUp.npz`).

8 *Localization index* (`LI`) refers to the ratio of service flows within an HSA divided by the total service flows originated from the same HSA. `POPU` is the total population in a region. `Compactness` is the corrected perimeter-to-area ratio, or *PAC* $= P / (3.54 * A^{0.5})$, where P denotes perimeter and A denotes area. `EstTime` is the weighted average travel time for patients originated from the HSA.

9 The default values for the parameter of power, exponential, square-root exponential, and log-normal functions are set at 3, 0.03, 0.3, and 0.3, respectively. For the Gaussian function, the coefficient is the value for σ (with a default value = 80), since $\mu = 0$ for a monotonic distance decay pattern when assuming no crater around $x = 0$. The distance decay coefficient alpha appears only when a two-parameter function is chosen. The default values for both parameters (alpha and beta) in the log-logistic and compound power-exponential functions are set at 1.

10 In our case, we typed: `cd   C:\Users\WH\AppData\Local\ESRI\conda\envs\arcgispropy3-clone`.

5 GIS-Based Measures of Spatial Accessibility and Application in Examining Healthcare Disparity

Accessibility refers to the relative ease by which the locations of activities, such as work, school, shopping, recreation, and healthcare, can be reached from a given location. Accessibility is an important issue for several reasons. Resources or services are scarce, and their efficient delivery requires adequate access by people. The spatial distribution of resources or services is not uniform and needs effective planning and allocation to match demands. Disadvantaged population groups (e.g., low-income and minority residents) often suffer from poor access to certain activities or opportunities because of their lack of economic or transportation means. Access can thus become a social justice issue, which calls for careful planning and public policies by government agencies, as discussed in Chapter 11.

Accessibility is determined by the distributions of supply and demand and how they are connected in space, and thus a classic issue for location analysis well suited for GIS to address. This chapter focuses on how spatial accessibility is measured by GIS-based methods and uses the application in assessing healthcare disparity to illustrate the value of those methods. The COVID-19 (coronavirus disease 2019) pandemic has expanded telehealth utilization in unprecedented ways and raised the urgency of refining existing methods for measuring accessibility from physical visits to virtual care delivered online.

The chapter is structured as follows. Section 5.1 overviews the issues on accessibility, followed by two GIS-based methods for defining spatial accessibility: the floating catchment area methods, including the popular two-step floating catchment area (2SFCA) method in Section 5.2 and the generalized 2SFCA (G2SFCA) method in Section 5.3. Sections 5.4 and 5.5 introduce the newly developed "inverted 2SFCA (i2SFCA)" and "two-step virtual catchment area (2SVCA)" methods, respectively. The i2SFCA method estimates potential crowdedness in facilities, and the 2SVCA method measures accessibility via internet or virtual accessibility, for example, telehealth. Section 5.6 illustrates how the methods are implemented in a case study of various accessibility measures for primary care physicians in Baton Rouge metropolitan region. Section 5.7 concludes the chapter with a brief summary.

5.1 ISSUES ON ACCESSIBILITY

Access may be classified according to two dichotomous dimensions (potential vs. revealed, and spatial vs. aspatial) into four categories, such as potential spatial

DOI: 10.1201/9781003292302-7

access, potential aspatial access, revealed spatial access, and revealed aspatial access (Khan, 1992). *Revealed accessibility* focuses on actual use of a service, whereas *potential accessibility* signifies the probable utilization of a service. The revealed accessibility may be reflected by frequency or satisfaction level of using a service and thus obtained in a survey or an electronic health record (EHR) system. Most studies examine potential accessibility, based on which planners and policy analysts evaluate the existing system of service delivery and identify strategies for improvement. *Spatial access* emphasizes the importance of spatial separation between supply and demand as a barrier or a facilitator, whereas *aspatial access* stresses non-geographic barriers or facilitators (Joseph and Phillips, 1984). Aspatial access is related to many demographic and socioeconomic variables. In the study on job access, Wang (2001b) examined how workers' characteristics such as race, sex, wage, family structure, educational attainment, and home ownership status affect commuting time and thus job access. In the study on healthcare access, Wang and Luo (2005) included several categories of aspatial variables: demographics, such as age, sex, and ethnicity; socioeconomic status, such as population under the poverty line, female-headed households, home ownership, and income; environment, such as residential crowdedness and housing units' lack of basic amenities; linguistic barrier and service awareness, such as population without a high school diploma and households linguistically isolated; and transportation mobility, such as households without vehicles. Since these variables are often correlated to each other, they may be consolidated into a few independent factors by using the principal components and factor analysis techniques (see Chapter 7).

This chapter focuses on measuring *potential spatial accessibility*, an issue particularly interesting to geographers and location analysts.

If the capacity of supply is less a concern, one can use simple *supply-oriented accessibility measures* that emphasize the proximity to supply locations. For instance, Onega et al. (2008) used *minimum travel time* to the closest cancer care facility to measure accessibility to cancer care service. Distance or time from the nearest provider can be obtained using the techniques illustrated in Chapter 2. Scott and Horner (2008) used *cumulative opportunities* within a distance (travel time) range to measure job accessibility. Hansen (1959) used a simple gravity-based *potential model* to measure accessibility to jobs. The model is written as:

$$A_i = \sum_{j=1}^{n} S_j d_{ij}^{-\beta}, \tag{5.1}$$

where A_i is the accessibility at location i, S_j is the supply capacity at location j, d_{ij} is the distance or travel time between the demand (at location i) and a supply location j, β is the travel friction coefficient, and n is the total number of supply locations. The potential model values supplies at all locations, each of which is discounted by a distance term. The model does not account for the demand side. That is to say, the amount of population competing for the supplies is not considered to affect accessibility. The model is the foundation for a more advanced method that accounts for the distance decay effect in Section 5.3.

In most cases, accessibility measures need to account for both supply and demand because of scarcity of supply. Prior to the widespread use of GIS, the *simple supply–demand ratio method* computes the ratio of supply vs. demand in an area (usually an administrative unit, such as township or county) to measure accessibility. For example, Cervero (1989) and Giuliano and Small (1993) measured job accessibility by the ratio of jobs vs. resident workers across subareas (central city and combined suburban townships) and used the ratio to explain intraurban variations of commuting time. In the literature on job access and commuting, the method is commonly referred to as the *jobs–housing balance approach*. The US Department of Health and Human Services (DHHS) uses the population-to-physician ratio within a "rational service area" (most as large as a whole county or a portion of a county or established neighborhoods and communities) as a basic indicator for defining physician shortage areas (GAO, 1995; Lee, 1991). In the literature on healthcare access and physician shortage area designation, the method is referred to as the *regional availability measure* (vs. the *regional accessibility measure* based on a gravity model) (Joseph and Phillips, 1984).

The simple supply–demand ratio method has at least two shortcomings. First, it cannot reveal the detailed spatial variations within an area unit (usually large). For example, the job–housing balance approach computes the jobs–resident workers ratio in each city and uses it to explain commuting across cities but cannot explain the variation within a city. Secondly, it assumes that the boundaries are impermeable, that is, demand is met by supply only within the areas. For instance, in physician shortage area designation by the DHHS, the population-to-physician ratio is often calculated at the county level, implying that residents do not visit physicians beyond county borders.

The next two sections discuss the floating catchment area (FCA) methods and some more advanced models. All consider both supply and demand and overcome the shortcomings mentioned earlier.

5.2 FLOATING CATCHMENT AREA METHODS

5.2.1 Earlier Versions of Floating Catchment Area (FCA) Method

Earlier versions of the *floating catchment area (FCA) method* are very much like the one discussed in Section 3.1 on spatial smoothing. For example, in Peng (1997), a catchment area is defined as a square around each location of residents, and the jobs–residents ratio within the square measures the job accessibility for that location. The catchment area "floats" from one residential location to another across the study area and defines the accessibility for all locations. The catchment area may also be defined as a circle (Immergluck, 1998; Wang, 2000) or a fixed travel time range (Wang and Minor, 2002), and the concept remains the same.

Figure 5.1 uses an example to illustrate the method. For simplicity, assume that each demand location (e.g., tract) has only one resident at its centroid, and the capacity of each supply location is also one. Assume that a circle around the centroid of a residential location defines its catchment area. *Accessibility* in a tract is defined as the supply-to-demand ratio within its catchment area. For instance, within the catchment area of tract 2, total supply is 1 (i.e., only at site *a*), and total demand is 7. Therefore, accessibility at tract 2 is the supply–demand ratio, that is, 1/7. The circle floats from

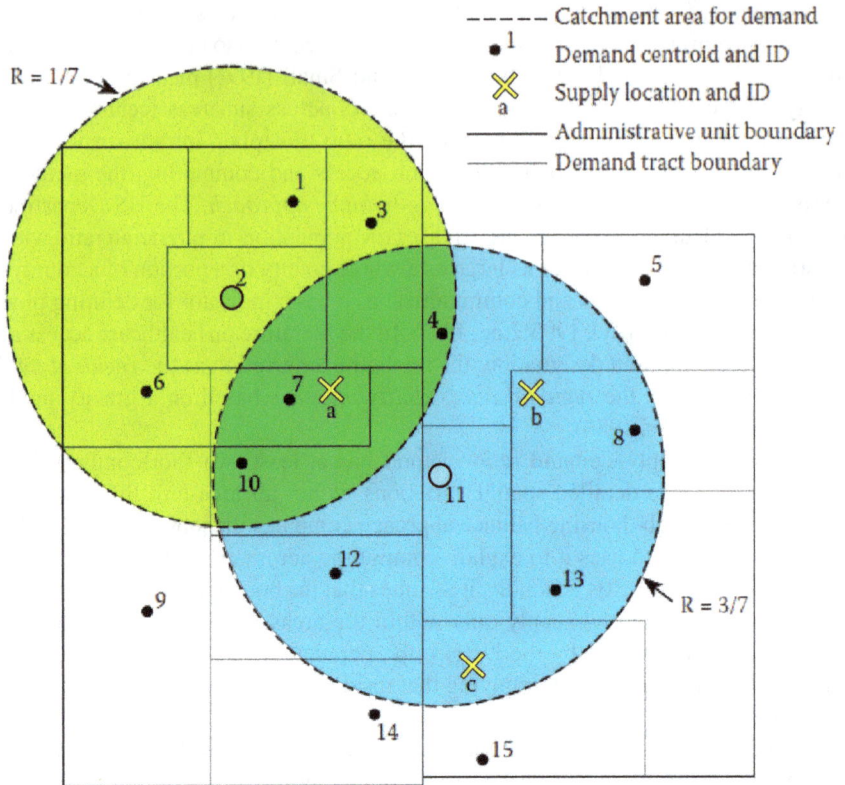

FIGURE 5.1 The floating catchment area (FCA) method in Euclidean distance.

one centroid to another, while its radius remains the same. The catchment area of tract 11 contains a total supply of 3 (i.e., at sites *a, b,* and *c*) and a total demand of 7, and thus the accessibility at tract 11 is 3/7. Here the ratio is based on the floating catchment area and not confined by the boundary of an administrative unit.

The previous example can also be used to explain the fallacies of this simple FCA method. It assumes that services within a catchment area are fully available to residents within that catchment area, and residents only use those services. However, the distance between a supply and a demand within the catchment area may exceed the threshold distance (e.g., in Figure 5.1, the distance between 13 and *a* is greater than the radius of the catchment area of tract 11). Furthermore, the supply at *a* is within the catchment of tract 2 but may not be fully available to serve demands within that catchment, as it is also reachable by tract 11. This points out the need to discount the availability of a supplier by the intensity of competition for its service of surrounding demand.

5.2.2 Two-Step Floating Catchment Area (2SFCA) Method

A method developed by Luo and Wang (2003) overcomes the previous fallacies. It repeats the process of "floating catchment" twice (once on supply locations and another on demand locations) and is therefore referred to as the *two-step floating catchment area (2SFCA) method.*

First, for each supply location j, search all demand locations (k) that are within a threshold travel distance (d_0) from location j (i.e., catchment area j), and compute the supply-to-demand ratio R_j within the catchment area:

$$R_j = \frac{S_j}{\sum_{k \in \{d_{kj} \leq d_0\}} D_k},$$

where d_{kj} is the distance between k and j, D_k is the demand at location k that falls within the catchment (i.e., $d_{kj} \leq d_0$), and S_j is the capacity of supply at location j.

Next, for each demand location i, search all supply locations (j) that are within the threshold distance (d_0) from location i (i.e., catchment area i), and sum up the supply-to-demand ratios R_j at those locations to obtain the accessibility A_i at demand location i:

$$A_i = \sum_{j \in \{d_{ij} \leq d_0\}} R_j = \sum_{j \in \{d_{ij} \leq d_0\}} \left(\frac{S_j}{\sum_{k \in \{d_{kj} \leq d_0\}} D_k} \right), \qquad (5.2)$$

where d_{ij} is the distance between i and j, and R_j is the supply-to-demand ratio at supply location j that falls within the catchment centered at i (i.e., $d_{ij} \leq d_0$). A larger value of A_i indicates a better accessibility at a location.

The first step earlier assigns an initial ratio to each service area centered at a supply location as a measure of supply availability (or crowdedness). The second step sums up the initial ratios in the overlapped service areas to measure accessibility for a demand location, where residents have access to multiple supply locations. The method considers interaction between demands and supplies across areal unit borders and computes an accessibility measure that varies from one location to another. Equation 5.2 is basically the ratio of supply to demand (filtered by a threshold distance).

Figure 5.2 uses the same example to illustrate the 2SFCA method. Here we use travel time instead of Euclidean distance to define catchment area. The catchment area for supply a has one supply and eight residents and thus carries a supply-to-demand ratio of 1/8. Similarly, the ratio for catchment b is 1/4, and for catchment c is 1/5. The resident at tract 3 has access to a only, and the accessibility at tract 3 is equal to the supply-to-demand ratio at a (the only supply location), that is, $R_a = 0.125$. Similarly, the resident at tract 5 has access to b only, and thus its accessibility is $R_b = 0.25$. However, the resident at 4 can reach both supplies a and b (shown in an area overlapped by catchment areas a and b) and therefore enjoys better accessibility (i.e., $R_a + R_b = 0.375$).

The catchment drawn in the first step is centered at a supply location, and thus the travel time between the supply and any demand within the catchment does not exceed the threshold. The catchment drawn in the second step is centered at a demand location, and all supplies within the catchment contribute to the supply–demand ratios at that demand location. The method overcomes the fallacies in the earlier FCA

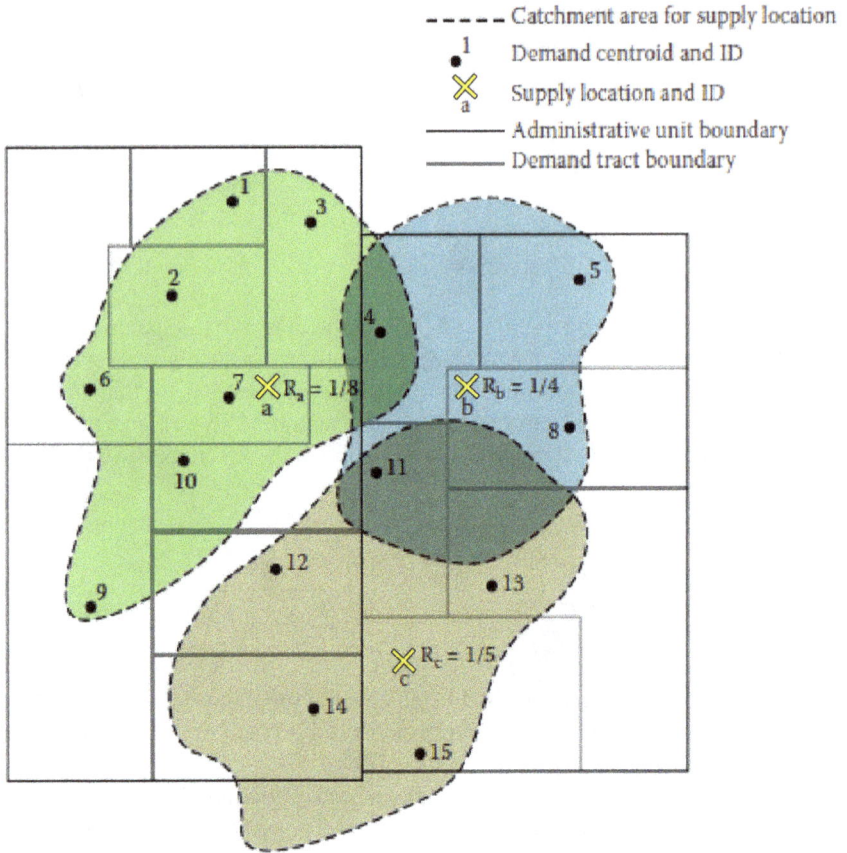

FIGURE 5.2 The two-step floating catchment area (2SFCA) method in travel time.

methods. Equation 5.2 is basically a ratio of supply to demand with only selected supplies and demands entering the numerator and the denominator, and the selections are based on a threshold distance or time within which supplies and demands interact. If a supply location falls within the catchment area from a demand site, conversely the demand site is also within the catchment area from that supply facility. The distance or time matrix is calculated once but used twice in the search in both steps. Travel time should be used if distance is a poor measure of travel impedance.

The method can be implemented in ArcGIS. The detailed procedures are explained in Section 5.6.

5.3 FROM GRAVITY-BASED TO GENERALIZED 2SFCA METHOD

5.3.1 GRAVITY-BASED 2SFCA METHOD

The 2SFCA method draws an artificial line (say, 15 miles or 30 minutes) between an accessible and inaccessible location. Supplies within that range are counted equally regardless of the actual travel distance or time (e.g., 2 miles vs. 12 miles).

Similarly, all supplies beyond that threshold are considered inaccessible, regardless of any difference in travel distance or time. The gravity model rates a nearby supply more accessible than a remote one and thus reflects a continuous decay of access in distance.

The potential model in equation 5.1 considers only the supply, not the demand side (i.e., competition for available supplies among demands). Weibull (1976) improved the measurement by accounting for competition for services among residents (demands). Joseph and Bantock (1982) applied the method to assess healthcare accessibility. Shen (1998) and Wang (2001b) used the method for evaluating job accessibility. The gravity-based accessibility measure at location i can be written as:

$$A_i = \sum_{j=1}^{n} \frac{S_j d_{ij}^{-\beta}}{V_j}, \text{where } V_j = \sum_{k=1}^{m} D_k d_{kj}^{-\beta}, \tag{5.3}$$

where n and m are the total numbers of supply and demand locations, respectively, and the other variables are the same as in equations 5.1 and 5.2. Compared to the primitive accessibility measure based on the Hansen model, equation 5.3 discounts the availability of a physician by the service competition intensity at that location, V_j, measured by its population potential. A larger A_i implies better accessibility.

This accessibility index can be interpreted like the one defined by the 2SFCA method. It is essentially the ratio of supply S to demand D, each of which is weighted by travel distance or time to a negative power. The total accessibility scores (i.e., sum of "individual accessibility indexes multiplied by corresponding demand amounts"), either by the 2SFCA or the gravity-based method, are equal to the total supply. Alternatively, the weighted average of accessibility in all demand locations is equal to the supply-to-demand ratio in the whole study area (see Appendix 5A for proof).

A careful examination of the two methods further reveals that the 2SFCA method is merely a special case of the gravity-based method.

The 2SFCA method treats distance (time) impedance as a dichotomous measure, that is, any distance (time) within a threshold is equally accessible, and any distance (time) beyond the threshold is equally inaccessible. Using d_0 as the threshold travel distance (time), distance or time can be recoded as:

1. d_{ij} (or d_{kj}) = ∞ if d_{ij} (or d_{kj}) > d_0
2. d_{ij} (or d_{kj}) = 1 if d_{ij} (or d_{kj}) ≤ d_0

For any $\beta > 0$ in equation 5.3, we have:

1. $d_{ij}^{-\beta}$ (or $d_{kj}^{-\beta}$) = 0 when d_{ij} (or d_{kj}) = ∞
2. $d_{ij}^{-\beta}$ (or $d_{kj}^{-\beta}$) = 1 when d_{ij} (or d_{kj}) = 1

In case 1, S_j or P_k will be excluded by being multiplied by 0, and in case 2, S_j or P_k will be included by being multiplied by 1. Therefore, equation 5.3 is regressed to equation 5.2, and thus the 2SFCA measure is just a special case of the gravity-based measure. Therefore, *accessibility* defined in equation 5.3 is termed *gravity-based 2SFCA method*. Considering that the two methods have been developed in different fields for a variety of applications, this proof validates their rationale for capturing the essence of accessibility measures.

In the 2SFCA method, a larger threshold distance or time reduces variability of accessibility across space and thus leads to stronger spatial smoothing (Fotheringham et al., 2000, p. 46; also see Section 3.1). In the gravity-based method, a lower value of travel friction coefficient β leads to a lower variance of accessibility scores, and thus stronger spatial smoothing. The effect of a larger threshold travel time in the 2SFCA method is equivalent to that of a smaller travel friction coefficient in the gravity-based method. Indeed, a lower β value implies that travel distance or time matters less, and people are willing to travel farther to seek a service.

The gravity-based method seems to be more conceptually sound than the 2SFCA method. However, the 2SFCA method may be a better choice in some cases for two reasons. First, the gravity-based method tends to inflate accessibility scores in poor-access areas than the 2SFCA method (Luo and Wang, 2003), but the poor-access areas are usually the areas of most interest to many public policymakers. In addition, the gravity-based method also involves more computation and is less intuitive. In particular, finding the value of the distance friction coefficient β requires additional data and work to define and may be region-specific (Huff, 2000).

5.3.2 GENERALIZED 2SFCA METHOD

Despite the popularity of 2SFCA, the method's major limitation remains its dichotomous approach that defines a doctor being accessible or inaccessible by a cut-off distance (time). Many studies have attempted to improve it, and most are from healthcare-related applications. A kernel density function (Guagliardo, 2004) or a Gaussian function (Dai, 2010; Shi et al., 2012) is proposed to model the distance decay effect (i.e., a continuously gradual decay within a threshold distance and no effect beyond). The catchment radius may also vary by provider types or neighborhood types (Yang et al., 2006) and by geographic settings (e.g., shorter in urban and longer in rural areas) (McGrail and Humphreys, 2009a). Weights can be assigned to different travel time zones so that the supply–demand interaction drops with travel time by multiple steps, referred to as *"expanded 2SFCA (E2SFCA)"* (Luo and Qi, 2009). McGrail and Humphreys (2009b) proposed a constant weight within 10 minutes, a zero weight beyond 60 minutes, and a weight of gradual decay between, and their method can be termed "three-zone hybrid approach." Most recently, in measuring healthcare accessibility, Shao and Luo (2022) proposed the "supply–demand adjusted two-steps floating catchment area (SDA-2SFCA)" that further adjusted the supply side by the insurance plans accepted by doctors and refined the demand side by demography.

These methods can be synthesized together in one framework and vary in their conceptualization of distance decay in patient–physician interactions (Wang, 2012). Representing the distance decay effect in a function *f(d)*, we can write the *generalized 2SFCA method*, such as:

$$A_i = \sum_{j=1}^{n} \left[S_j f\left(d_{ij}\right) / \left(\sum_{k=1}^{m} D_k f\left(d_{kj}\right)\right) \right]. \tag{5.4}$$

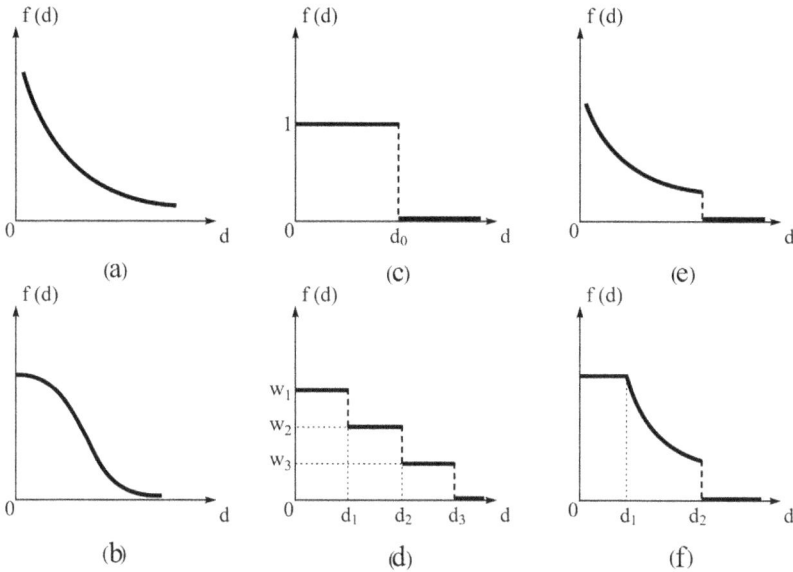

FIGURE 5.3 Conceptualizing distance decay in G2SFCA: (a) gravity function, (b) Gaussian function, (c) 2SFCA, (d) E2SFCA, (e) kernel, and (f) three-zone hybrid approach.

In equation 5.4, a general distance decay function $f(d_{ij})$ or $f(d_{kj})$, replacing the power function $d_{ij}^{-\beta}$ or $d_{kj}^{-\beta}$ in equation 5.3, can take various forms. It is simply referred to as *G2SFCA*.

Figure 5.3a–f summarizes various ways of conceptualizing the distance decay function $f(d)$: a continuous function, such as a power function (Figure 5.3a) or a Gaussian function (Figure 5.3b); a discrete variable, such as binary in 2SFCA (Figure 5.3c) or multiple weights in E2SFCA (Figure 5.3d); or a hybrid of the two, such as a kernel function (Figure 5.3e) or a three-zone hybrid approach (Figure 5.3f).

In Case Study 5, Subsection 5.6.1 implements the conventional 2SFCA method step-by-step to help readers understand the calibration process and model construction, and Subsection 5.6.2 implements the gravity-based 2SFCA method step-by-step to highlight its difference from the conventional 2SFCA. One may also use the Generalized 2SFCA/I2SFCA Pro tool to automate the process with options to define various distance decay behaviors, as discussed in more detail in Appendix 5B.

What is the appropriate size for the catchment area(s)? Which is the best function to capture the distance decay effect? Choosing which function and using what parameter values should be based on analyzing the actual travel behavior in a study area. See Section 2.3 of Chapter 2 for various methods for deriving the best-fitting continuous function and related parameter(s). However, many researchers do not have access to such a dataset and often conduct sensitivity analysis by experimenting with different functions and employing a set of parameter values (e.g., Bauer and Groneberg, 2016).

5.4   INVERTED 2SFCA METHOD

The 2SFCA (or generalized 2SFCA) is residents- or demand-oriented and assesses their relative ease of getting a service from a residential location. By turning the regular 2SFCA upside down, this section introduces a method for examining an equally important property on the supply side. In other words, by switching the supply and demand terms in the formulation of 2SFCA, we have an *inverted 2SFCA* (abbreviated as "*i2SFCA*") to measure the "potential crowdedness" (or busyness, saturation, or stress level) for a service provided at a facility.

In implementation, the *inverted 2SFCA* reverses the order of two steps (i.e., switches the variables for supply and demand) in the traditional 2SFCA. Here we use the generalized distance decay function as formulated in Subsection 5.3.2 to illustrate it. Specifically, in step 1, for each demand location i, sum up all surrounding supplies (S_l) discounted by their distance decay effects $(f(d_{il}))$, and compute the demand to total adjusted supplies ratio r_j:

$$r_j = D_i / [\sum_{l=1}^{n} S_l f(d_{il})]$$

In step 2, for each supply location j, sum up ratios r_j discounted by their distance decay effects $(f(d_{ij}))$ to yield the crowdedness C_j at supply location j:

$$C_j = \sum_{i=1}^{m} \left[r_j f(d_{ij}) \right] = \sum_{i=1}^{m} \left[D_i f(d_{ij}) / \sum_{l=1}^{n} (S_l f(d_{il})) \right] \qquad (5.5)$$

As summarized in Table 5.1, symmetry is evident when implementing equation 5.4 for the (generalized) 2SFCA and equation 5.5 for the i2SFCA by switching supply S and demand D in the formulations. Similar to the property of 2SFCA that the weighted mean of accessibility equals the ratio of total supply (S) to total demand (D) in a study area (as proved in Appendix 5A), the weighted mean of i2SFCA-derived crowdedness equals the ratio of total demand to total supply in the study area. A_i at demand location i or C_j at supply location j focuses on an element of "estimated," "projected," or "predicted" size of demand or supply, respectively. In 2SFCA, it is estimated supply per demand volume, whereas in i2SFCA, it is estimated demand per supply capacity.

The Generalized 2SFCA/I2SFCA Pro tool also has the capability of implementing the i2SFCA method with various distance decay formulation options. Its implementation is straightforward, by choosing "I2SFCA Method" under General Model, and thus not illustrated in Case Study 5.

Appendix 5C demonstrates the derivations of both methods and therefore strengthens the theoretical foundation for the methods. How good are the accessibility and crowdedness measures projected by the 2SFCA and i2SFCA methods, respectively? Given population size or demand D_i and its derived accessibility A_i, $W_i = A_i D_i$ is total accessibility for population at i (e.g., in a ZIP code area). Similarly, for facility capacity S_j with estimated crowded C_j, its total crowdedness is projected to be $V_j = C_j S_j$ (e.g., in a hospital). A study on hospitalization patterns in Florida (Wang, 2021) found that (1) total accessibility W_i was a good predictor of actual volume of

TABLE 5.1
Implementing 2SFCA versus i2SFCA

	2SFCA	i2SFCA
Objective	Measuring spatial accessibility of service by residents at location (i).	Measuring potential crowdedness at a facility location (j).
Step 1	For each supply location j, sum up surrounding demands D_k, discounted by distance decay function $f(d_{kj})$, across all demand locations k (= 1, 2, . . ., m), and compute the supply-to-demand ratio R_j: $R_j = S_j / \sum_{k=1}^{m} \left(D_k f\left(d_{kj}\right) \right)$.	For each demand location i, sum up supplies S_l, discounted by distance decay function $f(d_{il})$, across all supply locations l (= 1, 2, . . ., n), and compute the demand-to-supply ratio r_i: $r_i = D_i / \sum_{l=1}^{n} \left(S_l f\left(d_{il}\right) \right)$.
Step 2	For each demand location i, sum up ratios R_j, discounted by distance decay function $f(d_{ij})$, across all supply locations j (= 1, 2, . . ., n), to obtain the accessibility A_i at demand location i: $A_i = \sum_{j=1}^{n} \left[R_j f\left(d_{ij}\right) \right]$ $= \sum_{j=1}^{n} \left[S_j f\left(d_{ij}\right) / \sum_{k=1}^{m} \left(D_k f\left(d_{kj}\right) \right) \right]$.	For each supply location j, sum up ratios r_i, discounted by distance decay function $f(d_{ij})$, across all demand locations i (= 1, 2, . . ., m), to obtain the crowdedness C_j at supply location j: $C_j = \sum_{i=1}^{m} \left[r_i f\left(d_{ij}\right) \right] = \sum_{i=1}^{m} \left[D_i f\left(d_{ij}\right) / \sum_{l=1}^{n} \left(S_l f\left(d_{il}\right) \right) \right]$.
Property	Weighted mean of accessibility (using the demand amount as weight) is equal to the ratio of total supply to total demand in the study area.	Weighted mean of crowdedness (using the supply capacity as weight) is equal to the ratio of total demand to total supply in the study area.
	Weighted means of accessibility and crowdedness are reciprocal of each other.	

patients generated by ZIP code area i (with $R^2 = 0.68$ or a correlation coefficient of 0.82), and (2) total crowdedness V_j is a good estimator of actual number of patients discharged by hospitals (with $R^2 = 0.62$ or a correlation coefficient of 0.78).

5.5   TWO-STEP VIRTUAL CATCHMENT AREA (2SVCA) METHOD

Most recently, the COVID-19 pandemic has brought renewed awareness of new healthcare delivery patterns, such as telehealth. A new method is proposed to measure virtual accessibility of telehealth to account for the availability and quality of under-lying broadband network that supports telemedicine and e-health (Alford-Teaster et al., 2021). Its formulation is modeled on the basis of 2SFCA and is also composed of two steps, each of which is confined to a virtual catchment area. Therefore, it is termed "*two-step virtual catchment area (2SVCA) method*." While the method is conceptualized with telehealth access as a primary application field, it can be used for measuring accessibility of any resource or service delivered via internet.

Here, the formulation of 2SVCA method is based on Wang et al. (2023) and Liu et al. (2023), an improved version over the earlier method proposed in Alford-Teaster et al. (2021). Take telehealth as an example in measuring virtual accessibility. It rarely works completely independently of physical visits to healthcare providers, and rather as supplementary consultation to reduce travel burdens for patients (Sorensen et al., 2020), and thus takes effect within the provider's physical catchment area.

The 2SVCA for virtual accessibility is formulated as:

$$VA_i = a_i \sum_{j \in (d_{ij} \leq d_0)}^{n} \left[a_j S_j f\left(b_j\right) / \sum_{k \in (d_{kj} \leq d_0)}^{m} \left(a_k D_k f\left(b_k\right)\right) \right], \quad (5.6)$$

where the degree of a facility S_j and a demand location D_k participating in telehealth services between them is a function of their broadband strengths $f(b_j)$ and $f(b_k)$, in place of the distance decay function $f(d_{kj})$ or $f(d_{ij})$ in equation 5.4 for the G2SFCA. There are three additional parameters associated with broadband subscription rates: (1) a_j is applied on facility S_j so that only a portion of the facility with commercial broadband subscription provides telehealth service, (2) a_k is applied on demand D_k to capture that only those residents with consumer broadband subscription would contribute to telehealth services offered by S_j, and (3) a_i is applied to discount the initial virtual accessibility score assigned to demand location i because only this portion of residents has the consumer broadband subscription. Case Study 5 in Section 5.6 assumes that broadband is affordable for any primary care physician (PCP) facilities and thus a uniform $a_j = 1$. All other notations remain the same as in equation 5.4.

The model captures two issues associated with virtual accessibility. The broadband quality variable b is associated with locations and reflects where it is available, termed *availability factor*, and its subscription rate a is associated with population or facility and captures to whom it is affordable and termed *affordability factor*. Similar to the notion of assigning a binary 0–1 value to the distance decay function $f(d_{kj})$ or $f(d_{ij})$ in equation 5.4, a baseline 2SVCA method (as adopted in the Case Study 5) sets $f(b_j) = 1$ or $f(b_k) = 1$ in equation 5.6 for simplicity when broadband is available at supply location j or demand location k, respectively, and set their value as 0 otherwise. Similar to the E2SFCA method for refining 2SFCA by Luo and Qi (2009), a recent paper by Shao and Luo (2023) proposes the E2SVCA method by using a stepwise function in place of the binary broadband availability factor $f(b_j)$ or $f(b_k)$ in equation 5.6.

Subsection 5.6.3 in the case study illustrates how to implement the calibration of 2SVCA method.

In equation 5.6, the formulation of 2SVCA method focuses on the notion that the virtual interaction between supply and demand takes place within a fixed physical catchment area d_0. It is possible that the physical and virtual interactions are independent of each other, or they intersect with each other with more complexity. Modeling virtual accessibility is an emerging theme, and our effort only represents a first attempt at this increasingly important and complex issue.

5.6 CASE STUDY 5: MEASURING ACCESSIBILITY OF PRIMARY CARE PHYSICIANS IN BATON ROUGE

This case study extends the work reported in Wang et al. (2020) and uses the same study area: nine parishes in the Baton Rouge Metropolitan Statistical Area

(MSA). Updates include using more recent population data and adding a virtual accessibility measure. The case study uses Euclidean distances to measure travel impedance d_{ij} for simplicity so that we can focus on implementing the accessibility measures in GIS.

The population data is extracted from the 2020 census as illustrated in Case Study 1 in Section 1.3 of Chapter 1. Data of physicians (including specialties and geographic locations) in Louisiana in 2022 are extracted from the Doctors and Clinicians National Downloadable File released by the Centers for Medicare and Medicaid Services (CMS). There are 594 primary care physicians (PCP) aggregated to 172 locations. The internet infrastructure data is based on the Fixed Broadband Deployment Block Data (dated December 31, 2020) via the FCC website (www.fcc. gov/general/broadband-deployment-data-fcc-form-477). The FCC broadband data contains maximum upload and download speed records by different service providers (consumer vs. business).

The following features in the geodatabase BRMSA.gdb under the data folder BRMSA are used for this project:

1. Polygon features MSAbkg (n = 574) and MSAcounty for 2020 (n = 9) census block groups and counties, respectively, for reference and mapping.
2. Point feature MSAbkgPt (n = 570)[1] for 2020 census block group centroids with the field GEOID for unique geography identifier ID, POP100 for population, and MaxAdDown, MaxAdUp, and bandR for mean download speed, mean upload speed, and subscription rate of broadband of consumer service providers, respectively.
3. Point feature PCPmsa with the field PCP for number of FTE (full-time equivalent) primary care physicians (PCP), and MaxAdDown and MaxAdUp for mean fixed broadband download and upload speeds by business service providers, in 2020 census block centroids.
4. OD Cost Matrix table MSAtime for travel time between MSAbkgPt and PCPmsa based on the road network feature dataset MSAhigh, with fields OriginID and DestinationID representing population and physician location IDs, respectively, and Minutes representing the drive time between them in minutes.

5.6.1 Implementing the 2SFCA Method

Step 1. Computing Distances between Population and Physician Locations: In the Geoprocessing Toolboxes pane, use the Generate Near Table tool to compute the Euclidean distances between population locations (MSAbkgPt as Input Features) and physician locations (PCPmsa as Near Features) and save the Output Table as DistAll. Refer to step 3 in Section 2.2.1 if needed. The distance table has $570 \times 172 = 98,040$ records.

Step 2. Attaching Population and Physician Data to the Distance Table: Join the attribute table of population MSAblkpop (based on the field OBJECTID) to the distance table DistAll (based on the field IN_FID), and then join the attribute table of physicians PCPmsa (based on the field OBJECTID) to the distance table

DistAll (based on the field NEAR _ FID). Refer to step 3 in Subsection 3.3.1 of Chapter 3 for attribute join if needed.

Step 3. Extracting Distances within a Catchment Area: On the distance table DistAll, select the records with "NEAR _ DIST is less than or equal to 32180" meters (i.e., 20 miles), and export to a new table Dist20mi, which has 69,546 records. Refer to step 4 in Subsection 3.3.1 of Chapter 3 for attribute query if needed. The new distance table only includes those distances within 20 miles² and thus implements the selection conditions $i \in \{d_{ij} \leq d_0\}$ and $k \in \{d_{kj} \leq d_0\}$ in equation 5.2.

Step 4. Summing Up Population around Each Physician Location: On the table Dist20mi, use the Summarize tool to sum up population POP100 by physician locations NEAR _ FID, and save the result in a new table DocAvl20mi, which has 172 records (i.e., number of physician locations in blocks). Refer to step 5 in Section 3.3.1 of Chapter 3 for the Summarize tool if needed. The field Sum _ POP100 is the total population within the threshold distance from each physician location and therefore implements calibration of the term $\sum_{k \in \{d_{kj} \leq d_0\}} D_k$ in equation 5.2.

Step 5. Computing Initial Physician-to-Population Ratio at Each Physician Location: Join the updated table DocAvl20mi back to the distance table Dist20mi (based on their common field NEAR _ FID).

On the table Dist20mi, add a field docpopR (Data Type as Double), and compute it as docpopR = 1000*!Dist20mi.PCP!/!DocAvl20mi.SUM _ POP100! This assigns an initial physician-to-population ratio to each physician location. This step computes the term $\dfrac{S_j}{\sum_{k \in \{d_{kj} \leq d_0\}} D_k}$ in equation 5.2. The ratio is inflated 1,000 times to indicate the physician availability per 1,000 residents.

Step 6. Summing Up Physician-to-Population Ratios by Population Locations: On the updated table Dist20mi, remove the joins, and then use the Summarize tool again to sum up the initial physician-to-population ratios docpopR by population locations IN_FID, and save the result in a new table BkgAcc20mi, which has 570 records (i.e., number of population locations in census tracts). The field Sum_docpopR is the summed-up availability of physicians that are reachable from each residential location and thus returns the accessibility score A_i in equation 5.2.

Figure 5.4 illustrates the process of 2SFCA implementation in ArcGIS Pro from steps 1 to 6. The layer/table names with critical field names are in a box, and the actions (tools) are numbered according to the corresponding steps.

Step 7. Mapping Physical Accessibility: In the Geoprocessing Toolboxes pane, use the tool "Join Field" to implement a permanent attribute join by joining the table BkgAcc20mi with accessibility scores Sum _ docpopR (based on the field IN _ FID) to the attribute table of population layer MSAbkgPt (based on the field OBJECTID). This is to ensure that the accessibility scores are carried over in the next join (see Subsection 1.1.2). Join the updated attribute table of MSAbkgPt to the census block group polygon layer MSAbkgAll (based on their common fields GEOID) for mapping.

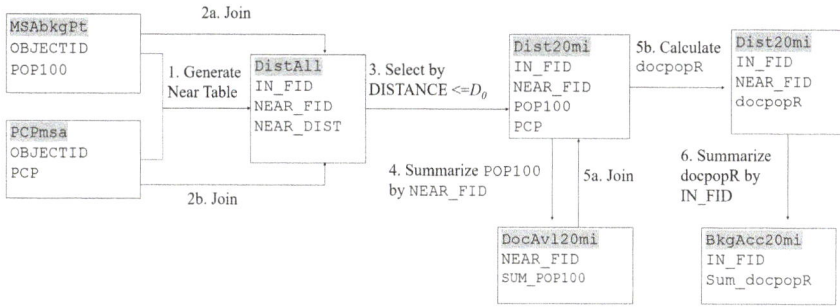

FIGURE 5.4 Flowchart for implementing the 2SFCA in ArcGIS.

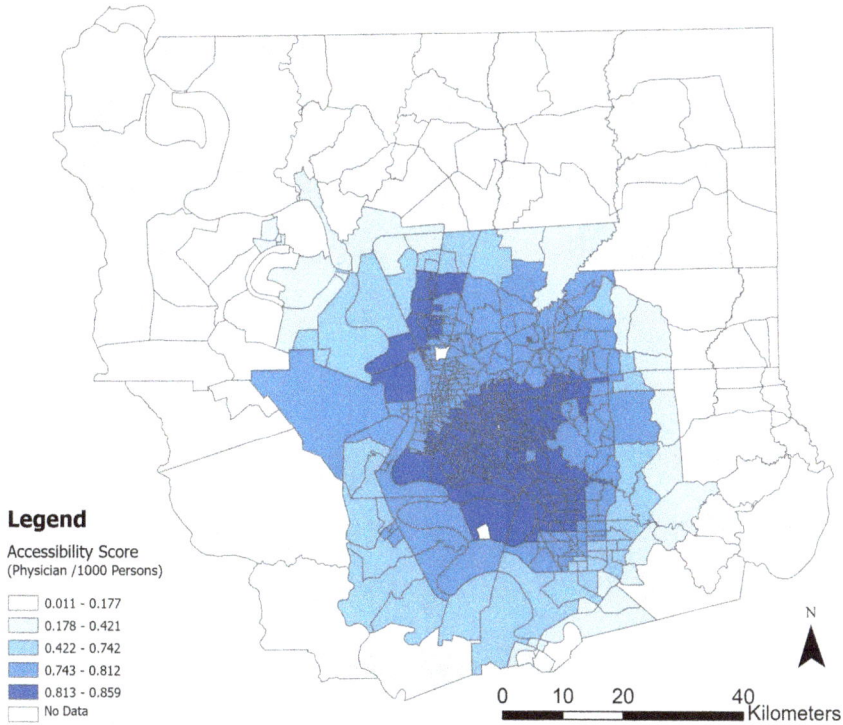

FIGURE 5.5 Spatial accessibility for primary care physicians in Baton Rouge MSA by 2SFCA (20-mile catchment).

As shown in Figure 5.5, the accessibility scores are the highest in two areas (one north of downtown and another southeast of downtown) and decline outward. Like any accessibility studies, results near the borders of the study area need to be interpreted with caution because of *edge effects* discussed in Section 3.1. In other words, physicians outside of the study area may contribute to accessibility of residents near the borders, and residents may also seek services beyond the MSA border. Neither is accounted for.

Step 8. Optional: Sensitivity Analysis Using Various Threshold Distances: A sensitivity analysis can be conducted to examine the impact of using different threshold distances. For instance, the study can be repeated through steps 3–7 by using threshold distances of 15, 10, and 5 miles, and results can be compared.

As shown Figure A5.1a, this process is automated in the tool "Generalized 2SFCA/I2SFCA Pro" in the toolkit `Accessibility Pro.tbx` (available under the folder `ArcPy-tools`). Choose Distance Type as `Euclidean` and Distance Decay Function Category `Discrete`. See Appendix 5B for its usage.

As shown Figure A5.1b for the same tool, one may instead choose Distance Type as `External Table` and use the predefined travel time matrix between census block group centroids and PCP points, that is, table `MSAtime` under the same `BRMSA.gdb`.

Figure 5.6 shows the result by the 2SFCA using a catchment area size of 30-minute travel time. While the general pattern is consistent with Figure 5.5 based on Euclidean distances, it shows the areas of high accessibility stretching along expressways. Different threshold travel times can also be tested for sensitivity analysis. Table 5.2 (top part) shows the results by 2SFCA using the threshold time ranging from 20 to 40 minutes. As the threshold time increases, a larger catchment smooths the variability of accessibility scores to a narrower range and a smaller standard deviation.

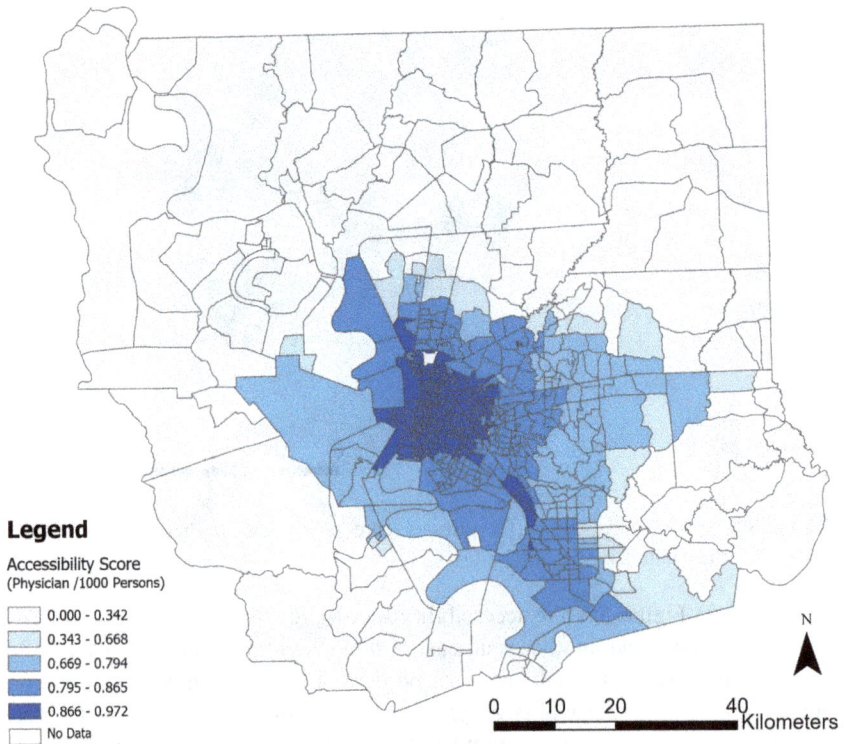

Legend

Accessibility Score
(Physician /1000 Persons)

- 0.000 - 0.342
- 0.343 - 0.668
- 0.669 - 0.794
- 0.795 - 0.865
- 0.866 - 0.972
- No Data

0   10   20   40 Kilometers

N

FIGURE 5.6 Spatial accessibility for primary care physicians in Baton Rouge MSA by 2SFCA (30-minute catchment).

TABLE 5.2

Comparison of 2SFCA Accessibility Measures

Method	Parameter	Min.	Max.	Std. Dev.	Mean*
Conventional	$d_0 = 10$ min	0.0	2.223	0.721	0.721
2SFCA	$d_0 = 15$ min	0.0	1.397	0.502	0.727
	$d_0 = 20$ min	0.0	1.258	0.409	0.716
	$d_0 = 25$ min	0.0	1.160	0.348	0.707
	$d_0 = 30$ min	0.0	0.972	0.289	0.700
	$d_0 = 35$ min	0.004	0.903	0.243	0.692
	$d_0 = 40$ min	0.020	0.846	0.215	0.691
Gravity-based	$\beta = 0.2$	0.506	1.004	0.087	0.702
2SFCA	$\beta = 0.4$	0.362	1.519	0.176	0.705
	$\beta = 0.6$	0.255	2.429	0.275	0.709
	$\beta = 0.8$	0.177	4.074	0.402	0.713
	$\beta = 1.0$	0.121	6.848	0.587	0.718
	$\beta = 1.2$	0.082	11.025	0.837	0.724
	$\beta = 1.4$	0.057	17.267	1.145	0.733

Note: All based on travel time as spatial impedance.

*While the mean value differs slightly, the weighted mean is a constant.

5.6.2 IMPLEMENTING THE GRAVITY-BASED 2SFCA METHOD

The process of implementing the gravity-based 2SFCA method is similar to that of the conventional 2SFCA method in the previous subsection. The differences are highlighted in the following text.

The gravity-based method may utilize all distances. Therefore, step 3 for extracting distances within a threshold is skipped, and subsequent computation are based on the original distance table DistAll.

Step 9. Computing Population Potential for Each Physician Location: Similar to step 4, make sure the fields POP100 and PCP are still attached to DistAll, add a field PPotent (Data Type as Double) to the table DistAll, and compute it as !MSAbkgPt.POP100!*!DistAll.NEAR _ DIST!**(-1) (assuming a distance friction coefficient $\beta = 1.0$). Based on the updated distance table, summarize the potential PPotent by physician locations NEAR _ FID, and save the result in a new table DocAvlg. The field SUM _ DistAll.PPotent calibrates the term $V_j = \sum_{k=1}^{m} D_k d_{kj}^{-\beta}$ in equation 5.3, which is the population potential for each physician location.

Step 10. Computing Accessibility Attributable to Individual Physician Locations: Similar to step 5, join the updated table DocAvlg to the distance table DistAll by physician locations (i.e., based on the common field NEAR _ FID). On the table DistAll, add a field R (Data Type as Double) and compute it as 1000*!PCPmsa. PCP! * !DistAll.NEAR _ DIST! ** (-1) / !DocAvlg. SUM _ Dist

All _ PPotent! This computes the term $\dfrac{S_j d_{ij}^{-\beta}}{V_j}$ in equation 5.3. Again, the multiplier

1,000 is applied in computing R to avoid small values.

Step 11. Summing Up Accessibility to All Physician Locations: Similar to step 6, on the updated DistAll, remove the joins, sum up R by population locations IN _ FID, and name the output table BkgAccg. The field Sum _ R is the gravity-based accessibility in equation 5.3.

Similarly, the sensitivity analysis for the gravity-based method can be conducted by changing the β value. The process can also be automated in the tool "Generalized 2SFCA/I2SFCA Pro" by setting the continuous distance decay function as the power function (Figure A5.1c). One may use the external travel time matrix MSAtime to measure travel impedance.

Table 5.2 (bottom part) shows the results with various β in the range 0.2–1.4 by the gravity-based method when travel time is used for measuring spatial impedance. As the β value increases, the distance friction effect is stronger, equivalent to a shorter catchment area size in the conventional 2SFCA. Therefore, the variation range for the accessibility scores expands and the standard deviation value increases. There are also some subtle differences in distribution patterns within each method. Figure 5.7a shows the distribution by the gravity-based method (a more evenly distributed bell shape). Figure 5.7b shows the distribution by the 2SFCA method (skewed toward high scores). Figure 5.7c plots them against each other in one graph, showing that the gravity-based method tends to inflate the scores in low-accessibility areas. Overall, the improvement of transportation (e.g., via road network improvement or a faster transportation mode) leads to declining β or large d_0 and thus helps close some accessibility gap.

5.6.3 IMPLEMENTING THE 2SVCA METHOD

The FCC fixed-broadband file contains multiple records (providers) in a single census block group. In this case study, if the mean download and download speeds in a block group are $\geq$ 25 and 3 mbps, respectively, it is considered with broadband access (i.e., b_i or $b_k = 1$), and otherwise, no access (i.e., b_i or $b_k = 0$).

Step 12. Defining Broadband Availability for Population and Physician Locations: Open the attribute table of MSAbkgPt, and add a new field MSAbb (Data Type as short) to indicate whether a block group has broadband. Select those block groups with MaxAdUp >= 3 and MaxAdDown >= 25, and use Calculate Field to assign the value 1 to the field MSAbb; otherwise, assign the value 0 to the field MSAbb.

Add a new field PCPbb to the attribute table PCPmsa, and follow the same rule to calculate its value as 1 or 0 for broadband availability based on the fields MaxAdUp and MaxAdDown.

Remember to clear the selections afterward.

Step 13. Attaching Broadband Availability Data to the OD Matrix: Similar to step 2, join the table MSAbkgPt (common field OBJECTID) to the OD table DistAll (common field IN _ FID), and join the table PCPmsa (common field OBJECTID) also to the OD table DistAll (common field Near _ FID).

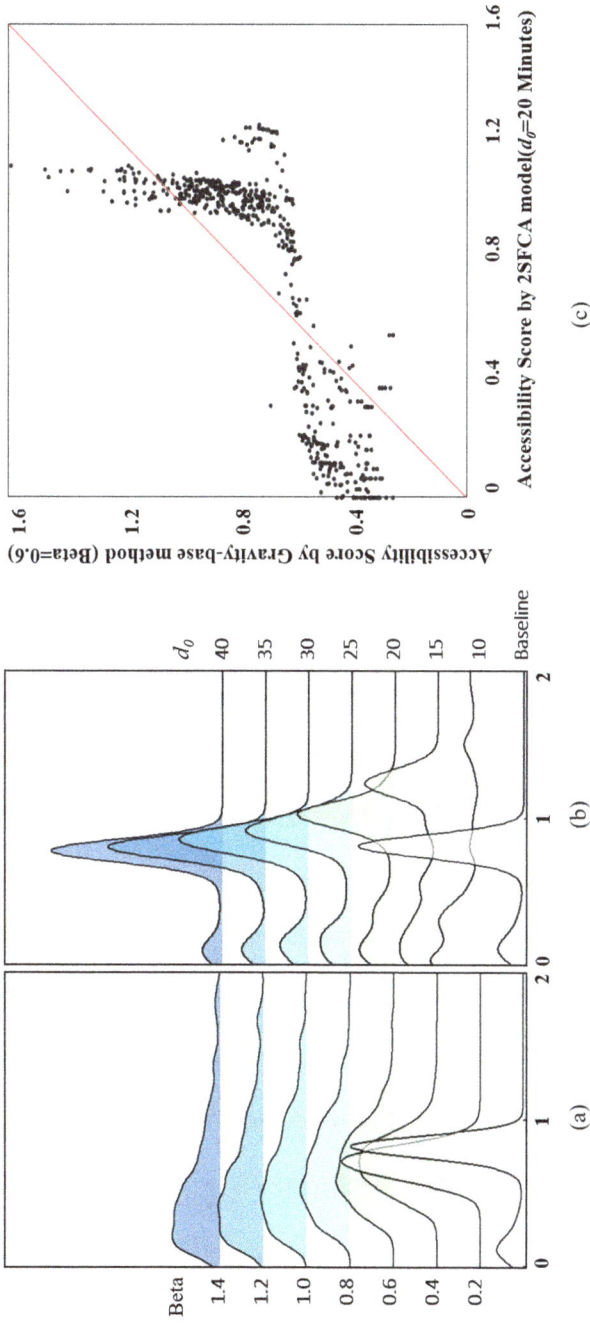

FIGURE 5.7 Probability density function curves of accessibility scores based on (a) gravity-based (β = 0.2–1.4) vs. (b) conventional 2SFCA (d0 = 10–40 minutes), and (c) accessibility scores of gravity-based (β = 0.6) vs. conventional 2SFCA (d0 = 20 minutes).

Step 14. Extracting Records with Threshold Distance: Similar to step 3, on the distance table `DistAll`, select the records with "NEAR _ DISTANCE is less than 32180" meters (i.e., 20 miles), and export to a new table `Dist20miB`, which has 69,546 records. The distance table includes only those distance pairs within 20 miles and thus implements the selection conditions $j \in \{d_{ij} \le d_0\}$ and $k \in \{d_{kj} \le d_0\}$ in equation 5.6.

Step 15. Summing Up Population with Broadband Subscription in Areas with Available Broadband Services around Each Physician Location: On the table `Dist20miB`, add a new field wpop (Data Type as Double) and calculate it as wpop =!bandR!*!POP100!*!MSAbb!, which is the value $a_k D_k f(b_k)$ in equation 5.6. Here, only locations with broadband services are eligible for participating in telehealth, and the population there is further discounted by (limited to) those with broadband subscriptions.

Use the Summarize tool to sum up population wpop by physician locations NEAR _ FID, and save the result in a new table `DocAvl20miB`, which has 172 records, similar to step 4. The field Sum _ wpop is the total population having adequate broadband access within the threshold distance from each physician location and therefore implements calibration of the term $\sum_{k \in (d_{kj} \le d_0)}^{m} (a_k D_k f(b_k))$ in equation 5.6.

Step 16. Computing Initial Physician-to-Population Ratio at Each Physician Location: Join the updated table `DocAvl20miB` back to the distance table `Dist20miB` (based on their common field NEAR _ FID). On the table `Dist20miB`, add a field docpopR (Data Type as Double) and compute it as 1000*!Dist20miB. PCP! *!Dist20miB.PCPbb!/!DocAvl20miB.SUM _ wpop! This assigns an initial physician-to-population ratio to each physician location, where only physicians in locations with broadband services and with subscription to broadband participate in the calibration. This step computes the term $\dfrac{a_j S_j f(b_j)}{\sum_{k \in (d_{kj} \le d_0)}^{m} (a_k D_k f(b_k))}$ in equation 5.6.[3]

Again, the ratio is inflated 1,000 times to indicate the physician availability per 1,000 residents.

Step 17. Summing Up Physician-to-Population Ratios by Population Locations: On the updated table `Dist20miB`, remove the joins, then use the Summarize tool again to sum up the initial physician-to-population ratios docpopR by population locations IN_FID, and save the result in a new table `BkgAcc20miB`, which has 570 records. The field Sum_docpopR is the virtual accessibility for physicians from each residential location, and it includes five locations with zero values due to inade-quate access to internet. This computes the term $\sum_{j \in (d_{ij} \le d_0)}^{n} \dfrac{a_j S_j f(b_j)}{\sum_{k \in (d_{kj} \le d_0)}^{m} (a_k D_k f(b_k))}$ in equation 5.6, which is a preliminary virtual accessibility, as it is limited to those with broadband subscription.

Join the table `BkgAcc20miB` (based on the field IN _ FID) to the attribute table of population layer `MSAbkgPt` (based on the field OBJECTID). Add a field BkgAccV to the attribute table of `MSAbkgPt`, and calculate it as !BkgAcc20miB.

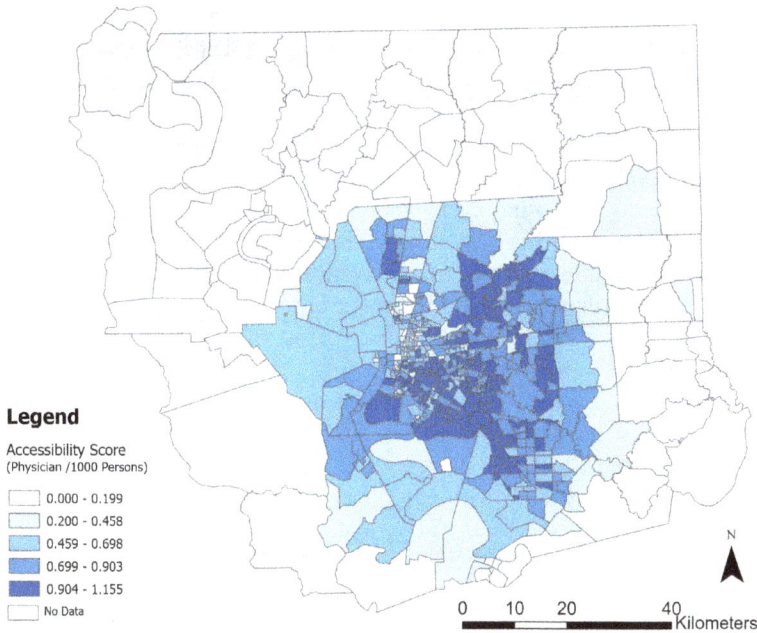

FIGURE 5.8 Telehealth accessibility for primary care physicians in Baton Rouge MSA by 2SVCA (20-mile catchment).

Sum _ docpopR! *!MSAbkgPt.bandR! By multiplying the broadband subscription rates, it returns the overall virtual accessibility score VA_i associated with residential area i in equation 5.6.

Step 18. Mapping Virtual Accessibility: Join the updated attribute table of MSAbkgPt to the census block group polygon layer MSAbkg (based on their common field GEOID) for mapping.

As shown in Figure 5.8, the virtual accessibility pattern is more scattered than the monocentric physical accessibility pattern in Figure 5.6. It is largely attributable to the complex interaction between broadband availability and affordability factors. Noticeably, a good number of census block groups have no access to telehealth.

This process is automated in a tool "2SVCA Pro" in the toolkit Accessibility Pro.tbx under the folder ArcPy-tools (Figure 5A.2). See Appendix 5B for its usage. As the development of 2SVCA method remains at an exploratory stage, we refrain from making further discussion on possible effects of parameter settings and differences between 2SFCA-derived physical accessibility and 2SVCA-derived virtual accessibility.

5.7 SUMMARY

Accessibility is a common issue in many studies at various scales. For example, a local community may be interested in examining the accessibility to children's playgrounds, identifying underserved areas and designing a plan of constructing new playgrounds or expanding existing ones (Talen and Anselin, 1998). It is also important to assess the accessibility of public parks, as good access may encourage physical activity and promote a healthy lifestyle (Zhang X et al., 2011). Research also suggests that spatial access to schools also matters in student achievement (Talen, 2001; Williams and Wang, 2014). Poor job accessibility tied to one's residential location not only has an adverse effect on employment prospects but also incurs high monetary and psychological costs for workers already in the labor force and increases their willingness to risk losing their jobs through involvement in deviant or criminal behavior (Wang and Minor, 2002). Accessibility of primary care physicians (as demonstrated in Case Study 5) affects timely utilization of cancer screening services by patients and contributes to risk of late-stage cancer diagnosis (Wang et al., 2008). The expansion of accessibility research along the application domain can be summarized as "accessibility for anything."

This chapter focuses on various GIS-based measures of spatial accessibility. Aspatial factors as discussed in Section 5.1 also play important roles in affecting accessibility. Indeed, the US Department of Health and Human Services designates two types of Health Professional Shortage Areas (HPSA): *geographic areas* and *population groups*. Generally, geographic-area HPSA is intended to capture spatial accessibility, and population-group HPSA accounts for aspatial factors (Wang and Luo, 2005). Examining the disparity of accessibility across various demographic groups is of great interest to public policy and guides planning and mitigation strategies for closing the gaps. For example, Section 11.4 of Chapter 11 discusses a new planning problem of explicitly maximizing accessibility equality. Research along this line of work can be grouped under the theme of "accessibility to anyone."

In short, the broad applications of spatial accessibility studies can be summarized as accessibility matters for "4As" (Wang, 2022):

1. For *anything* (e.g., healthcare, job, education, recreation, food)
2. To *anyone* (e.g., on disparity in access between socio-demographic groups)
3. By *any means* (e.g., via different transportation modes, challenges by the handicapped)
4. At *any time* (e.g., accounting for temporal variability of supply, demand, and transportation between daytime vs. nighttime, seasonally, normal vs. in the event of natural disasters) (Li et al., 2022)

APPENDIX 5A: A PROPERTY OF ACCESSIBILITY MEASURES

The accessibility index measured in equation 5.2 or 5.3 has an important property: the total accessibility score is equal to the total supply, or the weighted mean of accessibility is equal to the ratio of total supply to total demand in a study area. The following uses the gravity-based measure to prove the property (also see Shen, 1998, pp. 363–364). As shown in Section 5.3, the two-step floating catchment area (2SFCA) method in equation 5.2 is only a special case of the gravity-based method in equation 5.3, and therefore the proof also applies to the index defined by the 2SFCA method.

Recall the gravity-based accessibility index for demand site i, written as:

$$A_i = \sum_{j=1}^{n} \frac{S_j d_{ij}^{-\beta}}{\sum_{k=1}^{m} D_k d_{kj}^{-\beta}} \tag{A5.1}$$

The total accessibility (TA) is:

$$TA = \sum_{i=1}^{m} D_i A_i = D_1 A_1 + D_2 A_2 + \ldots + D_m A_m \tag{A5.2}$$

Substituting A5.1 into A5.2 and expanding the summation terms yield:

$$TA = D_1 \sum_j \frac{S_j d_{1j}^{-\beta}}{\sum_k D_k d_{kj}^{-\beta}} + D_2 \sum_j \frac{S_j d_{2j}^{-\beta}}{\sum_k D_k d_{kj}^{-\beta}} + \cdots + D_m \sum_j \frac{S_j d_{mj}^{-\beta}}{\sum_k D_k d_{kj}^{-\beta}}$$

$$= \frac{D_1 S_1 d_{11}^{-\beta}}{\sum_k D_k d_{k1}^{-\beta}} + \frac{D_1 S_2 d_{12}^{-\beta}}{\sum_k D_k d_{k2}^{-\beta}} + \cdots + \frac{D_1 S_n d_{1n}^{-\beta}}{\sum_k D_k d_{kn}^{-\beta}}$$

$$+ \frac{D_2 S_1 d_{21}^{-\beta}}{\sum_k D_k d_{k1}^{-\beta}} + \frac{D_2 S_2 d_{22}^{-\beta}}{\sum_k D_k d_{k2}^{-\beta}} + \cdots + \frac{D_2 S_n d_{2n}^{-\beta}}{\sum_k D_k d_{kn}^{-\beta}} + \cdots$$

$$+ \frac{D_m S_1 d_{m1}^{-\beta}}{\sum_k D_k d_{k1}^{-\beta}} + \frac{D_m S_2 d_{m2}^{-\beta}}{\sum_k D_k d_{k2}^{-\beta}} + \cdots + \frac{D_m S_n d_{mn}^{-\beta}}{\sum_k D_k d_{kn}^{-\beta}}$$

Rearranging the terms, we obtain:

$$TA = \frac{D_1 S_1 d_{11}^{-\beta} + D_2 S_1 d_{21}^{-\beta} + \cdots + D_m S_1 d_{m1}^{-\beta}}{\sum_k D_k d_{k1}^{-\beta}} + \frac{D_1 S_2 d_{12}^{-\beta} + D_2 S_2 d_{22}^{-\beta} + \cdots + D_m S_2 d_{m2}^{-\beta}}{\sum_k D_k d_{k2}^{-\beta}} +$$

$$\cdots + \frac{D_1 S_n d_{1n}^{-\beta} + D_2 S_n d_{2n}^{-\beta} + \cdots + D_m S_n d_{mn}^{-\beta}}{\sum_k D_k d_{kn}^{-\beta}}$$

$$= \frac{S_1 \sum_k D_k d_{k1}^{-\beta}}{\sum_k D_k d_{k1}^{-\beta}} + \frac{S_2 \sum_k D_k d_{k2}^{-\beta}}{\sum_k D_k d_{k2}^{-\beta}} + \cdots + \frac{S_n \sum_k D_k d_{kn}^{-\beta}}{\sum_k D_k d_{kn}^{-\beta}}$$

$$= S_1 + S_2 + \cdots + S_n$$

Denoting the total supply in the study area as S (i.e., $S = \sum_{i=1}^{n} S_i$), the previous equation shows that $TA = S$, that is, total accessibility is equal to total supply.

Denoting the total demand in the study area as D (i.e., $D = \sum_{i=1}^{m} D_i$), the weighted average of accessibility is:

$$W = \sum_{i=1}^{m} \left(\frac{D_i}{D} \right) A_i = (1/D)(D_1 A_1 + D_2 A_2 + \ldots + D_m A_m) = TA/D = S/D,$$

which is the ratio of total supply to total demand.

APPENDIX 5B: A TOOLKIT FOR AUTOMATED SPATIAL ACCESSIBILITY AND VIRTUAL ACCESSIBILITY MEASURES

An all-in-one toolkit is developed to automate the spatial accessibility and virtual accessibility measures discussed in this chapter.

The interface of Generalized 2SFCA/I2SFCA Pro tool for measuring spatial accessibility is shown in Figure 5A.1. It includes three definitions for distance type: Euclidean, geodesic, and external table. When Euclidean or geodesic is chosen, it uses the tool Generate Near Table to calculate the Euclidean or geodesic distance matrix between the Customer and Facility layers (or their centroids if the features are in polygons), preferably in projected coordinate systems. The Distance Unit Conversion Factor transforms the unit from meters to kilometers, or vice versa.

Much of the improvement of this tool over the previous version focuses on a rich set of choices for modeling the distance decay behaviors as discussed in Subsection 5.3.2. Specifically, for distance decay, three function categories are available: discrete, continuous, and hybrid, sharing a weight range of 0–1.

1. *Discrete.* The input of Discrete Value accepts a list of values ranging from a single value to multiple ones and assigns the arithmetic sequence of 0–1 as corresponding weights. For example, in the Case Study 5 for conventional 2SFCA (Figure A5.1a–b, a single value of 32180 (meters) or 30 (minutes) is used, with assigned weights (1, 0) for those within the catchment vs. beyond. If one inputs 15;40 (minutes), a two-stage stepwise series of weights (1, 0.5, 0) is assigned to those in three travel time ranges (0, 15), (15, 40), and (40, ∞). Similarly, a three-stage stepwise uses a weight series (1, 0.67, 0.33, 0), a four-stage stepwise uses a weight series (1, 0.75, 0.50, 0.25 0), and so on.
2. *Continuous.* The continuous functions include one-parameter (power, exponential, square-foot exponential, Gaussian, and log-normal) or two-parameter (log-logistic and compound power-exponential) functions and their associated coefficients. A min–max normalization function is then applied to the result, such as *(x-min)/(max-min)*, to convert the values to a 0–1 range. For example, Figure A5.1c shows the definitions used in the Huff model in Case Study 5.
3. *Hybrid.* (a) *One-stage hybrid* uses only one threshold value. A weight 1 is assigned to any records within the threshold, and the remaining records are applied with the continuous function and then normalized between 0 and 1. (b) *Two-stage hybrid* (as shown in Figure A5.1d) inputs two threshold values separated by ";" as delimiter. A weight 1 is assigned to any records within the first threshold, a weight 0 is assigned to any records beyond the second threshold, and the remaining records between are applied with the continuous function and then normalized between 0 and 1.

Finally, the weight scale factor is designed to inflate the values for accessibility scores to avoid too small values (e.g., when using a weight scale factor of 1,000 in Case Study 5, the scores indicate the numbers of accessibly physicians per 1,000 persons).

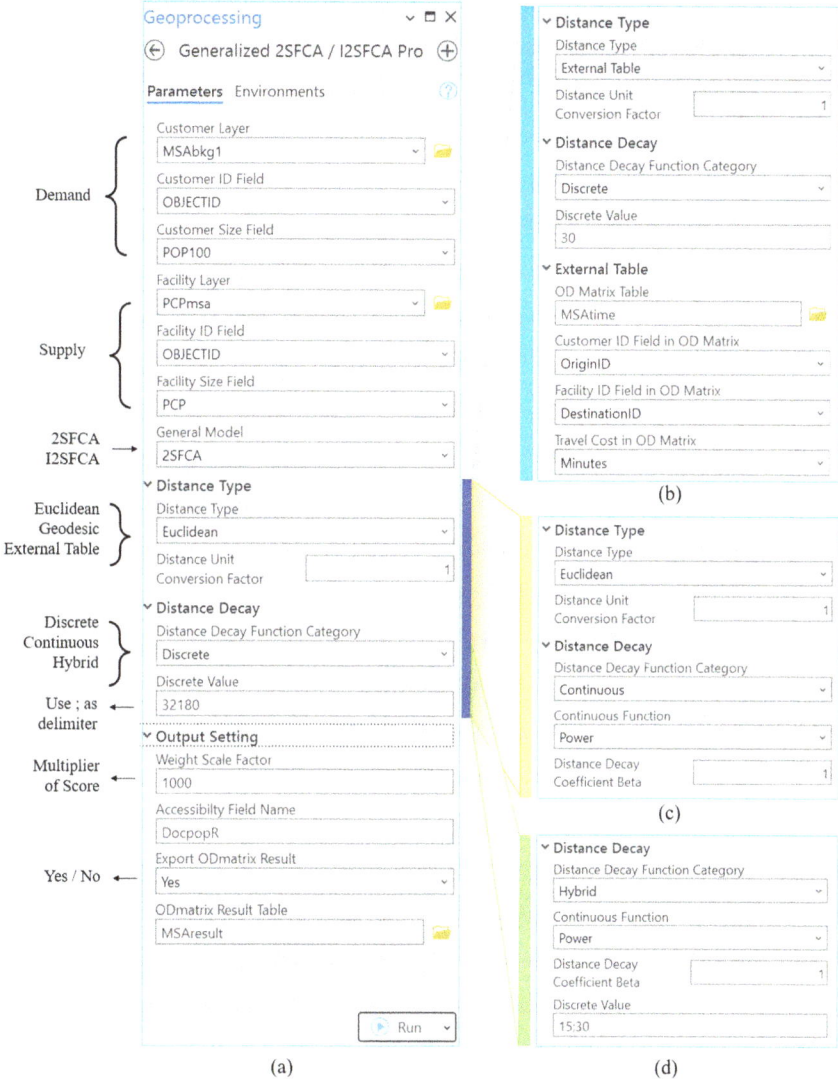

FIGURE A5.1 Interface for the Generalized 2SFCA/I2SFCA Pro tool: discrete decay function with (a) Euclidean distances and (b) external distance table, (c) continuous decay function, and (d) hybrid decay function.

Other inputs, such as Customer layer, Facility layer, and their associated fields, are routine and not discussed here. For results, when 2SFCA is chosen for General Model, it saves accessibility scores in the field DocpopR associated with the Customer layer, and when I2SFCA is chosen for General Model, it saves crowdedness scores also in the field DocpopR associated with the Facility layer. The option of "Export ODmatirx Result" may be turned on for users to save the OD matrix table (e.g., named as `MSAresult`) and review the intermedium result, such as distance decay effect.

The interface of 2SVCA Pro tool for measuring virtual accessibility is shown in Figure A5.2. The parameter settings are similar to those of the Generalized 2SFCA/

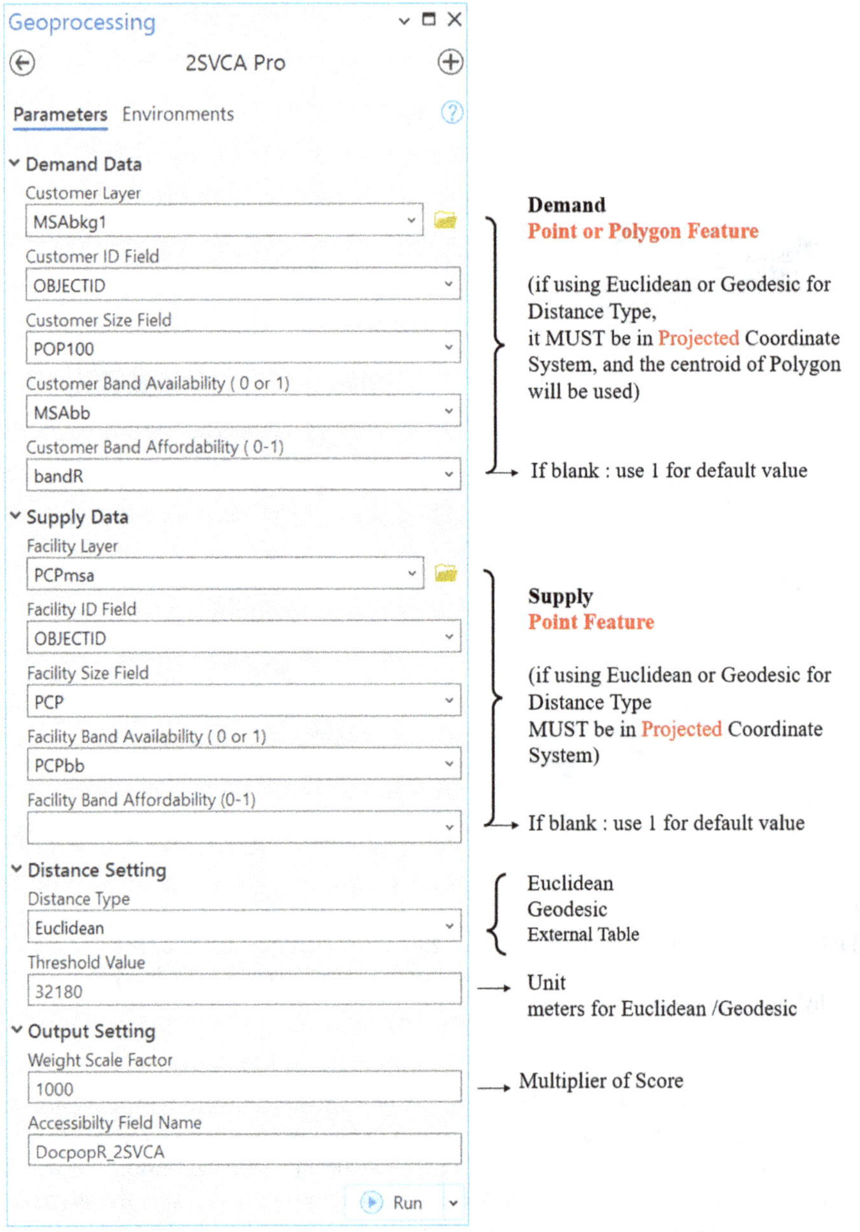

FIGURE A5.2 Interface for the 2SVCA Pro tool.

I2SFCA Pro tool and add two items to both Demand (Customer) and Supply (Facility) layers: (1) band availability defines parameter $f(b)$ in equation 5.6, currently a binary choice 0 or 1 (e.g., whether broadband services are available in an area), and (2) band affordability defines a in equation 5.6, for values ranging from 0 to 1 (e.g., broadband subscription rate). Since the distance decay setting in the current formulation of 2SVCA in equation 5.6 adopts the conventional 2SFCA in equation 5.2, users only need to input the distance type (Euclidean, geodesic, or external table) and a threshold value. All others are the same as in the Generalized 2SFCA/I2SFCA Pro tool.

APPENDIX 5C: DERIVING THE 2SFCA AND I2SFCA METHODS

Readers may refer to Figure A5.3 for the graphic illustration of the derivations of both methods.

This section begins with a brief review of deriving i2SFCA as in Wang (2018). According to the Huff (1963) model, residents at location i (denoted by D_i) choose a particular facility j with capacity S_j among a set of alternatives S_l ($l = 1, 2, \ldots, n$), with a probability such as:

$$Prob_{Di\_j} = S_j f\left(d_{ij}\right) / \sum_{l=1}^{n}\left(S_l f\left(d_{il}\right)\right) \qquad (A5.3)$$

Note that a generic distance decay function $f(d)$ is adopted in place of Huff's power (gravity-based) function (Jia et al., 2017).

Multiplying the probability $Prob_{Di\_j}$ by the demand size (e.g., population) D_i at i yields the number of visitors (customers, patients, etc.) from residential location i to facility j, or service volume, denoted by F_{ij}. It is projected as:

$$F_{ij} = D_i Prob_{Di\_j}$$

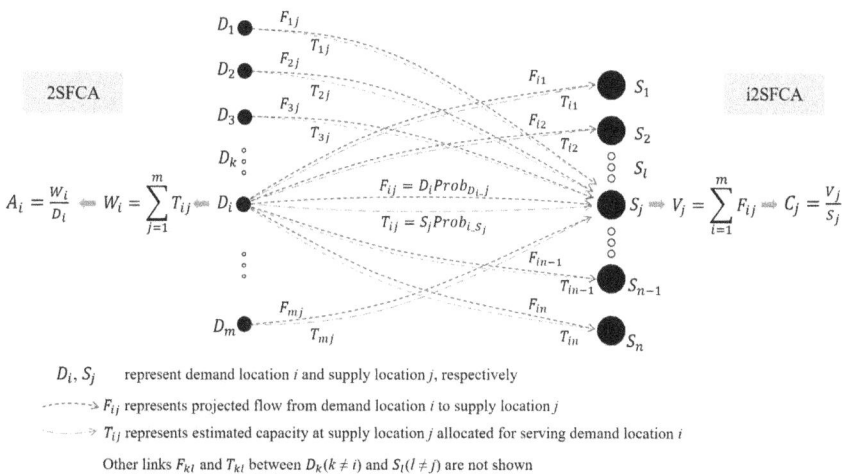

D_i, S_j represent demand location i and supply location j, respectively

$\dashrightarrow F_{ij}$ represents projected flow from demand location i to supply location j

$\dashrightarrow T_{ij}$ represents estimated capacity at supply location j allocated for serving demand location i

Other links F_{kl} and T_{kl} between $D_k (k \neq i)$ and $S_l (l \neq j)$ are not shown

FIGURE A5.3 Deriving the 2SFCA and i2SFCA methods.

Therefore, the total estimated number of visitors to facility j, denoted by V_j, is:

$$V_j = \sum_{i=1}^{m} F_{ij} \qquad (A5.4)$$

Finally, normalizing V_j by its capacity S_j yields:

$$C_j = V_j / S_j = \sum_{i=1}^{m} F_{ij} / S_j = (\sum_{i=1}^{m} D_i Prob_{Di\_j}) / S_j$$
$$= \sum_{i=1}^{m} [D_i f(d_{ij}) / \sum_{l=1}^{n} (S_l f(d_{il}))] \qquad (A5.5)$$

Note that the term S_j is eliminated in both the numerator and denominator for simplification in equation A5.5. C_j is the final formulation of i2SFCA, termed "potential crowdedness" (e.g., patients per bed in a hospital, clients per clerk, etc.). A higher C_j value corresponds to a service facility being more crowded.

The derivation of 2SFCA imitates the preceding process for deriving i2SFCA (Wang, 2021). Similar to the notion of the Huff model but beginning with a focus on facility location S_j, the proportion of visitors from residential location D_i, denoted by $Prob_{i\_Sj}$, out of alternative origins (other residential locations) D_k ($k = 1, 2, \ldots, m$), is:

$$Prob_{i\_Sj} = D_i f(d_{ij}) / \sum_{k=1}^{m} (D_k f(d_{kj})) \qquad (A5.6)$$

Note the symmetry between equations A5.3 and A5.6, where demand size D and facility capacity S switch with each other. In other words, it is the Huff model upside down. While the Huff model predicts that the probability visiting a facility out of alternative destinations is proportional to the facility's size and reversely related to distance, the argument made here is that the probability coming from a residential area out of alternative origins is proportional to the residential area's population size and reversely related to distance. One may relate this symmetry to the derivation of the Garin–Lowry model in Chapter 10, where "the proportion of service employment in an area owing to the influence of population in another area out of its impacts on all areas" is symmetrical to "the proportion of population in an area owing to the influence of employment in another area out of its impacts on all areas" (Wang, 2015, p. 221).

With the total capacity at facility j defined as S_j, the portion dedicated (allocated, attributable) for serving the visitors from i out of all demand locations is:

$$T_{ij} = S_j Prob_{i\_Sj}$$

If a facility j uses capacity T_{ij} for serving residential location i, summing up these capacities across all facilities ($j = 1, 2, \ldots, n$) with their respective portions dedicated for i yields total resource commanded by demand location i, denoted as W_i:

$$W_i = \sum_{j=1}^{n} T_{ij} = \sum_{j=1}^{n} (S_j Prob_{i\_Sj}) = \sum_{j=1}^{n} [S_j D_i f(d_{ij}) / \sum_{k=1}^{m} (D_k f(d_{kj}))] \qquad (A5.7)$$

Finally, normalizing the total resource W_i available for i by its demand size D_i yields:

$$A_i = W_i / D_i = \sum_{j=1}^{n} [S_j f(d_{ij}) / \sum_{k=1}^{m} (D_k f(d_{kj}))] \qquad (A5.8)$$

where D_i is eliminated in both the numerator and the denominator for simplification. Equation A5.8 is the traditional 2SFCA, which is now derived. It can be interpreted as the projected resource available for each resident at location i.

NOTES

1 With comparison to the polygon feature MSAbkg (n = 574), MSAbkgPt (n = 570) excludes four census block groups with no residents, and its corresponding polygon feature is MSAbkg1 (n = 570).

2 A reasonable threshold distance for visiting a primary care physician is 15 miles in travel distance. We set the search radius at 20 miles in Euclidean distance, roughly equivalent to a travel distance of 30 miles, that is, twice the reasonable travel distance. This is perhaps an upper limit.

3 Here, for lack of data on broadband subscription rates for physicians, we assume $a_j = 1$. In a rare case, if the denominator $\sum_{k \in (d_{kj} \le d_0)}^{m} (a_k D_k f(b_k)) = 0$, a physician facility S_j is surrounded with no valid demand via virtual connectivity, the term is invalid, coded as 0, and does not contribute to the overall accessibility.

6 Function Fittings by Regressions and Application in Analyzing Urban Density Patterns

Urban and regional studies begin with analyzing the spatial structure, particularly population density patterns. As population serves as both supply (labor) and demand (consumers) in an economic system, the distribution of population represents that of economic activities. Analysis of changing population distribution patterns over time is a starting point for examining economic development patterns in a city or region. Urban and regional density patterns mirror each other: the *central business district* (CBD) is the center of a city, whereas the whole city itself is the center of a region, and densities decline with distances both from the CBD in a city and from the central city in a region. While the theoretical foundations for urban and regional density patterns are different, the methods for empirical studies are similar and closely related.

This chapter discusses how we find the best-fitting function to capture the density pattern, and what we can learn about urban and regional growth patterns from this approach. The methodological focus is on function fittings by regressions and related statistical issues. Section 6.1 explains how density functions are used to examine urban and regional structures. Section 6.2 presents various functions for a monocentric structure. Section 6.3 discusses some statistical concerns on monocentric function fittings and introduces nonlinear regression and weighted regression. Section 6.4 examines various assumptions for a polycentric structure and corresponding function forms. Section 6.5 uses a case study in Chicago urban area to illustrate the techniques (monocentric vs. polycentric models, linear vs. nonlinear, and weighted regressions). The chapter is concluded in Section 6.6 with discussions and a brief summary.

6.1 THE DENSITY FUNCTION APPROACH TO URBAN AND REGIONAL STRUCTURES

6.1.1 URBAN DENSITY FUNCTIONS

Since the classic study by Clark (1951), there has been great interest in empirical studies of urban population density functions. This cannot be solely explained by the easy availability of data. Many researchers are fascinated by the regularity of urban density pattern and its solid foundation in economic theory. McDonald (1989, p. 361)

DOI: 10.1201/9781003292302-8

considers the population density pattern as "a critical economic and social feature of an urban area."

Among all functions, the *exponential function* or the *Clark's model* is the one used most widely:

$$D_r = ae^{br}, \tag{6.1}$$

where D_r is the density at distance r from the city center (i.e., central business district or CBD), a is a constant or the CBD intercept, and b is another constant or the density gradient. Since the density gradient b is often a negative value, the function is also referred to as the *negative exponential function*. Empirical studies show that it is a good fit for most cities in both developed and developing countries (Mills and Tan, 1980).

The economic model by Mills (1972) and Muth (1969), often referred to as the *Mills–Muth model*, is developed to explain the empirical finding of urban density pattern as a negative exponential function. The model assumes a *monocentric structure*: a city has only one center, where all employment is concentrated. Intuitively, as everyone commutes to the city center for work, a household farther away from the CBD spends more on commuting and is compensated by living in a larger-lot house that is cheaper in terms of housing price per area unit. The resulting population density exhibits a declining pattern with distance from the city center. Appendix 6A shows how the negative exponential urban density function is derived by the economic model. In the deriving process, the parameter b in equation 6.1 is the same as the unit cost of transportation or commuting. Therefore, the declining transportation cost over time, as a result of improvements in transportation technologies and road networks, is associated with a flatter density gradient (i.e., more gradual or less sharp decline of density with increasing distance from the city center). This helps us understand that *urban sprawl* or *suburbanization* (characterized by an increasingly flatter density gradient over time) is mainly attributable to transportation improvements (Wang and Zhou, 1999).

However, economic models are "simplification and abstractions that may prove too limiting and confining when it comes to understanding and modifying complex realities" (Casetti, 1993, p. 527). The main criticisms lie in its assumptions of the monocentric city and unit price elasticity for housing, neither of which is fully supported by empirical studies. Wang and Guldmann (1996) developed a *gravity-based model* to explain the urban density pattern (see Appendix 6A). The basic assumption of the gravity-based model is that population at a particular location is proportional to its accessibility to all other locations in a city, measured as a gravity potential. Simulated density patterns from the model conform to the negative exponential function when the distance friction coefficient β falls within a certain range ($0.2 \leq \beta \leq 1.0$ in the simulated example). The gravity-based model does not make the same restrictive assumptions as the economic model and thus implies wide applicability. It also explains two important empirical findings: (1) flattening density gradient over time (corresponding to smaller β) and (2) flatter gradients in larger cities. The economic model explains the first finding well, but not the second (McDonald, 1989, p. 380). Both the economic model and the gravity-based model explain the change of density gradient over time through transportation improvements since the distance friction coefficient β in the gravity model or the unit cost of transportation in the economic model declines over time.

Another perspective that helps us understand the urban population density pattern is the *space syntax theory* (Hillier and Hanson, 1984), which emphasizes that urban structure is "determined by the structure of the urban grid itself rather than by the presence of specific attractors or magnets" (Hillier et al., 1993, p. 32). In other words, it is the configuration of a city's street network that shapes its structure in terms of land use intensity (employment and population density patterns). Based on the approach, various centrality measures derived from the street network capture a location's (un)attractiveness and thus explain the concentration (or lack of concentration) of economic (including residential) activities. Appendix 6B illustrates how centrality indices are measured and implemented and how the indices are associated with urban densities.

Earlier empirical studies of urban density patterns are based on the monocentric model, that is, how population density varies with distance from the city center. It emphasizes the impact of the primary center (CBD) on citywide population distribution. Since 1970s, more and more researchers recognize the changing urban form from *monocentricity* to *polycentricity* (Ladd and Wheaton, 1991; Berry and Kim, 1993). In addition to the major center in the CBD, most large cities have secondary centers or subcenters and thus are better characterized as polycentric cities. In a polycentric city, assumptions of whether residents need to access all centers or some of the centers lead to various function forms. Section 6.4 examines the polycentric models in detail.

6.1.2 REGIONAL DENSITY FUNCTIONS

The study of regional density patterns is a natural extension to that of urban density patterns when the study area is expanded to include rural areas. The urban population density patterns, particularly the negative exponential function, are empirically observed first and then explained by theoretical models (e.g., the economic model and the gravity-based model). Even considering *Alonso's* (1964) *urban land use model* as the precedent of the Mills–Muth urban economic model, the theoretical explanation lags behind the empirical finding on urban density patterns. In contrast, following the rural land use theory by Von Thünen (1966, English version), economic models for the regional density pattern by Beckmann (1971) and Webber (1973) were developed before the work of empirical models for regional population density functions by Parr (1985), Parr et al. (1988), and Parr and O'Neill (1989). The city center used in the urban density models remains as the center in regional density models. The declining regional density pattern has a different explanation. In essence, rural residents farther away from a city pay higher transportation costs for the shipment of agricultural products to the urban market and for gaining access to industrial goods and urban services in the city and are compensated by occupying cheaper and, hence, more land (Wang and Guldmann, 1997).

Similarly, empirical studies of regional density patterns can be based on a monocentric or a polycentric structure. Obviously, as the territory for a region is much larger than a city, it is less likely for physical environments (e.g., topography, weather, and land use suitability) to be uniform across a region than a city. Therefore, population density patterns in a region tend to exhibit less regularity than in a city. An ideal study area for empirical studies of regional density functions would be an area with uniform physical environments, like the "isolated state" in the von Thünen model (Wang, 2001a, p. 233).

Analyzing the function change over time has important implications for both urban and regional structures. For urban areas, we can examine the trend of *urban polarization* vs. *suburbanization*. The former represents an increasing percentage of population in the urban core relative to its suburbia, and the latter refers to a reverse trend with an increasing portion in the suburbia. For regions, we can identify the process as *centralization* vs. *decentralization*. Similarly, the former refers to the migration trend from peripheral rural to central urban areas, and the latter is the opposite. Both can be synthesized into a framework of core vs. periphery. According to Gaile (1980), economic development in the core (city) impacts the surrounding (suburban and rural) region through a complex set of dynamic spatial processes (i.e., intraregional flows of capital, goods and services, information and technology, and residents). If the processes result in an increase in activity (e.g., population) in the periphery, the impact is *spread*. If the activity in the periphery declines while the core expands, the effect is *backwash*. Such concepts help us understand core–hinterland interdependencies and the various relationships between them (Barkley et al., 1996). If the exponential function is a good fit for regional density patterns, the changes can be illustrated as in Figure 6.1, where $t + 1$ represents a more recent time than t. In a monocentric model, we can see the relative importance of the city center; in a polycentric model, we can understand the strengthening or weakening of various centers.

In the remainder of this chapter, the discussion focuses on urban density patterns. However, similar techniques can be applied to studies of regional density patterns.

6.2 FUNCTION FITTINGS FOR MONOCENTRIC MODELS

In addition to the exponential function (equation 6.1) introduced earlier, three other simple bivariate functions for the monocentric structure have often been used:

$$D_r = a + br \tag{6.2}$$

$$D_r = a + b \ln r \tag{6.3}$$

$$D_r = ar^b \tag{6.4}$$

Equation 6.2 is a *linear function*, equation 6.3 is a *logarithmic function*, and equation 6.4 is a *power function*. The parameter b in all the aforementioned four functions is expected to be negative, indicating declining densities with distances from the city center.

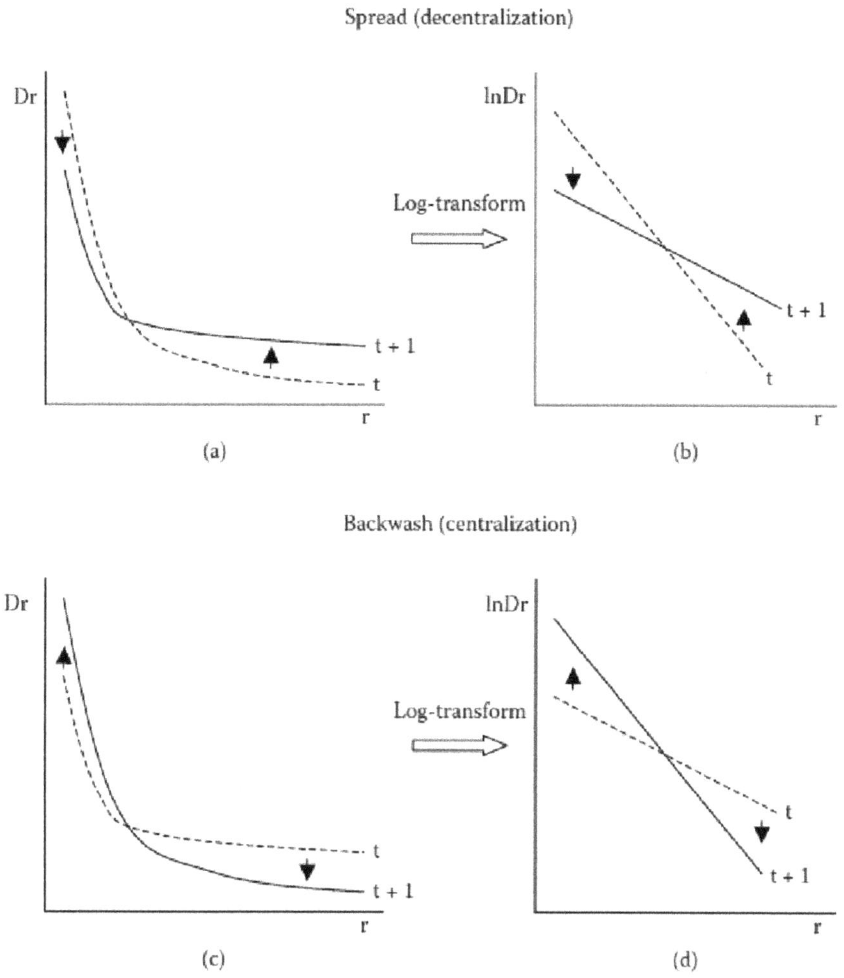

Spread (decentralization)

(a)   (b)

Backwash (centralization)

(c)   (d)

FIGURE 6.1 Regional growth patterns by the density function approach.

Equations 6.2 and 6.3 can be easily estimated by an *ordinary least square (OLS) linear regression*. Equation 6.1 and 6.4 can be transformed to linear functions by taking the logarithms on both sides, such as:

$$lnD_r = A + br \qquad (6.5)$$

$$lnD_r = A + blnr \qquad (6.6)$$

Equation 6.5 is the log-transform of equation 6.1, and equation 6.6 is the log-transform of equation 6.4. The intercept A in both equations 6.5 and 6.6 is just the log-transform of constant a (i.e., $A = lna$) in equations 6.1 and 6.4, respectively. The value of a can be easily recovered by taking the reverse of logarithm, that is, $a = e^A$. Equations 6.5 and 6.6 can also be estimated by a linear OLS regression. In regressions for equations 6.3 and 6.6 containing the term lnr, samples should not include observations where $r = 0$, to avoid taking logarithms of zero. Similarly, in equations 6.5 and 6.6 containing the term lnD_r, samples should not include those where $D_r = 0$ (with zero population).

Take the log-transform of exponential function in equation 6.5 for an example. The two parameters, intercept A and gradient b, characterize the density pattern in a city. A smaller value of A indicates a lower density around the central city, and a lower value of b (in terms of absolute value) represents a flatter density pattern. Many cities have experienced a lower intercept A and a flatter gradient b over time, representing a common trend of urban sprawl and suburbanization. The changing pattern is similar to Figure 6.1a, which also depicts decentralization in the context of regional growth patterns.

In addition to the four simple bivariate functions discussed earlier, three other functions are also used widely in the literature. One was proposed by Tanner (1961) and Sherratt (1960) independently from each other, commonly referred to as the *Tanner–Sherratt model*. The model is written as:

$$D_r = ae^{br^2},$$ (6.7)

where the density D_r declines exponentially with distance squared r^2. This function is referred to as the simplified Gaussian or normal function in Section 2.3 of Chapter 2 when the distance decay functions are discussed. By extension, one could also test whether other distance decay functions (e.g., square-root exponential, log-normal, log-logistic, etc.) could be a better fit.

Newling (1969) incorporated both Clark's model and the Tanner–Sherratt model and suggested the following model:

$$D_r = ae^{b_1r+b_2r^2},$$ (6.8)

where the constant term b_1 is most likely to be positive and b_2 negative and other notations remain the same. In *Newling's model*, a positive b_1 represents a *density crater* around the CBD, where population density is relatively low due to the presence of commercial and other non-residential land uses. According to Newling's model, the highest population density does not occur at the city center but rather at a certain distance away from the city center.

Some researchers (e.g., Anderson, 1985; Zheng, 1991) use the *cubic spline function* to capture the complexity of urban density pattern. The function is written as:

$$D_x = a_1 + b_1\left(x - x_0\right) + c_1(x - x_0)^2 + d_1(x - x_0)^3 + \sum_{i=1}^{k} d_{i+1}(x - x_i)^3 Z_i^*,$$ (6.9)

where x is the distance from the city center, D_x is the density there, x_0 is the distance of first density peak from the city center, x_i is the distance of ith knot from the city center (defined by either the second, third . . . density peak or simply even intervals across the whole region), Z_i^* is a dummy variable (= 0, if x is inside the knot; =1, if x is outside of the knot).

The cubic spline function intends to capture more fluctuations of the density pattern (e.g., local peaks in suburban areas) and thus cannot be strictly defined as a monocentric model. However, it is still limited to examining density variations related to distance from the city center regardless of directions and thus assumes a *concentric* density pattern.

The density function analysis only uses two variables: one is Euclidean distance r from the city center, and the other is the corresponding population density D_r. Euclidean distances from the city center can be obtained using the techniques explained in Section 2.1. Identifying the city center requires knowledge of the study area and is often defined as a commonly recognized landmark site by the public. In the absence of a commonly recognized city center, one may use the location with the highest job concentration to define it. Density is simply computed as population divided by area size. Area size is saved in the field Shape_Area in any polygon feature in a geodatabase. Once the two variables are obtained in GIS, the dataset can be exported to an external file for regression analysis.

Both the Tanner–Sherratt model (equation 6.7) and Newling's model (equation 6.8) can be estimated by linear OLS regression on their log-transformed forms (see Table 6.1). In the Tanner–Sherratt model, the X variable is distance squared (r^2). In Newling's model, there are two X variables (r and r^2). Newling's model has one more explanatory variable (i.e., r^2) than Clark's model (exponential function) and thus always generates a higher R^2 regardless of the significance of the term r^2. In this sense, Newling's model is not comparable to the other five bivariate models in terms of fitting power. Table 6.1 summarizes the models.

Fitting the cubic spline function (equation 6.9) is similar to that of other monocentric functions with some extra work in preparing the data. First, sort the data by the variable "distance" in an ascending order. Second, define the constant x_0, and calculate the terms $(x-x_0)$, $(x-x_0)^2$, and $(x-x_0)^3$. Third, define the constants x_i (i.e., $x_1, x_2 \ldots$),

---

TABLE 6.1

Linear Regressions for a Monocentric City

Models	Function Used in Regression	Original Function	X Variable(s)	Y Variable	Restrictions
Linear	$D_r = a + br$	Same	r	D_r	none
Logarithmic	$D_r = a + blnr$	Same	lnr	D_r	$r \neq 0$
Power	$lnD_r = A + blnr$	$D_r = ar^b$	lnr	lnD_r	$r \neq 0$ and $D_r \neq 0$
Exponential	$lnD_r = A + br$	$D_r = ae^{br}$	r	lnD_r	$D_r \neq 0$
Tanner–Sherratt	$lnD_r = A + br^2$	$D_r = ae^{br^2}$	r^2	lnD_r	$D_r \neq 0$
Newling's	$lnD_r = A + b_1r + b_2r^2$	$D_r = ae^{b_1r+b_2r^2}$	r, r^2	lnD_r	$D_r \neq 0$

and compute the terms $(x - x_i)^3 Z_i^*$. Take one term $(x - x_1)^3 Z_1^*$ as an example: (1) set the values = 0 for those records with $x \le x_1$, and (2) compute the values = $(x - x_1)^3$ for those records with $x > x_1$. Finally, run a multivariate regression, where the Y variable is density D_x, and the X variables are $(x\text{-}x_0)$, $(x\text{-}x_0)^2$, $(x\text{-}x_0)^3$, $(x - x_1)^3 Z_1^*$, $(x - x_2)^3 Z_2^*$, and so on. The cubic spline function contains multiple X variables, and thus its regression R^2 tends to be higher than in other models.

Linear OLS regression is available in many software packages, including ArcGIS Pro. Case Study 6 in Section 6.5 illustrates the use of a customized R-ArcGIS tool "RegressionR" to implement the OLS regression in addition to a regular OLS regression tool in ArcGIS. Note that equations 6.5 and 6.6 are the log-transformations of exponential function (equation 6.1) and power function (equation 6.4), respectively. Based on the results, equation 6.1 or 6.4 can be easily recovered by computing the coefficient $a = e^A$ and the coefficient b is unchanged.

6.3   NONLINEAR AND WEIGHTED REGRESSIONS IN FUNCTION FITTINGS

In function fittings for the monocentric structure, two statistical issues merit more discussion. One is the choice between nonlinear regression directly on the exponential or power function vs. linear regression on its log-transformation (as discussed in Section 6.2). Generally, they yield different results since the two have different dependent variables (D_r in nonlinear regression vs. lnD_r in linear regression) and imply different assumptions of the error term (Greene and Barnbrock, 1978).

We use the exponential function (equation 6.1) and its log-transformation (equation 6.5) to explain the differences. The linear regression on equation 6.5 has the errors ε, such as:

$$lnD_r = A + br + \varepsilon. \tag{6.10}$$

That implies multiplicative errors and weights equal *percentage errors* equally, such as:

$$D_r = ae^{br+\varepsilon}.$$

The nonlinear regression on the original exponential function (equation 6.1) assumes additive errors and weights all equal *absolute errors* equally, such as:

$$D_r = ae^{br} + \varepsilon. \tag{6.11}$$

The *ordinary least square (OLS) regression* seeks the optimal values of A (or a) and b so that residual sum of squares (RSS) is minimized. For the linear regression based on equation 6.10, it is to minimize:

$$RSS = \sum_i \left(\ln D_i - A - br_i \right)^2.$$

For the nonlinear regression based on equation 6.11, it is to minimize:

$$RSS = \sum_i (D_i - ae^{br_i})^2,$$

where i indexes individual observations.

For linear regression, a popular measure for a model's goodness of fit is R^2 (*coefficient of determination*). Appendix 6C illustrates how the parameters in a bivariate linear OLS regression are estimated and how R^2 is defined. In essence, R^2 is the portion of dependent variable's variation (termed "total sum squared (TSS)") explained by a regression model, termed "explained sum squared (ESS)," that is, $R^2 = ESS/TSS = 1- RSS/TSS$.

There are several ways to estimate the parameters in a nonlinear regression (Griffith and Amrhein, 1997, p. 265), and all use iterations to gradually improve guesses. For example, the *modified Gauss–Newton method* uses linear approxima-tions to estimate how RSS changes with small shifts from the present set of parame-ter estimates. Good initial estimates (i.e., those close to the correct parameter values) are critical in finding a successful nonlinear fit. The initialization of parameters is often guided by experience and knowledge. Note that R^2 is no longer applicable to nonlinear regression, as the identity "ESS + RSS = TSS" no longer holds and the residuals do not add up to 0. However, one may use a *pseudo-R^2* (defined similarly as *1- RSS/TSS*) as an approximate measure of goodness of fit.

Which is a better method for estimating density functions, linear or nonlinear regression? The answer depends on the emphasis and objective of a study. Linear regression is based on log-transformation. By weighting equal *percentage errors* equally, the errors generated by high-density observations are scaled down (in terms of percentage). However, the differences between the estimated and observed values in those high-density areas tend to be much greater than those in low-density areas (in terms of absolute value). As a result, the total estimated population in the city could be off by a large margin. On the contrary, nonlinear regression is to mini-mize the residual sum of squares (RSS) directly based on densities instead of their logarithms. By weighting all equal *absolute errors* equally, the regression limits the errors (in terms of absolute value) contributed by high-density samples. As a result, the total estimated population in the city is often closer to the actual value than the one based on linear regression, but the estimated densities in low-density areas may be off by high percentages.

Another issue in estimating urban density functions concerns *randomness of sample* (Frankena, 1978). A common problem for census data (not only in the United States) is that high-density observations are many and are clustered in a small area near the city center, whereas low-density ones are fewer and spread in remote areas. In other words, high-density samples may be overrepresented as they are concen-trated within a short distance from the city center, and low-density samples may be underrepresented as they spread across a wide distance range from the city center. A plot of density vs. distance show many more observations in short distances and fewer in long distances. This is referred to as *non-randomness of sample* and causes biased (usually upward) estimators. A weighted regression can be used to mitigate the problem. Frankena (1978) suggests weighting observations in proportion to their areas. In the regression, the objective is to minimize the weighted RSS. For the same reason explained for nonlinear regression, R^2 is no longer valid for a weighted regression, and a *pseudo-R^2* may be constructed from calculated RSS and TSS. See Wang and Zhou (1999) for an example. Some researchers favor samples with uniform area sizes. In Case Study 6 (Section 6.5.3 in particular), we will also analyze population density functions based on survey townships of approximately the same area size.

Case Study 6 in Section 6.5 uses the same R-ArcGIS tool "RegressionR" to implement the weighted and nonlinear OLS regressions.

6.4 FUNCTION FITTINGS FOR POLYCENTRIC MODELS

Monocentric density functions simply assume that densities are uniform at the same distance from the city center regardless of direction. Urban density patterns in some cities may be better captured by a polycentric structure. In a polycentric city, residents and business value access to multiple centers, and therefore population densities, are functions of distances to these centers (Small and Song, 1994). Centers other than the primary or major center at the CBD are called subcenters.

A polycentric density function can be established under several alternative assumptions:

1. If the influences from different centers are perfectly substitutable so that only the nearest center (CBD or subcenter) matters, the city is composed of multiple monocentric subregions. Each subregion is the proximal area for a center (see Section 4.1), within which various monocentric density functions can be estimated. Taking the exponential function as an example, the model for the subregion around the ith center is:

$$D = a_i e^{b_i r_i}, \tag{6.12}$$

where D is the density of an area, r_i is the distance between the area and its nearest center i, and a_i and b_i ($i = 1, 2 \ldots$) are parameters to be estimated.

2. If the influences are complementary so that some access to all centers is necessary, then the polycentric density is the product of those monocentric functions (McDonald and Prather, 1994). For example, the log-transformed polycentric exponential function is written as:

$$\ln D = a + \sum_{i=1}^{n} b_i r_i, \tag{6.13}$$

where D is the density of an area, n is the number of centers, r_i is the distance between the area and center i, and a and b_i ($i = 1, 2 \ldots$) are parameters to be estimated.

3. Most researchers (Griffith, 1981; Small and Song, 1994) believe that the relationship among the influences of various centers is between assumptions 1 and 2, and the polycentric density is the sum of center-specific functions.

For example, a polycentric model based on the exponential function is expressed as:

$$D = \sum_{i=1}^{n} a_i e^{b_i r_i}. \tag{6.14}$$

The preceding three assumptions are based on Heikkila et al. (1989).

4. According to the central place theory, the major center at the CBD and the subcenters play different roles. All residents in a city need access to the major center for higher-order services; for other lower-order services, residents only need to use the nearest subcenter (Wang, 2000). In other words, everyone values access to the major center and access to the nearest center. Using the exponential function as an example, the corresponding model is:

$$ln\,D = a + b_1 r_1 + b_2 r_2, \tag{6.15}$$

where r_1 is the distance from the major center, r_2 is the distance from the nearest center, and a, b_1, and b_2 are parameters to be estimated.

Figure 6.2 illustrates the different assumptions for a polycentric city. Residents need access to all centers under assumption 2 or 3, but effects are multiplicative in 2 and additive in 3. Table 6.2 summarizes the preceding discussion.

Analysis of polycentric models requires the identification of multiple centers first. Ideally, these centers should be based on the distribution of employment (e.g., Gordon et al., 1986; Giuliano and Small, 1991; Forstall and Greene, 1998). In addition to traditional choropleth maps, Wang (2000) used surface modeling techniques to generate isolines (contours) of employment density[1] and identified employment centers based on both the contour value (measuring the threshold employment density) and the size of area enclosed by the contour line (measuring the base value of

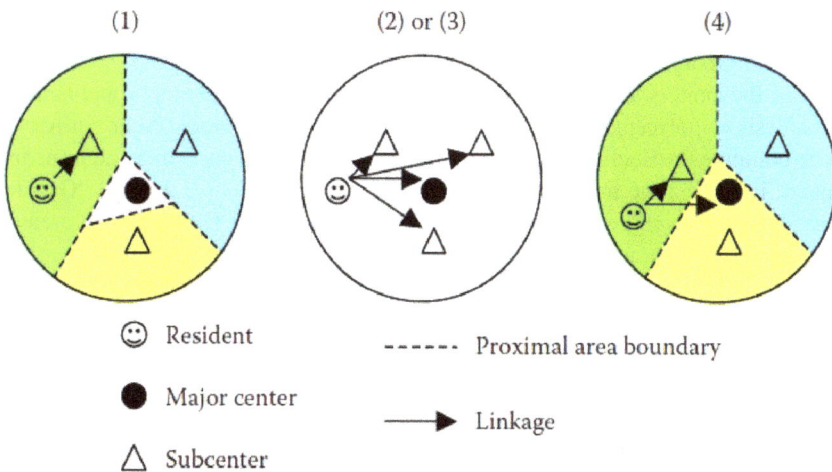

FIGURE 6.2 Illustration of polycentric assumptions.

TABLE 6.2

Polycentric Assumptions and Corresponding Functions

Label	Assumption	Model (Exponential as an Example)	X Variables	Sample	Estimation Method
1	Only access to the nearest center is needed.	$\ln D = A_i + b_i r_i$	Distances r_i from the nearest center i (1 variable)	Areas in subregion i	Linear regression (or nonlinear regression on $D = a_i e^{b_i r_i}$)
2	Access to all centers is necessary (multiplicative effects).	$\ln D = a + \sum_{i=1}^{n} b_i r_i$	Distances from each center (n variables r_i)	All areas	Linear regression
3	Access to all centers is necessary (additive effects).	$D = \sum_{i=1}^{n} a_i e^{b_i r_i}$	Distances from each center (n variables r_i)	All areas	Nonlinear regression
4	Access to CBD and the nearest center is needed.	$\ln D = a + b_1 r_1 + b_2 r_2$	Distances from the major and nearest center (2 variables)	All areas	Linear regression (or nonlinear regression on $D = a_1 e^{b_1 r_1} + a_2 e^{b_2 r_2}$)

total employment). The monocentric urban economic model explains urban population density functions while assuming a single center at the CBD where all jobs are located. Polycentric models extend the assumption to multiple job centers in a city. Therefore, centers should be based on the employment instead of population distribution pattern. For spatial data on employment in cities in the United States, visit the Census Transportation Planning Products (CTPP) program website ctpp. transportation.org.

Once the centers are identified, GIS prepares the data of distances and densities for analysis of polycentric models. For assumption 1, use the tool "Near" in ArcGIS Pro to compute the Euclidean or geodesic distances between each area and its nearest centers (including the major center). For assumption 2 or 3, use the tool "Generate Near Table" to compute the Euclidean or geodesic distances between each area and every center. For assumption 4, it requires two distances: the distance between each area and the major center and the distance between each area and its nearest center. The two distances are obtained by using the Near tool in ArcGIS Pro twice. See Section 6.5.2 for detail.

Based on assumption 1, the polycentric model is degraded to monocentric functions (equation 6.12) within each center's proximal area, which can be estimated by the techniques explained in Sections 6.2 and 6.3. Equation 6.13 for assumption 2 and equation 6.15 for assumption 4 can also be estimated by simple multivariate linear regressions. However, equation 6.14 based on assumption 3 needs to be estimated by a nonlinear regression.

6.5 CASE STUDY 6: ANALYZING URBAN DENSITY PATTERNS IN CHICAGO URBAN AREA

Chicago has been an important study site for urban studies. The classic urban concentric model by Burgess (1925) was based on Chicago and led to a series of studies on urban structure, forming the so-called *Chicago School*. This case study uses the 2000 census data to examine the urban density patterns in the Chicago urban area. The study area is limited to the core six-county area in Chicago CMSA (Consolidated Metropolitan Statistical Area) (county codes in parentheses): Cook (031), DuPage (043), Kane (089), Lake (097), McHenry (111), and Will (197) (see Figure 6.3). In order to examine the possible *modifiable areal unit problem (MAUP)*, the project analyzes the density patterns at both the census tract and survey township levels. The MAUP refers to sensitivity of results to the analysis units for which data are collected or measured and is well-known to geographers and spatial analysts (Openshaw, 1984; Fotheringham and Wong, 1991).

The following features in geodatabase `ChiUrArea.gdb` under the folder `Chicago` for the study area are provided:

1. Census tract feature `trt2k` for the Chicago ten-county region is used to extract census tracts in this study area (field "popu" is the population data in 2000).
2. Feature `polycent15` contains 15 centers identified as employment concentrations from a previous study (Wang, 2000).
3. Feature `twnshp` contains 115 survey townships in the study area, providing an alternative areal unit that is relatively uniform in area size.
4. Feature `cnty6` defines the study area.

6.5.1 FUNCTION FITTINGS FOR MONOCENTRIC MODELS AT THE CENSUS TRACT LEVEL

Step 1. Data Preparation in ArcGIS Pro: Extracting Study Area and CBD Location: Add all data to the map. Click `trt2k` > use a spatial query ("Selection by Location" in the Map tab) to select features from `trt2k` that have their center in `cnty6` > export the selected features to a new layer `cnty6trt`. It contains 1,837 census tracts in the six-county area.

Add a field `popden` with Data Type as Double to the attribute table of `cnty6trt`, and calculate it as `1000000*!POPU!/!Shape _ Area!`, which is population density in persons per square kilometer. We limit the analysis of population density patterns to tracts with nonzero population. Therefore, use an attribute query to select from `cnty6trt` with popden > 0, and export the result to a new feature `cnty6trt1`. This eliminates five tracts with $Dr = 0$.

Use the tool "Feature to Point" to generate a point feature `cnty6trtpt` from the polygon feature `cnty6trt1`. Refer to step 2 of Section 2.2.1 if needed.

Based on the feature `polycent15`, select the point with `cent15_` is equal 15 and export it to a new layer `monocent`, which identifies the location of CBD.

Step 2. Mapping Population Density Surface: Use the surface modeling techniques learned in Section 3.3.2 to map the population density surface of cnty-6trtpt in the study area. A sample map using the IDW method with cnty6 as Mask is shown in Figure 6.3. Note that job centers are not necessarily the peak points of population density. As we learned from Newling's model, commercial and other

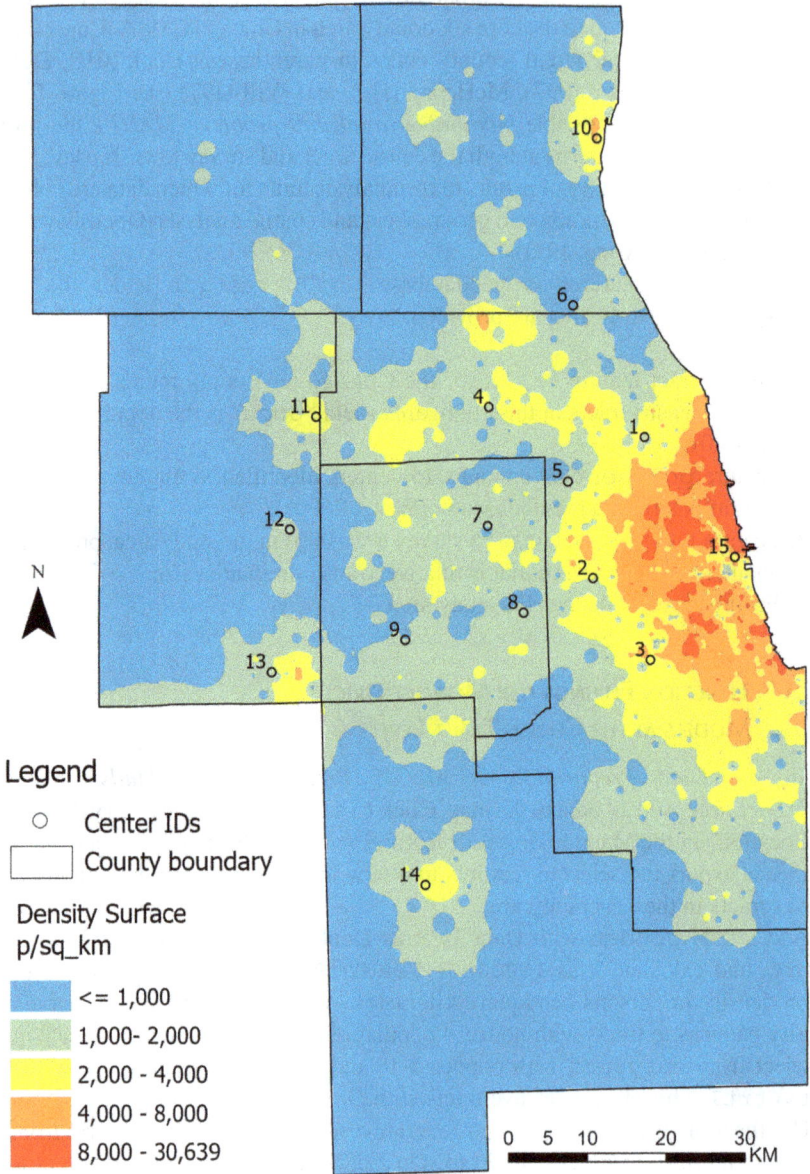

FIGURE 6.3 Population density surface and job centers in Chicago 2000.

business entities often dominate the land uses in an employment center with relatively lower population density. However, some suburban job centers attract residents around them and are near the local density peaks.

Step 3. Computing Distances between Tract Centroids and CBD: Search and activate the Near tool in the Geoprocessing pane to compute Euclidean distances between tract centroids (cnty6trtpt) and CBD (monocent). In the updated attribute table of cnty6trtpt, the field NEAR _ FID identifies the nearest center (in this case the only point in monocent), and the field NEAR _ DIST is the distance of each tract centroid from the center. Add a field DIST _ CBD with Data Type as Double and calculate it as !NEAR _ DIST!/1000, which is the distances from the CBD in kilometers.[2]

Step 4. Optional: Exploring Bivariate Linear Regression with Scatter Plot and Ordinary Least Squares (OLS): Click cnty6trtpt in the Map pane > Data tab > Create Chart > Scatter Plot. In the dialogue of Chart Properties, input DIST _ CBD as X-axis Number and popden as Y-axis Number. The chart displays the linear trend and R^2 value.

In the Geoprocessing pane, search and activate the tool "Ordinary Least Squares (OLS)." In the dialog, choose cnty6trtpt as Input Data, set popden as Dependent Variable and DIST _ CBD as Explanatory Variable, input file name cnty _ OLS for Output Feature Class or Output Features and ORIG _ FID as Unique ID Field, and keep other default settings. Click Run. The report can be accessed by clicking on View Details. The output includes estimated coefficients, corresponding standard errors, *t* statistics and *p*-values, and R^2. The output feature class also includes residuals.

Step 5. Implementing Bivariate Linear Regression and Weighted Regression with RegressionR: The Scatter Plot or the OLS tool in ArcGIS Pro implements an OLS regression. This step illustrates the customized R-ArcGIS tool "RegressionR" with an option to run a weighted regression. For more information on R-ArcGIS tools, refer to Appendix 2C of Chapter 2.

In the Catalog pane, activate the RegressionR tool under the folder CMGIS-V3-Toolbox > R _ ArcGIS _ tools, choose cnty6trtpt as Input Data, choose popden under Dependent Variable and DIST _ CBD for Independent Variables, keep Linear for Regression Model, choose Off for Weighted Model, set Yes for Export Predict Data, input OLS.csv for Regression Summary and OLS _ Data for Output Prediction Data, keep No for Filter, and click Run.

Click on View Details or open the table OLS.csv to check the regression result. The other table OLS _ Data contains predicted values by the regression model.

For a weighted OLS regression, as discussed in Section 6.3, repeat the preceding process of the RegressionR tool but choose On for Weighted Model and then AREA under Weight Field, and use different file names for Regression Summary and Output Prediction Data. The result is reported in Table 6.3.

Step 6. Additional Monocentric Function Fittings with RegressionR: In the Map pane, choose cnty6trtpt, then add and compute new fields dist _ sq, lndist, and lnpopden, with Data Type all as Double, representing distance squared, logarithm of distance, and logarithm of population density, respectively. As noted

TABLE 6.3

Regressions Based on Monocentric Functions

Regression techniques and functions	Census Tracts (n = 1,832)			Townships (n = 115)		
	a (or A)	b	Adjusted R^2	a (or A)	b	Adjusted R^2
Linear						
$D_r = a + br$	7225.62	−120.92	0.328	3535.46	−48.17	0.484
$D_r = a + b\ln r$	12167.03	−1383.68	0.353	9330.86	−2175.69	0.684
$\ln D_r = A + b\ln r$	8.3245	−0.0006	0.390	14.0869	−2.1571	0.601
$\ln D_r = A + br$	**8.8448**	**−0.0428**	**0.437**	**9.0288**	**−0.0607**	**0.692**
$\ln D_r = A + br^2$	10.1983	−0.4222	0.349	7.5831	−0.0005	0.625
$\ln D_r =$ $A + b_1 r + b_2 r^2$	8.9070	$b_1 = -0.0487$; $b_2 = 0.0004$	0.438	9.4048	$b_1 = -0.0784$; $b_2 = 0.0002$	0.693
Nonlinear						
$D_r = ar^b$	12247.45	−0.3823	0.262	16558.26	−0.7281	0.508
$D_r = ae^{br}$	10110.03	−0.0474	0.397	9433.93	−0.0580	0.772
$D_r = ae^{br^2}$	**8233.2972**	**−0.0019**	**0.419**	**7042.93**	**−0.0019**	**0.799**
$D_r = ae^{b_1 r + b_2 r^2}$	6407.33	$b_1 = 0.057$; $b_2 = -0.0044$	0.434	8455.80	$b_1 = -0.0384$; $b_2 = -0.0005$	0.780
Weighted						
$\ln D_r = A + br$	8.5851	−0.0561	0.556	8.9696	− 0.0599	0.674

Note: Results in bold indicate the best-fitting models (Newling's model has one more explanatory variable and thus is not comparable to others); adjusted $R^2 = 1$- RSS/TSS based on reported RSS and TSS.

previously, for density functions with logarithmic terms (`lnr` or `lnDr`), the distance (`r`) or density (`Dr`) value cannot be 0.

The corresponding inputs in Calculate field are:

```
dist_sq = math.pow(!DIST_CBD!,2)
lndist= math.log(!DIST_CBD!)
lnpopden = math.log(!popden!)
```

Repeat step 5 to fit the logarithmic, power, exponential, Tanner–Sherratt, and Newling's functions, as outlined in Table 6.1. All regressions, except for Newling's model, are simple bivariate models. Regression results are summarized in Table 6.3. For those bivariate models sharing the same dependent variable, we can run the regression models at once by the tool RegressionR, as shown in step 3 of Subsection 2.4.1 in Chapter 2.

Step 7. Implementing Nonlinear Regressions with RegressionR: Reactivate the RegressionR tool. In the dialog as shown in Figure 6.4, choose `cnty6trtpt` as Input Data; choose `popden` under Dependent Variable, `DIST _ CBD` for Independent Variables, `Nonlinear` for Regression Model, `a*exp(b*x1)` for

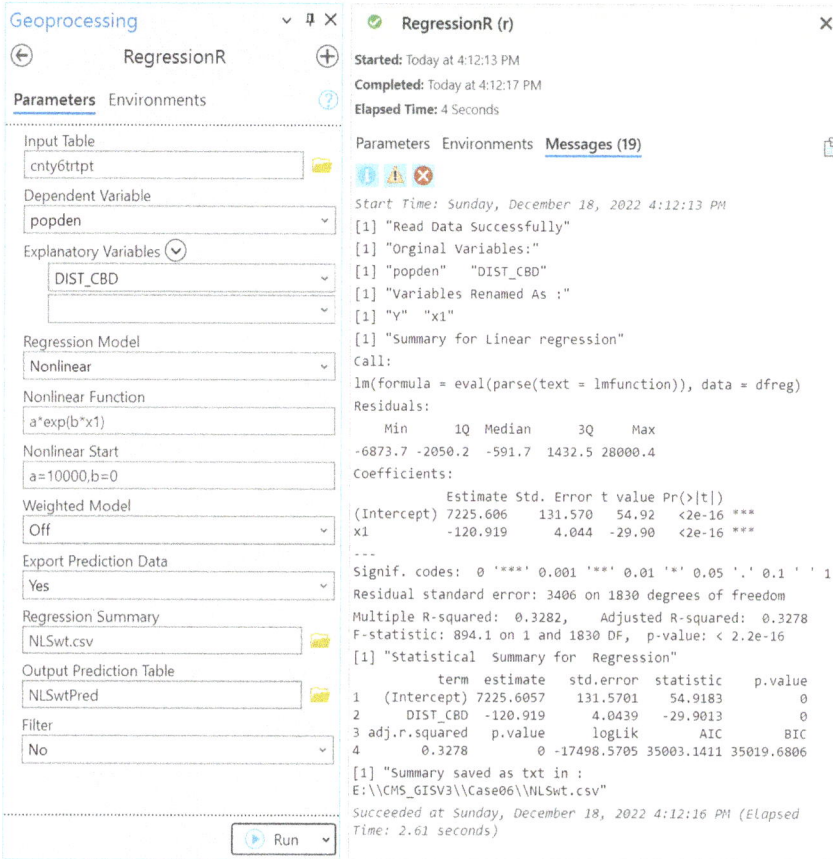

FIGURE 6.4 Dialog window for running a nonlinear regression in RegressionR.

Nonlinear Function, "a=10000, b=0" for Nonlinear Start; input NLSwt.csv for Regression Summary; set Export Prediction Data as Yes; input NLSwtPred as Output Prediction Table; keep No for Filter; and click Run. This fits the exponential function by nonlinear regression.

Note that for the nonlinear function to fit, x1, x2, . . . , xn denote Independent Variables (here only one variable x1 for DIST _ CBD). Under Nonlinear Start, define starting values for any parameters in the Nonlinear Function to be estimated (here a and b).

Repeat the same tool to fit the power, Tanner–Sherratt and Newling's functions, by nonlinear regression, defined as:

1. Power function: a*x1^b, Independent Variable DIST _ CBD for x1
2. Tanner–Sherratt model: a*exp(b*x1), Independent Variable dist_sq for x1
3. Newling's model: a*exp(b1*x1+b2*x1^2), Independent Variable DIST _ CBD for x1

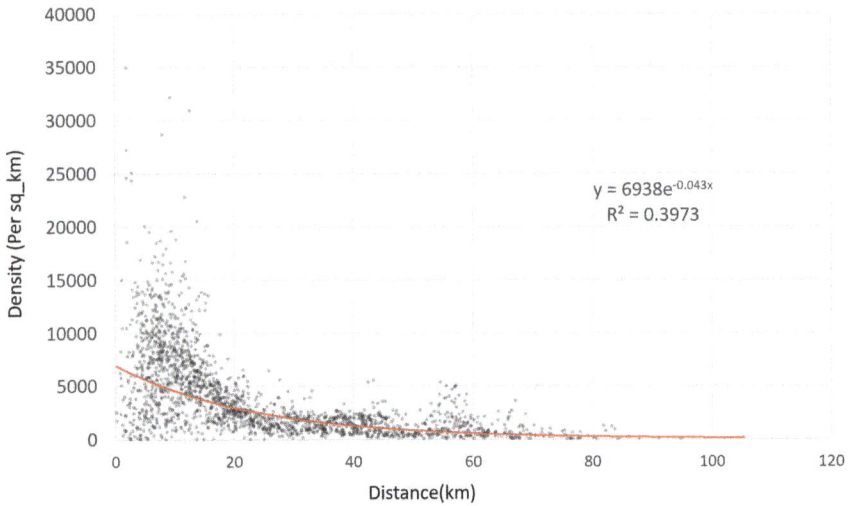

FIGURE 6.5 Density vs. distance with the exponential trend line across census tracts in Chicago.

For the power function and Tanner–Sherratt model, one may set Nonlinear Start as "a=10000, b=0.0." For Newling's model, set Nonlinear Start as "a=10000, b1 = 0, b2 = 0."

All regression results are summarized in Table 6.3. Among comparable functions, the ones with the highest R^2 are highlighted. Both the linear and nonlinear regressions indicate that the exponential function has the best fit among the bivariate models. Newling's model has two explanatory terms and thus is not comparable. Figure 6.5 shows the exponential trend line superimposed on the X-Y scatter graph of density versus distance.

6.5.2 Function Fittings for Polycentric Models at the Census Tract Level

Step 8. Computing Distances between Tract Centroids and Their Nearest Centers: Use the Near tool to compute distances between tract centroids (cnty6trtpt) and their nearest centers (polycent15). In the updated attribute table for cnty6trtpt, the fields NEAR _ FID and NEAR _ DIST identify the nearest center from each tract centroid and the distance between them, respectively. The Near tool is already executed once in step 3, and the fields NEAR _ FID and NEAR _ DIST are updated here. Add a field D _ NEARC with Data Type as Double and calculate it as !NEAR _ DIST! / 1000. This distance field will be used to test the polycentric assumptions 1 and 4.

Step 9. Exploring Polycentric Assumption 1 with RegressionR: Activate the RegressionR tool. In the dialog, choose cnty6trtpt for Input Data, lnpopden for Dependent Variable, and D _ NEARC for Independent Variables; input PolyAssump1 _ 1.csv under Regression Summary; choose Yes Under Filter; then input NEAR _ FID==1 under Filter Condition. Choose linear for Regression Model, keep other default settings, and click Run. Click View Details or open PolyAssump1 _ 1.csv to view the regression result based on the center ID = 1, which captures the distance decay pattern of population density in the proximal area around and anchored by center 1.

One may repeat the process to explore the density functions in proximal areas around other centers by changing the Filter Condition as NEAR_FID==2 or 3, . . . , 15. This implements the regression models under assumption 1, with results shown in Table 6.4.

Step 10. Exploring Polycentric Assumption 4 with RegressionR: Activate the RegressionR tool again. In the dialog, choose cnty6trtpt for Input Data, lnpopden for Dependent Variable, and DIST _ CBD and D _ NEARC for Independent Variables; input PolyAssump4.csv under Regression Summary; keep other default setting; and click Run.

TABLE 6.4

Regressions Based on Polycentric Assumptions 1 and 2 (Census Tracts)

Center Index i	Assumption 1: $lnD = A_i + b_i r_i$ for Center i's Proximal Area				Assumption 2: $lnD = a + \sum_{i=1}^{n} b_i r_i$ for the Whole Study Area	
	No. Tracts	A_i	b_i	R^2	b_i	
1	185	7.2999***	0.1577***	0.1774	−0.1079***	$a = 11.23$***
2	105	7.4100***	0.1511***	0.2594	0.0689**	*Adjusted*
3	398	8.4746***	−0.0514***	0.1603	−0.0181	$R^2 = 0.484$
4	76	7.6874***	−0.0513**	0.1020	−0.066**	$n = 1,832$
5	50	5.444***	0.3810***	0.2565	0.1439***	
6	68	7.0739***	−0.0339	0.0304	0.1130***	
7	51	7.2447***	0.0000	0.0000	−0.0555*	
8	47	7.5770***	−0.0700*	0.1075	0.1141***	
9	60	7.1746***	−0.0175	0.0000	−0.0344	
10	103	7.4754***	−0.0675***	0.3419	−0.0558***	
11	87	7.3185***	−0.0588***	0.2830	−0.0395*	
12	22	6.6123***	−0.0286	0.0000	0.0817***	
13	27	8.1252***	−0.2557***	0.6223	−0.0492**	
14	65	7.1786***	0.0814***	0.4477	−0.0047	
15	488	8.3384***	0.0347*	0.0074	−0.0291***	

Note:

*** Significant at 0.001.

** Significant at 0.01.

* Significant at 0.05.

The regression model implements the polycentric assumption 4, and the result shows how the population density in a tract is jointly influenced by its distance from the CBD and the distance from its nearest subcenter:

```
lnD = 8.914 - 0.0406 DIST_CBD -0.0137D_NEARC with R² = 0.4414
      (218.08)  (-31.81)        (-3.90)
```

The corresponding t-values in parentheses imply that both distances from the CBD and the nearest center are statistically significant, but the distance from the CBD is far more important.

Step 11. Computing Distances between Tract Centroids and All Centers: Use the analysis tool Generate Near Table to compute distances between tract centroids (cnty6trtpt) and all 15 centers (polycent15); remember to uncheck Location, Angle, and Find only closet feature, and name the output table Trt2PolyD, which has $1832 \times 15 = 27,480$ records. Update the value NEAR _ DIST with !NEAR _ DIST!/1000 to set its units as kilometers.

As the distance to each center is an explanatory variable in the polycentric assumptions 2 and 3, the long table needs to be pivoted to a wide table. The Pivot Table tool can be applied to flatten the OD list to an OD Matrix. Search and activate Pivot Table in the Geoprocessing pane. In the dialog, input Trt2PolyD as Input Table; select IN_FID under Input Fields, NEAR_FID for Pivot Field, NEAR_DIST for Value Field, Trt2PolyD_Pivot for Output Table; and click Run. In the resulting table Trt2PolyD_Pivot, fields ranging from NEAR_FID1 to NEAR_FID15 represent the distances between census tract and job centers 1 to 15, respectively. See step 10 in Subsection 11.5.2 of Chapter 11 for another example using the Pivot Table tool.

Join the attribute table of cnty6trtpt to Trt2PolyD_Pivot and export it to a new table PolyDist, which will be used to test the polycentric assumptions 2 and 3.

Step 12. Exploring Polycentric Assumption 2 with RegressionR: Activate the RegressionR tool. In the dialog, input PolyDist as Input Data and lnpopden as Dependent Variable, add all fields ranging from NEAR _ FID1 to NEAR _ FID15 under Independent Variables, select Linear under Regression Model, input PolyAssump2.csv as Regression Summary, keep other default settings, and click Run. This implements the regression model under polycentric assumption 2, with the results reported in Table 6.4. With exceptions of five centers (2, 5, 6, 8, and 12), most of the centers exert influence on region-wide population density patterns as expected, with negative coefficients for corresponding distances, and such influences from centers 1, 4, 7, 10, 11, 13, and 15 are significant.

Step 13. Exploring Polycentric Assumption 3 with RegressionR: The function based on assumption 3 is a complex nonlinear function with 30 parameters to estimate (two for each center). After numerous trials, the model does not converge, and no regression results may be obtained. For illustration, this step experiments with implementing a model of two largest centers (center 15 at the CBD and center 5 at the O'Hare airport).

Activate the RegressionR tool. In the dialog, input PolyDist as Input Data and popden as Dependent Variable, add fields NEAR _ FID15 and NEAR _ FID5 under Independent Variables, select NonLinear under Regression Model, and input Nonlinear Function as:

```
a1*exp(b1 * x1)+a2*exp(b2 * x2)
```

Input Nonlinear Start Parameter as:

```
a1 = 1, b1 = 0, a2 = 1, b2 = 0
```

Choose No for Export Prediction Data, input PolyAssmp3.csv for Regression Summary, select No for Filter, and click Run.

The nonlinear regression result can be written as:

$$D = 6.7575e^{-0.0078\,\text{NEAR\_FID15}} + 2.2905e^{-0.0024\text{NEAR\_FID5}} \text{ with pseudo-}R^2 = 0.438.$$
(3.21) (-2.50)   (1.06) (-1.09)

The corresponding t-values in parentheses indicate that the CBD is more significant than the O'Hare airport center.

6.5.3 FUNCTION FITTINGS FOR MONOCENTRIC MODELS AT THE TOWNSHIP LEVEL

Step 14. Using Area-Based Interpolation to Estimate Population in Townships: This step uses the simple areal weighting interpolator in Section 3.5.1 of Chapter 3 to transform population at the census tract level to population at the township level.

Search and activate the analysis tool Intersect in the Geoprocessing pane to overlay cnty6trt and twnshp, name the output layer twntrt, and click Run. Add a field EstPopu to the attribute table of twntrt and calculate it as !POPU!/!AREA!*!Shape _ Area! Here, population in a census tract is distributed proportionally to the area size of split area unit in twntrt. On the attribute table of twntrt, summarize EstPopu by the field RNGTWN (township IDs), and name the output table twn _ pop. The field Sum _ EstPopu in the table twn _ pop is the estimated population in townships.

Join the table twn _ pop to the attribute table of twnshp based on their common field RNGTWN. Add a field popden to the joined table, and calculate it as 1000000*!twn _ pop.SUM _ EstPopu!/!twnshp.Shape _ Area!

Step 15. Computing Distances between Townships and CBD: Similar to step 1, use the tool Feature to Point to generate a point feature twnshppt from the polygon feature twnshp, and click Run. Use the tool Near to obtain the distances between townships (twnshppt) and the CBD (monocent). The attribute table of twnshppt now contains the variables needed for function fittings: population density (popden), distance (NEAR _ DIST).

Step 16. Fitting Monocentric Functions with RegressionR: Repeat steps 5–7 on twnshppt to implement the linear and nonlinear regressions. Results are also reported in Table 6.3. Given the small sample size, polycentric functions are not tested.

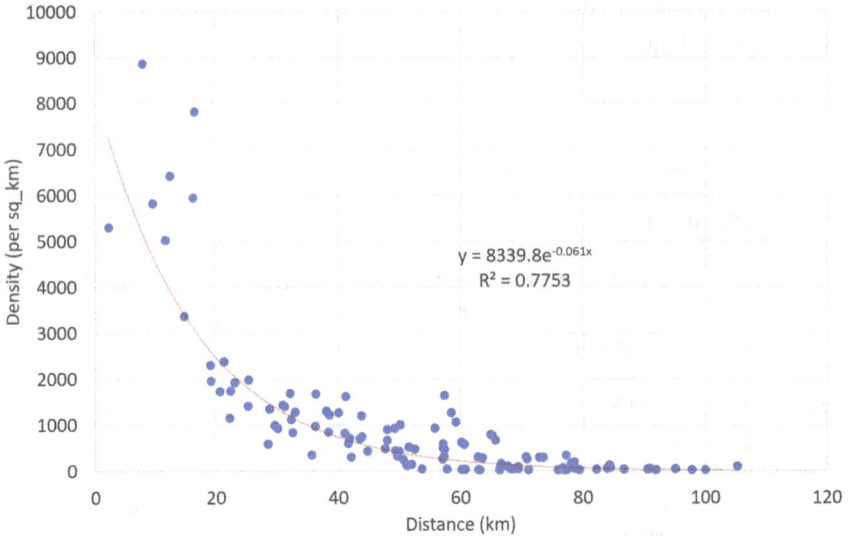

$$y = 8339.8e^{-0.061x}$$
$$R^2 = 0.7753$$

FIGURE 6.6 Density vs. distance with the exponential trend line across survey townships in Chicago.

Figure 6.6 shows the fitted exponential function curve based on the interpolated population data for survey townships. Survey townships are much larger than census tracts and have far fewer observations. It is not surprising that the function is a better fit at the township level (as shown in Figure 6.6) than at the tract level (as shown in Figure 6.5).

6.6 DISCUSSION AND SUMMARY

Based on Table 6.3, among the five bivariate monocentric functions (linear, logarithmic, power, exponential, and Tanner–Sherratt), the exponential function has the best fitting overall. It generates the highest R^2 by linear regressions using both census tract data and survey township data. Only by nonlinear regressions does the Tanner–Sherratt model have R^2 slightly higher than the exponential model. Newling's model has two terms (distance and distance squared), and thus its R^2 is not comparable to the bivariate functions. In fact, Newling's model is not a very good fit here since the term "distance squared" is not statistically significant by linear regression on either census tract data or survey township data. Data aggregated at the survey township level smooth out variations in both densities and distances. As expected, R^2 is higher for each function obtained from the survey township data than that from the census tract data.

We here use the exponential function as an example to compare the regression results by linear and nonlinear regressions. The nonlinear regression generates a higher intercept than the linear regression ($10110.03 > e^{8.8448} = 6938.22$ for census tracts and $9433.93 > e^{9.0288} = 8339.8$ for survey townships). Recall the discussion in Section 6.3 that nonlinear regression tends to value high-density areas more than low-density areas in minimizing residual sum squared (RSS). In other words, the logarithmic transformation in linear regression reduces the error contributions by high-density areas, and thus its fitting intercept tends to swing lower. The trend for the slope is unclear: steeper by nonlinear than by linear regression when using the census tract data, but flatter by nonlinear than by linear regression when using the survey township data.

In weighted regression, observations are weighted by their area sizes. The area size for survey townships is considered comparable except for some incomplete townships on the east border. Therefore, the results between weighted and unweighted regressions (using the exponential function as an example) are similar based on the survey township data but are more different based on the census tract data. In other words, when area sizes are comparable, the concern of unfair sampling and the need for running a weighted regression are alleviated. In Section 12.2 of Chapter 12, a uniform area unit is designed to address the issue.

Based on Table 6.4, the regression results based on the first polycentric assumption reveal that within most of the proximal areas, population densities decline with distances from their nearest centers. This is particularly true for suburban centers. But there are areas with a reversed pattern with densities increasing with distances, particularly in the central city (e.g., centers 1, 2, 5, and 15; refer to Figure 6.3 for their locations). This clearly indicates density craters around these downtown or near-downtown job centers because of significant non-residential land uses (commercial, industrial, or public) around these centers. The analysis of density function fittings based on polycentric assumption 1 enables us to examine the effect of centers on the local areas surrounding the centers. The regression result based on assumption 2 indicates distance decay effects for seven centers, with their coefficients b_i being negative and statistically significant. The regression based on assumption 3 is most difficult to implement, particularly when the number of centers is large. The regression result based on assumption 4 indicates that the CBD exerts the dominant effect on the density pattern, and the effect from the nearest subcenters is also significant.

Function fitting is commonly encountered in many quantitative analysis tasks to characterize how an activity or event varies with distance from a source of influence. Urban and regional studies suggest that in addition to population density, land use intensity (reflected in its price), land productivity, commodity prices, and wage rate may all experience some "distance decay effect" (Wang and Guldmann, 1997), and studies of their spatial patterns may benefit from the function fitting approach. Furthermore, various growth patterns (backwash vs. spread, centralization vs. decentralization) can be identified by examining the changes over time and analyzing whether the changes vary in directions (e.g., northward vs. southward, along certain transportation routes, etc.).

APPENDIX 6A: DERIVING URBAN DENSITY FUNCTIONS

This appendix discusses the theoretical foundations for urban density functions: an economic model following Mills (1972) and Muth (1969) (also see Fisch, 1991), and a gravity-based model based on Wang and Guldmann (1996).

Mills–Muth Economic Model

Each urban resident intends to maximize his/her utility, that is,

$$Max \ U(h,x)$$

by consuming h amount of land (i.e., housing lot size) and x amount of everything else.

The budget constraint y is given as:

$$y = p_h h + p_x x + tr,$$

where p_h and p_x are the prices of land and everything else, respectively, t is unit cost for transportation, and r is the distance to the city center, where all jobs are located.

The utility maximization yields a first-order condition given by:

$$\frac{dU}{dr} = 0 = \frac{dp_h}{dr} h + t. \tag{A6.1}$$

Assume that the price elasticity of the land demand is -1 (i.e., often referred to as the assumption of "negative unit elasticity for housing demand"), such as:

$$h = p_h^{-1}. \tag{A6.2}$$

Combining A6.1 and A6.2 yields the negative exponential rent gradient:

$$\frac{1}{p_h} \frac{dp_h}{dr} = -t. \tag{A6.3}$$

As population density $D(r)$ is given by the inverse of lot size h (i.e., $D(r) = 1/h$), we have $D(r) = 1/(p_h^{-1}) = p_h$. Substituting into A6.3 and solving the differential equation yield the negative exponential density function:

$$D(r) = D_0 e^{-tr}. \tag{A6.4}$$

Gravity-Based Model

Consider a city composed of n equal-area tracts. The population in tract j, x_j, is proportional to the potential there, such as:

$$kx_j = \sum_{i=1}^{n} \frac{x_i}{d_{ij}^{\beta}}, \tag{A6.5}$$

where d_{ij} is the distance between tract i and j, β is the distance friction coefficient, and n is the total number of tracts in the city. This proposition assumes that population at a particular location is determined by its location advantage or accessibility to all other locations in the city, measured as a gravity potential.

Equation A6.5 can also be written, in matrix notation, as:

$$kX = AX, \tag{A6.6}$$

where X is a column vector of n variables $(x_1, x_2, \ldots, x_n)$, A is an $n \times n$ matrix with terms involving the constants d_{ij}'s and β, and k is an unknown scalar. Normalizing one of the population terms, say, $x_1 = 1$, A6.6 becomes a system of n equations with n unknown variables that can be solved by numerical analysis.

Assuming a transportation network made of circular and radial roads (see Section 10.3 in Chapter 10) that define the distance term d_{ij} and given a β value, the population distribution pattern can be simulated in the city. The simulation indicates that the density pattern conforms to the negative exponential function when β is within a particular range (i.e., $0.2 \le \beta < 1.5$). When β is larger (i.e., $1.5 \le \beta \le 2.0$), the log-linear function becomes a better fit. Therefore, the model suggests that the best-fitting density function may vary over time and in different cities.

APPENDIX 6B: CENTRALITY MEASURES AND ASSOCIATION WITH URBAN DENSITIES

Among various centrality indices (Kuby et al., 2005), three are most widely used: *closeness* (C^C), *betweenness* (C^B), and *straightness* (C^S), measuring a location being *close* to all others, being the *intermediary* between others, and being accessible via a *straight* route to all others. All are based on a street network that is composed of nodes and edges connecting them. The distance between nodes can be measured in a dual graph that does not differentiate edge lengths (i.e., topological distance, as discussed in Section 2.1) or by a *primal* approach that represents road segments (edges) with lengths (Jiang and Claramunt, 2004). The latter has been increasingly adopted in recent studies.

Closeness centrality C^C measures how close a node is to all the other nodes along the shortest paths of the network. C^C for a node i is defined as:

$$C_i^C = \frac{N-1}{\sum_{j=1; j \neq i}^{N} d_{ij}}, \tag{A6.7}$$

where N is the total number of nodes in the network, and d_{ij} is the shortest distance between nodes i and j. In other words, C^C is the inverse of average distance from this node to all other nodes.

Betweenness centrality C^B measures how often a node is traversed by the shortest paths connecting all pairs of nodes in the network. C^B is defined as:

$$C_i^B = \frac{1}{(N-1)(N-2)} \sum_{j=1; k=1; j \neq k \neq i}^{N} \frac{n_{jk}(i)}{n_{jk}}, \tag{A6.8}$$

where n_{jk} is the number of shortest paths between nodes j and k, and $n_{jk}(i)$ is the number of these shortest paths that contain node i. C^B captures a special property for a place: it does not act as an origin or a destination for trips but as a pass-through point.

Straightness centrality C^S measures how much the shortest paths from a node to all others deviate from the virtual straight lines (Euclidean distances) connecting them. C^S is defined as:

$$C_i^S = \frac{1}{N-1}\sum_{j=1; j\neq i}^{N} \frac{d_{ij}^{Eucl}}{d_{ij}}, \tag{A6.9}$$

where d_{ij}^{Eucl} is the Euclidean distance between nodes i and j. C^S measures the extent to which a place can be reached directly, like on a straight line, from all other places in a city.

The global centrality indices are calculated on the street network of the whole study area, and local centrality indices for any location are based on the street network within a radius (catchment area) around the location.

One may use the ArcGIS Network Analyst module to prepare a street network dataset (see step 5 in Subsection 2.2.2) and download and install the Urban Network Analysis tool by Sevtsuk et al. (2013) to implement the computation of the centrality indices. To examine the association between centrality and land use intensity (e.g., population density), one needs to convert the two features (e.g., centrality values at nodes of a street network vs. population density in census tracts) into one data frame. One approach is to convert both to a raster dataset by kernel density estimation, as discussed in step 9 in Subsection 3.3.2, and use the Band Collection Statistics tool in ArcGIS to examine the correlation between them. In a study reported in Wang et al. (2011), the centrality indices outperform the economic monocentric model in explaining the intraurban variation of population density significantly.

APPENDIX 6C: OLS REGRESSION FOR A LINEAR BIVARIATE MODEL

A linear bivariate regression model is written as:

$$y_i = a + bx_i + e_i,$$

where x is a predictor (independent variable) of y (dependent variable), e is the error term or residual, i indexes individual cases or observations, and a and b are parameter estimates (often referred to as "intercept" and "slope," respectively). The predicted value for y by the linear model is:

$$\hat{y}_i = a + bx_i.$$

Residuals e measure prediction error, that is:

$$e_i = y_i - \hat{y}_i = y_i - (a + bx_i).$$

When $e > 0$, the actual y_i is higher than predicted $\hat{y}_i$ (i.e., under-prediction). When $e < 0$, the actual y_i is lower than predicted $\hat{y}_i$ (i.e., over-prediction). A perfect prediction results in a zero residual ($e = 0$). Either under-prediction or over-prediction contributes to inaccuracy, and therefore, the total error or the *residual sum squared (RSS)* is:

$$RSS = \sum_i e_i^2 = \sum_i (y_i - a - bx_i)^2. \tag{A6.10}$$

Denoting the mean of Y_i as $\overline{Y}$, *total sum squared (TSS)*, defined as $TSS = \sum_i (Y_i - \overline{Y})^2$, represents the total variation of dependent variable Y_i, and *explained sum squares (ESS)*, defined as $ESS = \sum_i (\hat{Y}_i - \overline{Y})^2$, reflects the variation explained by the model. The difference between them is RSS (i.e., $RSS = TSS\text{-}ESS$). The *ordinary least square (OLS) regression* seeks the optimal values of a and b so that RSS is minimized. Here x_i and y_i are observed, and a and b are the only variables. The optimization conditions for minimizing (A6.10) are:

$$\partial RSS \,/\, \partial a = -2\sum_i (y_i - a - bx_i) = 0, \tag{A6.11}$$

$$\partial RSS \,/\, \partial b = 2\sum_i (y_i - a - bx_i)(-x_i) = 0. \tag{A6.12}$$

Assuming that n is the total number of observations, we have $\sum_i a = na$. Solve equation A6.11 for a:

$$a = \frac{1}{n}\sum_i y_i - \frac{b}{n}\sum_i x_i. \tag{A6.13}$$

Substituting A6.13 into A6.12 and solving for b yield:

$$b = \frac{n\sum_i x_i y_i - \sum_i x_i \sum_i y_i}{n\sum_i x_i^2 - \left(\sum_i x_i\right)^2}. \tag{A6.14}$$

Substituting A6.14 back to A6.13 yields the solution to a.
How good is the regression model? A popular measure for goodness of fit is:

$$R^2 = ESS \,/\, TSS = 1 - RSS \,/\, TSS$$

It captures the portion of Y's variation from its mean explained by the model.

NOTES

1 The contour map of employment density was based on the logarithm of density. The direct use of density values would lead to crowded contour lines in high-density areas.
2 Note that the value of DIST_CBD for the CBD tract is nonzero, as it measures the distance from the CBD tract centroid (a tract with residents) from the CBD location monocent (based on the employment pattern). This ensures the validity of ln(DIST_CBD).

7 Principal Components, Factor Analysis, and Cluster Analysis and Application in Social Area Analysis

This chapter discusses three important multivariate statistical analysis methods: principal components analysis (PCA), factor analysis (FA), and cluster analysis (CA). PCA and FA are often used together for data reduction by structuring many variables into a limited number of components (factors). The techniques are particularly useful for eliminating variable collinearity and uncovering latent variables. Applications of the methods are widely seen in socioeconomic studies (e.g., Subsection 8.8.1 in Chapter 8). While the PCA and FA consolidate variables, the CA classifies observations into categories according to similarity among their attributes. In other words, given a dataset as a table, the PCA and FA reduce the number of columns and the CA reduces the number of rows.

By reducing the data dimension, benefits of PCA and FA include uncovering latent variables for easy interpretation and removing multicollinearity for subsequent regression analysis. In many socioeconomic applications, variables extracted from census data are often correlated with each other and thus contain duplicated information to some extent. PCA and FA use fewer components or factors to represent the original variables and thus simplify the structure for analysis. Resulting component or factor scores are uncorrelated to each other (if not rotated or orthogonally rotated) and thus can be used as independent explanatory variables in regression analysis. Despite the commonalities, PCA and FA are "both conceptually and mathematically very different" (Bailey and Gatrell, 1995, p. 225). PCA uses the same number of components to simply transform the original data and thus is strictly a mathematical transformation. FA uses fewer factors to capture most of variation among the original variables with error terms and thus is a statistical analysis process. PCA attempts to explain the variance of observed variables, whereas FA intends to explain their intercorrelations (Hamilton, 1992, p. 252).

Social area analysis is used to illustrate the techniques as it employs all three methods. The interpretation of social area analysis results also leads us to a review and comparison of three classic models on urban structure, namely, the concentric zone model, the sector model, and the multi-nuclei model. The analysis demonstrates how analytical statistical methods synthesize descriptive models into one

DOI: 10.1201/9781003292302-9

framework. Beijing, the capital city of China, in the midst of forming its social areas after decades under a socialist regime, is chosen as the study area for a case study.

Sections 7.1–7.3 discuss principal components analysis, factor analysis, and cluster analysis, respectively. Section 7.4 introduces social area analysis. A case study on the social space in Beijing is presented in Section 7.5. The chapter is concluded with discussion and a brief summary in Section 7.6.

7.1 PRINCIPAL COMPONENTS ANALYSIS (PCA)

For convenience, the original data of observed variables X_k are first standardized so that variables do not depend on measurement scale and thus are comparable to each other. Denote the mean and the standard deviation for a series of data X_k as $\bar{X}$ and σ, respectively. *Data standardization* involves the process of converting the data series X_k to a new series Z_k, such as $Z_k = \left(X_k - \bar{X} \right) / \sigma$. By doing so, the resulting data series Z_k has the mean equal to 0 and the standard deviation equal to 1.

Principal components analysis (PCA) transforms data of K observed variables Z_k (likely correlated with each other to some degree) to data of K principal components F_k that are independent from each other:

$$Z_k = l_{k1}F_1 + l_{k2}F_2 + \ldots + l_{kj}F_j + \ldots + l_{kK}F_K. \qquad (7.1)$$

Reversely, the components F_j can also be expressed as a linear combination of the original variables Z_k:

$$F_j = a_{1j}Z_1 + a_{2j}Z_2 + \ldots + a_{kj}Z_j + \ldots + a_{Kj}Z_K. \qquad (7.2)$$

The components F_j are constructed to be uncorrelated with each other and are ordered such that the first component F_1 has the largest sample variance (λ_1), F_2 the second largest, and so on. The variances λ_j corresponding to various components are termed *eigenvalues*, and $\lambda_1 > \lambda_2 > \ldots$. A larger eigenvalue represents a larger share of the total variance in all the original variables captured by a component and thus indicates a component of more importance. That is to say, the first component is more important than the second, and the second is more important than the third, and so on.

For readers familiar with the terminology of linear algebra, the eigenvalues λ_j $(j = 1, 2, \ldots, K)$ are based on the correlation matrix **R** (or, less commonly, the covariance matrix **C**) of the original variables **Z**. The correlation matrix **R** is written as:

$$R = \begin{pmatrix} 1 & r_{12} & \cdots & r_{1K} \\ r_{21} & 1 & \cdots & r_{2K} \\ . & . & \cdots & . \\ r_{K1} & r_{K2} & \cdots & 1 \end{pmatrix},$$

where r_{12} is the correlation between Z_1 and Z_2, and so on. The coefficients $(a_{1j}, a_{2j}, \ldots, a_{Kj})$ for component F_j in equation 7.2 comprise the eigenvector associated with the jth largest eigenvalue λ_j.

Since standardized variables have variances of 1, the total variance of all variables also equals to the number of variables, such as:

$$\lambda_1 + \lambda_2 + \ldots + \lambda_k = K.$$

Therefore, the proportion of total variance explained by the jth component is λ_j/K. In PCA, the number of components equals the number of original variables, so no information is lost in the process. The newly constructed components are a linear function of the original variables and have two important and useful properties: (a) they are independent of each other, and (b) their corresponding eigenvalues reflect their relative importance. The first property ensures that multicollinearity is avoided if resulting components are used in regression, and the second property enables us to use the first fewer components to capture most of the information in a multivariate dataset.

By discarding the less important components and keeping only the first few components, we achieve data reduction. If the J largest components ($J < K$) are retained, equation 7.1 becomes:

$$Z_k = l_{k1}F_1 + l_{k2}F_2 + \ldots + l_{kJ}F_J + v_k, \tag{7.3}$$

where the discarded components are represented by the residual term v_k, such as:

$$v_k = l_{k,J+1}F_{J+1} + l_{k,J+2}F_{J+2} + \ldots + l_{kK}F_K. \tag{7.4}$$

Equations 7.3–7.4 are sometimes termed a *principal components factor* model. It is not a true *factor analysis* (FA) model.

7.2 FACTOR ANALYSIS (FA)

Factor analysis (FA) is based on a different assumption than PCA. FA assumes that each variable has common and unique variance, and the method seeks to create factors that account for the common, not the unique, variance. FA uses a linear function of J unobserved factors F_j, plus a residual term u_k, to model each of K observed variables Z_k:

$$Z_k = l_{k1}F_1 + l_{k2}F_2 + \ldots + l_{kJ}F_J + u_k. \tag{7.5}$$

The same J factors, F_j, appear in the function for each Z_k and thus are termed *common factors*. Equation 7.5 looks like the principal components factor model in equation 7.3, but the residuals u_k in equation 7.5 are unique to variables Z_k, termed *unique factors*, and uncorrelated. In other words, after all the common factors are controlled for, correlations between the unique factors become zero. On the other side, residuals v_k, defined as in equation 7.4 for PCA, are linear functions of the same discarded components and thus correlated.

In order to understand this important property of FA, it is helpful to introduce the classical example used by psychologist Charles Spearman, who invented the method about 100 years ago. Spearman hypothesized that a variety of tests of mental ability, such as scores in mathematics, vocabulary, and other verbal skills, artistic skills, and logical reasoning ability, could all be explained by one underlying factor of general intelligence (g). If g could be measured and controlled for, there would be no correlations among any tests of mental ability. In other words, g was the only factor common to all the scores, and the remainder was the unique factor, indicating the special ability in that area.

When both Z_k and F_j are standardized, the l_{kj} in equation 7.5 are standardized coefficients in the regression of variables Z_k on common factors F_j, also termed *factor loadings*. In other words, the loadings are correlation coefficients between the standardized Z_k and F_j and thus range from -1 to $+1$. For example, l_{k1} is the loading of variables Z_k on standardized component F_1. Factor loading reflects the strength of relations between variables and factors.

Estimates of the factors are termed *factor scores*. For instance, the estimate for F_j ($j = 1, 2, \ldots, J$, and $J < K$) is:

$$\hat{F}_j = c_{1j}Z_1 + c_{2j}Z_2 + \ldots + c_{Kj}Z_K, \tag{7.6}$$

where c_{kj} is the *factor score coefficient* for the kth variable on the jth factor. Equation 7.6 is used to generate factor scores for corresponding observations.

The computing process of FA is more complex than that of PCA. A commonly used procedure for FA is *principal factor analysis*, which utilizes the results from PCA in iterations to obtain the final factor structure and factor scores. The PCA extracts the principal components based on the eigenvalues of correlation matrix **R**, and its computation is straightforward. The principal factor analysis extracts principal factors of a modified correlation matrix **R***. Whereas elements on the major diagonal of **R** are 1, these elements in **R*** are replaced by estimates of *communality* (a variable's variance shared by the common factors). For instance, the communality of variable Z_k in equation 7.5 equals the proportion of Z_k's variance explained by the common factors, that is, $\sum_{j=1}^{J} l_{kj}^2$. After an initial number of factors is decided, the communality and the modified correlation matrix **R*** are estimated through an iterative process until there is little change in the estimates. Based on the final modified correlation matrix **R*** and extracted principal components, the first principal component defines the first factor, the second component defines the second factor, and so forth. In addition to the principal factor analysis method, there are also several alternative methods for FA based on *maximum-likelihood estimation*.

How do we decide the number of factors to use? Eigenvalues provide a basis for judging which factors are important and which are not. In deciding the number of factors to include, one has to make a tradeoff between the total variance explained (higher by including more factors) and interpretability of factors (better with fewer factors). A rule of thumb is to include only factors with eigenvalues greater than 1. Since the variance of each standardized variable is 1, a factor with $\lambda < 1$ accounts for less variance than an original variable's variance and thus does not serve the purpose of data reduction.

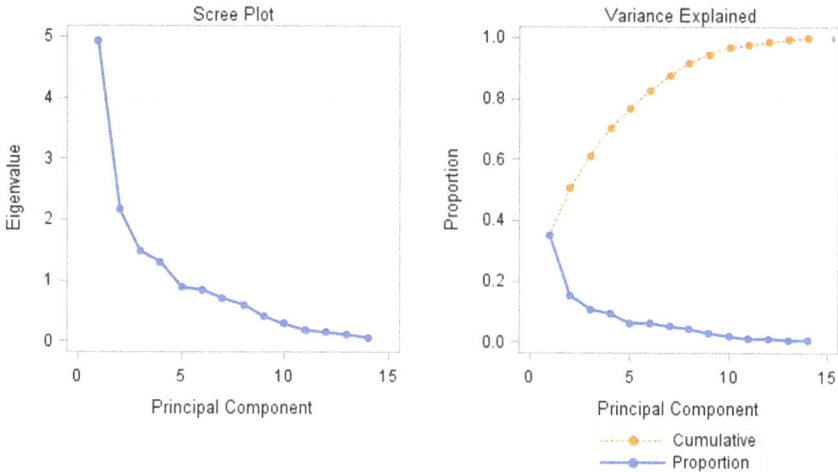

FIGURE 7.1 Scree plot and variance explained in principal components analysis (PCA).

Using one as a cutoff value to identify "important" factors is arbitrary. A *scree graph* plots eigenvalues against factor numbers and provides more useful guidance. For example, Figure 7.1 shows the scree graph of eigenvalues as reported in step 1 of Case Study 7 in Section 7.5. The graph levels off after component 4, indicating that components 5–14 account for relatively little additional variance. Therefore, four components (factors) are retained to account for a cumulative 70% variance. Other techniques, such as statistical significance testing based on the maximum-likelihood FA, may also be used to choose the number of factors. However, most importantly, selecting the number of factors must be guided by understanding the true forces underlying the research issue and supported by theoretical justifications. There is always a tradeoff between the total variance explained (higher by using more factors) and interpretability of factors (simpler with fewer factors).

Initial results from FA are often hard to interpret as variables load across factors. While fitting the data equally well, *rotation* generates simpler structure by attempting to find factors that have loadings close to ±1 or 0. This is done by maximizing the loading (positive or negative) of each variable on one factor and minimizing the loadings on the others. As a result, we can detect more clearly which factors mainly capture what variables (not others) and therefore label the factors adequately. The derived factors are considered latent variables representing underlying dimensions that are combinations of several observed variables.

Orthogonal rotation generates independent (uncorrelated) factors, an important property for many applications (e.g., avoidance of multicollinearity in regression). A widely used orthogonal rotation method is *varimax rotation*, which maximizes the variance of the squared loadings for each factor and thus polarizes loadings (either high or low) on factors. *Oblique rotation* (e.g., *promax rotation*) generates even greater polarization but allows correlation between factors.

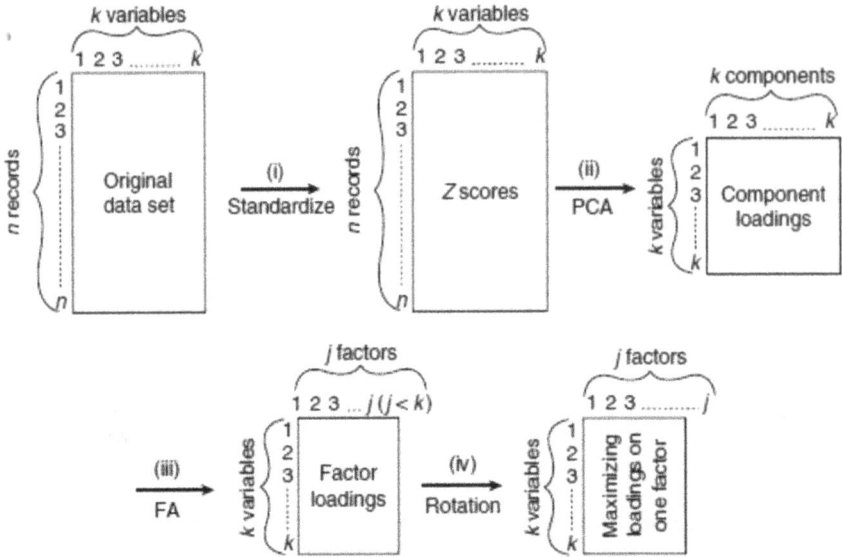

FIGURE 7.2 Major steps in principal components analysis (PCA) and factor analysis (FA).

As a summary, Figure 7.2 illustrates the major steps in PCA and FA:

1. The original dataset of K observed variables with n records is first standardized.
2. PCA then generates K uncorrelated components to account for all the variance of the K variables and suggests the first J components accounting for most of the variance.
3. FA uses only J ($J < K$) factors to model interrelations of the original K variables.
4. A rotation method can be used to maximize a variable's loading on one factor (closer to ±1) and minimize its loading on the others (closer to 0) for easier interpretation.

In essence, PCA explains all the variance in the original variables, whereas FA tries to explain their intercorrelations (covariances). Therefore, PCA is used to construct composite variables that reproduce the maximum variance of the original variables, and FA is used to reveal the relationships between the original variables and unobserved underlying dimensions. Despite the differences, PCA and FA share much common ground. Both are primarily used for data reduction, uncovering latent variables, and removing multicollinearity in regression analysis. Often, PCA results are reported as part of the intermediate results of FA.

Case Study 7 in Section 7.5 implements the PCA and FA in the R-ArcGIS tool "PCAandFA" from the R package *psych* (Revelle, 2022). There are several options for factor analysis method, such as `minres`, `pa`, `wls`, `gls`, and `ml`, representing minimum residual, principal axis, weighted least squares, generalized weighted least squares, and maximum likelihood, respectively. The default option `minres` attempts to minimize the off diagonal residual correlation matrix by adjusting the eigen values of the original correlation matrix.

7.3 CLUSTER ANALYSIS (CA)

Cluster analysis (CA) groups observations according to similarity among their attributes. As a result, the observations within a cluster are more similar than observations between clusters as measured by the clustering criterion. Note the difference between CA and another similar multivariate analysis technique—discriminant function analysis (DFA). Both group observations into categories based on the characteristic variables. The difference is: categories are unknown in CA but known in DFA. See Appendix 7 for further discussion on DFA. Geographers have a long-standing interest in cluster analysis (CA) and have used it in applications such as regionalization and city classification.

A key element in deciding assignment of observations to clusters is "*attributive distance*," in contrast to the various spatial distances discussed in Chapter 2. The most commonly used attributive distance measure is Euclidean distance:

$$d_{ij} = (\sum_{k=1}^{K} (x_{ik} - x_{jk})^2)^{1/2}, \tag{7.7}$$

where x_{ik} and x_{jk} are the kth variable of the K-dimensional observations for individuals i and j, respectively. When $K = 2$, Euclidean distance is simply the straight-line distance between observations i and j in a two-dimensional space. Other distance measures include Manhattan distance and others (e.g., Minkowski distance, Canberra distance) (Everitt et al., 2001, p. 40).

The most widely used clustering method is the *agglomerative hierarchical method* (AHM). The method produces a series of groupings: the first consists of single-member clusters, and the last consists of a single cluster of all members. The results of these algorithms can be summarized with a *dendrogram*, a tree diagram showing the history of sequential grouping process. See Figure 7.3 for the example illustrated in the following. In the diagram, the clusters are nested, and each cluster is a member of a larger and higher-level cluster.

For illustration, an example is used to explain a simple AHM, the *single linkage method* or the *nearest-neighbor method*. Consider a dataset of four observations with the following distance matrix:

$$D_1 = \begin{matrix} 1 \\ 2 \\ 3 \\ 4 \end{matrix} \begin{bmatrix} 0 & & & \\ 3 & 0 & & \\ 6 & 5 & 0 & \\ 9 & 7 & 4 & 0 \end{bmatrix}.$$

The smallest no-zero entry in the preceding matrix D_1 is $(2 \rightarrow 1) = 3$, and therefore individuals 1 and 2 are grouped together to form the first cluster C1. Distances

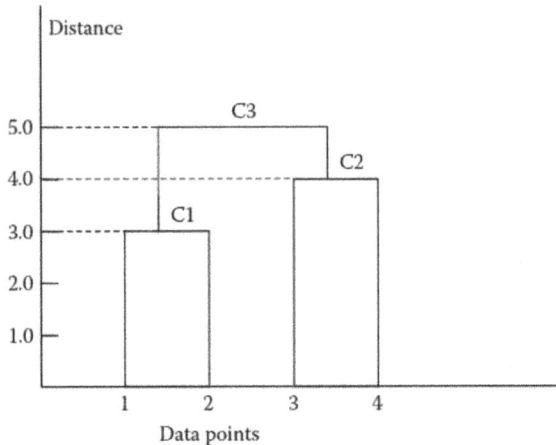

FIGURE 7.3 Dendrogram for a cluster analysis example.

between this cluster and the other two individuals are defined according to the nearest-neighbor criterion:

$$d_{(12)3} = \min\{d_{13}, d_{23}\} = d_{23} = 5$$

$$d_{(12)4} = \min\{d_{14}, d_{24}\} = d_{24} = 7.$$

A new matrix is now obtained, with cells representing distances between cluster C1 and individuals 3 and 4:

$$D_2 = \begin{matrix} (12) \\ 3 \\ 4 \end{matrix} \begin{bmatrix} 0 & & \\ 5 & 0 & \\ 7 & 4 & 0 \end{bmatrix}.$$

The smallest no-zero entry in D_2 is $(4{\to}3) = 4$, and thus, individuals 3 and 4 are grouped to form a cluster C2. Finally, clusters C1 and C2 are grouped together, with distance equal to 5, to form one cluster C3 containing all four members. The process is summarized in the dendrogram in Figure 7.3, where the height represents the distance at which each fusion is made.

Similarly, the *complete linkage (furthest-neighbor) method* uses the maximum distance between pair of objects (one object from one cluster and another object from the other cluster), the *average linkage method* uses the average distance between pair of objects, and the *centroid method* uses squared Euclidean distance between individuals and cluster means (centroids).

Another commonly used AHM is *Ward's method*. The objective at each stage is to minimize the increase in the total within-cluster error sum of squares, given by:

$$E = \sum_{c=1}^{C} E_c,$$

where:

$$E_c = \sum_i^{n_c} \sum_{k=1}^{K} (x_{ck,i} - \overline{x_{ck}})^2,$$

in which $x_{ck,i}$ is the value for the kth variable for the ith observation in the cth cluster, and $\overline{x_{ck}}$ is the mean of kth variable in the cth cluster.

Case Study 7 in Section 7.5 uses a popular method, k-means, available in ArcGIS Pro, to implement clustering analysis. It is a method of vector quantization that aims to partition n observations into k clusters in which each observation belongs to the cluster with the nearest mean (cluster centers or cluster centroid). The algorithm randomly selects seed locations for the Initialization Method parameter and incorporates heuristics to grow the clusters. Therefore, it may return a slightly different result each time the tool is run.

Each clustering method has its advantages and disadvantages. A desirable clustering should produce clusters of similar size, densely located, compact in shape, and internally homogeneous (Griffith and Amrhein, 1997, p. 217). The single linkage method tends to produce unbalanced and straggly clusters and should be avoided in most cases. If outlier is a major concern, the centroid method should be used. If compactness of clusters is a primary objective, the complete linkage method should be used.

The choice for the number of clusters depends on objectives of specific applications. The pseudo-F statistic is a ratio of the between-cluster variation to the within-cluster variation, and thus, a larger pseudo-F statistic is generally more desirable. An analyst often experiments with various numbers of clusters and selects the number of clusters that not necessarily corresponds to the maximum pseudo-F statistic, rather helps tell the best story.

7.4   SOCIAL AREA ANALYSIS

The social area analysis was developed by Shevky and Williams (1949) in a study of Los Angeles and was later elaborated on by Shevky and Bell (1955) in a study of San Francisco. The basic thesis is that the changing social differentiation of society leads to residential differentiation within cities. The studies classified census tracts into types of social areas based on three basic constructs: economic status (social rank), family status (urbanization), and segregation (ethnic status). Originally, the three constructs were measured by six variables: economic status was captured by occupation and education; family status by fertility, women labor participation, and single-family houses; and ethnic status by percentage of minorities (Cadwallader, 1996, p. 135). In factor analysis, an idealized factor loadings matrix probably looks like Table 7.1. Subsequent studies using a large number and variety of measures generally confirmed the validity of the three constructs (Berry, 1972, p. 285; Hartschorn, 1992, p. 235).

Geographers made an important advancement in social area analysis by analyzing the spatial patterns associated with these dimensions (e.g., Rees, 1970; Knox, 1987). The socioeconomic status factor tends to exhibit a sector pattern (Figure 7.4a): tracts

TABLE 7.1

Idealized Factor Loadings in Social Area Analysis

	Economic Status	Family Status	Ethnic Status
Occupation	I	O	O
Education	I	O	O
Fertility	O	I	O
Female labor participation	O	I	O
Single-family house	O	I	O
Minorities	O	O	I

Note: *I* denotes a number close to 1 or −1; *O* denotes a number close to 0.

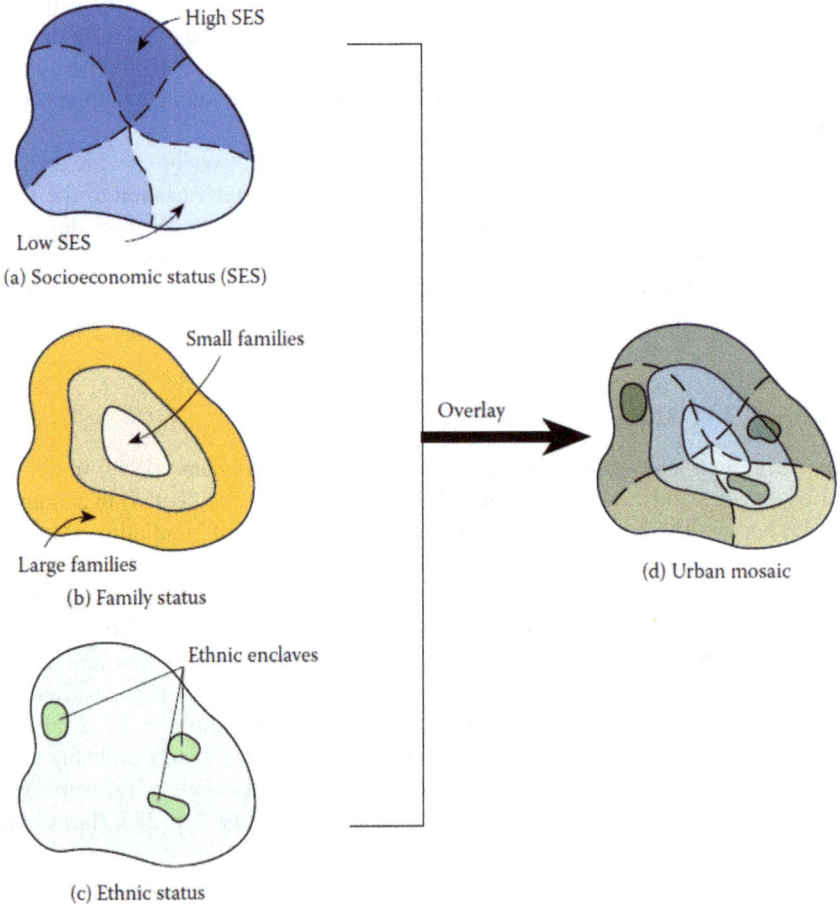

FIGURE 7.4 Conceptual model for urban mosaic.

with high values for variables such as income and education form one or more sectors, and low-status tracts form other sectors. The family status factor tends to form concentric zones (Figure 7.4b): inner zones are dominated by tracts with small families with either younger or older household heads, and tracts in outer zones are mostly occupied by large families with middle-aged household heads. The ethnic status factor tends to form clusters, each of which is dominated by a particular ethnic group (Figure 7.4c). Superimposing the three constructs generates a complex urban mosaic, which can be grouped into various social areas by cluster analysis (Figure 7.4d). By studying the spatial patterns from social area analysis, three classic models for urban structure, Burgess's (1925) concentric zone model, Hoyt's (1939) sector model, and Ullman–Harris (Harris and Ullman, 1945) multi-nuclei model, are synthesized into one framework. In other words, each of the three models reflects one specific dimension of urban structure and is complementary to each other.

There are at least three criticisms of the factorial ecological approach to understanding residential differentiation in cities (Cadwallader, 1996, p. 151). First, the results from social area analysis are sensitive to research design, such as variables selected and measured, analysis units, and factor analysis methods. Secondly, it is still a descriptive form of analysis and fails to explain the underlying process that causes the pattern. Thirdly, the social areas identified by the studies are merely homogeneous, but not necessarily functional regions or cohesive communities. Despite the criticisms, social area analysis helps us understand residential differentiation within cities and serves as an important instrument for studying intraurban social spatial structure. Applications of social area analysis can be seen on cities in developed countries, particularly rich on cities in North America (see a review by Davies and Herbert, 1993), and also on cities in developing countries (e.g., Berry and Rees, 1969; Abu-Lughod, 1969).

7.5 CASE STUDY 7: SOCIAL AREA ANALYSIS IN BEIJING

This case study is based on a study reported in Gu et al. (2005). Detailed research design and interpretation of the results can be found in the original paper. This section shows the procedures to implement the study, with emphasis on illustrating the three statistical methods discussed in Sections 7.1–7.3. In addition, the study illustrates how to test the spatial structure of factors by regression models with dummy explanatory variables.

The study area was the contiguous urbanized area of Beijing, including four inner-city districts (Xicheng, Dongcheng, Xuanwu, and Chongwen)[1] and four suburban districts (Haidian, Chaoyang, Shijingshan, and Fengtai), with a total of 107 subdistricts (*jiedao*) in 1998. Some subdistricts on the periphery of those four suburban districts were mostly rural and were thus excluded from the study (see Figure 7.5). The study area had a total population of 5.9 million, and the analysis unit "subdistrict" had an average population of 55,200 in 1998.

The following datasets are provided under the data folder `Beijing`:

1. Feature `subdist` in geodatabase `BJSA.gdb` contains 107 urban subdistricts.
2. Text file `bjattr.csv` is the attribute dataset.

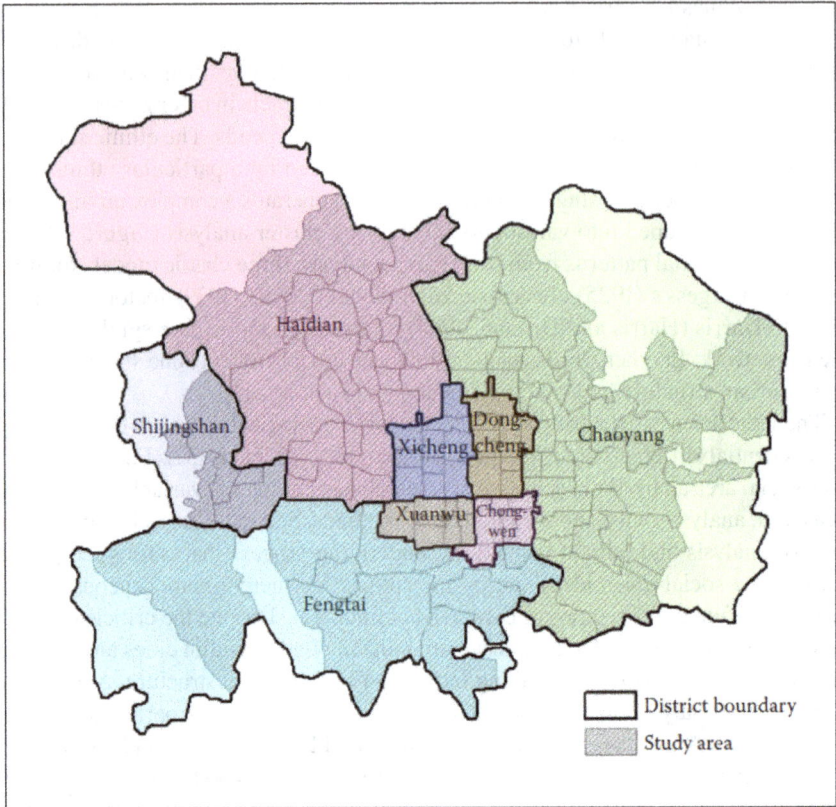

FIGURE 7.5 Districts and subdistricts in Beijing.

In the attribute table of subdist, the field sector identifies four sectors (= 1 for NE, 2 for SE, 3 for SW, and 4 for NW), and the field ring identifies four zones (= 1 for the most inner zone, 2 for the next, and so on). Sectors and zones are needed for testing spatial structures of social space. The text file bjattr.csv has 14 socio-economic variables (X1 to X14) for social area analysis (see Table 7.2). Both the attribute table of subdist and the text file bjattr.csv contain the common field ref _ id identifying the subdistricts.

The study uses the principal component analysis (PCA) to derive the factors for social area analysis. The factor analysis (FA) is also implemented for illustration but not adopted for subsequent social area analysis.

Step 1. Implementing Principal Components Analysis (PCA) with R-ArcGIS Tool: Add the table bjattr.csv to the map. Activate the PCAandFA tool under the folder R _ ArcGIS _ tools. In the dialog (Figure 7.6a), choose bjattr.csv as Input Table; choose ref _ id under Excluded Variables; input 4 for Factor Number; set Principal components analysis for Model, varimax for PCA Rotate,

TABLE 7.2

Basic Statistics for Socioeconomic Variables in Beijing (n = 107)

Index	Variables	Mean	Std. Dev.	Minimum	Maximum
X1	Population density (persons/km^2)	14,797.09	13,692.93	245.86	56,378.00
X2	Natural growth rate (‰)	−1.11	2.79	−16.41	8.58
X3	Sex ratio (M/F)	1.03	0.08	0.72	1.32
X4	Labor participation ratio (%)[a]	0.60	0.06	0.47	0.73
X5	Household size (persons/ household)	2.98	0.53	2.02	6.55
X6	Dependency ratio[b]	1.53	0.22	1.34	2.14
X7	Income (yuan/person)	29,446.49	127,223.03	7,505.00	984,566.00
X8	Public service density (no./ km^2)[c]	8.35	8.60	0.05	29.38
X9	Industry density (no./km^2)	1.66	1.81	0.00	10.71
X10	Office/retail density (no./km^2)	14.90	15.94	0.26	87.86
X11	Ethnic enclave (0, 1)[d]	0.10	0.31	0.00	1.00
X12	Floating population ratio (%)[e]	6.81	7.55	0.00	65.59
X13	Living space (m^2/person)	8.89	1.71	7.53	15.10
X14	Housing price (yuan/m^2)	6,686.54	3,361.22	1,400.00	18,000.00

[a] Labor participation ratio is percentage of persons in the labor force out of total population at the working ages, that is, 18–60 years old for males and 18–55 years old for females.

[b] Dependency ratio is the number of dependents divided by number of persons in the labor force.

[c] Public service density is the number of governmental agencies, non-profit organizations, educational units, hospitals, postal and telecommunication units per square kilometer.

[d] Ethnic enclave is a dummy variable identified whether a minority (mainly Muslims in Beijing) or migrant concentrated area was present in a subdistrict.

[e] Population includes permanent residents (those with a registration status in the hukou system) and floating population (those migrants without a permanent resident status). The ratio is the floating population ratio out of total population.

None for Plot, bjsaPCA4 for Output Table; keep TRUE for Sort Loading; and click Run.

The report on View Details (Figure 7.6b) contains a summary on standardized loadings, eigenvalues, loading scores and their proportions, cumulative values, etc. The first four components (here simply referred to as *factors*) with eigenvalues greater than 1 are retained and account for a cumulative 70.4% of total variance explained.

The first part in the report is the standardized loadings (pattern matrix) after the varimax rotation as shown in Table 7.3, where the variables are reordered to highlight the factor loading structure.

(a)

(b)

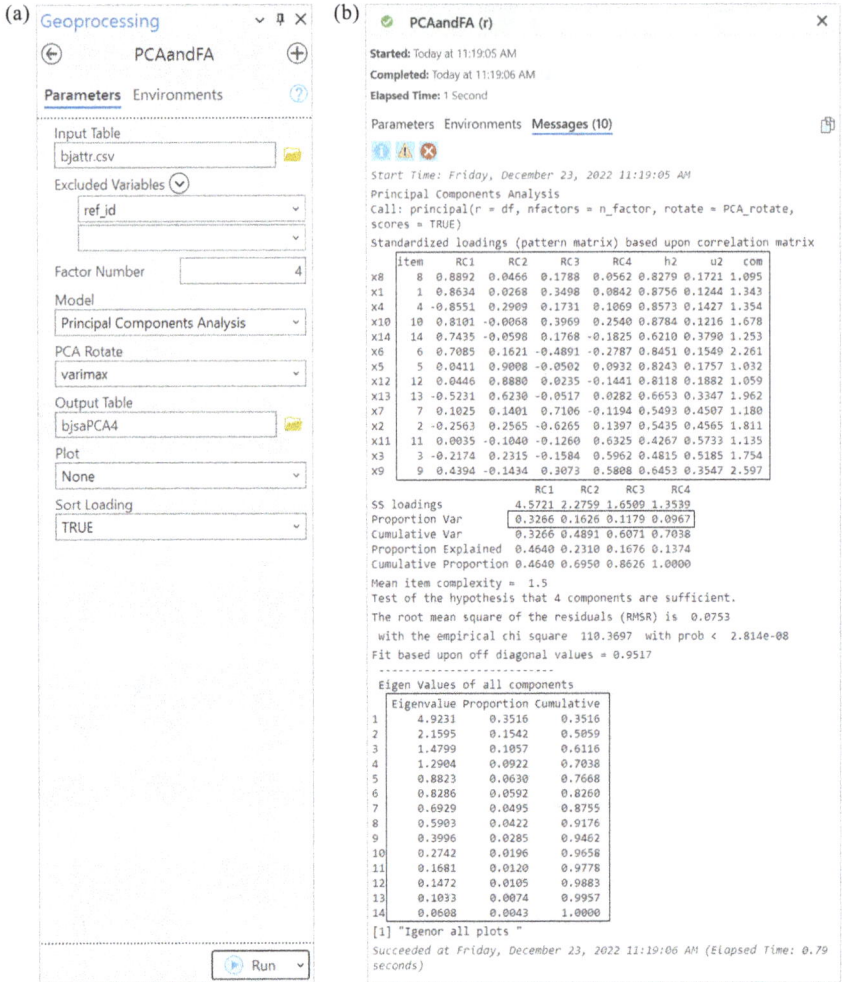

FIGURE 7.6 Principal component analysis by R-ArcGIS tool: (a) interface and (b) result.

TABLE 7.3
Factor Loadings in Social Area Analysis

Variables		Land Use Intensity	Neighborhood Dynamics	Socioeconomic Status	Ethnicity
		RC1	RC2	RC3	RC4
Public service density	x8	**0.8892**	0.0466	0.1788	0.0562
Population density	x1	**0.8634**	0.0268	0.3498	0.0842
Labor participation ratio	x4	**−0.8551**	0.2909	0.1731	0.1069

TABLE 7.3 (*Continued*)
Factor Loadings in Social Area Analysis

Variables		Land Use Intensity	Neighborhood Dynamics	Socioeconomic Status	Ethnicity
		RC1	RC2	RC3	RC4
Office/retail density	x10	**0.8101**	−0.0068	0.3969	0.254
Housing price	x14	**0.7435**	−0.0598	0.1768	−0.1825
Dependency ratio	x6	**0.7085**	0.1621	−0.4891	−0.2787
Household size	x5	0.0411	**0.9008**	−0.0502	0.0932
Floating population ratio	x12	0.0446	**0.888**	0.0235	−0.1441
Living space	x13	−0.5231	**0.623**	−0.0517	0.0282
Income	x7	0.1025	0.1401	**0.7106**	−0.1194
Natural growth rate	x2	−0.2563	0.2565	**−0.6265**	0.1397
Ethnic enclave	x11	0.0035	−0.104	−0.126	**0.6325**
Sex ratio	x3	−0.2174	0.2315	−0.1584	**0.5962**
Industry density	x9	0.4394	−0.1434	0.3073	**0.5808**

The factors are labeled to reflect major variables loaded:

a. "Land use intensity" is by far the most important factor, explaining 32.66% of the total variance, and captures mainly six variables: three density measures (population density, public service density, and office and retail density), housing price, and two demographic variables (labor participation ratio and dependency ratio).

b. "Neighborhood dynamics" accounts for 16.26% of the total variance and includes three variables: floating population ratio, household size, and living space.

c. "Socioeconomic status" accounts for 11.79% of the total variance and includes two variables: average annual income per capita and population natural growth rate.

d. "Ethnicity" accounts for 9.67% of the total variance and includes three variables: ethnic enclave, sex ratio, and industry density.

Step 2. Optional: Implementing Factor Analysis (FA): Activate the PCAandFA tool again. In the dialog, choose bjattr.csv as Input Table; choose ref _ id under Excluded Variables; input 4 for Factor Number, Factor Analysis for Model, varimax for FA Rotate, pa (principal axis factoring)[2] for FA method, bjattrFA4 for Output Table; choose Scree plot under Plot; and keep FALSE for Sort Loading. Click Run.

Experiment with other selections for FA Rotate or FA method to explore more options in FA. The scree plot in the R Graphics window is similar to Figure 7.1.

Step 3. Implementing K-Means Cluster Analysis in ArcGIS Pro: In ArcGIS Pro, add and open the layer `subdist`, and then join the text file `bjsaPCA4` to it based on the common key `ref _ id`. Under Geoprocessing toolboxes > Spatial Statistics Tools > choose Multivariate Clustering.[3] In the dialog, choose `subdist` for Input Features, input `subdist _ CL` for Output Features, check RC1, RC2, RC3, and RC4 under Analysis Fields, set `K means` for Clustering Method, keep `Optimized seed locations` for Initialization Method and blank for Number of Clusters, input `subdist _ CLre` for Output Table for Evaluating Number of Clusters, and click Run. As stated previously, the K-means method may generate slightly different results every time it runs.

The Output Table `subdist _ CLre` reports how the pseudo-F statistic corresponds to the number of clusters in a chart. The pseudo-F statistic is a ratio of the between-cluster variation to the within-cluster variation, and thus, the curve shows how the clustering scenario performs at maximizing both within-cluster similarities and between-cluster differences. As shown in Figure 7.7, the first decline of pseudo-F statistic occurs at nine clusters and corresponds to an "elbow" or "knee" point of variance explained curve. Therefore, we set the number of clusters at 9.

Rerun the Multivariate Clustering tool by choosing 9 for Number of Clusters, naming `subdist _ CL9` for Output Features, and leaving it blank under Output Table for Evaluating Number of Clusters.

Step 4. Mapping Factor Patterns and Social Areas: For understanding the characteristics of each social area, use the "Summarize" tool on the table of `subdist _ CL9` to compute the mean values of factor scores within each cluster. The results are reported in Table 7.4. The clusters are labeled by analyzing the factor scores and their locations relative to the city center.

Based on the fields RC1, RC2, RC3, and RC4 in the layer `bjsaPCA4`, map the factor scores of land use intensity, neighborhood dynamics, socioeconomic status, and ethnicity, as shown in Figure 7.8a–d.

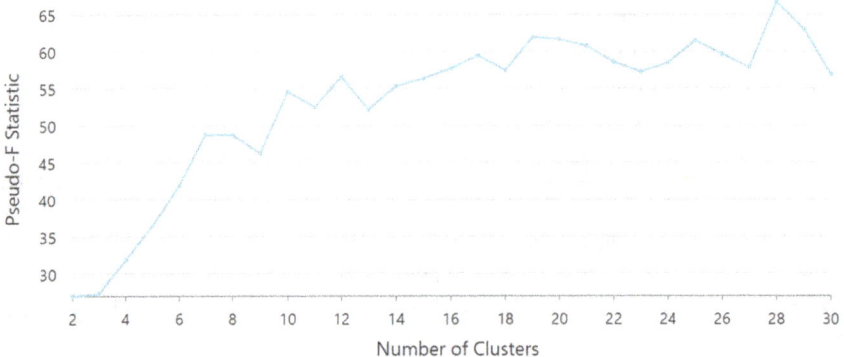

FIGURE 7.7 Pseudo-F statistic vs. number of clusters by the k-means clustering method.

TABLE 7.4
Characteristics of Social Areas

Clusters			Averages of Factor Scores			
ID	Features	No. Subdistricts	Land Use Intensity	Neighborhood Dynamics	Socioeconomic Status	Ethnicity
1	Outer suburban with highest floating population	1	0.1016	5.7971	–0.2524	–1.8756
2	Outer suburban manufacturing with high floating population	6	–1.4853	2.0669	0.3645	0.1866
3	Inner-city low-income	10	2.0533	0.0333	–1.1476	–0.7611
4	Inner-city moderate-income	22	0.8802	–0.1912	0.5521	0.1707
5	Inner-city high-income	2	0.7280	0.9623	5.1487	–0.8146
6	Suburban moderate-density	21	–0.2071	0.6729	–0.6926	0.3590
7	Inner-city ethnic enclave	1	1.8829	–0.0153	1.8282	4.3565
8	Inner-suburban moderate-income	23	–0.4917	–0.5159	–0.0507	0.4149
9	Outer-suburban moderate-income	21	–0.8936	–0.8809	0.0467	–0.7237

Based on the field `Cluster` ID in the layer `subdist _ CL9`, map the nine types of social areas as shown in Figure 7.9.

Step 5. Optional: Implementing Other Clustering Methods in R-ArcGIS Tool: The tool `BasicClustering` under the folder `R _ ArcGIS _ tools` can be used to implement more clustering methods, including k-means, as in step 3 by ArcGIS Pro. Here, the complete hierarchical clustering method is used as an example. Activate the tool "BasicClustering." In the interface, choose `subdist` for Input Data and RC1, RC2, RC3, and RC4 for Cluster Variables, input 9 for Clustering Number, choose `hierarchical` for Model Type and `complete` for Hierarchical Clustering Method, choose `Yes` for Plot Dendrogram, keep No for Plot Clustering Statistics, input `subdist _ h1` for Output Data, and click Run. It generates a dendrogram like Figure 7.3. In the resulting table `subdist _ h1`, field `ClusterID` represents the clustering group ID.

Step 6. Testing Urban Spatial Structure by Regression with Dummy Explanatory Variables: Regression models can be constructed to test whether the spatial pattern of a factor is better characterized as a zonal or sector model (Cadwallader, 1981). Based on the circular ring roads, Beijing is divided into four zones, coded by three *dummy variables* (x_2, x_3, and x_4). Similarly, three additional dummy variables (y_2, y_3,

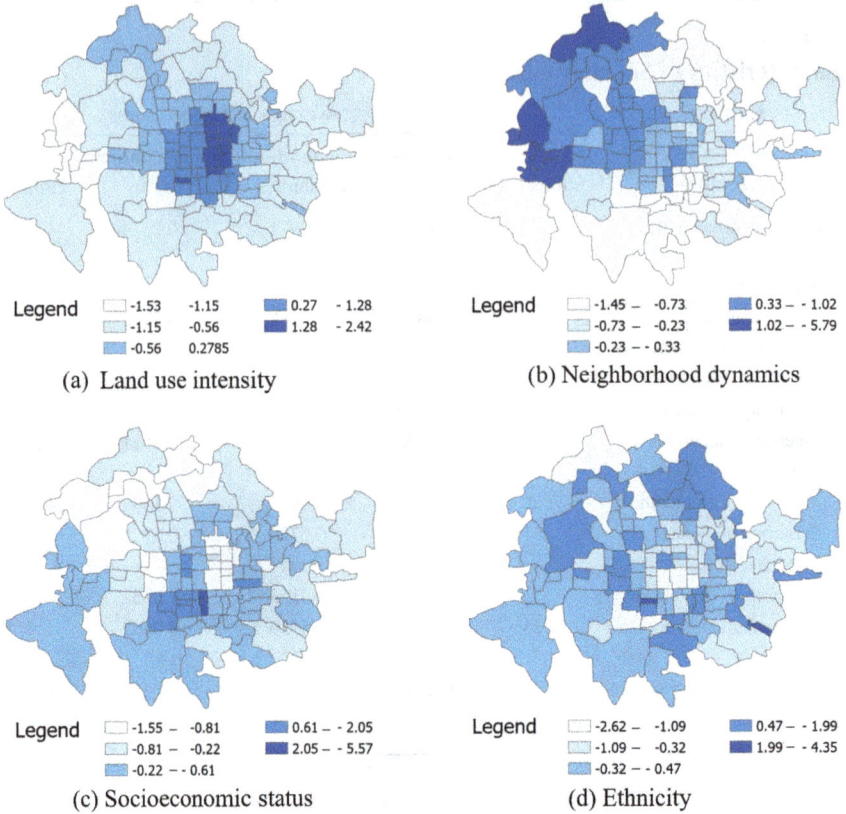

Legend ☐ -1.53   -1.15   ■ 0.27 - 1.28
        ☐ -1.15   -0.56   ■ 1.28 - 2.42
        ■ -0.56   0.2785

(a) Land use intensity

Legend ☐ -1.45 – -0.73   ■ 0.33 – - 1.02
        ☐ -0.73 – -0.23   ■ 1.02 – - 5.79
        ■ -0.23 – - 0.33

(b) Neighborhood dynamics

Legend ☐ -1.55 – -0.81   ■ 0.61 – - 2.05
        ☐ -0.81 – -0.22   ■ 2.05 – - 5.57
        ■ -0.22 – - 0.61

(c) Socioeconomic status

Legend ☐ -2.62 – -1.09   ■ 0.47 – - 1.99
        ☐ -1.09 – -0.32   ■ 1.99 – - 4.35
        ■ -0.32 – - 0.47

(d) Ethnicity

FIGURE 7.8 Spatial patterns of factor scores in Beijing.

and y_4) are used to code the four sectors (NE, SE, SW, and NW). Table 7.5 shows how the zones and sectors are coded by the dummy variables. A simple linear regression model for testing the zonal structure can be written as:

$$F_i = b_1 + b_2 x_2 + b_3 x_3 + b_4 x_4, \tag{7.8}$$

where F_i is the score of a factor ($i = 1, 2, 3,$ and 4), the constant term b_1 is the average factor score in zone 1 (when $x_2 = x_3 = x_4 = 0$, also referred to as *reference zone*), and the coefficient b_2 or b_3 or b_4 is the difference of average factor scores between zone 1 and zone 2, 3, or 4, respectively. Similarly, a regression model for testing the sector structure can be written as:

$$F_i = c_1 + c_2 y_2 + c_3 y_3 + c_4 y_4, \tag{7.9}$$

where notations have similar interpretations to those in equation 7.8.

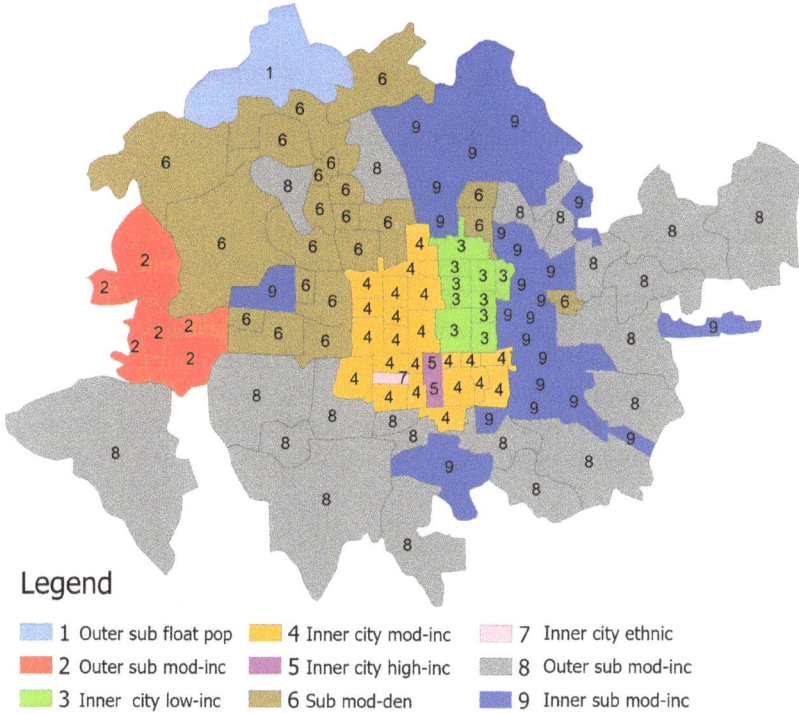

Legend

▦ 1 Outer sub float pop	▦ 4 Inner city mod-inc	▦ 7 Inner city ethnic
▦ 2 Outer sub mod-inc	▦ 5 Inner city high-inc	▦ 8 Outer sub mod-inc
▦ 3 Inner city low-inc	▦ 6 Sub mod-den	▦ 9 Inner sub mod-inc

FIGURE 7.9 Social areas in Beijing.

TABLE 7.5

Zones and Sectors Coded by Dummy Variables

Index and Location	Zones		Index and Location	Sectors	
	Codes	Select By		Codes	Select By
1. Inside 2nd ring	$x_2=x_3=x_4=0$		1. NE	$y_2=y_3=y_4=0$	
2. Between 2nd and 3rd ring	$x_2=1, x_3=x_4=0$	Ring=2	2. SE	$y_2=1, y_3=y_4=0$	Sector=2
3. Between 3rd and 4th ring	$x_3=1, x_2=x_4=0$	Ring=3	3. SW	$y_3=1, y_2=y_4=0$	Sector=3
4. Outside 4th ring	$x_4=1, x_2=x_3=0$	Ring=4	4. NW	$y_4=1, y_2=y_3=0$	Sector=4

In ArcGIS Pro, add layer `subdist` with joined table `bjsaPCA4` to the map. Note that fields of `RING` and `SECTOR` indicate corresponding rings and sectors. Open the attribute table of `subdist`, add six new fields, `x2`, `x3`, `x4`, `y2`, `y3`, and `y4`, and initiate their values by calculating them as 0. Then update their values, such as:

1. Select by Attribute with "`SECTOR is equal 2`" and update the selected values in `x2` as 1.

2. Repeat on sector = 3 and 4, and update the selected values in x3 and x4 as 1, respectively.
3. Similarly, select by Attribute with "RING is equal 2" to update the selected value in y2 as 1.
4. Repeat on ring = 3 and 4, and update the selected values in y3 and y4 as 1, respectively.

Activate the RegressionR tool and run regression models in equations 7.8 and 7.9. Refer to Section 6.5 in Chapter 6 if needed. The results are presented in Table 7.6.

R^2 in Table 7.6 indicates whether the zonal or sector model is a good fit, and an individual t statistic (in parentheses) indicates whether a coefficient is statistically significant (i.e., whether a zone or a sector is significantly different from the reference zone or sector). Clearly, the land use intensity pattern fits the zonal model well, and the negative coefficients b_2, b_3, and b_4 are all statistically significant and indicate declining land use intensity from inner to outer zones. The neighborhood dynamics pattern is better

TABLE 7.6
Regressions for Testing Zonal versus Sector Structures (n = 107)

Factors		Land Use Intensity	Neighborhood Dynamics	Socioeconomic Status	Ethnicity
Zonal model	b_1	1.2980***	−0.1365	0.4861**	−0.0992
		(12.07)	(−0.72)	(2.63)	(−0.51)
	b_2	−1.2145***	0.0512	−0.4089	0.1522
		(−7.98)	(0.19)	(−1.57)	(0.56)
	b_3	−1.8009***	−0.0223	−0.8408**	−0.0308
		(−11.61)	(−0.08)	(−3.16)	(−0.11)
	b_4	−2.1810***	0.4923	−0.7125**	0.2596
		(−14.47)	(1.84)	(−2.75)	(0.96)
	R^2	*0.697*	*0.046*	*0.105*	*0.014*
Sector model	c_1	0.1929	−0.3803**	−0.3833**	−0.2206
		(1.14)	(−2.88)	(−2.70)	(−1.32)
	c_2	−0.1763	−0.3511	0.4990*	0.6029*
		(−0.59)	(−1.52)	(2.01)	(2.06)
	c_3	−0.2553	0.0212	1.6074***	0.4609
		(−0.86)	(0.09)	(6.47)	(1.58)
	c_4	−0.3499	1.2184***	0.1369	0.1452
		(−1.49)	(6.65)	(0.69)	(0.63)
	R^2	*0.022*	***0.406***	***0.313***	*0.051*

Note:
*Significant at 0.05.
**Significant at 0.01.
***Significant at 0.001. R^2 in bold indicates models with an overall statistical significance.

characterized by the sector model, and the positive coefficient c_4 (statistically significant) confirms high portions of floating population in the northwest sector of Beijing. The socioeconomic status factor displays both zonal and sector patterns, but a stronger sector structure. The negative coefficients b_3 and b_4 (statistically significant) in the socioeconomic status model imply that factor scores decline toward the third zone and fourth zone, and the positive coefficient c_3 (statistically significant) indicates higher factor scores in the southwest sector, mainly because of two high-income subdistricts in Xuanwu District. The ethnicity factor does not conform to either the zonal or sector model. Ethnic enclaves scatter citywide and may be best characterized by a multiple nuclei model.

Land use intensity is clearly the primary factor forming the concentric social spatial structure in Beijing. From the inner city (clusters 4, 6, 8, and 9) to inner suburbs (clusters 1 and 2) and to remote suburbs (clusters 3, 5 and 7), population densities as well as densities of public services, offices, and retails declined along with land prices. The neighborhood dynamics, mainly the influence of floating population, is the second factor shaping the formation of social areas. Migrants are attracted to economic opportunities in the fast-growing Haidian District (cluster 1) and manufacturing jobs in Shijingshan District (cluster 3). The effects of the third factor (socioeconomic status) can be found in the emergence of the high-income areas in two inner-city subdistricts (cluster 8) and the differentiation between middle-income (cluster 1) and low-income areas in suburbs (clusters 2, 3, and 5). The fourth factor of ethnicity does not come to play until the number of clusters is expanded to nine.

7.6   SUMMARY

This chapter introduces three multivariate statistical methods, principal components analysis, factor analysis, and cluster analysis. Principal components and factor analyses can be considered as data-reduction methods to reduce the dimensions (columns) of data by structuring many correlated variables into a smaller number of components (factors). Cluster analysis may also be regarded as a data consolidation method to reduce the observations (rows) of data by grouping numerous similar data points into fewer clusters. The methods all simplify and organize the data in a way for better understanding. As the derived variables from principal component or factor analysis are usually independent from each other, they become truly independent variables in many statistical analyses, such as OLS regression, that are free from concern of collinearity. The two methods also help uncover latent variables from observed variables. Data of latent variables cannot be collected because they are hard to measure or could not be collected due to privacy concerns.

Correlation between observations may also arise from location, as similar or dissimilar observations tend to cluster together, and analysis of such data merits the use of some spatial statistical methods to be discussed in Chapter 8. Clustering analysis may also be extended to account for spatial adjacency, in addition to attributive distance that is the focus of cluster analysis discussed in this chapter. Chapter 9 will cover spatial clustering (or regionalization) methods that group spatial features together that are both similar in attributes and close in spatial distance.

To demonstrate the three multivariate statistical analysis methods, we choose the application in social area analysis, a popular topic of great interest to researchers in sociology and urban geography. Social area analysis uses the principal components or factor analysis to consolidate many socio-demographic variables collected from census or field surveys into a few factors (constructs) and then employs the cluster analysis to group small analysis areas measured in the constructs into larger social areas. Moreover, when the results of social areas are mapped, these constructs exhibit distinctive spatial patterns that can be characterized, such as the concentric (zonal) model, the sector model, or the multi-nuclei model. The regression model with dummy variables is designed to test whether a factor conforms to a concentric or sector model with statistical significance.

APPENDIX 7: DISCRIMINANT FUNCTION ANALYSIS

Certain categorical objects bear some characteristics, each of which can be measured in a quantitative way. The goal of *discriminant function analysis (DFA)* is to find a linear function of the characteristic variables and use the function to classify future observations into the preceding known categories. DFA is different from cluster analysis, in which categories are unknown. For example, we know that females and males bear different bone structures. Now, some body remnants are found, and we can identify the genders of the bodies by DFA.

Here we use a two-category example to illustrate the concept. Say, we have two types of objects, A and B, measured by p characteristic variables. The first type has m observations, and the second type has n observations. In other words, the observed data are:

$$X_{ijA} \left(i = 1,2,\ldots,m; j = 1,2,\ldots,p \right),$$

$$X_{ijB} \left(i = 1,2,\ldots,n; j = 1,2,\ldots,p \right).$$

The objective is to find a discriminant function R such that:

$$R = \sum\nolimits_{k=1}^{p} c_k X_k - R_0, \tag{A7.1}$$

where c_k $(k = 1, 2, \ldots, p)$ and R_0 are constants.

After substituting all m observations X_{ijA} into R, we have m values of $R(A)$. Similarly, we have n values of $R(B)$. The $R(A)$ values have a statistical distribution; so do the $R(B)$s. The goal is to find a function R such that the distributions of $R(A)$ and $R(B)$ are most remote to each other. This goal is met by two conditions, namely:

1. The mean difference $Q = \overline{R(A)} - \overline{R(B)}$ is maximized.
2. The variances $F = S_A^2 + S_B^2$ are minimized (i.e., with narrow bands of distribution curves).

That is equivalent to maximizing $V = Q/F$ by selecting coefficients c_k. Once the c_ks are obtained, we simply use the pooled average of estimated means of $R(A)$ and $R(B)$ to represent R_0:

$$R_0 = \left[m\overline{R(A)} + n\overline{R(B)} \right] / (m+n). \tag{A7.2}$$

For any given sample, we can calculate its R value and compare it to R_0. If it is greater than R_0, it is category A; otherwise, it is category B.

The discriminant analysis can be implemented by in R, for example, using the lda() function of the package MASS for linear discriminant analysis.

NOTES

1 Xuanwu and Chongwen are now merged into Xicheng and Dongcheng, respectively.
2 Principal axis factor analysis iteratively applies an eigen value decomposition of the correlation matrix with the diagonal replaced by the values estimated by the factors of the previous iteration. It is not as sensitive to improper matrices as other methods and may produce more interpretable results (Revelle, 2022).
3 The other choice is spatially constrained multivariate clustering, which generates spatially contiguous clusters based on spatial constraints, such as polygon contiguity or point distance range, a topic covered in Chapter 9.

8 Spatial Statistics and Applications

Spatial statistics is statistics that analyzes the pattern, process, and relationship in spatial (geographic) data. While built upon statistical concepts and methods, spatial statistics has some unique capacities that are beyond regular statistics. Some major spatial statistical concepts were raised several decades ago, but related applications were initially limited because of their requirements of intensive computation. Recent advancements in GIS and the development of several free packages on spatial statistics have stimulated greater interests and wide usage. This chapter serves an overview of fundamental tasks in spatial statistics: measuring geographic distributions, spatial cluster analysis, and spatial regression models.

Measuring geographic distributions intends to capture the characteristics of a feature's distribution, such as its center, compactness, and orientation, by some descriptive statistics. For example, by mapping the historical trend of population center in the United States (https://www2.census.gov/geo/maps/DC2020/PopCenter/CenterPop_Mean_1790-2020.pdf), one can tell that population in the West has grown faster than the rest of the country, and to a less degree, the South has outpaced the North. By drawing the standard distances for various crimes in a city, we may detect that one type of crime tends to be more geographically confined than others. Based on the ellipses of an ethnic group's settlement in different eras, a historian can examine the migration trend and identify whether the trend is related to a transportation route or some environmental barriers along certain directions. The techniques computing the center, standard distance, and ellipse are also referred to as "*centrographic measures*."

The centrographic measures are intuitive and easy to interpret, but merely descriptive, and do not put any statistical significance on a revealed pattern. *Spatial cluster analysis* detects unusual concentrations or non-randomness of events in space with a rigorous statistical test. One type of spatial cluster analysis focuses on the pattern of *feature locations* (usually discrete features) only, and another examines *feature values* (usually for analyzing contiguous areas). The former requires exact locations of individual occurrences, whereas the latter uses aggregated rates in areas. Therefore, they are also referred to as point-based and area-based cluster analysis, respectively. Data availability dictates which methods to use. The common belief that point-based methods are better than area-based methods is not well grounded (Oden et al., 1996). It is also of interest to examine whether two types of points tend to be clustered together or dispersed apart, termed "*colocation analysis*."

Two application fields utilize spatial cluster analysis extensively. In crime studies, it is often referred to as "hotspot" analysis. Concentrations of criminal activities or hotspots in certain areas may be caused by (1) particular activities, such as drug trading (e.g., Weisburd and Green, 1995), (2) specific land uses, such as skid row areas and bars, or (3) interaction between activities and land uses, such as thefts at bus stops and transit stations (e.g., Block and Block, 1995). Identifying hotspots is useful for police and crime prevention units to target their efforts on limited areas. Health-related research is another field with wide usage of spatial cluster analysis. Does the disease exhibit any spatial clustering pattern? What areas experience a high or low prevalence of disease? Elevated disease rates in some areas may arise simply by chance alone and thus have no public health significance. The pattern generally warrants study only when it is statistically significant (Jacquez, 1998). Spatial cluster analysis is an essential and effective first step in any exploratory investigation. If the spatial cluster patterns of a disease do exist, case-control, retrospective cohort, and other observational studies can follow up.

Non-randomness of events in spatial distribution indicates the existence of *spatial autocorrelation*, a common issue encountered in analysis of geographic data. Spatial autocorrelation violates the assumption of independent observations in ordinary least squares (OLS) regression discussed in Chapter 6 and necessitates the usage of *spatial regression* models. Spatial regression models include global models, such as *spatial lag and spatial error models*, with constant coefficients for explanatory variables, and local models, such as *geographically weighted regression* (GWR), with coefficients varying across a study area.

This chapter begins with a discussion of centrographic measures in Section 8.1 and then point-based spatial clustering analysis in Section 8.2, followed by a case study of place-names in Yunnan, China, in Section 8.3. Section 8.4 introduces colocation analysis of two types of points, and Section 8.5 uses a case study to demonstrate the methods for detecting global and location colocation patterns. Section 8.6 examines spatial cluster analysis based on feature values, and Section 8.7 presents spatial regression, including global and local models, followed by a case study of homicide patterns in Chicago in Section 8.8. Section 8.9 concludes the chapter with a brief summary.

8.1   THE CENTROGRAPHIC MEASURES

As fundamental as mean and standard deviation in basic statistics, centrographic measures describe the geographic distribution of point features in terms of central tendency and spatial dispersion. Similar to the mean in regular statistics that represents the average value of observations, the *mean center* may be interpreted as the average location of a set of points. As a location has x- and y-coordinates, the mean center's coordinates $(\bar{X}, \bar{Y})$ are the average x-coordinate and average y-coordinate of all points (X_i, Y_i) for i = 1, 2, . . . , n, respectively, such as:

$$\bar{X} = \sum_i X_i / n, \ \bar{Y} = \sum_i Y_i / n.$$

If the points are weighted by an attribute variable w_i, the *weighted mean center* has coordinates such as:

$$\bar{X} = \sum_i (w_i X_i) / \sum_i w_i, \ \bar{Y} = \sum_i (w_i Y_i) / \sum_i w_i.$$

Two other measures of center are used less often: median center and central feature. *Median center* is the location having the shortest total distance to all points. *Central feature* is the point (among all input points) that has the shortest total distance to all other points. Median center or central feature finds the location that is most accessible. The difference is that the central feature must be one from the input points, and the median center does not need to be. Computing the central feature is straightforward, by selecting the point with the lowest total distance from others. However, the location of median center can only be found approximately by experimenting with locations around the mean center and identifying one with the lowest total distance from all points (Mitchell, 2005).

Similar to the standard deviation in regular statistics that captures the degree of total variations of observations from the mean, the *standard distance* is the average difference in distance between the points and their mean center, such as:

$$SD = \sqrt{\sum_i \frac{\left(X_i - \bar{X}\right)^2}{n} + \sum_i \frac{\left(Y_i - \bar{Y}\right)^2}{n}}.$$

For the weighted standard distance, it is written as:

$$SD_w = \sqrt{\frac{\sum_i w_i \left(X_i - \bar{X}\right)^2}{\sum_i w_i} + \frac{\sum_i w_i \left(Y_i - \bar{Y}\right)^2}{\sum_i w_i}}.$$

In graphic representation, it is a circle around the mean center with a radius equal to the standard distance. The larger the standard distance is, the more widely dispersed the features are from the mean center. Therefore, the standard distance measures the compactness of the features.

The ellipse further advances the standard distance measure by identifying the orientation of the features and the average differences in distance along the long axis and the short axis. In GIS, the orientation of the long axis (i.e., the new y-axis) is determined by rotating from 0^0 so that the sum of the squares of the distance between the features and the axis is minimal, and the short axis (i.e., the new x-axis) is then vertical to the derived long axis. The lengths of the two axes are twice the standard deviation along each axis, which is written as:

$$SD_x = \sqrt{\sum_i \frac{\left(X_i - \bar{X}\right)^2}{n}} \, , \, SD_y = \sqrt{\sum_i \frac{\left(Y_i - \bar{Y}\right)^2}{n}}.$$

Therefore, it is also termed the *"standard deviational ellipse."*

In graphic representation, an ellipse indicates that features are more dispersed along the long axis than the short axis from the mean center.

In ArcGIS Pro, all the centrographic measures are grouped under Spatial Statistics Tools > Measuring Geographic Distributions. If lines or areas are used as input features, they are represented by the coordinates of the center points of the lines or the centroids of the areas, respectively.

8.2 INFERENTIAL MEASURES OF POINT DISTRIBUTION PATTERNS

Section 8.1 uses the centrographic measures to summarize location information of points (and their associated attribute values, if available) on where the points are generally located and to what extent they are dispersed and if they exhibit a directional bias. But the measures cannot indicate if the points are clustered or dispersed with any statistical significance. This section covers inferential statistics to characterize the spatial pattern of points.

8.2.1 SPATIAL CLUSTERING PATTERN OF ONE TYPE OF POINTS

Two general approaches have been proposed to analyze point patterns. The density-based approach examines whether the observed points exhibit a density pattern (number of points per unit area) that is different from that of a random pattern, and the distance-based approach considers if distances between points are larger (dispersed) or smaller (clustered) than those of a random pattern. Here, two popular methods are introduced: Ripley's (1977) *K-function* represents the former, and the *nearest neighbor index* represents the latter.

The K-function, *K(h)*, is constructed by counting the number of points within different distance rings, *h* (spatial lag), around each point. Therefore, it may be regarded as the observed point density levels at different spatial lags. Highly clustered patterns have high density levels for small *h*, and dispersed or evenly distributed points have similar density levels with different *h*'s. By comparing this observed distribution to an expected distribution (i.e., a random pattern), one can detect the degree of clustering or dispersion. A reasonable density estimate of random pattern is πh^2. The difference between the *K(h)* and πh^2 is the difference function *L(h)*, written as:

$$L(h) = \sqrt{\frac{K(h)}{\pi}} - h.$$

High (and positive) values in the *L*-function indicate clustering at the corresponding lags, while low (and negative) values reflect dispersion. Therefore, plotting the *L*-function against *h* shows the different degrees of clustering–dispersion over different spatial lags. The *L*-function of a random point pattern should be close to 0 at all spatial lags.

Figure 8.1 shows the *L(h)* plot for an empirical case (Wong and Wang, 2018). The expected *L*-function is horizontal at 0, with a 95% confidence upper and lower bound around it. The observed *L*-function is way above the expected one and the bounds, indicating a highly clustered pattern. At the local scale, the clustering is mild; as the lag increases, the degree of clustering increases first and then diminishes slightly but stays significantly different from a random pattern.

In ArcGIS Pro, the Ripley's K-Function can be accessed in Geoprocessing Toolboxes > Spatial Statistics Tools > Analyzing Patterns > Multi-Distance Spatial Cluster Analysis (Ripley's K-function).

The *nearest neighbor (NN) index (statistic)* computes the average distance of each point from its nearest point (r_{obs}) and compares that to the expected nearest

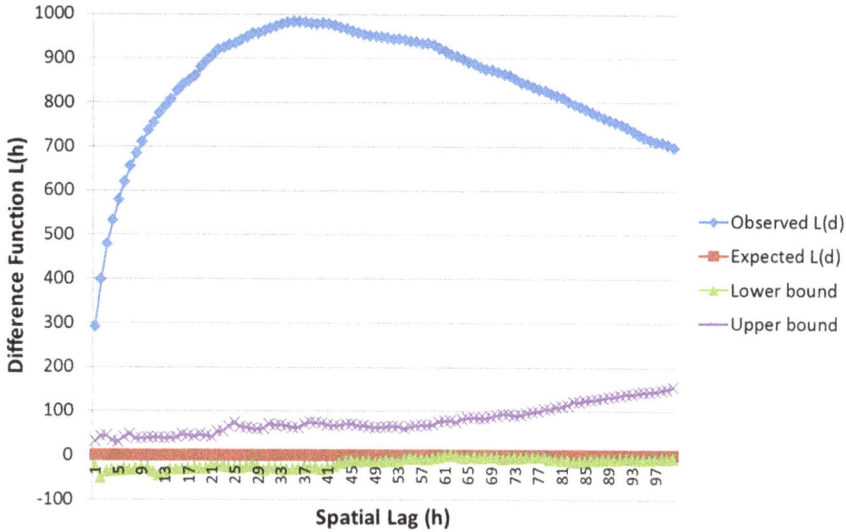

FIGURE 8.1 Sample L(h) function: observed, expected (= 0), and upper and lower bounds.

neighbor distance (r_{exp}). The latter is derived analytically for a random pattern such that $r_{exp} = 0.5\sqrt{A/r}$, where A is the area size and r is the number of points in a study area. If the observed average nearest distance is significantly longer than that of a random pattern, it is dispersed; if the observed average nearest distance is significantly shorter than that of a random pattern, it is clustered. Any pattern between is considered random. In addition to the detected pattern, the nearest neighbor index reports a z-score to indicate the statistical significance level.

The NN statistic relies entirely on nearest (first-order) neighbor distances. Similar to the change of *L(h)* function in response to the spatial lag *h*, the NN statistic can be extended to *higher-ordered neighbor statistics* when second- or higher-ordered neighbors are considered. The clustering–dispersion pattern may change as the order of neighbors (or scale) changes. Considering higher-ordered neighbors can reveal the impact of spatial scale on the detected spatial pattern.

In ArcGIS Pro, the tool Average Nearest Neighbor is accessed under Spatial Statistics Tools > Analyzing Patterns. The tool returns five values: observed mean distance, expected mean distance, nearest neighbor index, z-score, and p-value. The tools for higher-ordered neighbor statistics are not yet available in ArcGIS Pro.

The preceding analysis of spatial pattern is to investigate whether there is clustering or dispersion throughout a study region and is thus termed *global clustering*. Most studies of using the K-function and NN statistics detect global clustering. Conceivably, we can divide a study region into subregions, apply these measures to each, and thus detect variability of clustering across the region. Alternatively, we can employ the floating catchment area (FCA) method introduced in Subsection 3.1.1 to identify a subset of points around each point, use these spatial statistics to determine the clustering pattern in each FCA, and then examine the varying pattern across the region. Those analyses can identify *local clusters* in certain areas even when global clustering is absent.

8.2.2   CLUSTER ANALYSIS OF CASE-CONTROL POINTS

In many applications (e.g., medical fields), researchers are interested in identifying whether one type of points is concentrated (clustered) among other points, either across the entire study area (global clustering) or in certain areas (local clusters). *Cases* represent individuals with the disease (or the events in general) being studied, and *controls* represent individuals without the disease (or the non-events in general). The test by Whittemore et al. (1987) computes the average distance between all cases and the average distance between all individuals (including both cases and controls). If the former is lower than the latter, it indicates clustering. The method is useful if there are abundant cases in the central area of the study area but not good if there is a prevalence of cases in peripheral areas (Kulldorff, 1998, p. 53). The method by Cuzick and Edwards (1990) examines the K-nearest neighbors to each case and tests whether there are more cases (not controls) than what would be expected under the null hypothesis of a purely random configuration. Other tests for global clustering include Diggle and Chetwynd (1991), Grimson and Rose (1991), and others.

In addition to global clustering, it is also important to identify cluster locations or *local clusters*. The geographical analysis machine (GAM) developed by Openshaw et al. (1987) first generates grid points in a study region, then draws circles of various radii around each grid point, and finally searches for circles containing a significantly high prevalence of cases. One shortcoming of the GAM method is that it tends to generate a high percentage of "false positive" circles (Fotheringham and Zhan, 1996). Since many significant circles overlap and contain the same cluster of cases, the Poisson tests that determine each circle's significance are not independent and thus lead to the problem of multiple testing.

The test by Besag and Newell (1991) only searches for clusters around cases. Say, k is the minimum number of cases needed to constitute a cluster. The method identifies the areas that contain the $k-1$ nearest cases (excluding the centroid case), then analyzes whether the total number of cases in these areas[1] is large relative to total risk population. Common values for k are between 3 and 6 and may be chosen based on sensitivity analysis using different k. As in the GAM, clusters identified by Besag and Newell's test often appear as overlapping circles. But the method is less likely to identify false-positive circles than the GAM and also less computationally intensive (Cromley and McLafferty, 2002, p. 153). Other point-based spatial cluster analysis methods include Rushton and Lolonis (1996) and others.

The following discusses the *spatial scan statistic* by Kulldorff (1997). Its main usage is to evaluate reported spatial or space–time disease clusters and to see if they are statistically significant. Like the GAM, the spatial scan statistic uses a circular scan window to search the entire study region but considers the problem of multiple testing. The radius of the window varies continuously in size from 0 to 50% of the total population at risk. For each circle, the method computes the likelihood that the risk of disease is higher inside the window compared to outside the window. The spatial scan statistic uses either a Poisson-based model or a Bernoulli model to assess statistical significance. When the risk (base) population is available as aggregated area data, the Poisson-based model is used, and it requires case and population counts by areal units and the geographic coordinates of the points. When binary event data for case-control studies are available, the Bernoulli model is used, and it requires the geographic coordinates of all individuals. The cases are coded as 1s, and the controls are coded as 0s.

For instance, under the Bernoulli model, the likelihood function for a specific window z is:

$$L(z,p,q) = p^c (1-p)^{n-c} q^{C-c} (1-q)^{(N-n)-(C-c)},$$

where N is the combined total number of cases and controls in the study region, n is the combined total number of cases and controls in the window, C is the total number of cases in the study region, c is the number of cases in the window, $p = c/n$ (probability of being a case within the window), and $q = (C-c)/(N-n)$ (probability of being a case outside the window).

The likelihood function is maximized over all windows, and the "most likely" cluster is one that is least likely to have occurred by chance. The likelihood ratio for the window is reported and constitutes the *maximum likelihood ratio test* statistic. Its distribution under the null hypothesis and its corresponding *p*-value are determined by a *Monte Carlo simulation* approach (see Chapter 12). The method also detects secondary clusters with the highest likelihood function for a particular window that do not overlap with the most likely cluster or other secondary clusters.

The spatial scan statistic and its associated significance test are implemented in a SaTSCanR tool in Case Study 8A.

8.3 CASE STUDY 8A: SPATIAL DISTRIBUTION AND CLUSTERS OF PLACE-NAMES IN YUNNAN, CHINA

This project is based on a study reported in Wang, Zhang, et al. (2014). Similar to the toponymical (place-name) study of Zhuang in Guangxi of China introduced in Section 3.3 in Chapter 3, this case study shifts the study area to Guangxi's neighboring province, Yunnan. The place-names in Guangxi are largely binary and grouped into Zhuang and Han. Yunnan has many ethnic minorities, and thus a mix of multiethnic toponyms. Analysis of multiple types of point data presents unique challenges and affords us an opportunity to showcase the values of spatial statistics in historical and cultural geography. The case study has two parts: Subsection 8.3.1 implements the centrographic measures to describe the spatial patterns of toponyms, and Subsection 8.3.2 identifies whether concentrations of one type of toponyms in some areas are random or statistically significant.

Among a total population of 44.5 million in Yunnan, two-thirds are the Chinese majority Han and one-third are ethnic minority groups. More than a quarter of current toponyms in Yunnan are derived from ethnic minority languages. The rich set of multiethnic toponyms has earned Yunnan the nickname of "museum of toponyms" (Wang, Zhang, et al., 2014). The multiethnic toponyms in Yunnan are grouped into three language branches: Zang–Mian (Tibeto–Burman), Zhuang–Dong (Kam–ai), and Han, all of which belong to the large Sino-Tibetan language family.

Data for the case study is organized in geodatabase YN.gdb under the folder Yunnan that includes:

1. Three reference layers: Yunnan _ pro (boundary of Yunnan Province), Prefectures1994 (prefectures in Yunnan), Majorcity (two major cities Kunming and Dali in Yunnan).
2. Layer PlaceNames with (1) a field Large_grou identifying place-names of three language branches (han, zang-mian, and zhuang-dong) and (2) two fields PointX and PointY representing the XY coordinates.
3. Four layers for Han place-names corresponding to four different eras: Han _ PreYuan (before 1279), Han _ Yuan (1279–1368), Han _ Ming (1368–1644), Han _ Qing _ postQ (after 1644).

8.3.1 ANALYSIS OF MULTIETHNIC PLACE-NAMES BY CENTROGRAPHIC MEASURES

Toponym entries in the four layers for different eras contain a field Dynasty indicating their historical origins, for example, a value "Ming" in the layer Han _ Ming means that the place was first named by Han people in the Ming dynasty and has been preserved till present. This timestamp sheds an interesting light on the historical evolution of multiethnic settlement in the region. The centrographic analysis approach is adopted here to reveal the overall spatial trend of historical evolution of Han toponyms. Specifically, as Han gradually moved into the region historically inhabited by native ethnic minorities, it led to *sinification*, a process whereby those minorities came under the influence of dominant Han Chinese. One of the outcomes was that many toponyms of minority language origins got obscured (translated or recorded in Han) or renamed, and it led to a dispersion of Han toponyms over time.

Historically, major agriculture development and population settlements were anchored by two cities, Kunming and Dali (Figure 8.2). Both were major military posts established by the Han rulers to control and exert influence in the region. Distance from whichever of the nearer of the two captures this effect. We begin the analysis by separating the toponyms into two regions, either closer to Dali in the west or closer to Kunming in the east. By doing so, we can examine how the distribution of Han toponyms evolved around the two major cities as springboards of Han's incursion to Yunnan over time. Either subset is then further divided to four layers, corresponding to the Han toponyms named in each of four eras: (1) pre-Yuan (before 1279), (2) Yuan dynasty (1279–1368), (3) Ming dynasty (1368–1644), and (4) Qing dynasty (1644–1912) and post-Qing (after 1912). The centrographic measures are repeated on each of four eras in either of two regions, that is, eight scenarios.

Step 1. Data Preparation: Activate ArcGIS Pro and add all data layers to the project. Access the tool "Near" by searching in the Geoprocessing pane. In its dialog window, select Han_Qing_postQ as Input Features, choose Majorcity under Near Features, keep other default settings, and click Run. Open the attribute table of Han_Qing_postQ, where field NEAR_FID indicates the nearest city for each place; and select the records with "NEAR_FID=1." In the Map pane, right-click the layer Han_Qing_postQ > Data > Export Features, and name the output feature Han_Qing1, which includes the place-names in the east region (anchored by Kunming).

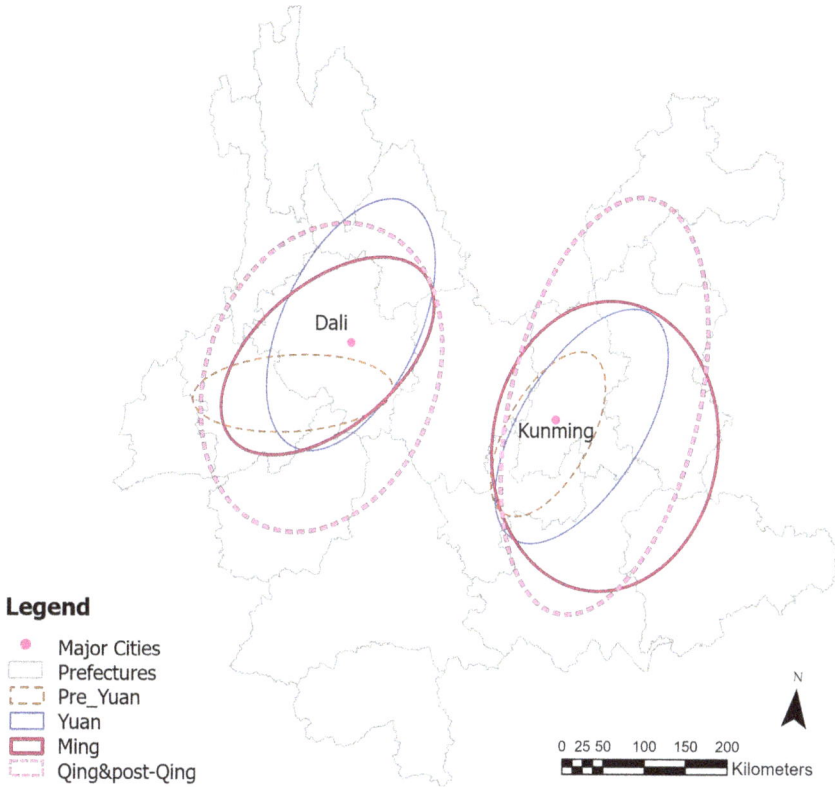

FIGURE 8.2 Ellipses of Han toponyms around Dali and Kunming in various eras.

Similarly, on the attribute table of Han _ Qing _ postQ again, select the records with "NEAR _ FID=0," and export the selected features to a new layer Han _ Qing0, which includes the place-names in the west region (anchored by Dali).

Repeat the previous process to generate (1) three new layers for place-names in the east region, Han_PreYuan1, Han_Yuan1, and Han_Ming1, for the other three eras, and (2) three new layers for place-names in the west region, Han_PreYuan0, Han_Yuan0, and Han_Ming0, also for the other three eras. In summary, a total of eight layers are prepared and correspond to four eras and two regions.

Step 2. Generating the Mean Center: Search and activate the tool Mean Center in the Geoprocessing pane. In its dialog window, select Han _ Qing1 as Input Feature Class, keep the default name Han _ Qing1 _ MeanCenter for the Output Feature Class, keep other default settings, and click OK. The result is the mean center for the Han place-names in the east region since the Qing dynasty.

Step 3. Generating the Standard Distance: Similarly, search and activate the Standard Distance tool in the Geoprocessing pane. Select the same Input Feature Class, use the default name Han_Qing1_StandardDistance for the Output Feature Class, keep other default settings, and click OK to generate the standard distance.

Step 4. Generating the Ellipse: Similarly, use the tool Directional Distribution (Standard Deviational Ellipse) to derive the ellipse Han _ Qing1 _ DD.

Repeat it on the other seven layers prepared from step 1.

Overlay the eight layers generated by the tool Directional Distribution in one map frame, as shown in Figure 8.2. In the pre-Yuan era, most Han toponyms were limited in their spatial ranges around Dali (more in the west–east direction) and Kunming (more in southwest–northeast direction). In the Yuan dynasty, the Han toponyms diffused gradually outward around the two cities, both along the southwest–northeast direction. The expansion continued throughout the Ming and Qing dynasties into the modern days. The only areas with minimal incursion of Han toponyms were the northwest corner, populated by the Zang–Mian toponyms, and the southwest corner, dominated by the Zhuang–Dong toponyms, both known for their difficult terrains. In short, the spatial expansion of Han toponyms reflects the impact of increasing migration and settlement of Han population in Yunnan.

8.3.2 Cluster Analysis of Multiethnic Place-Names by SaTScanR

This part of the analysis focuses on the distinctive pattern of toponyms of either minority ethnic group (Zang–Mian or Zhuang–Dong) in comparison to the majority Han by cluster analysis.

Step 5. Preparing Data in ArcGIS Pro: Add the dataset PlaceNames containing the three toponym categories. Add a new field Count and calculate it as Count = 1 for all rows. Use Select by Attribute to select those Zang–Mian toponyms with "Large _ grou=zang-mian" and export to a new feature ZMCase; select those Zhuang–Dong toponyms with "Large _ grou=zhuang-dong" and export to a new feature ZDCase; and select those Han toponyms with "Large _ grou=han" and export to a new feature HanControl.

The feature ZMCase or ZDCase contains the cases, and the feature HanControl contains the controls.

Step 6. Spatial Cluster Analysis by SaTScanR: In the Catalog, browse to the folder R-ArcGIS-Tools > double-click the tool SaTScanR under R-ArcGIS-Tools. tbx to activate the dialog (Figure 8.3).

In the dialog, choose Hancontrol for Control Feature, ZMcase for Case Feature, Yunnan _ pro for Boundary Feature; input 49 for Simulation Times, ZMcluster for Output Cluster, and ZMclusterCircle for Cluster Circle; then click Run to execute the tool. Feature ZMcluster updates feature ZMcase with an additional field ClusterID identifying the Zang–Mian toponym clusters of statistical significance. For those not significant, the value of ClusterID is Null. Another feature ZMclusterCircle contains the scan circles for the clusters, which also have fields to record the centers, radius, p-value, etc.

Repeat the step by defining the Case Feature as ZDcase and naming the Output Cluster as ZDcluster and the Cluster Circle as ZDclusterCircle. It generates the spatial scan results for Zhuang–Dong toponyms.

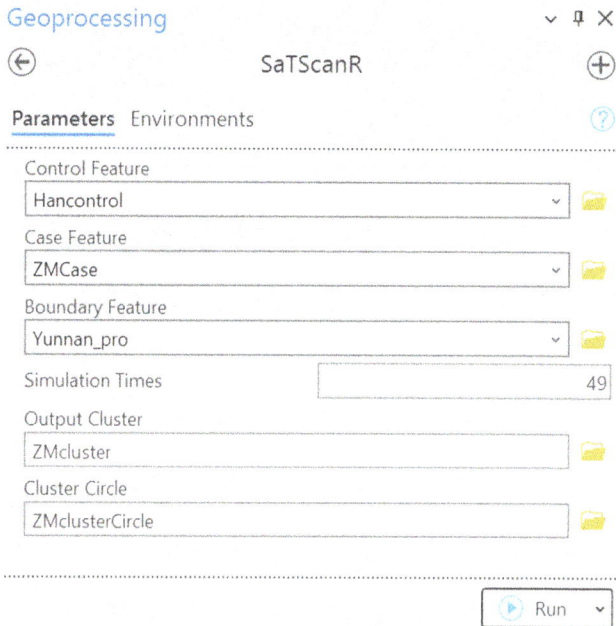

FIGURE 8.3 SaTScanR dialog window for cluster analysis of case-control points.

Step 7. Mapping Spatial Cluster Analysis Results: Add the resulting features from step 6 if they have not been added to the project.

Map the layers ZMcluster and ZMclusterCircle by using unique colors to highlight Zang–Mian toponyms with distinctive ClusterID values in the feature ZMcluster. Their values range from 1 to 5 to identify five clusters for Zang–Mian toponyms (Figure 8.4, left).

Repeat the mapping on the layers ZDcluster and ZDclusterCircle, with two clusters for those Zhuang–Dong toponyms (Figure 8.4, right).

In Figure 8.4, any circle defines a cluster when the ratio of minority (Zang–Mian or Zhuang–Dong) toponyms against the Han toponyms is higher inside the window than that outside it and such a difference is statistically significant. The largest cluster of Zang–Mian toponyms is in the northwest corner, then the second largest in the northeast, and the next one immediately southward of the largest cluster. All three border the Tibet region with historical and modern-day concentrations of Tibetan (or Zang) settlement. The two minor Zang–Mian toponym clusters in the south are associated with Mian ethnic groups. The primary cluster of Zhuang–Dong toponyms is in the southwest, where Yunnan borders with Myanmar and Laos, and a secondary cluster is in the southeast corner, where Yunnan borders with Vietnam to the south and its neighbor Guangxi to the east. Both regions are known for ethnic groups of Zhuang–Dong linguistic branch.

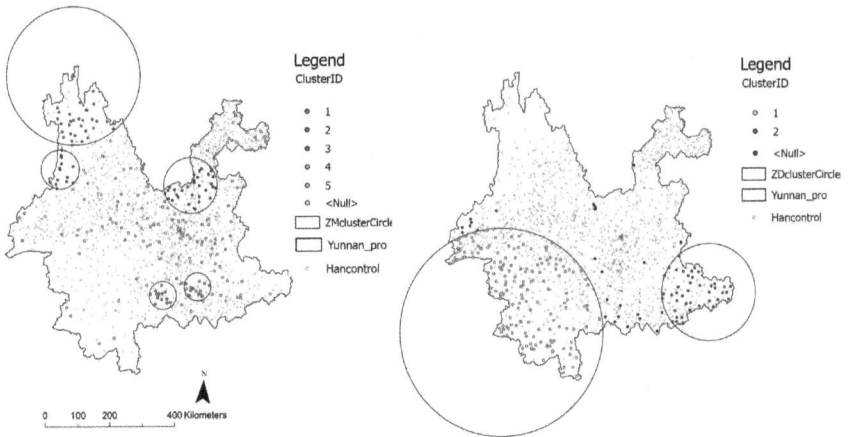

FIGURE 8.4 Clusters of Zang–Mian (left) and Zhuang–Dong (right) toponyms among Han toponyms.

8.4 COLOCATION ANALYSIS OF TWO TYPES OF POINTS

Colocation analysis examines the extent that two types of objects are within the vicinity of each other. This section focuses on colocation analysis of points.

8.4.1 SPATIAL WEIGHTS DEFINITION AND CROSS K-FUNCTION ANALYSIS

Colocation analysis begins with a definition of neighboring relationship between observations, or a *spatial weights matrix*.

As introduced in Appendix 1 of Chapter 1, defining spatial weights can be based on polygon (rook or queen) contiguity, where $w_{ij} = 1$ if area j is adjacent to i, and 0 otherwise. Defining spatial weights can be based on distance (d), such as inverse distance ($1/d$), inverse distance squared ($1/d^2$), distance band (= 1 within a specified critical distance and = 0 outside of the distance), or a continuous weighting function of distance, such as:

$$w_{ij} = \exp(-d_{ij}^2 / h^2),$$

where d_{ij} is the distance between areas i and j, and h is referred to as the *bandwidth* (Fotheringham et al., 2000, p. 111).

In ArcGIS, the spatial weights matrix is critical for most of the tools under the category "Spatial Statistics Tool" in the Geoprocessing tools, defined as "Conceptualization of Spatial Relationships." It provides the options of "Inverse Distance," "Inverse Distance Squared," "Fixed Distance Band," "Zone of Indifference," "Contiguity Edges Only" (i.e., rook contiguity), "Contiguity Edges Corners" (i.e., queen contiguity), and "Get Spatial Weights from File." One may also use the tool "Generate Spatial Weights Matrix" or "Generate Network Spatial Weights" to prepare a spatial weights matrix separately (see steps 2–3 in section 8.5).

Once the neighboring relationship is defined in a spatial weights matrix, we are ready to discuss colocation analysis. The *cross K (or bivariate K) function*, an extension of Ripley's K-function in Subsection 8.2.1, is frequently used to investigate the relationship between two spatial processes under the assumption of independence. It measures the ratio of observed overall density of type B points within a specified distance of a type A point over what we would expect by chance (Cressie, 1991). A significance test is fulfilled by a series of Monte Carlo simulations to examine whether the two types of points are clustered or dispersed. Here, the neighboring relationship is defined as "Fixed Distance Band," and distance can be measured in Euclidean, geodesic, or network distance, as discussed in Section 2.1 of Chapter 2.

Results for cross K-function analysis are presented as a graph illustrating the change of cross K value (y-axis) as distance (x-axis) increases, similar to Figure 8.1. Based on the graph, one can detect at what distance range the two types are spatially correlated (i.e., *colocated*) or independent from each other. For example, statistically significant colocation exists at a certain distance range if the cross K value at that distance falls above an upper envelop of a given level of statistical significance; significant dispersion of two point sets is suggested if it lies below the lower envelop line, and the spatial distribution of type A points has no impact on that of type B points if it falls within the upper and lower envelops. Note that the cross K-function is sensitive to the direction of colocation association. In other words, the cross K-function for $A{\rightarrow}B$, which stands for the extent to which type A points are attracted to type B points, differs from $B{\rightarrow}A$, which indicates the extent that type B points are drawn by type A points.

In ArcGIS Pro, the tool "Multi-Distance Spatial Cluster Analysis (Ripley's K-function) in Analyzing Pattern" is under Spatial Statistics Tools. It examines whether objects exhibit statistically significant clustering or dispersion over a range of distances.

8.4.2 Global Colocation Quotient (CLQ)

As Leslie and Kronenfeld (2011) point out, the cross K-function of two types of point features may be significantly biased by the distribution of the underlying population if they collectively exhibit a significant spatial clustering pattern. They proposed a *colocation quotient (CLQ)* measure, based on the concept of location quotient commonly used in economic geography studies, to mitigate the problem. Different from the cross K-function, the global CLQ is designed on the basis of nearest neighbors rather than actual metrical distance in order to account for the aforementioned effect of the joint population distribution.

The *global CLQ*, which examines the overall association between observed and expected numbers of type B points in proximity to type A points, is formulated as:

$$CLQ_{A \rightarrow B} = \frac{N_{A \rightarrow B} / N_A}{N_B / (N-1)}, \tag{8.1}$$

where N represents the total number of points, N_A denotes the number of type A points, N_B depicts the number of type B points, and $N_{A \rightarrow B}$ denotes the number of type A points that have type B points as their nearest neighbors. The numerator calculates the observed proportion of type A points that has type B points as the nearest neighbor, while the denominator estimates the expected proportion by chance. Given that a point itself is not an eligible candidate for its nearest neighbor, N-1 rather than N is used in the measure of expected proportion. Like the cross K-function, the global CLQ takes the direction of colocation relationship into account.

In practice, point A may have multiple nearest neighbors within a bandwidth. Under this circumstance, the global CLQ treats each nearest neighbor of point A equally in the calculation of $N_{A \to B}$. It is formulated as:

$$N_{A \to B} = \sum_{i=1}^{N_A} \sum_{j=1}^{nn_i} \frac{f_{ij}}{nn_i}, \tag{8.2}$$

where i denotes each type A point, nn_i represents the number of nearest neighbors of point i, j indicates each of point i's nearest neighbors nn_i, and f_{ij} is a binary variable indicating whether or not point j is of type B under investigation (1 indicates yes, and 0 otherwise). Here the notation nn_i, also termed "spatial weights," highlights how the definition of neighbors influences the index. Steps 1–3 in Section 8.5 discuss various definitions of spatial weights in detail.

The global CLQ is proven to have an expected value of one when all points are randomly re-labeled, given the frequency distribution of each point set (Leslie and Kronenfeld, 2011). Consequently, a $CLQ_{A \to B}$ larger than 1 indicates that type A points tend to be more spatially colocated with B points, and a larger CLQ value indicates a stronger colocation association. On the contrary, a $CLQ_{A \to B}$ less than 1 suggests that type A points tend to be isolated from B points. A Monte Carlo simulation-based statistical test is used to determine whether such spatial colocation or isolation behavior is significantly nonrandom. The Monte Carlo simulation process reassigns the category of each point in a random manner, but subject to the frequency distribution of each point category. A distribution of CLQ from the randomization is obtained by repeating this process many times. Then the distribution of observed CLQ is compared with the distribution derived from randomization to derive a test statistic and significance level.

8.4.3 Local Indicator of Colocation Quotient (LCLQ)

More recently, Cromley et al. (2014) developed a local version of CLQ, the *local indicator of colocation quotient (LCLQ)*, in order to reveal the spatial variability of correlation between two point sets. The LCLQ is formulated as:

$$LCLQ_{A_i \to B} = \frac{N_{A_i \to B}}{N_B / (N-1)}, \tag{8.3}$$

$$N_{A_i \to B} = \sum_{j=1(j \neq i)}^{N} \left(\frac{w_{ij} f_{ij}}{\sum_{j=1(j \neq i)}^{N} w_{ij}} \right), \tag{8.4}$$

$$w = \exp\left(-0.5 * \frac{d_{ij}^2}{d_{ib}^2}\right), \tag{8.5}$$

where the LCLQ for point A_i relative to type B points has a similar expression as the global CLQ in equation 8.1; $N_{Ai \to B}$ denotes the weighted average number of type B points that are the nearest neighbors of point A_i; f_{ij} still represents a binary variable indicating if point j is a marked B point (1 for yes, and 0 otherwise); w_{ij} denotes the weight of point j, indicating the importance of point j to the ith A point; d_{ij} is the distance between the ith A point and point j; d_{ib} denotes the bandwidth distance around the ith A point. Equation 8.5 defines the Gaussian kernel density function that is used to assign geographic weights to each neighbor of point A_i, and basically the farther a neighbor locates beyond point A_i, the less important it is to point A_i. One may also consider other density functions to define a point's weight, such as the box kernel density function, which treats each point within the prescribed bandwidth distance from a marked A point identically as one, regardless of their distances.

Wang, Hu, et al. (2017) used the Monte Carlo simulation technique to develop a statistical significance test for the LCLQ. According to the "restricted random labeling approach" (Leslie and Kronenfeld, 2011), at a given location, one simulation trial randomly assigns the categories of points by keeping the frequency distribution of each category fixed in this relabeling step. In other words, the simulation process at point i re-labels other points in a random fashion until all points are re-categorized and the number of elements in each point set remains unchanged after the simulation. This simulation trial is repeated for a predetermined number of times, and a sample distribution of LCLQs can be obtained at each point of interest (say, point i) by recalculating a LCLQ for each trial. Then the significance of calculated LCLQ at point i is determined by comparing it with the sample distribution generated from simulation.

In ArcGIS Pro, the colocation quotient tool is available in Geoprocessing Tools > Spatial Statistics toolbox > Modeling Spatial Relationships > Colocation Analysis.

8.5 CASE STUDY 8B: DETECTING COLOCATION BETWEEN CRIME INCIDENTS AND FACILITIES

This case study is based on the work reported in Wang, Hu, et al. (2017) and detects colocation between motorcycle thefts and three types of facilities, namely, entertainment entities, retail shops, and schools, in a city in Jiangsu Province, China. It implements the global and local colocation quotients introduced in Subsections 8.4.2 and 8.4.3 and focuses on various ways of defining spatial weights.

Data for the case study is organized in geodatabase `JS.gdb` under the folder `Jiangsu` that includes:

1. A reference layer `District` for the study area boundary and administrative districts
2. A network feature dataset `Cityroad` for the road network of the city

3. A point layer MotorTheft with a field Category, whose values = Motorcycle _ theft, Retail _ shop, School, or Entertainment, corresponding to four types of points

Step 1. Implementing Colocation Analysis Tool with K Nearest Neighbors: Activate ArcGIS Pro and add all data to the project. In the Geoprocessing pane, search for the tool "Colocation Analysis" and activate it. In the dialog (Figure 8.5), choose Single dataset for Input Type; select MotorTheft as input Features of Interest, Category for Field of Interest, Motorcycle _ theft for Category of Interest, Entertainment for Neighboring Category; input MotorTheft-knn for Output Features; keep the other default settings (e.g., K-nearest neighbors for Neighborhood Type, 8 for Number of Neighbors, 99 for Number of Permutations); click additional Options and keep the default setting "Gaussian" for Local Weighting

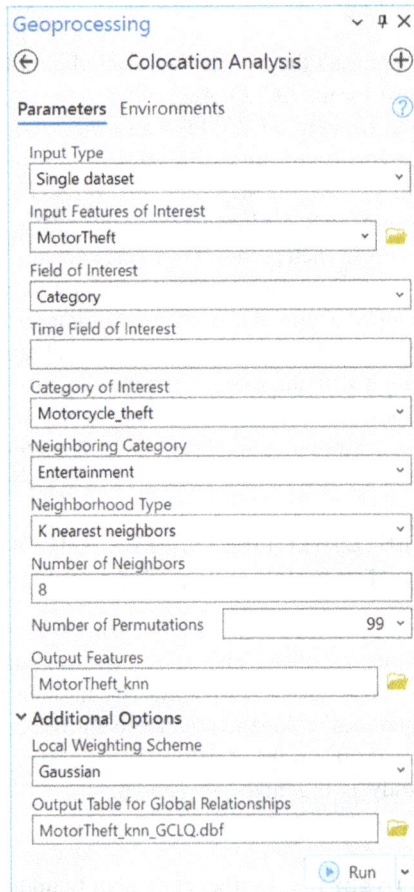

FIGURE 8.5 Dialog window for colocation analysis.

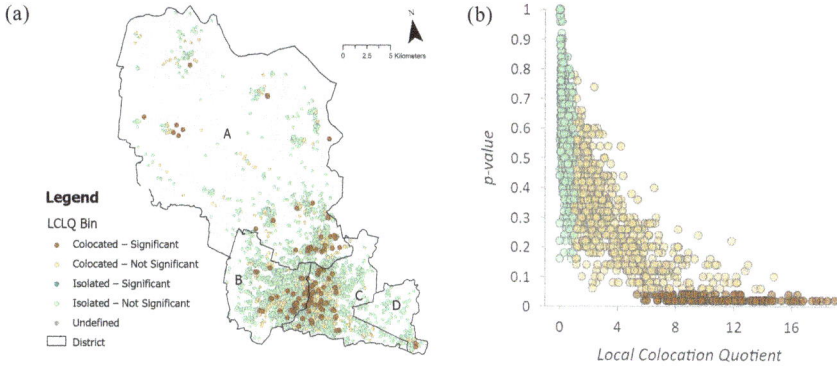

FIGURE 8.6 LCLQs between motorcycle thefts and entertainment facilities: (a) collocation of various significances and (b) LCLQ vs. p-value.

Scheme; input `MotorTheft _ knn _ GCLQ.dbf` for Output Table for Global Relationships; and click Run.

The GCLQ is reported in the table `MotorTheft _ knn _ GCLQ.dbf`, which contains 16 global colocation pairs for the four types of points. Note that the GCLQ of A to B is not equal to that of B to A. The top 3 highest GCLQ values are all from the internal colocations of retail (7.1370), school (4.2154), and entertainment (3.9350), indicating clustering patterns of such facilities. For motorcycle theft, the GCLQ values are 1.0361 (with motorcycle theft), 0.6669 (with entertainment), 0.6534 (with school), and 0.6255 (with retail shop), suggesting that motorcycle theft only globally collocates with itself.

Figure 8.6a–b shows the LCLQ result in the study area and how the LCLQ values are related to their corresponding *p*-values, respectively. Check the attribute table of `MotorTheft-knn` for more information about the result. The LCLQs detect some areas where motorcycle theft crimes are significantly collocated with entertainment establishments. Significant colocations between them are mostly observed in the CBD area and the southeast corner of District A and are also spotted in small pockets in the north of District A and the southeast corner of District D. These findings indicate that the global CLQ results—a significant dispersion pattern between crimes and facilities, regardless of types—do not hold from place to place and thus demonstrate the usefulness of LCLQ, especially for areas with heterogeneous land use patterns.

Step 2. Defining Neighborhood Type by a Predefined Spatial Weights Matrix: Search for and activate the tool "Generate Spatial Weights Matrix." In the dialog, select `MotorTheft` as Input Feature Class and `ID` as Unique ID Field;[2] input `MotorTheftD2.swm` for Output Spatial Weights Matrix File; use `Inverse distance` for Conceptualization of Spatial Relationships, `Euclidean` for Distance Method, and 2 for Exponent; check Row Standardization; and click Run.

Reactivate the Colocation Analysis tool. In the dialog, choose `Get spatial weights from file` for Neighborhood Type, browse to select `MotorTheftD2.swm` under Weight Matrix File, input `MotorTheft_SWM_D2` for Output Features and `MotorTheft_SWM_D2_GCLQ.dbf` for Output Table for Global Relationships, and click Run. The result shows a more clustered pattern than that from step 1.

Step 3. Defining Neighborhood Type by Network Spatial Weights: To define a network spatial weight matrix, users need to build the network dataset first. Refer to step 5 in Subsection 2.2.2 of Chapter 2. Here we use a prepared network dataset.

Activate the tool "Generate Network Spatial Weights." In the dialog, select `MotorTheft` as Input Feature Class, `ID` as Unique ID Field, `MotorTheftNW.swm` for Output Spatial Weights Matrix File, `Cityroad` for Input Network Data Source, then keep other default settings and click Run.

Similar to step 2, in the dialog of the Colocation Analysis tool, choose `Get spatial weights from file` for Neighborhood Type, browse to `MotorTheftNW.swm` under Weight Matrix File, input `MotorTheft_SWM_NW` for Output Features and `MotorTheft_SWM_NW.dbf` for Output Table for Global Relationships, and click Run. The result also differs slightly from those by steps 1 and 2.

Step 4. Optional: Implementing Colocation Analysis with CLQ Tools: Users may also use the global CLQ tool developed by Leslie and Kronenfeld (2011) and the localized CLQ tool developed by Wang, Hu, et al. (2017). Visit http://seg.gmu.edu/clq to download the global CLQ tool. Access the localized CLQ tool under the project folder `CLQ`. Consider setting the Monte Carlo simulations for 1,000 times and the bandwidth of 10 nearest neighbors based on Euclidean distances.

8.6 SPATIAL CLUSTER ANALYSIS BASED ON FEATURE VALUES

Similar to global and local colocation quotients in Section 8.4, spatial cluster analysis methods based on feature values include tests for global clustering and corresponding tests for local clusters. The former is usually developed earlier than the latter.

8.6.1 Tests for Global Clustering Based on Feature Values

Moran's I statistic (Moran, 1950) is one of the oldest indicators that detect global clustering (Cliff and Ord, 1973). It detects whether nearby areas have similar or dissimilar attributes overall, that is, positive or negative *spatial autocorrelation*, respectively. Moran's *I* is calculated as:

$$I = \frac{N \sum_i \sum_j w_{ij} \left(x_i - \bar{x} \right)\left(x_j - \bar{x} \right)}{\left(\sum_i \sum_j w_{ij} \right) \sum_i (x_i - \bar{x})^2}, \tag{8.6}$$

where N is the total number of areas, w_{ij} is the spatial weight linking area i and j, x_i and x_j are the attribute values for area i and j respectively, and $\bar{x}$ is the mean of the attribute values.

It is helpful to interpret Moran's I as the correlation coefficient between a variable and its *spatial lag*. The spatial lag for variable x is the average value of x in neighboring areas j, defined as:

$$x_{i,-1} = \sum_j w_{ij} x_j / \sum_j w_{ij}. \tag{8.7}$$

Therefore, Moran's I varies between -1 and 1. A value near 1 indicates that similar attributes are clustered (either high values near high values or low values near low values), and a value near -1 indicates that dissimilar attributes are clustered (either high values near low values or low values near high values). If a Moran's I is close to 0, it indicates a random pattern or absence of spatial autocorrelation.

Getis and Ord (1992) developed *general G* statistic. General G is a multiplicative measure of overall spatial association of values, which fall within a critical distance (d) of each other, defined as:

$$G(d) = \frac{\sum_i \sum_{j \neq i} w_{ij}(d) x_i x_j}{\sum_i \sum_{j \neq i} x_i x_j}, \tag{8.8}$$

where x_i and x_j can only be positive variables. A large G value (a positive z-score) indicates clustering of high values. A small G (a negative z-score) indicates clustering of low values. Z-score value is a measure of statistical significance. For example, a z-score larger than 1.96, 2.58, and 3.30 (or smaller than -1.96, -2.58, and -3.30) corresponds to a significance level at 0.05, 0.01, and 0.001, respectively. A z-score near 0 indicates no apparent clustering within the study area.

In addition, *Geary's C* (Geary, 1954) also detects global clustering. Unlike Moran's I, using the cross product of the deviations from the mean, Geary's C uses the deviations in intensities of each observation with one another. It is defined as:

$$C = \frac{(N-1) \sum_i \sum_j w_{ij} (x_i - x_j)^2}{2(\sum_i \sum_j w_{ij}) \sum_i (x_i - \bar{x})^2}. \tag{8.9}$$

The values of Geary's C typically vary between 0 and 2 (although 2 is not a strict upper limit), with $C = 1$, indicating that all values are spatially independent from each other. Values between 0 and 1 typically indicate positive spatial autocorrelation, while values between 1 and 2 indicate negative spatial autocorrelation, and thus, Geary's C is inversely related to Moran's I.

For either the Getis–Ord general G or the Moran's I, the statistical test is a normal z-test, such as $z = (Index - Expected) / \sqrt{variance}$. If z is larger than 1.96 (critical value), it is statistically significant at the 0.05 (5%) level; if z is larger than 2.58, it is statistically significant at the 0.01 (1%) level; and if z is larger than 3.29, it is statistically significant at the 0.001 (0.1%) level. For a general overview of significance testing for Moran's I, $G(d)$, and Geary's C, interested readers may refer to Wong and Lee (2005). For detailed theoretical explanations, readers should consult Cliff and Ord (1973) and Getis and Ord (1992).

In ArcGIS Pro, choose Spatial Statistics Toolbox > Analyzing Patterns to access the tools implementing the computations of Moran's I and Getis–Ord General G.

8.6.2 Tests for Local Clusters Based on Feature Values

Anselin (1995) proposed a local Moran index or Local Indicator of Spatial Association (LISA) to capture local pockets of instability or local clusters. The local Moran index for an area i measures the association between a value at i and values of its nearby areas, defined as:

$$I_i = \frac{(x_i - \bar{x})}{s_x^2} \sum_j \left[w_{ij} (x_j - \bar{x}) \right],$$ (8.10)

where $s_x^2 = \sum_j (x_j - \bar{x})^2 / n$ is the variance, and other notations are the same as in equation 8.2. Note that the summation over j does not include the area i itself, that is, $j \neq i$. A positive I_i means either a high value surrounded by high values (high–high) or a low value surrounded by low values (low–low). A negative I_i means either a low value surrounded by high values (low–high) or a high value surrounded by low values (high–low).

Similarly, Getis and Ord (1992) developed the G_i statistic, a local version of the global or general G statistic, to identify local clusters with statistically significant high or low attribute values. The G_i statistic is written as:

$$G_i = \frac{\sum_j \left(w_{ij} x_j \right)}{\sum_j x_j},$$ (8.11)

where the summations over j may or may not include i. When the target feature (i) is not included, it is the G_i statistic which captures the effect of the target feature on its surrounding ones. When the target feature (i) is included, it is called the G_i^* statistic, which detects hotspots or cold spots. ArcGIS computes G_i^* statistic. A high G_i value indicates that high values tend to be near each other, and a low G_i value indicates that low values tend to be near each other. The G_i statistic can also be used for spatial filtering in regression analysis (Getis and Griffith, 2002).

For an overview of significance testing for the local Moran's and local G_i, one may refer to Wong and Lee (2005). For in-depth theoretical formulations of the tests, please refer to Anselin (1995) and Getis and Ord (1992).

In ArcGIS Pro, ArcToolbox > Spatial Statistics Tools > Mapping Clusters, choose Cluster and Outlier Analysis (Anselin Local Morans I) for computing the local Moran, or Hot Spot Analysis (Getis-Ord Gi^*) for computing the local G_i^*.

In analysis for disease or crime risks, it may be interesting to focus only on local concentrations of high rates or the high–high areas. In some applications, all four types of associations (high–high, low–low, high–low, and low–high) have important implications. For example, Shen (1994, p. 177) used the Moran's I to test two hypotheses on the impact of growth-control policies in San Francisco area. The first is that residents who are not able to settle in communities with growth-control policies would find the second-best choice in a nearby area, and consequently, areas of population loss (or very slow growth) would be close to areas of population growth. This leads to a negative spatial autocorrelation. The second is related to the so-called NIMBY (Not in My Backyard) syndrome. In this case, growth-control communities tend to cluster together; so do the pro-growth communities. This leads to a positive spatial autocorrelation.

8.7   SPATIAL REGRESSION

8.7.1   SPATIAL LAG MODEL AND SPATIAL ERROR MODEL

The spatial cluster analysis detects *spatial autocorrelation*, in which values of a variable are systematically related to geographic location. In the absence of spatial autocorrelation or spatial dependence, the ordinary least square (OLS) regression model can be used. It is expressed in matrix form:

$$y = X\beta + \varepsilon \tag{8.12}$$

where y is a vector of n observations of the dependent variable, X is an $n \times m$ matrix for n observations of m independent variables, β is a vector of m regression coefficients, and ε is a vector of n random errors or residuals, which are independently distributed with a mean of 0.

When spatial dependence is present, the residuals are not independent from each other, and the OLS regression is no longer applicable. This section discusses two commonly used models of *maximum likelihood estimator.* The first is a *spatial lag model* (Baller et al., 2001) or *spatially autoregressive model* (Fotheringham et al., 2000, p. 167). The model includes the mean of the dependent variable in neighboring areas (i.e., *spatial lag*) as an extra explanatory variable. Denoting the weights matrix by W, the spatial lag of y is written as Wy. The element of W in the ith row and jth column is $w_{ij} / \sum_j w_{ij}$, as defined in equation 8.3. The model is expressed as:

$$y = \rho Wy + X\beta + \varepsilon, \tag{8.13}$$

where ρ is the regression coefficient for the spatial lag, and other notations are the same as in equation 8.12.

Rearranging equation 8.13 yields:

$$(I - \rho W)y = X\beta + \varepsilon.$$

Assuming the matrix $(I - \rho W)$ is invertible, we have:

$$y = (I - \rho W)^{-1} X \beta + (I - \rho W)^{-1} \varepsilon. \tag{8.14}$$

This reduced form shows that the value of y_i at each location i is determined not only by x_i at that location (like in the OLS regression model) but also by the x_j at other locations through the spatial multiplier $(I - \rho W)^{-1}$ (not present in the OLS regression model). The model is also different from the autoregressive model in time series analysis and cannot be calibrated by the SAS procedures for time series modeling, such as AR or AMAR.

The second is a *spatial error model* (Baller et al., 2001) or *spatial moving average model* (Fotheringham et al., 2000, p. 169) or *simultaneous autoregressive (SAR) model* (Griffith and Amrhein, 1997, p. 276). Instead of treating the dependent variable as autoregressive, the model considers the error term as autoregressive. The model is expressed as:

$$y = X \beta + u, \tag{8.15}$$

where u is related to its spatial lag, such as:

$$u = \lambda W u + \varepsilon, \tag{8.16}$$

where λ is a vector of spatial autoregressive coefficients, and the second error term ε is independent.

Solving equation 8.16 for u and substituting into equation 8.15 yield the reduced form:

$$y = X \beta + (I - \lambda W)^{-1} \varepsilon. \tag{8.17}$$

This shows that the value of y_i at each location i is affected by the stochastic errors ε_j at all other locations through the spatial multiplier $(I - \lambda W)^{-1}$.

Estimation of either the spatial lag model in equation 8.14 or the spatial error model in equation 8.17 is implemented by the maximum likelihood (ML) method (Anselin and Bera, 1998). Case Study 8C in Section 8.8 illustrates how the spatial lag and the spatial error models are implemented in R-ArcGIS tools.

8.7.2 GEOGRAPHICALLY WEIGHTED REGRESSION

Geographically weighted regression (GWR) allows the regression coefficients to vary over space to investigate the spatial nonstationarity (Fotheringham et al., 2002). The model can be expressed as:

$$y = \beta_{0i} + \beta_{1i} x_1 + \beta_{2i} x_2 + \dots + \beta_{mi} x_m + \varepsilon \tag{8.18}$$

where the β's have subscript i, indicating that the coefficients are variant with location i.

GWR estimates the coefficients at each location, such as:

$$\beta_i = (X'W_iX)^{-1} X'W_iY, \qquad (8.19)$$

where β_i is the coefficient set for location i and W_i is the diagonal matrix with diagonal elements being weight of observations for location i. Equation 8.19 represents weighted regressions locally at every observation location using the weighted least square approach. Usually, a monotone decreasing function (e.g., Gaussian kernel) of the distance is used to assign weights to observations. Closer observations are weighted more than distant ones in each local regression.

One may revisit the concept of weighted regression by OLS introduced in Section 6.3 of Chapter 6 to understand the GWR conceptually. An OLS weighted regression seeks the best-fitting coefficients in order to minimize the residual sum of squares (RSS), where the contribution of each observation to RSS is weighted by an externally defined variable. For example, in the weighted OLS regression on a monocentric urban density model, the weights may be defined by the area size of each census tract. Here, in a GWR model, a weighted regression is constructed for each observation. Take census tract i as an example. We may implement a weighted regression for this census tract by using the data of all census tracts in the study area, and the weights associated with individual tracts j are defined a function of their distances from tract i (i.e., $1/d_{ij}$). The regression yields a set of estimated coefficients associated with tract i. Now we move to focus on another tract (say, $i + 1$). Once again, all tracts in the study area participate in the regression, but the weights have changed as their distances are from a different tract $i + 1$. Consequently, the weighted regression yields a different set of estimated coefficients. As we continue the process, we obtain a series of regression results for all tracts in the study area. In short, the same complete set of data points may participate in the regression,[3] and it is the varying weights assigned to those observations that change the regression results (including estimated coefficients, residuals, and R^2). As the weights are defined geographically, the method is thus termed "geographically weighted regression."

GWR yields an improved R^2 than OLS regression because it has more parameters to be estimated. This does not necessarily indicate a better model. The corrected Akaike's information criterion (AICc) index is used to measure the relative performance of a model since it builds in a tradeoff between the data fit of the model and the model's complexity (Joseph et al., 2012). A smaller AIC value indicates a better data fit and a less-complicated model. In addition to removing the spatial autocorrelation of residuals, the major benefit of GWR is that it reports local coefficients, local t values, and local R^2 values that can be mapped to examine their spatial variations.

In ArcGIS Pro, GWR is accessed in the Geoprocessing Toolboxes pane > Spatial Statistics Tools > Modeling Spatial Relationship > Geographically Weighted Regression (GWR).

8.8 CASE STUDY 8C: SPATIAL CLUSTER AND REGRESSION ANALYSES OF HOMICIDE PATTERNS IN CHICAGO

Most crime theories suggest, or at least imply, an inverse relationship between legal and illegal employment. The *strain theory* (e.g., Agnew, 1985) argues that crime results from the inability to achieve desired goals, such as monetary success, through conventional means, like legitimate employment. The *control theory* (e.g., Hirschi, 1969) suggests that individuals unemployed or with less-desirable employment have less to lose by engaging in crime. The *rational choice* (e.g., Cornish and Clarke, 1986) and *economic theories* (e.g., Becker, 1968) argue that people make rational choices to engage in a legal or illegal activity by assessing the cost, benefit, and risk associated with it. Research along this line has focused on the relationship between unemployment and crime rates (e.g., Chiricos's, 1987). According to economic theories, job market probably affects economic crimes (e.g., burglary) more than violent crimes, including homicide (Chiricos, 1987). Support for the relationship between job access and homicide can be found in the *social stress theory*. According to the theory, "high stress can indicate the lack of access to basic economic resources and is thought to be a precipitator of . . . homicide risk" (Rose and McClain, 1990, pp. 47–48). Social stressors include any psychological, social, and economic factors that form "an unfavorable perception of the social environment and its dynamics," particularly unemployment and poverty, which are explicitly linked to social problems, including crime (Brown, 1980).

Most literature on the relation between job market and crime has focused on the link between unemployment and crime using large areas, such as the whole nation, states, or metropolitan areas (Levitt, 2001). There may be more variation *within* such units than *between* them, and intraurban variation of crime rates needs to be explained by examining the relationship between local job market and crime (e.g., Bellair and Roscigno, 2000). Wang and Minor (2002) argued that not every job was an economic opportunity for all, and only an accessible job was meaningful. They proposed that *job accessibility*, reflecting one's ability to overcome spatial and other barriers to employment, was a better measure of local job market conditions. Their study in Cleveland suggested a reverse relationship between job accessibility and crime and stronger (negative) relationships with economic crimes (including auto theft, burglary, and robbery) than violent crimes (including aggravated assault, homicide, and rape). Wang (2005) further extended the work to focus on the relationship between job access and homicide patterns with refined methodology, based on which this case study is developed.

The following datasets are provided under the study area folder `Chicago`:

1. A polygon feature `citytrt` in the geodatabase `ChiCity.gdb` contains 846 census tracts in the city of Chicago (not including the O'Hare airport tract because of its unique land use and non-contiguity with other tracts).
2. A text file `cityattr.csv` contains tract IDs and ten corresponding socio-economic attribute values based on census 1990.

In the attribute table of `citytrt`, the field CNTYBNA is each tract's unique ID, the field POPU is population in 1990, the field JA is job accessibility measured by the methods discussed in Chapter 5 (a higher JA value corresponds to better job accessibility),

TABLE 8.1
Principal Components of Socioeconomic Variables in Chicago

	1. Concentrated Disadvantages	2. Concentrated Latino Immigration	3. Residential Instability
Public assistance	**0.93**	0.18	−0.01
Female-headed households	**0.89**	0.15	0.17
Black	**0.87**	−0.23	−0.15
Poverty	**0.84**	0.31	0.25
Unemployment	**0.77**	0.19	−0.06
Non–high school diplomas	0.40	**0.81**	−0.12
Crowded households	0.25	**0.83**	−0.13
Latino	−0.51	**0.79**	0.19
New residents	−0.21	−0.02	**0.91**
Renters-occupied	0.45	0.20	**0.77**
Portion of variances	*0.45*	*0.23*	*0.16*

Note: Values in bold indicate the largest loading of each variable on one component among the three and in a descending order by their loadings.

and the field CT89 _ 91 is total homicide counts for a three-year period around 1990 (i.e., 1989, 1990, and 1991). Homicide data for the study area are extracted from the 1965–1995 Chicago homicide dataset compiled by Block et al. (1998), available through the National Archive of Criminal Justice Data (NACJD) at https://www.icpsr.umich.edu/web/NACJD/%20studies/6399. Homicide counts over a period of three years are used to help reduce measurement errors and stabilize rates. Note that the job market for defining job accessibility is based on a much wider area (mostly urbanized six counties: Cook, Lake, McHenry, Kane, DuPage, and Will) than the city of Chicago.

Data for defining the ten socioeconomic variables and population are based on the STF3A files from the 1990 census and are measured in percentage. In the text file cityattr.csv, the first column is tract IDs (i.e., identical to the field cntybna in the GIS layer citytrt), and the variables are listed in Table 8.1, such as (1) families below the *poverty* line, (2) families receiving *public assistance*, (3) *female-headed households* with children under 18, (4) *unemployment*, (5) *new residents* who moved in the last five years, (6) *renter-occupied* homes, (7) residents with *no high school diplomas*, (8) *crowded households* with an average of more than one person per room, (9) *Black* residents, and (10) *Latino* residents.

8.8.1 SPATIAL CLUSTER ANALYSIS OF HOMICIDE RATES

Step 1. Optional: Principal Components Analysis on Socioeconomic Variables in R-ArcGIS Tool: Add the file cityattr.txt to a new map, and export it to a table Cityattr under geodatabase ChiCity.gdb. Activate the tool "PCAandFA" under the folder R _ ArcGIS _ tools. In the dialog, choose Cityattr as Input

Table, choose `OBJECTID` and `Field1` under Excluded Variables, input 3 for Factor Number, select `Principal Components Analysis` for Model, enter `CityattrFA` for Output Table, keep other default settings, and click Run. Refer to step 1 in Section 7.5 of Chapter 7 if needed.

Join `CityattrFA` (principal component scores) to the GIS layer `citytrt` (based on their respective common fields `Field1` and `CNTYBNA`) to view the distribution patterns of three component scores.

This step is optional and provides another opportunity to implement the PCA method discussed in Chapter 7. The result (component scores RC1, RC2, and RC3) is already provided in the attribute table of `citytrt` in fields FACTOR1, FACTOR2, and FACTOR3. As shown in Table 8.1, these three components capture 84% of the total variance of the original ten variables. Component 1 (accounting for 45% variance) is labeled "concentrated disadvantages" and captures mostly five variables (public assistance, female-headed households, Black, poverty, and unemployment). Component 2 (accounting for 23% variance) is labeled "concentrated Latino immigration" and captures mostly three variables (residents with no high school diplomas, households with more than one person per room, and Latinos). Component 3 (accounting for 16% variance) is labeled "residential instability" and captures mostly two variables (residential instability and renters-occupied homes). The three components are independent from each other and used as control variables (socioeconomic covariates) in the regression analysis of job access and homicide rate. The higher the value of a component, the more disadvantageous a tract in terms of socioeconomic structure.

Step 2. Computing Homicide Rates in ArcGIS Pro: In ArcGIS Pro, on the layer `citytrt`, use an attribute query to select features with `POPU > 0` (845 tracts selected), and export it to a new feature layer `citytract`. This excludes a tract with no population at the O'Hare airport and thus avoids 0 denominator in computing homicide rates.

Because of rarity of the incidence, homicide rates are usually measured as homicides per 100,000 residents. On the attribute table of `citytract`, add a field `HomiRate` and calculate it as `!CT89 _ 91!*100000/!POPU!`, which is the homicide rates per 100,000 residents during 1989–1991. In the subsequent regression analysis in Section 8.8.2, the logarithmic transformation of homicide rates (instead of the raw homicide rate) is used to measure the dependent variable (see Land et al., 1990, p. 937), and 1 is added to the rates to avoid taking logarithm of 0.[4] Add another field `LnHomiRate` to the attribute table of `citytract`, and calculate it as `math.log(!HomiRate!+1)`.

Step 3. Computing Getis–Ord General G and Moran's I: Search and activate the tool High-Low Clustering (Getis–Ord General *G*) in the Geoprocessing pane. In the dialog, choose `citytract` as Input Feature Class and `HomiRate` as Input Field, check the option Generate Report, leave other settings as default (e.g., `INVERSE _ DISTANCE` for Conceptualization of Spatial Relationships and `EUCLIDEAN _ DISTANCE` for Distance Method, `Row` for Standardization), and click Run.

The result reports that the observed and expected General G's are 0.002090 and 0.001185, respectively. The z-score (= 17.94) indicates a strong clustering of high homicide rates. One may repeat the analysis by selecting "CONTIGUITY_EDGES_ONLY" for Conceptualization of Spatial Relationships and obtain a similar result, suggesting a statistically significant clustering of high homicide rates (the observed and expected General G's are 0.002125 and 0.001185, respectively, with the z-score = 8.316).

Similarly, execute another analysis by using the tool "Spatial Autocorrelation (Morans I)." The result suggests even a stronger clustering of similar (either high–high or low–low) homicide rates (the observed and expected Moran's I's are 0.167664 and −0.001185, respectively, with the z-score = 17.92 when using the Euclidean distances for conceptualization of spatial relationships).

Step 4. Mapping Local Moran's Ii and Local Gi*: Search and activate the tool "Cluster and Outlier Analysis (Anselin Local Morans I)" in the Geoprocessing pane. In the dialog, define Input Feature Class, Input Field, Conceptualization of Spatial Relationships, and Distance Method similar to those in step 3 and name the Output Feature Class HomiR _ LISA. As shown in Figure 8.7a, most of the tracts belong to the type "not significant," there are two major areas of "high–high cluster," only several tracts in the north are "high–low outlier," about a dozen tracts of "low–high outlier" are interwoven with the high–high clusters, and the type "low–low cluster" concentrates in two major areas. The attribute table of HomiR _ LISA contains four new fields: LMiIndex for the local Moran's I_i, LmiZScore the corresponding z-score, LmiPValue the p-value, and COType taking the values of NULL (blank), HH, HL, LH, and LL for the five types of spatial autocorrelation as explained.

Repeat the analysis using the tool "Hot Spot Analysis (Getis-Ord Gi^*)." This time, experiment with another setting for "Conceptualization of Spatial Relationships," such as FIXED_DISTANCE_BAND, and save the output feature as HomiR _ Gi. As shown in Figure 8.7b, it clearly differentiates those hotspots (clusters of high homicide rates) versus cold spots (clusters of low homicide rates) and otherwise not significant. Among the hotspots or the cold spots, it is further divided into three types with different levels of

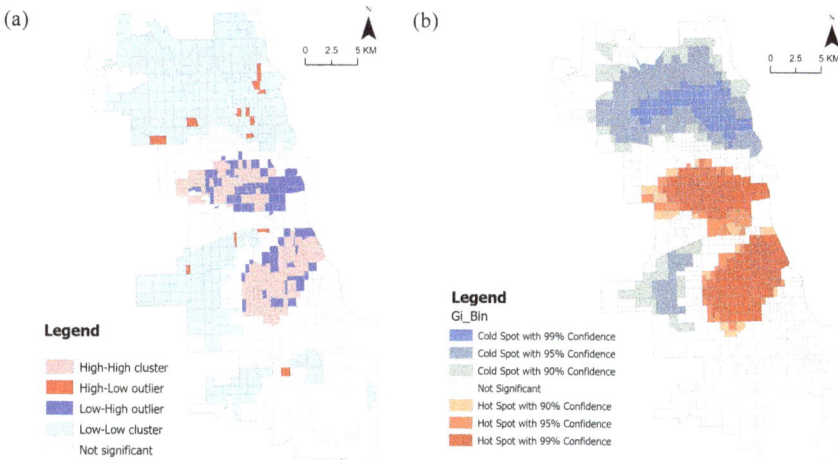

FIGURE 8.7 Spatial cluster analysis of homicide rates in Chicago: (a) local Moran's I and (b) local Gi*.

confidence (90%, 95%, and 99%). Note that the two hotspots of high homicide rates are largely consistent with the two major clusters of high–high clusters identified by the local Moran's I. The attribute table of `HomiR_Gi` contains three new fields: `GiZScore` the z-score, `GiPValue` the p-value, and `GiBin` taking the values of –3, –2, –1, 0, 1, 2, and 3 for 99% confidence cold spot, 95% confidence cold spot, 90% confidence cold spot, not significant, 90% confidence hotspot, 95% confidence hotspot, and 99% confidence hotspot, respectively. It does not report the exact $Gi*$ values, whose statistical significances, as indicated in the field `GiBin`, are derived from the corresponding z-scores.

8.8.2   Regression Analysis of Homicide Patterns

This part implements several regression models of homicide patterns in Chicago. Three global models include the OLS model, the spatial lag model, and the spatial error model, and the local one is the GWR. The OLS and GWR models are implemented in ArcGIS, and the spatial lag and spatial error models are implemented in R-ArcGIS tools. For the reason explained previously, the logarithm of homicide rate is used as the dependent variable in all regression models.

Step 5. Implementing the OLS Regression in ArcGIS Pro: Search and activate the tool Ordinary Least Squares. In the dialog window, choose `citytract` as the Input Feature Class and `IDINDEX` as the Unique ID Field, name the Output Feature Class `Homi_OLS`, choose `LnHomiRate` as Dependent Variable, check four fields (`FACTOR1`, `FACTOR2`, `FACTOR3`, and `JA`) as explanatory variables, name the Output Report File `OLSRpt`, and click Run.

The result is summarized in Table 8.2. Except for factor 3, both factor 1 and factor 2 are positive and significant, indicating that more concentrated disadvantage and concentrated Latino immigration contribute to higher homicide rates. Poorer job accessibility also contributes to higher homicide rates.

TABLE 8.2
OLS and Spatial Regressions of Homicide Rates in Chicago (n = 845)

Independent Variables	OLS Model	Spatial Lag Model	Spatial Error Model
Intercept	6.1324	4.5431	5.8325
	(10.87) ***	(7.53) ***	(8.98) ***
Factor1	1.2200	0.9669	1.1783
	(15.43) ***	(10.93) ***	(12.91) ***
Factor2	0.4989	0.4053	0.4779
	(7.41) ***	(6.02) ***	(6.02) ***
Factor3	–0.1230	–0.0995	–0.0858
	(–1.83)	(–1.53)	(–1.09)
Job access	–2.9143	–2.2096	–2.6340
	(–5.41) ***	(–4.14) ***	(–4.26) ***
Spatial lag (ρ)		0.2734	
		(5.86) ***	

TABLE 8.2 (*Continued*)
OLS and Spatial Regressions of Homicide Rates in Chicago (n = 845)

Independent Variables	OLS Model	Spatial Lag Model	Spatial Error Model
Spatial error (λ)			0.2606
			(4.79) ***
R^2	0.395	0.417	0.408

Note: Values in parentheses indicate t-values.
***Significant at 0.001.

Step 6. Implementing the Spatial Regression Model in R-ArcGIS Tools: In the Catalog, activate the SpatialRegressionModel tool under the folder R _ ArcGIS _ tools. In the dialog, choose citytract as Input Table, choose LnHomiRate under Dependent Variable and FACTOR1, FACTOR2, FACTOR3, and JA for Independent Variables, keep the default setting Queen under Spatial Weight, input citytractSPDq for Output Table, and click Run.

Click on View Details to view the report. It contains the Moran's I test to detect the presence or absence of spatial autocorrelation on the errors from the OLS model and the Lagrange Multiplier (LM) test to compares the model's fit relative to the OLS model. The table citytractSPDq summarizes the results of Spatial Lag Model and Spatial Error Model with coefficients, standard error, t-value, p-value, pseudo-R squared, AIC, and BIC.

The results from the spatial lag and spatial error models are largely consistent with the OLS regression result in step 5. The results are also summarized in Table 8.2. Note that both spatial regression models are obtained by the maximum likelihood (ML) estimation. Both t statistics for the spatial lag (ρ) and the spatial error (λ) are very significant and thus indicate the necessity of using spatial regression models over the OLS regression to control for spatial autocorrelation.

Step 7. Implementing the Geographically Weighted Regression (GWR) Model in ArcGIS: Search and activate the tool Geographically Weighted Regression (GWR) in the Geoprocessing pane. In the dialog (Figure 8.8), choose citytract as the Input Features, select LnHomiRate as Dependent Variable, keep Continuous (Gaussian) as Model Type, check four variables (FACTOR1, FACTOR2, FACTOR3, and JA) as Explanatory Variables, and name the Output Features Homi_GWR. Under Neighborhood Type, choose Number of Neighbors, Golden search for Neighborhood Selection Method, and keep the empty value under Minimum Number of Neighbor and Maximum Number of Neighbors.[5] Under Prediction Option, choose citytract for Prediction Locations, then the same variables are automatically added; input citytractGWR for Output Predicted Features, check Robust Prediction > Under Additional Options, keep Bisquare for Local Weighting Scheme and blank for Coefficient Raster Workspace, and click Run.

The results include maps of standard residuals (i.e., field StdResid in the output feature Homi _ GWR), GWR-predicted homicide rates, R-squared, and estimated coefficients. As shown in Figure 8.9, the estimated coefficients of factor 1, factor 2,

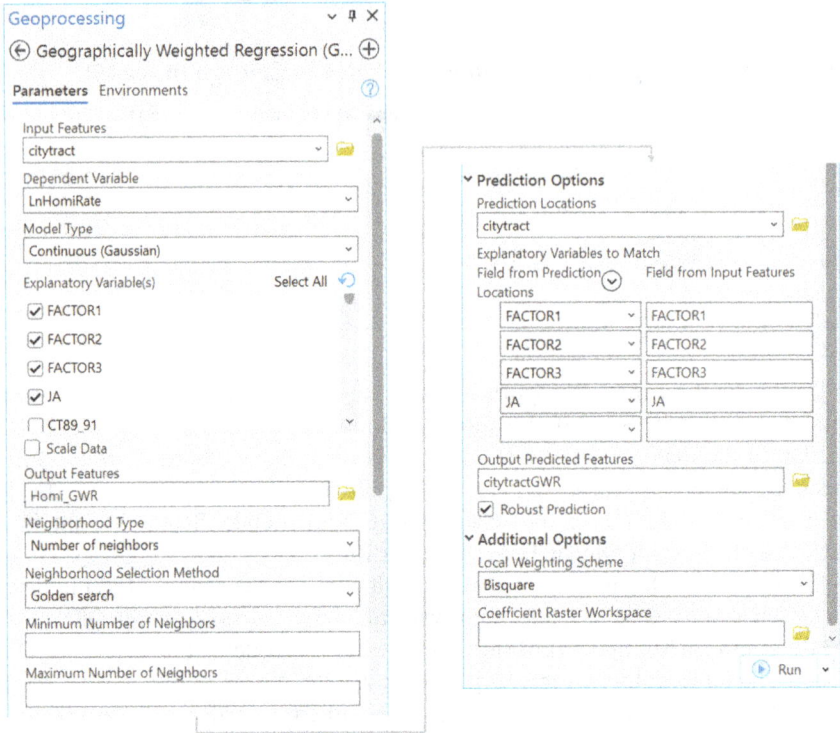

FIGURE 8.8 Dialog window for GWR in ArcGIS Pro.

factor 3, and job accessibility (JA) in fields C1 _ FACTOR1, C2 _ FACTOR2, C3 _
FACTOR3, and C4 _ JA vary across the study area. The coefficients are positive for
factor 1 (concentrated disadvantage) and factor 2 (concentrated Latino immigration)
and negative for job accessibility but vary from −0.36 to 0.32 for factor 3 (residential
instability). These are consistent with the findings from the three global regression
models reported in Table 8.2 (i.e., consistent with signs for factor 1, factor 2, and job
accessibility that are statistically significant and nonsignificant coefficient for factor
3). In addition, the positive effect of concentrated disadvantage on homicide rates is
the strongest on the "south side" (where the area has long been plagued with socioeconomic disadvantages) and declines gradually toward north and downtown; the positive
effect of concentrated Latino immigration is the strongest in the southwest (with the
highest concentration of Latino immigrants) and declines toward north and southeast,
and the negative effect of job accessibility is the strongest across an arc linking the
southwest-downtown-northeast and declines toward southeast and northwest.

8.9   SUMMARY

The centrographic measures provide a basic set of descriptive measures of geographic pattern in terms of the mean center, degree of dispersion from the center,
and possible orientation. Spatial cluster analysis detects non-randomness of spatial

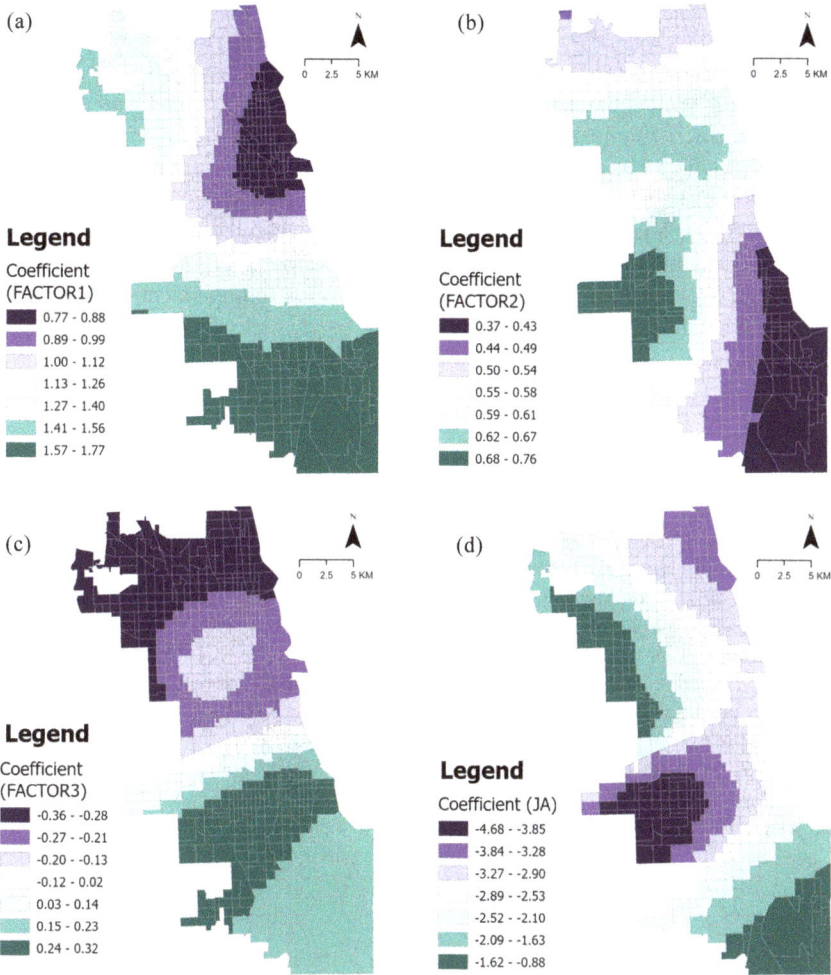

FIGURE 8.9 GWR estimated coefficients for (a) factor 1, (b) factor 2, (c) factor 3, and (d) job accessibility.

patterns or existence of spatial autocorrelation. The spatial cluster method, such as the nearest neighbor index, is based on locations only and identifies whether data points are clustered, dispersed, or random. Other spatial cluster methods also focus on feature locations, but features are distinctively classified as cases and controls; and the methods analyze whether events of cases within a radius exhibit a higher level of concentration than a random pattern would suggest. Colocation analysis examines whether two types of points tend to be clustered together or dispersed apart. Spatial cluster analysis methods based on feature values examine whether objects in proximity or adjacency are related (similar or dissimilar) to each other. The existence of spatial autocorrelation necessitates the usage of spatial regression in regression analysis. Global spatial regression models are the spatial lag model and the spatial error model. Both are estimated by the maximum likelihood (ML) method. The local model is the geographically weighted regression (GWR) that yields regression coefficients and corresponding t statistics varying across a study area.

Most of the spatial statistical methods covered in this chapter, such as the centrographic measures, the nearest neighbor index, the colocation quotient, spatial cluster analysis indices based on feature values, and the GWR, can be implemented in ArcGIS Pro. Others, such as the cluster analysis on case-and-control features and spatial lag and spatial error regressions, can be implemented in specific R-ArcGIS tools. The expanded capacity of ArcGIS in spatial statistics and access to related R-ArcGIS tools helps bring down the technical barriers to implementing these methods. Social scientists, including those working in applied fields such as crime and health studies, have increasingly benefited from this perspective.

NOTES

1 The number may be slightly larger than k since the last (farthest) area among those nearest areas may contain more than one case.
2 The value of Unique ID Field should be identical to the values of row index; otherwise, it will not be identified by the LCLQ tool.
3 If the spatial (geographic) weights are defined in such a way (e.g., with a threshold) that eliminates observations beyond a distance, only the observations within that threshold are included for the regression.
4 The choice of adding 1 is arbitrary and may bias the coefficient estimates. However, different additive constants have minimal consequence for significance testing as standard errors grow proportionally with the coefficients (Osgood, 2000, p. 36). In addition, adding 1 ensures that $log(r+1) = 0$ for $r = 0$ (zero homicide).
5 The minimum (maximum) number of neighbors cannot be too small. Otherwise, it may not estimate at least one local model due to multicollinearity (data redundancy).

9 Regionalization Methods and Application in Analysis of Cancer Data

Analysis of rare events (e.g., cancer, AIDS, homicide) often suffers from the *small population (numbers) problem*, which can lead to unreliable rate estimates, sensitivity to missing data and other data errors (Wang and O'Brien, 2005), and data suppression in sparsely populated areas. The spatial smoothing techniques, such as the floating catchment area method and the empirical Bayesian smoothing method, as discussed in Chapter 3, can be used to mitigate the problem. This chapter introduces a more advanced approach, namely, "regionalization," to this issue. *Regionalization* is to group many small areas into a smaller number of regions while optimizing a given objective function and satisfying certain constraints. Chapter 4 covers one type of regionalization that delineates *functional regions*, within which connection strengths (e.g., service flow, passenger volume, financial linkage, or communication) are the strongest. This chapter focuses on deriving *homogeneous regions*, within which areas are more similar in attributes than those beyond.

The chapter begins with an illustration of the small population problem and a brief survey of various approaches to the problem in Section 9.1. Section 9.2 discusses several GIS-automated regionalization methods, such as SCHC, SKATER, AZP, Max-P, and REDCAP methods, that construct a series of different numbers and thus different sizes of regions from small areas. Section 9.3 introduces the *mixed-level regionalization (MLR) method*, which decomposes areas of large population and merges areas of small population simultaneously to derive regions with comparable population size. Section 9.4 uses a case study of analyzing breast cancer rates in Louisiana to illustrate the implementation of these methods. Section 9.5 concludes with a summary.

9.1 THE SMALL POPULATION PROBLEM AND REGIONALIZATION

The small population problem is common for analysis of rare events. The following uses two fields, crime and health studies, to illustrate issues related to the problem and some attempts to mitigate the problem.

In criminology, the study of homicide rates across geographic units and for demographically specific groups often entails analysis of aggregate homicide rates in small populations. Research has found inconsistency between the two main sources of homicide data in the United States, that is, the Uniform Crime Reporting (UCR) and the National Vital Statistics System (NVSS) (Dobrin and Wiersema, 2000). In addition, a sizeable and growing number of unsolved homicides have to be excluded

DOI: 10.1201/9781003292302-11

from studies analyzing offender characteristics (Fox, 2000). For a small county with only hundreds of residents, data errors for one homicide could swing the homicide rate per 100,000 residents by hundreds of times (Chu et al., 2000). All raise the question of reliability of homicide rates.

Several non-geographic strategies have been attempted by criminologists to mitigate the problem. For example, Morenoff and Sampson (1997) used homicide counts instead of per capita rates or simply deleted outliers or unreliable estimates in areas with small population. Some used larger units of analysis (e.g., states, metropolitan areas, or large cities) or aggregated over more years to generate reliable homicide rates. Crime rates in small populations violate two assumptions of ordinary least squares (OLS) regressions, that is, homogeneity of error variance or homoscedasticity (because errors of prediction are larger for crime rates in smaller populations) and normal error distribution (because more crime rates of zero are observed as populations decrease). Land et al. (1996) and Osgood (2000) used *Poisson regression* to better capture the non-normal error distribution pattern in regression analysis of homicide rates in small population (see Appendix 9).

In health studies, another issue, such as *data suppression*, is associated with the small population problem. Figure 9.1, generated from the State Cancer Profiles website (statecancerprofiles.cancer.gov), shows age-adjusted death rates for female breast cancer in Illinois counties for 2003–2007. Rates for 37 out of 102 counties (i.e., 36.3%, mostly rural counties) are suppressed to "ensure confidentiality and stability of rate estimates" because counts were 15 cases or fewer. Cancer incidence in these

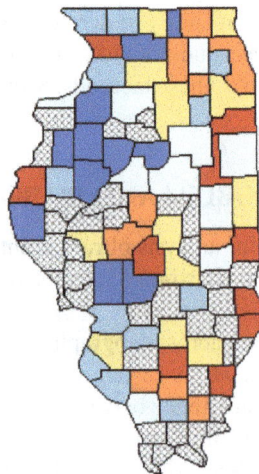

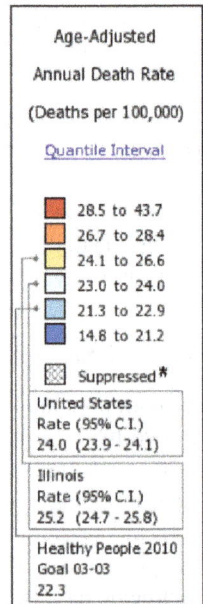

FIGURE 9.1 Female breast cancer death rates in Illinois, 2003–2007.

counties cannot be analyzed, leaving large gaps in our understanding of geographic variation in cancer and its social and environmental determinants (Wang et al., 2012).

Many researchers in health-related fields have used some spatial analytical or geographic methods to address the issue. For example, *spatial smoothing* computes the average rates in a larger spatial window to obtain more reliable rates (Talbot et al., 2000). Spatial smoothing methods include the floating catchment area method, kernel density estimation and empirical Bayes estimation as discussed in Chapter 3, locally weighted average (Shi, 2007), and adaptive spatial filtering (Tiwari and Rushton, 2004; Beyer and Rushton, 2009), among others. Another method, hierarchical Bayesian modeling (HBM), commonly used in spatial epidemiology, uses a nonparametric Bayesian approach to detect clusters of high risk and low risk with the prior model, assuming constant risk within a cluster (Knorr-Held, 2000; Knorr-Held and Rasser, 2000).

A geographic approach is to construct larger areas from the ones with a small population. The purpose is similar to that of aggregating over a longer period of time: to achieve a greater degree of stability in homicide rates across areas. The technique has much common ground with the long tradition of regional classification (*regionalization*) in geography (Cliff et al., 1975), which is particularly desirable for studies that require comparable analysis areas with a minimum base population. For instance, Black et al. (1996) developed the *ISD method* (named after the Information and Statistics Division of the Health Service in Scotland, where it was devised) to group a large number of census enumeration districts (ED) in the UK into larger analysis units of approximately equal population size. Lam and Liu (1996) used the *spatial order method* to generate a national rural sampling frame for HIV/AIDS research, in which some rural counties with insufficient HIV cases were merged to form larger sample areas. Both approaches emphasize spatial proximity, but neither considers within-area homogeneity of attributes.

Haining et al. (1994) attempted to consolidate many EDs in the Sheffield Health Authority Metropolitan District in the UK to a manageable number of regions for health service delivery. Commonly referred to as the *Sheffield method*, it started by merging adjacent EDs sharing similar deprivation index scores (i.e., complying with *within-area attribute homogeneity*) and then used several subjective rules and local knowledge to adjust the regions for spatial compactness (i.e., accounting for *spatial proximity*). The method attempted to balance the two criteria of attribute homogeneity and spatial proximity, a major challenge in regionalization analysis. In other words, only contiguous areas can be clustered together, and these areas must have similar attributes. If very different areas were grouped together, much of the geographic variation, which is of primary interest to spatial analysis, would be smoothed out in the regionalization process.

The Sheffield method explicitly strives to account for spatial and attributive similarity or distance in the process of aggregating or clustering small areas, but the process remains largely manual and incurs much uncertainty. Some early efforts to automate regionalization begin with spatializing classic clustering (SCC) methods (Anselin, 2020a). The classic clustering methods, such as k-means, k-medoids, hierarchical, and spectral clustering, some of which are discussed in Section 7.3 of Chapter 7, are applied to group areas by their spatial distance (i.e., locational dissimilarity) alone or a combination of spatial and attributive distances via assigned weighting factors. Such an

TABLE 9.1
Approaches to the Small Population Problem

Approach	Examples	Comments
Use counts instead of per capita rates	Morenoff and Sampson (1997)	Not applicable for most studies that are interested in the rates relative to population sizes.
Delete samples of small populations	Harrell and Gouvis (1994), Morenoff and Sampson (1997)	Omitted observations may contain valuable information and bias a study.
Aggregate over more years or to a higher geographic level	Messner et al. (1999), most studies surveyed by Land et al. (1990)	Infeasible for analysis of variations within the time period or within the larger areas.
Poisson regression	Osgood (2000), Osgood and Chambers (2000)	Not applicable to non-regression studies.
Spatial smoothing: floating catchment area, kernel density estimation, empirical Bayes estimation, locally weighted average, adaptive spatial filtering	Clayton and Kaldor (1987), Shi (2007), Tiwari and Rushton (2004)	Suitable for mapping trends, but smoothed rates are not true rates in analysis areas.
Regionalization: Sheffield method, spatial order, ISD, SCHC, SKATER, AZP, Max-P, REDCAP, MLR	Openshaw (1977), Haining et al. (1994), Lam and Liu (1996), Black et al. (1996), Assunção et al. (2006), Guo (2008), Duque et al. (2012), Mu et al. (2015)	Generating reliable rates for statistical reports, mapping, exploratory spatial analysis, regression analysis, and others.

approach does not enforce or ensure the polygon contiguity of constructed regions and thus is not strictly considered as spatial clustering or regionalization.

Table 9.1 summarizes the approaches to the small population problem. Regionalization shows the most promise. Constructing geographic areas by regionalization enables analysis to be conducted at multiple geographic levels and thus permits the test of the *modifiable areal unit problem* (*MAUP*) (Fotheringham and Wong, 1991). This issue is also tied to the *uncertain geographic context problem* (*UGCoP*) (e.g., in multilevel modeling), referring to sensitivity in research findings to different delineations of contextual units (Kwan, 2012). Furthermore, since similar areas are merged to form new regions, spatial autocorrelation is less of a concern in the new regions, which unburdens an analyst of the need of using advanced spatial regression methods for mitigating the effect of spatial autocorrelation as discussed in Section 8.7 of Chapter 8. Various automated regionalization methods in GIS will be discussed in the next two sections.

9.2 GIS-AUTOMATED REGIONALIZATION METHODS

Spatial regionalization methods can be divided into three types, spatializing classic clustering (SCC), spatially constrained hierarchical clustering (SCHC), and spatially constrained clustering with partitioning (SCCP) methods. As discussed briefly in Section 9.1, the SCC models can be seen as enforcing a soft contiguity constraint with attribute similarity and locational proximity and thus are not considered strictly regionalization. This section discusses other two methods, SCHC and SCCP, which impose contiguity as a hard constraint in a clustering procedure. A separate section (Section 9.3) is dedicated to the mixed-level regionalization (MLR) method, a rather-complex regionalization method.

9.2.1 THE SPATIALLY CONSTRAINED HIERARCHICAL CLUSTERING (SCHC) METHODS

The *spatially constrained hierarchical clustering (SCHC)* derives from unconstrained hierarchical clustering with the same linkage options as discussed in Section 7.3 of Chapter 7 (e.g., single linkage, complete linkage, average linkage, and Ward's method) but imposes an additional contiguity constraint (e.g., queen or rook). In other words, during the agglomerative (bottom-up) agglomeration process, the search for minimum attributive distance or dissimilarity measure is only limited to those pairs of contiguous observations (Anselin, 2020b). The following discusses two common methods: SKATER and REDCAP.

The *SKATER* (Spatial K'luster Analysis by Tree Edge Removal) algorithm was proposed by Assunção et al. (2006). The continuity structure between areas can be represented as a tree graph with contiguous observations as nodes connected via edges. The full graph is reduced to a minimum spanning tree (MST) so that a path connects all nodes, with each node being visited once. Define the overall dissimilarity measure as the total sum of squared deviations (SSD), $SSD = \sum_i (x_i - \bar{x})^2$, where x_i is attribute value for observation i, and $\bar{x}$ is overall mean. The MST is pruned (thus creating two graphs or regions) by removing an edge that maximizes the between-region SSD. The process continues until a desirable number of regions are derived.

The approach of SKATER is divisive instead of agglomerative, which starts with a single cluster and finds the optimal split into subclusters until the target group number is satisfied. Once the tree is cut at one point, all subsequent cuts are limited to the resulting subtrees, leading to possible limitations. As the name states, the *REDCAP (regionalization with dynamically constrained agglomerative clustering and partitioning)* method combines both the divisive (partitioning) and the agglomerative processes (Guo, 2008).

As illustrated in Figure 9.2, REDCAP is composed of two steps: (1) contiguity-constrained hierarchical clustering and (2) top-down tree partitioning. The color of each polygon represents its attribute value, and similar colors represent similar values. Two polygons are considered contiguous in space if they share a segment of boundary (i.e., based on the rook contiguity). In the first step, as shown in Figure 9.2a, REDCAP constructs a hierarchy of spatially contiguous clusters based on the attribute similarity under contiguity constraint. Two adjacent and most similar areas are grouped to form the first cluster, two adjacent and most similar clusters are grouped together to form a higher-level cluster, and so on until the whole study area is one cluster. It is a clustering process similar to the one explained in

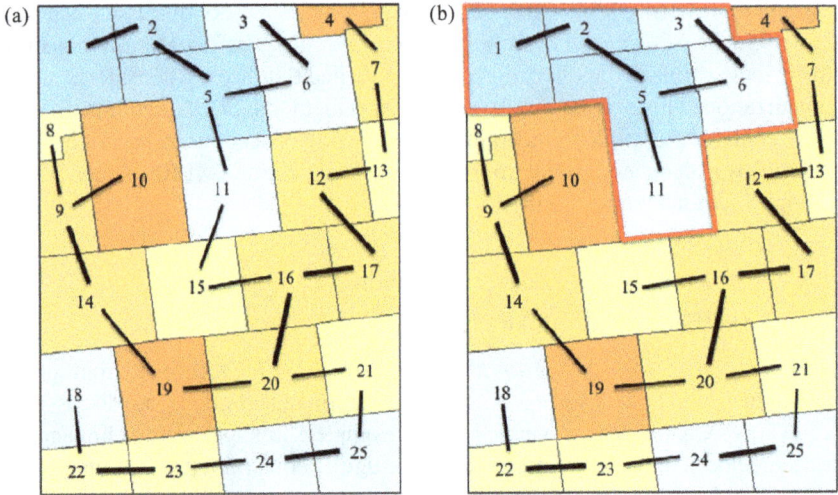

FIGURE 9.2 An example illustrating REDCAP: (a) a spatially contiguous tree is built with a hierarchal clustering method; (b) partitioning the tree by removing the edge that optimizes the SSD measure.

Section 7.3 and adds one new spatial constraint (i.e., only spatially contiguous polygons or clusters can be grouped). A clustering tree is generated to fully represent the cluster hierarchy (i.e., each cluster at any level is a sub-tree). In the second step, REDCAP partitions the tree to generate two regions by removing the best edge (i.e., link 11–15 in Figure 9.2b) that optimizes the homogeneity measure (i.e., SSD). In other words, the two regions are created in a way that the total within-region homogeneity is maximized. The partitioning continues until the desired number of regions is reached.

For the purpose of mitigating the small population problem, REDCAP is modified to accommodate one more constraint, such as a minimum regional size (Wang et al., 2012). It can be a threshold region population and/or the number of incidents (whichever is the denominator in a rate estimate). Such a constraint is enforced in the second step, that is, tree partitioning. For each potential cut, if it cannot produce two regions that both satisfy the constraints, the cut will not be considered as a candidate cut. Then the best of all candidate cuts is chosen to partition a tree into two regions. The process continues until no regions can be partitioned further. The resulting regions are all large enough and have the highest homogeneity within each region. Much of the discussion on applications is limited to crime and health studies in this chapter. Valuable applications of regionalization certainly go beyond. A recent study uses the REDCAP to divide China into two regions of nearly identical area size and greatest contrast in population and sheds new light on the scientific foundation of classic "Hu Line" (F Wang et al., 2019). Another uses mobile app data to define a hierarchical urbanization "source-sink" regions in China in terms of intensity of labor force import or output (Y Wang et al., 2019).

9.2.2   THE SPATIALLY CONSTRAINED CLUSTERING WITH PARTITIONING (SCCP) METHODS

The *spatially constrained clustering with partitioning (SCCP)* is also referred to as the *p-regions problem*, combining n spatial units into p larger regions that are made up of contiguous entities and maximize internal homogeneity based on heuristic algorithms (Anselin, 2020c). Two popular methods are discussed in this subsection: AZP and Max-P.

The *automatic zoning problem, or AZP* (Openshaw, 1977; Openshaw and Rao, 1995; Cockings and Martin, 2005; Grady and Enander, 2009), starts with an initial random regionalization and then iteratively refines the solution by reassigning objects to neighboring regions to improve an objective function value (e.g., maximum homogeneity within derived regions). The AZP method is a local optimization procedure, and therefore, its result varies, depending upon the initial randomization state. Therefore, the random seed, which is used to determine the initial positions, could be used to execute several rounds of Max-P algorithms for sensitive analysis.

The *Max-P model* groups a set of geographic areas into the maximum number of homogeneous regions by imposing a minimum size constraint for each region (Duque et al., 2012). It is formulated as an integer programming problem to solve the number of regions. As stated in Section 9.1, a minimum region size (e.g., population) is often desirable for reliable rate estimates or data dissemination. While meeting the constraint of a minimum region size, it strives to derive a maximum number of regions to retain the highest resolution.

Both the AZP and Max-P use various heuristic methods (e.g., greedy search, simulated annealing, or tabu search) to find a local optimum. In Case Study 9 in Section 9.4, the greedy algorithm is adopted for both and thus termed for *AZP_greedy* and *Maxp_greedy*.

All the five methods discussed in this section are included in the RegionalizationR tool introduced in Section 9.4. Only the SKATER method is available in ArcGIS Pro, via Spatial Statistics Tools > Mapping Clusters > Spatially Constrained Multivariate Clustering. SCHC, SKATER, REDCAP, or AZP_greedy needs to define the number of clusters, while the MaxP_greedy automatically generates an optimal number of clusters. The RegionalizationR tool can also specify a certain bound variable for newly formed regions with the support of the rgeoda package. Therefore, the methods can generate a series of different sizes of regions by changing the cluster number and/or bound variable.[1] A method with such a capacity is thus termed *"scale-flexible regionalization"* and enables an analyst to examine a study area at different scales and thus generate regions of sufficient size.

9.3   THE MIXED-LEVEL REGIONALIZATION (MLR) METHOD

The mixed-level regionalization (MLR) method decomposes areas of large population (to gain more spatial variability) and merges areas of small population (to mask privacy of data) in order to obtain regions of comparable and sufficient population. One key feature of the method is its capacity of balancing spatial compactness and attributive homogeneity in derived regions. Both are important properties in designing a regionalization method. In formulating the MLR, Mu et al. (2015) modified Peano curve algorithm (thus termed "MPC") to account for spatial connectivity and compactness, adopted the modified scale–space clustering (MSSC) to ensure attributive similarity, and integrated the two by assigning corresponding weighting factors. This section summarizes the design process.

9.3.1 MODIFIED PEANO CURVE ALGORITHM (MPC)

Italian mathematician Giuseppe Peano (1890) first developed *Peano curve* or the spatial order method. It uses *space-filling curves* to determine the nearness or spatial order of areas. Space-filling curves traverse space in a continuous and recursive manner to visit all areas and assign a spatial order from 0 to 1 to each area based on its relative positions in a two-dimensional space. As shown in Figure 9.3, the spatial order of each area is calculated and labeled (on the left). The centroids of these areas in the 2D space are then mapped onto the 1D line underneath. The Peano curve connects the centroids from the lowest spatial order to the highest (on the right). In general, areas that are close together have similar spatial order values, and areas that are far apart have dissimilar spatial order values. Based on their spatial order values, the areas can be grouped by using a clustering method discussed in Section 7.3 of Chapter 7.

A generic space-filling heuristic algorithm developed by Bartholdi and Platzman (1988) has been widely used in the GIS community (Mandloi, 2009). In ArcGIS Pro, the procedure is available in Data Management Tools > General > Sort > select Shape in the Field(s) parameter, and choose the option PEANO under Spatial Sort Method. The method has been used to construct larger areas from small ones in studies on health (Lam and Liu, 1996) and crime (Wang and O'Brien, 2005).

Some modifications are needed on the Peano or space-filling curve algorithm to address problems such as arbitrary jumps and predefined turns, and therefore the method is termed "*modified Peano curve (MPC) algorithm.*" The modifications

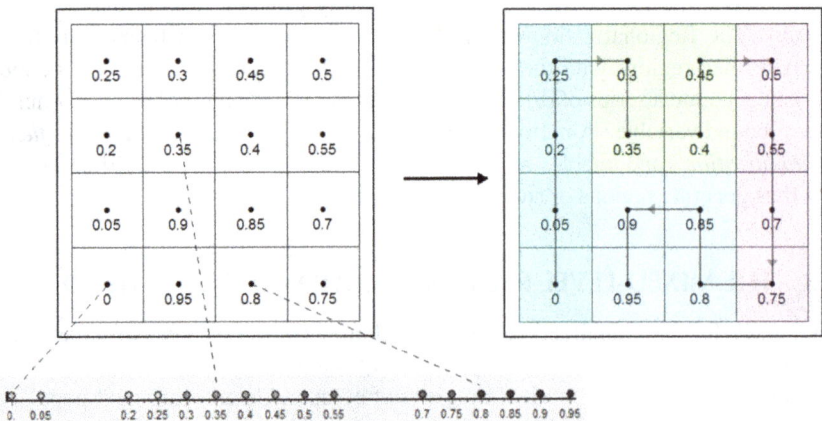

FIGURE 9.3 An example of assigning spatial order values to areas.

include (1) setting clustering boundaries with threshold population (e.g., lower limits of population and/or cancer count), (2) combining a spatial weight matrix (based on polygon adjacency) with spatial orders to enforce spatial contiguity, and (3) forcing a cluster membership for unclaimed or loose-end units.

9.3.2 MODIFIED SCALE–SPACE CLUSTERING (MSSC)

Modified scale–space clustering (MSSC) (Mu and Wang, 2008) is developed based on the scale–space theory (Koenderink, 1984). Using an analogy in viewing images, the scale–space theory treats scale like distance in viewing images. With the increase of scale (similar to the effect of an observer's distance increasing), the same image can reveal higher levels of generalizations and less details. Mathematically, a Gaussian function can be used to formulate a 2D scale–space in image analysis (Leung et al., 2000). In socioeconomic analysis, as it is our case, GIS data often come as polygons with multiple attributes. Mu and Wang (2008) simplified the algorithm to capture the attributes and neighboring relationships.

The idea is to melt each area to its most similar neighbor. The similarity is measured by the attribute distance between two neighboring areas, like Euclidean distance, as the sum of the (weighted) squares of multiple attribute differences of each pair between the unit itself and all its neighbors. In formula, a polygon i has t attributes standardized as $(x_{i1}, \ldots, x_{it})$, its adjacent polygons j have attributes also standardized as $(x_{j1}, \ldots, x_{jt})$, and the attributive distance between them is $D_{ij} = \sum_t \left(x_{it} - x_{jt}\right)^2$. By the minimum-distance criterion, we search for the shortest attribute distance (i.e., the highest attribute similarity) between i and its adjacent polygons j. By doing so, we merge each polygon to its most similar neighbor.

A polygon with the highest attribute value among its surrounding polygons serves as a local nucleus and searches outward for its most similar neighboring polygon for grouping. A branch of the outward-searching process continues until it reaches a polygon with the lowest attribute value among its surrounding polygons. The grouping process is much like our cognitive process of viewing a picture: the pattern or structure is captured by its brightest pixels, and the surrounding darker pixels serve as the background. In practice, for the purpose of operation, one needs to follow the direction from a local minimum to a local maximum and group all the polygons along the way.

9.3.3 INTEGRATING MPC AND MSSC

The modified Peano curve algorithm (MPC) accounts for the requirements of spatial connectivity and compactness with additional cluster bound constraints, such as population and cancer count. The modified scale–space clustering (MSSC) method considers attributive homogeneity in the clustering procedure. The two need to be integrated in order to balance spatial compactness and attributive homogeneity in clustering according to a user's definition. Similar efforts in explicitly formulating relative weights of spatial and attributive distances or similarities in spatial clustering can be found in Webster and Burrough (1972), Haining et al. (2000), and most recently, Chavent et al. (2018).[2]

Recall that the MPC algorithm yields a spatial order for each unit, o_{si}, and the MSSC generates an attributive order, o_{ai}. Both are normalized. When a user defines their corresponding weighting factors w_s and w_a, an integrated clustering order value o_i for unit i becomes:

$$o_i = w_s.o_{si} + w_a.o_{ai}. \tag{9.1}$$

In practice, set the value of w_s larger than 0 and preferably larger than 50% to emphasize entity connectivity, and $w_a = 1 - w_s$.

9.3.4 MIXED-LEVEL REGIONALIZATION (MLR)

The aforementioned integrated method is applied to areas of multiple levels and therefore termed *mixed-level regionalization (MLR)*. For instance, for rural counties with small population (or a low cancer count), it is desirable to group counties to form regions of similar population size; and for urban counties with a large population (or a high cancer count), it is necessary to segment each into multiple regions, also of similar size, and each region is composed of lower-level areas (e.g., census tracts). The method attempts to generate regions that are recognizable by preserving county boundaries as much as possible. One conceivable application of the method is the design and delineation of cancer data release regions that require a minimum population of 20,000 and the minimum cancer counts greater than 15 for the concern of privacy. The resulting regions are mixed-level, with some tract regions (sub-county level), some single-county regions, and some multi-county regions. The next section illustrates a detailed step-by-step implementation of the MLR method in a case study.

9.4   CASE STUDY 9: CONSTRUCTING GEOGRAPHICAL AREAS FOR MAPPING CANCER RATES IN LOUISIANA

Section 9.1 illustrates the small population problem that is commonly encountered in cancer data analysis. Areas of small population need to be merged to protect geo-privacy and/or generate reliable cancer rates. Section 9.2 covers five popular region-alization methods (SCHC, SKATER, REDCAP, AZP, and Max-P), all of which incorporate a threshold population constraint explicitly to construct regions of suf-ficient size. On the other side, areas of large population need to be decomposed to gain more spatial variability. The MLR method discussed in Section 9.3 is proposed to meet both challenges by creating regions of comparable population but made of different levels of area units (e.g., some are composed of census tracts, and others of one county or multiple whole counties).

This section illustrates the implementation of all these methods in analyzing can-cer rates in Louisiana. The case study is developed from the research reported in Mu et al. (2015). Due to the data use agreement, the cancer counts were modified by using an algorithm to introduce a randomized component while preserving the general pattern and total counts. The construction of new areas enables us to map reliable cancer rates.

Data used in this project are provided in the layer LA _ Mixtracts of geodatabase Louisiana.gdb under the folder Louisiana, which includes 1,132 census tracts in Louisiana. Its attribute table includes the following fields:

1. POPU is population, Count02 to Count06 stand for the yearly cancer counts from 2002 to 2006, and Count02 _ 06 is the five-year sum of cancer counts.
2. Fact1, Fact2, and Fact3 are the factor scores consolidated from 11 census variables, labeled "Socioeconomic disadvantages," "High healthcare needs," and "Language barrier," respectively (Wang et al., 2012). A higher score of any of the three factors corresponds to a more disadvantaged area. The three factors will be used for measuring attribute similarity in the regionalization methods.

Under the folder CMGIS-V3-Toolbox, the R-based tool RegionalizationR is under R _ ArcGIS _ tools.tbx in the subfolder R _ ArcGIS _ tools, and mixed-level regionalization (MLR) tools are under MLR Tools Pro.tbx in the subfolder ArcPY-Tools.

9.4.1 Implementing the SKATER, AZP, Max-P, SCHC and REDCAP Methods

Step 1. Using Spatially Constrained Multivariate Clustering in ArcGIS Pro: In ArcGIS Pro, build a new project and add all data to map. Search and activate the tool Spatially Constrained Multivariate Clustering (based on the SKATER method). In the dialog, choose LA _ Mixtracts as Input Features and input LA _ SCMC for Output Features; under Analysis Fields, check FACT1, FACT2, and FACT3; choose Attribute value for Cluster Size Constraints and COUNT06 for Constraint Fields; input 15 for Minimum Per Cluster, 100 for Number of Clusters; choose Contiguity edges corners for Spatial Constraints; choose 100 for Permutations to Calculate Membership Probabilities; and input LA _ SCMC _ T for Output Table for Evaluating Number of Clusters. Click Run.

The output feature contains two fields, Cluster ID and Membership Probability. The field Cluster ID identifies which cluster (region) a census tract is grouped into.

Step 2. Implementing the SKATER, AZP, and Max-P Methods with the RegionalizationR Tool: The RegionalizationR tool contains popular regionalization methods, such as SKATER, AZP_greedy, MaxP_greedy, SCHC, and REDCAP. We first use the SKATER method as an example. The other methods can be easily replicated by choosing a corresponding tool.

In the Catalog, activate the RegionalizationR tool under the folder R _ ArcGIS _ tools. In the dialog window (Figure 9.4a), choose LA _ Mixtracts as Input Feature; choose FACT1, FACT2, and FACT3 for Control Variables[3] and Count06 for Bound Variable; input 15 for Minimum Bound; choose SKATER for

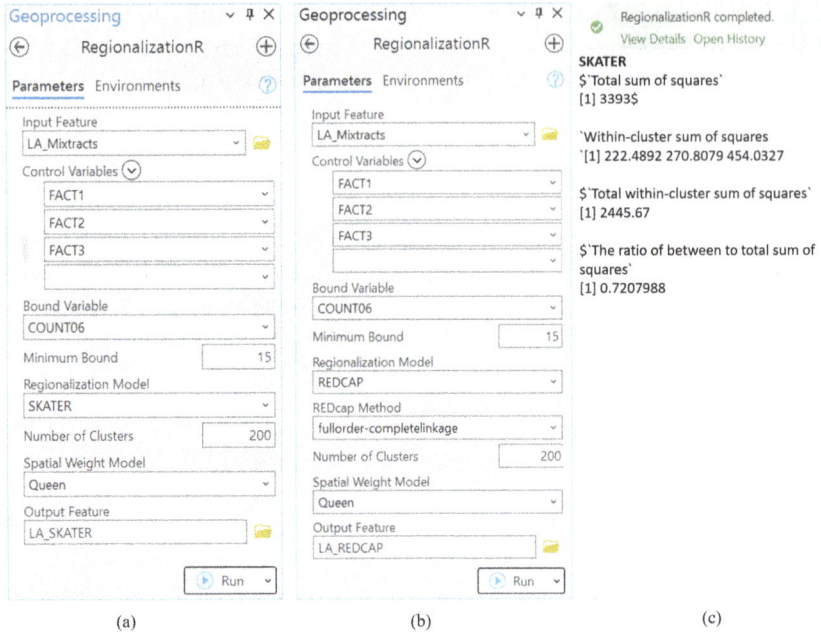

FIGURE 9.4 Implementing the RegionalizationR tool: (a) interface for SKATER, (b) interface for REDCAP, and (c) information in View Details.

Regionalization Model; input 200 for Number of Clusters; and choose Queen under Spatial Weight Model.[4] Input LA _ SKATER for Output Feature, and click Run.

The output feature LA_SKATER has one more field named "cluster," which indicates the new cluster IDs and can be used for aggregation. In View Details (Figure 9.4c), total sum of squares, within-cluster sum of squares, total within-cluster sum of squares, and the ratio of between to total sum of squares are reported.

The dialog interfaces for the AZP_greedy and MaxP_greedy have very similar input parameters but different corresponding regionalization model. As stated previously, the MaxP_greedy method does not have any parameter setting for "Number of Clusters" and automatically generates 548 clusters.

Step 3. Implementing the SCHC and REDCAP Methods with the RegionalizationR Tool: As explained in Section 9.2, both the SCHC and REDCAP use a linkage option (e.g., single linkage, average linkage, complete linkage, or Ward method) to enforce the hierarchical clustering procedure. Take REDCAP as an example, as shown in Figure 9.4b. Underneath the chosen Regionalization Method, define a linkage option (e.g., full-order complete linkage). The other parameters are similar to those for the SKATER method.

The output feature LA _ REDCAP has a new field cluster recording the cluster IDs. Based on the result reported in View Details, REDCAP has a more favorable result, that is, a higher ratio of between to total sum of squares than the SKATER method.

In other words, given the same scale (number) of regions, the REDCAP regions are more dissimilar from each other than the SKATER regions and thus more favorable.

Step 4. Mapping Cancer Rates in REDCAP-Derived Regions: In the Geoprocessing toolboxes pane, search and activate the Dissolve tool. In the dialog, choose LA _ REDCAP for Input Features; enter LA _ Ct15 for Output Feature Class; select (1) cluster under Dissolve Field(s) and (2) POPU and COUNT06 under Statistic Field(s) and (3) Sum for Statistic Type; and click Run. Add a new field BC _ rt to the attribute table of LA _ Ct15, and calculate it as 100000* SUM _ COUNT06/ SUM _ POPU, which is the crude breast cancer rate per 100,000 (Figure 9.5).

If desirable to enforce two constraints on region size, run the tool twice. For example, (1) implement the REDCAP method with the minimum cancer count of 15 and the cluster number of 400 to generate a layer LA _ REDCAP200, and then use the Dissolve tool to aggregate it in a layer LA _ Ct15400 and (2) run the REDCAP method on LA _ Ct15400 again by using MEAN _ FACT1, MEAN _ FACT1, and MEAN _ FACT1 for Control Variables, SUM _ POPU as Bound Variable, 20000 as Minimum Bound, and 200 for Number of Clusters to yield a layer LA _ REDCAP200, and then dissolve it to a final layer, LA _ Ct15200. The resulting regions meet both constraints of population and cancer counts. In the following subsection, the MLR method can accommodate double restraints simultaneously.

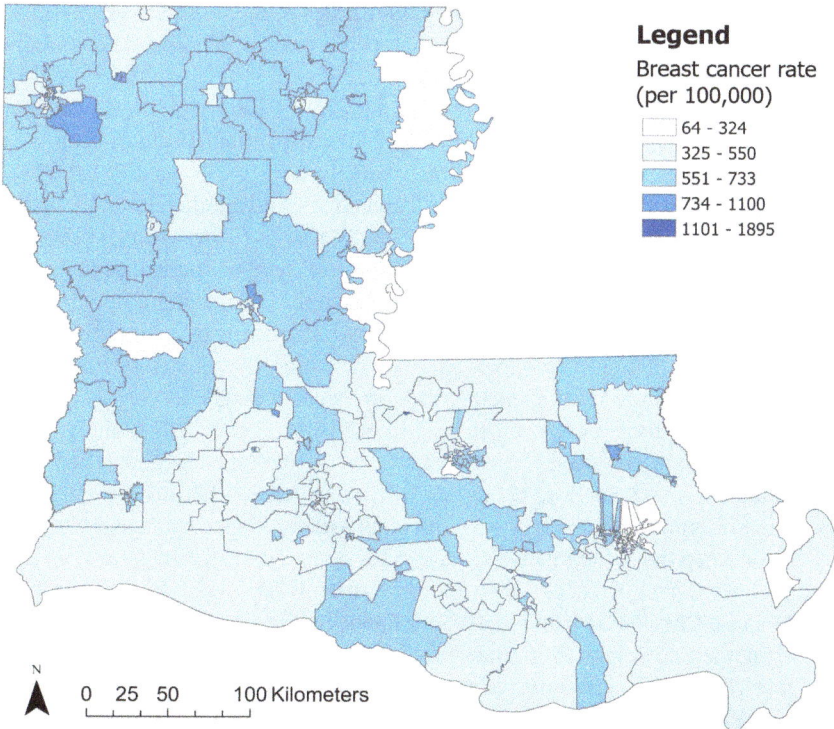

FIGURE 9.5 Crude breast cancer rates in REDCAP-derived regions in Louisiana, 2006.

9.4.2 Implementing the MLR Method

This subsection implements the MLR for two scenarios: one-level regionalization and mixed-level regionalization.

Step 5. Exploring One-Level Regionalization Step-by-Step with the MLR Tool: The one-level regionalization is decomposed into three sub-steps to illustrate the detailed procedures. In ArcGIS Pro, add LA_Mixtracts to the map, and then export it as a new feature LAtrtOne, which will be altered in this step as an experiment.

1. In the Catalog pane, under the toolbox MLR Tools.tbx, activate the tool "One level, step 1: clustering order (spatial and attribute)." In the dialog (Figure 9.6a), choose LAtrtOne for Input Feature. Keep the default column names for Spatial order field name, Attribute order field name, Normalized attribute order field name, and Integrated order value field name. Under list of attributes to be considered, check FACT1, FACT2, and FACT3. Under Weights of attributes, click "Add another" to input their corresponding weights, 0.5180, 0.1178, and 0.0902. Keep the default value 90 for "% of spatial connectivity in clustering" (i.e., $w_s = 0.9$ and $w_a = 0.1$ in equation 9.1), and click Run. It updates the input data LAtrtOne with four new columns: PeanoOrd, AttriOrd, NAttriOrd, and OrdVal, representing the normalized spatial orders o_{si}, attributive order, normalized attributive order o_{ai}, and integrated order o_i in equation 9.1, respectively. This sub-step calibrates spatial, attributive, and integrated clustering orders.

2. Activate the tool "One level, step 2: One-Level Clustering." In the dialog (Figure 9.6b), choose the updated LAtrtOne for Input Feature, keep the default value OrdVal under Order Value, check POPU and COUNT06 under List of Constraints (Variables), and click "Add another" to input their corresponding values 20000 and 16; under List of (Lower Limit) Capacities, keep the default column names for ClusterMember and Isolated cluster field name, and click Run.[5] This sub-step applies clustering and enforces the minimum-value constraints.[6] It updates LAtrtOne by adding two columns, SubClusID and isolate, representing the cluster IDs and the status of census tracts on whether they are isolated units.

3. Activate the tool "One level, step 3: Tackle isolated clusters." In the dialog (Figure 9.6c), choose the updated LAtrtOne for Input Feature, keep the default values SubClusID and isolate under Sub-Cluster ID and Isolated or Not, respectively. Leave the blanks under Upper Units and Cluster Type, keep the default name FinalClus under Name the Integrated Cluster, and input LAtrtOneFinal for Name of the Final Mixed Clusters' Feature. Check POPU and COUNT06 under List of Constraints (Variables), input their corresponding values of 20000 and 16 under the Capacity List, and click Run. This sub-step applies the dissolving method on the layer LAtrtOne based on FinalClus, adds a new column FinalClus for the new cluster IDs, and generates a new feature LAtrtOneFinal. In our

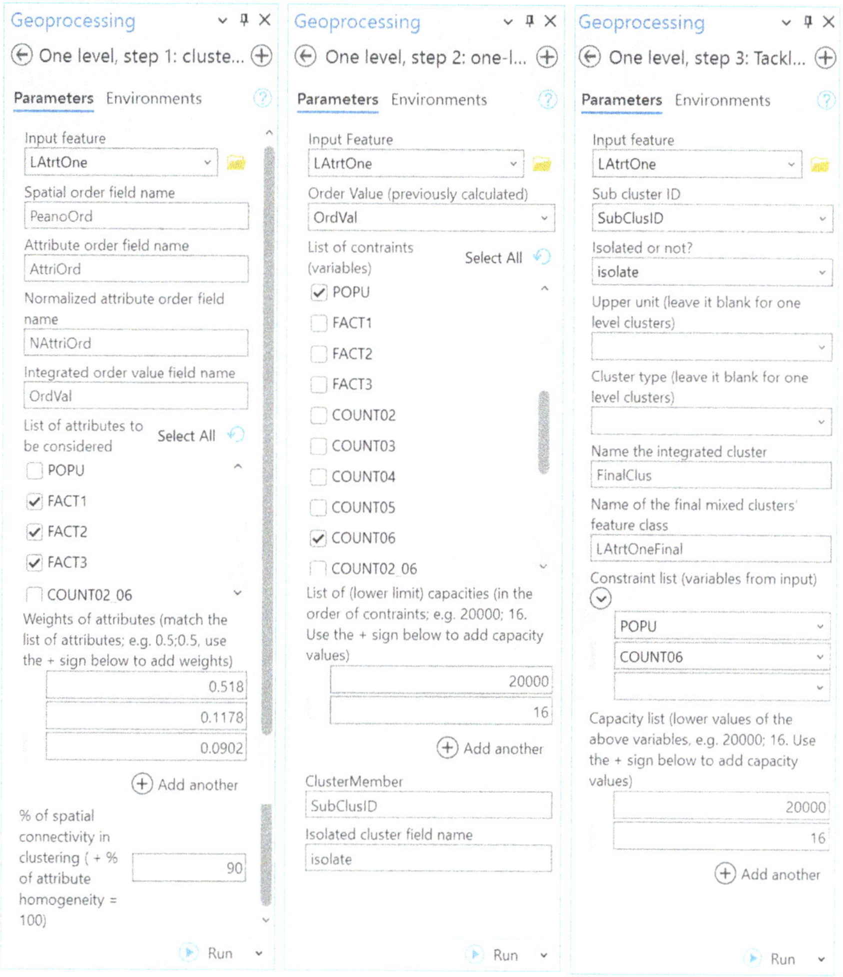

FIGURE 9.6 Interfaces for one-step regionalization in MLR: (a) sub-step 1, (b) sub-step 2, and (c) sub-step 3.

case, the result from the previous sub-step 2 does not contain any isolated clusters; LAtrtOneFinal has the same feature as LAtrtOne.

Step 6. Optional: Exploring the Three Scenarios in MLR: As shown in Figure 9.7, the one-level regionalization process in the previous step is applied to areas of multiple levels and leads to three scenarios. Once again, a parish in Louisiana is a county equivalent unit. There are three scenarios:

1. Disaggregation (color-coded as red). Both population and cancer count criteria overflow, and the values are more than twice the thresholds, so there is a need to break a parish into subregions.

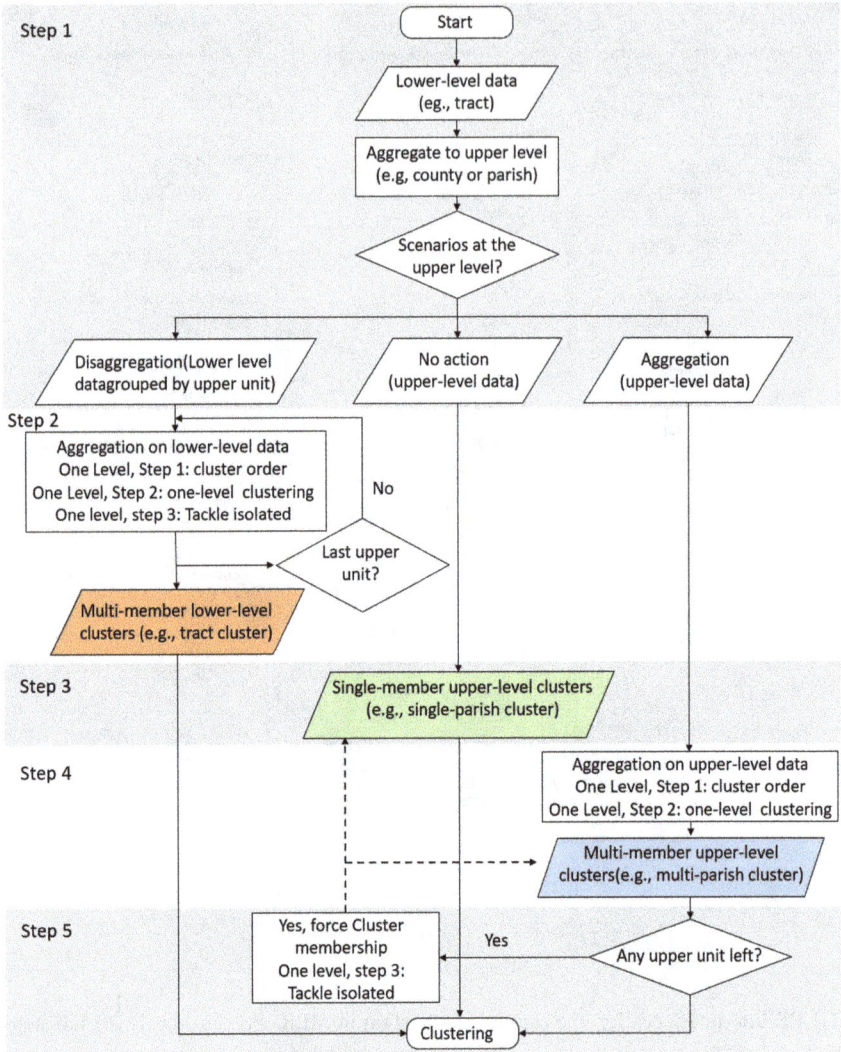

FIGURE 9.7 Flowchart of the mixed-level regionalization (MLR) method.

2. No action (color-coded as green). If a parish meets both criteria and the value for either criterion is not over twice the threshold, no further action is needed either.
3. Aggregation (color-coded as blue). If a parish has at least one unsatisfied criterion, it needs to be aggregated to adjacent parishes to reach the criteria.

To explore the different scenarios, a user can summarize the fields of population and cancer counts (POPU, COUNT06) based on the field COUNTY in the feature

LA _ Mixtracts, set SUM as Statistic Type, and COUNTY as Case field, and save the Output Table as LA _ CoutySum. Join the table LA _ CoutySum back to the feature LA _ Mixtracts. Add a field RegType to the layer LA _ Mixtracts, and calculate it, such as (Figure 9.8):

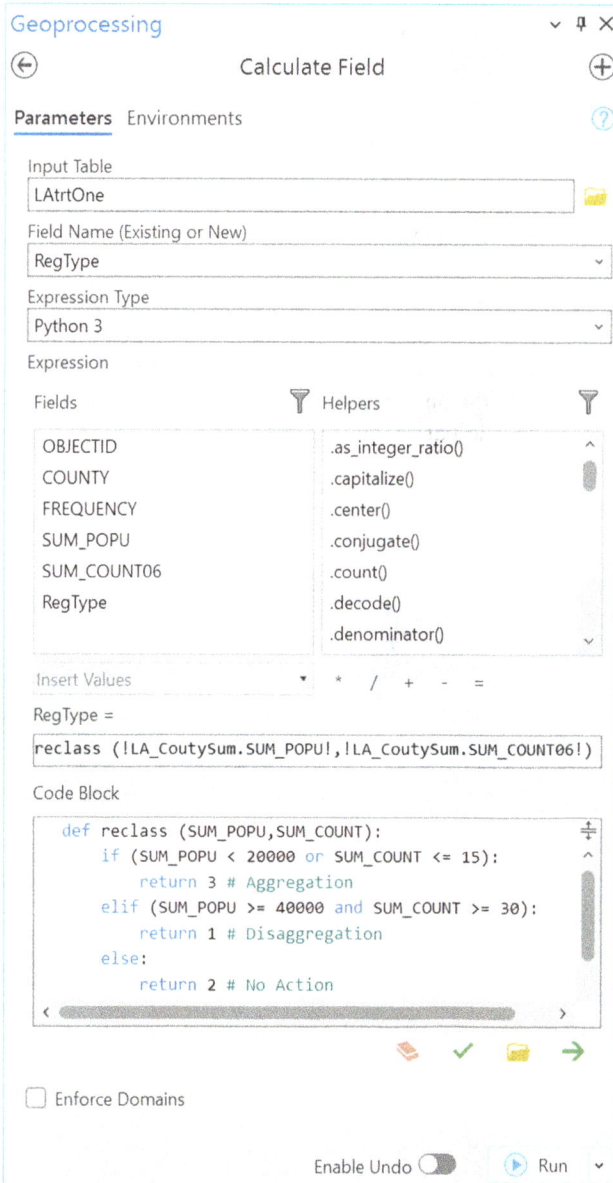

FIGURE 9.8 Interface for defining region types.

```
Under "RegType=", input
reclass (!LA_CoutySum.SUM_POPU!,!LA_CoutySum.SUM_COUNT06!)
Under "Code Block", input
def reclass (SUM_POPU,SUM_COUNT):
 if (SUM_POPU < 20000 or SUM_COUNT <= 15):
 return 3 # Aggregation
 elif (SUM_POPU >= 40000 and SUM_COUNT >= 30):
 return 1 # Disaggregation
 else:
 return 2 # No Action
```

Apply Summarize on LA _ Mixtracts with LA _ Mixtracts.RegType
for the Case field, any field for Field, and Count for Statistic Type, to check the
FREQUENCY in the result table. As shown in Figure 9.9, there are three scenarios for
the 64 parishes in Louisiana: 29 need to be divided into subregions (RegType = 1),
20 need no action (RegType = 2), and 15 need to be aggregated (RegType = 3). For
those with RegType = 1, apply the one-level regionalization on the census tracts
in each county. For those with RegType = 2 or 3, first aggregate the census tract
feature to the county level by summarizing the fields of POPU and COUNT06 and
calculating the weighted means FACT1, FACTT2, and FACT3, and then for those
with RegType = 3, apply the one-level regionalization to aggregate counties. Finally,
append the multi-tract, sole-parish, and multi-parish layers together to form a mixed-
level consolidated layer.

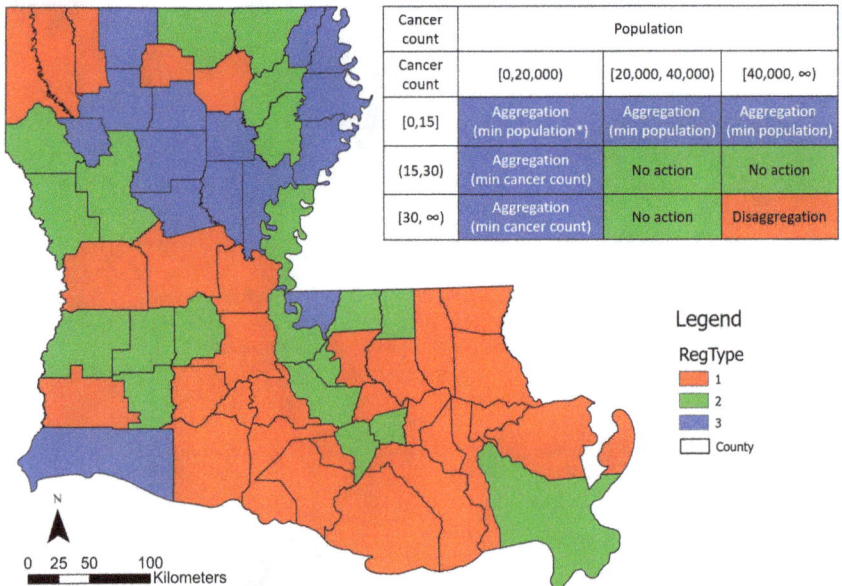

FIGURE 9.9 Three scenarios for parishes in Louisiana by mixed-level regionalization (MLR).

Step 7. Implementing All-in-One MLR: An all-in-one tool is prepared in the toolbox to automate the step-by-step mixed-level regionalization in step 6. Similarly, export LA_Mixtracts as a new feature LA_trtmix to preserve the input data.

Activate the tool "Mixed Level: All-in-One MLR." In the dialog (Figure 9.10), select the input Feature class LA _ trtmix; keep the default names in the next four boxes (PeanoOrd, AttriOrd, NAttriOrd, and OrdVal); check FACT1, FACT2, and FACT3 under Select a List of Attribute to Determine Attribute Order; and enter their corresponding weights: 0.5180, 0.1178, and 0.0902. Keep the default value 90 for Assign the Percent of Spatial Consideration. Check POPU and COUNT06 under Select a List of Clustering Constraints, and input their corresponding values, 20000 and 16. Keep the default names SubClusID and isolate for Name the Sub-Cluster Membership field and Name the Isolated Cluster field, respectively. Choose COUNTY for Select the Upper Geo Unit ID field, keep ClusType for Name the Final Mixed-Cluster Type field, select POPU for Select the Attribute Aggregation Weight field, keep FiClus for Name the Final Mixed-Cluster ID field, input LAtrtMixFinal for Name the Final Mixed-Cluster File name, and click Run.

As a result, the input layer LA _ trtmix is updated, and a new feature, LAtrtMixFinal, contains the fields ClusType and FiClus, indicating the type of the three scenarios (illustrated in step 5) and the cluster IDs, respectively.

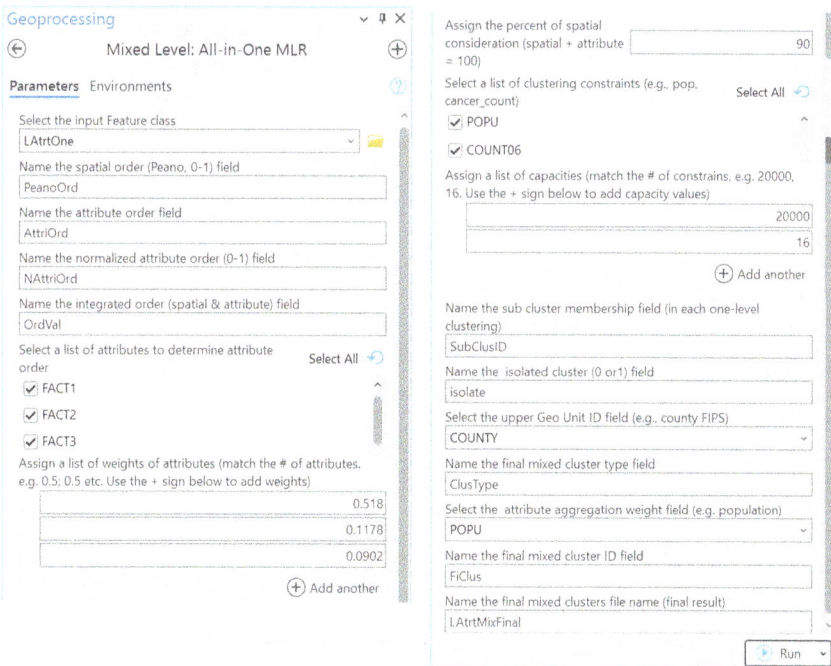

FIGURE 9.10 All-in-One MLR interface (top part on left, bottom part on right).

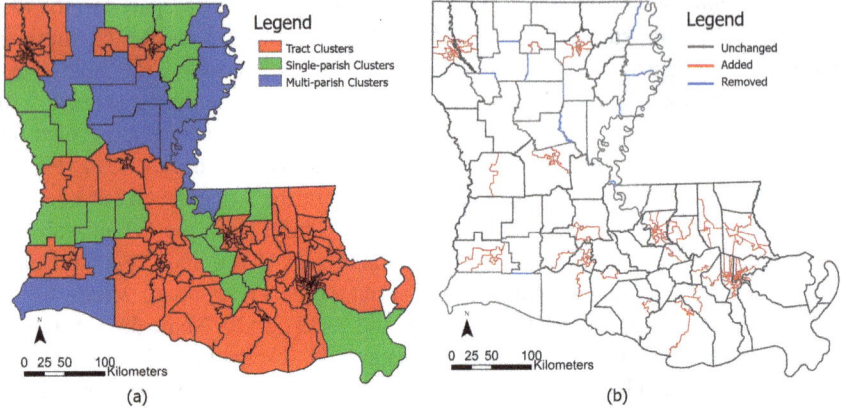

FIGURE 9.11 Constructed regions by mixed-level regionalization (MLR): (a) cluster types and (b) boundary changes between clusters and parishes.

Figure 9.11a shows the three types of constructed regions, and Figure 9.11b highlights the underneath parish boundary and the changes when the multi-level regions are formed. As summarized in Table 9.2, (1) 945 tracts in 29 type-1 parishes are aggregated to 144 sub-parish (or multi-tract) regions; (2) among the 20 initial no-action (type-2) parishes, 18 remain single-parish regions, and the other 2 are joined to type-3 parishes during the process of addressing loose-end parishes; and (3) the initial 15 type-3 parishes and 2 type-2 parishes are together clustered to form 7 multi-parish regions.

TABLE 9.2

Change in Numbers of Parishes and Census Tracts before and after the Mixed-Level Regionalization (MLR)

		Original Area Units (*n*)		MLR Clusters (*n*)	
		Initial Survey	After MPC	Number of Clusters	Type of Clusters
Scenario 1:	Parish #	29	29	144	Subparish (tract cluster)
Disaggregation	Tract #	945	945		
Scenario 2:	Parish #	20	18	18	Single parish
No action	Tract #	130	118		
Scenario 3:	Parish #	15	17	7	Multiparish
Aggregation	Tract #	57	69		
Total	Parish #	64		169	
	Tract #	1132			

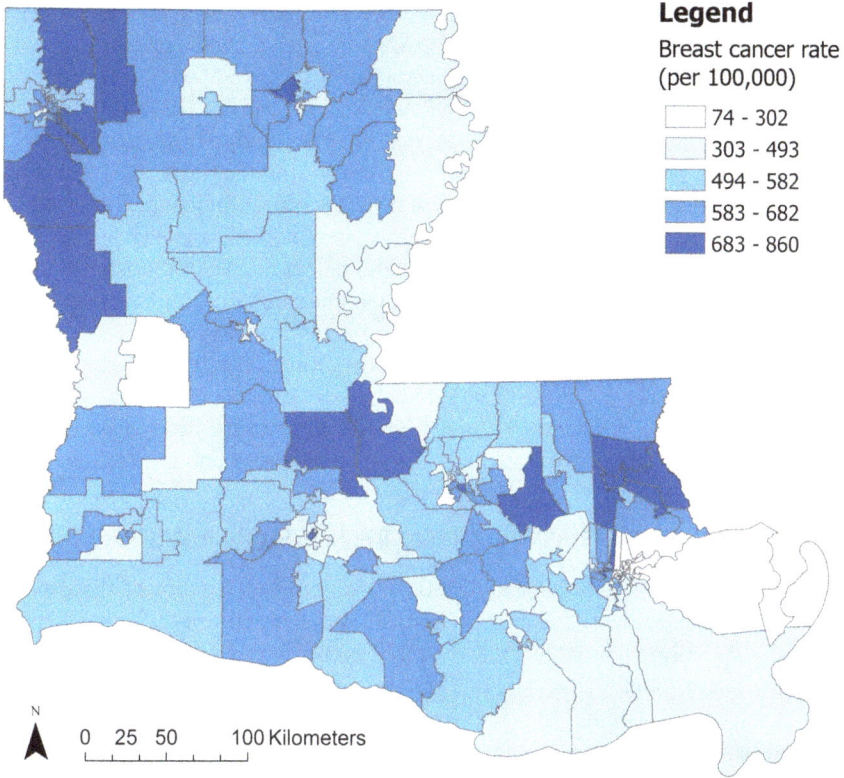

Legend

Breast cancer rate
(per 100,000)

☐ 74 - 302
☐ 303 - 493
☐ 494 - 582
☐ 583 - 682
■ 683 - 860

N

0 25 50   100 Kilometers

FIGURE 9.12 Crude breast cancer rates in MLR-derived regions in Louisiana, 2006.

Figure 9.12 shows crude cancer rates in Louisiana (2006) based on the MLR-derived mixed regions. The constructed regions have similar population sizes and are thus comparable, and each has a sufficient cancer count to avoid data suppression. This leads to more reliable cancer rate estimates, which permit meaningful mapping and subsequent exploratory spatial analysis.

9.5   SUMMARY

In geographic areas with a small number of events (e.g., cancer, AIDS, homicide), rate estimates are often unreliable because of high sensitivity to data errors associated with small numbers. Researchers have proposed various approaches to mitigate the problem. Applications are rich in criminology and health studies. Among various methods, geographic approaches seek to construct larger geographic areas, termed regionalization, so that more reliable rates may be obtained. Some of the earlier methods (e.g., the spatial order method) emphasize spatial distances in clustering. More recent models account for both spatial and attribute distances. On the one side, regionalization is like regular clustering methods in Chapter 7 and seeks to group similar areas to form homogenous regions (i.e., minimizing loss of information while generating larger analysis units). On the other side, only contiguous areas can be merged to form regions that are compact in shape. Such a challenge is addressed by various GIS-automated regionalization methods.

This chapter introduces the SCHC, SKATER, AZP, Max-P, REDCAP, and MLR methods, all automated in GIS and illustrated in a case study. By changing the desirable number of derived regions or imposing a constraint of a threshold region population, the methods can generate a series of different sizes of regions. A method with such a capacity is thus termed "scale-flexible regionalization" and enables an analyst to examine a study area at different scales and detect the MAUP raised in several places of this book. This issue is also tied to the uncertain geographic context problem (UGCoP) (e.g., in multilevel modeling), referring to sensitivity in research findings to different delineations of contextual units (Kwan, 2012). Furthermore, since similar areas are merged to form new regions, spatial autocorrelation is less of a concern in the new regions, which unburdens an analyst of the need to use advanced spatial statistical methods for mitigating the effect of spatial autocorrelation.

APPENDIX 9: THE POISSON-BASED REGRESSION ANALYSIS

This appendix is based on Osgood (2000). Assuming the timing of the events is random and independent, the *Poisson distribution* characterizes the probability of observing any discrete number (0, 1, 2, . . .) of events for an underlying mean count. When the mean count is low (e.g., in a small population), the Poisson distribution is skewed toward low counts. In other words, only these low counts have meaningful probabilities of occurrence. When the mean count is high, the Poisson distribution approaches the normal distribution, and a wide range of counts has meaningful probabilities of occurrence.

The *basic Poisson regression model* is:

$$\ln(\lambda_i) = \beta_0 + \beta_1 x_1 + \beta_2 x_2 + \ldots + \beta_k x_k \qquad (A9.1)$$

where λ_i is the mean (expected) number of events for case i, x's are explanatory variables, and β's are regression coefficients. Note that the left-hand side in equation A9.1 is the logarithmic transformation of the dependent variable. The probability of an observed outcome y_i follows the Poisson distribution, given the mean count λ_i, such as:

$$\Pr(Y_i = y_i) = \frac{e^{-\lambda_i} \lambda_i^{y_i}}{y_i!}. \qquad (A9.2)$$

Equation A9.2 indicates that the expected distribution of crime counts depends on the fitted mean count λ_i.

In many studies, it is the rates, not the counts of events, that are of most interest to analysts. Denoting the population size for case i as n_i, the corresponding rate is λ_i / n_i. The regression model for rates is written as:

$$ln(\lambda_i / n_i) = \beta_0 + \beta_1 x_1 + \beta_2 x_2 + \ldots + \beta_k x_k,$$

That is:

$$ln(\lambda_i) = ln(n_i) + \beta_0 + \beta_1 x_1 + \beta_2 x_2 + \ldots + \beta_k x_k. \qquad (A9.3)$$

Equation A9.3 adds the population size n_i (with a fixed coefficient of 1) to the basic Poisson regression model in equation A9.1 and transforms the model of analyzing counts to a regression model of analyzing rates. The model is a Poisson-based regression that is standardized for the size of base population, and solutions can be found in many statistical packages, such as SAS (PROC GENMOD), by defining the base population as the offset variable.

Note that the variance of the Poisson distribution is the mean count λ, and thus its standard deviation is $SD_\lambda = \sqrt{\lambda}$. The mean count of events λ equals the underlying per capita rate r multiplied by the population size n, that is, $\lambda = rn$. When a variable is divided by a constant, its standard deviation is also divided by the constant. Therefore, the standard deviation of rate r is:

$$SD_r = SD_\lambda / n = \sqrt{\lambda} / n = \sqrt{rn} / n = \sqrt{r} / \sqrt{n}. \qquad (A9.4)$$

Equation A9.4 shows that the standard deviation of per capita rate r is inversely related to the population size n, that is, the problem of heterogeneity of error variance discussed in Section 9.1. The Poisson-based regression explicitly addresses the issue by acknowledging the greater precision of rates in larger population.

NOTES

1 In terms of priority, the SCHC and AZP methods enforce the cluster number constraint first, and the REDCAP and SKATER methods enforce the bound variable constraint first.

2 In rgeoda (R Library for Spatial Data Analysis), the weighted clustering is invoked by checking the box next to "Use geometric centroids" and moving the weighting bar between 0 and 1. A weight of 1 implies clustering purely on spatial weights, and a weight of 0 implies clustering only based on attributive similarity.

3 The current setting for various spatial clustering (regionalization) tools in RegionalizationR assumes equal weights for all control variables in measuring attribute dissimilarity. The MLR method (Mu et al., 2015) assigns corresponding weights 0.5180, 0.1178, and 0.0902 for FACT1, FACT2, and FACT3, as derived from the factor analysis as portions of explained variances.

4 In addition to queen and rook contiguity, another option is K-nearest neighbor (KNN), which requires a user to input an integer for KNN number.

5 This step took 20 minutes on a laptop, described in the Preface.

6 A user can change the thresholds adaptable to specific applications. Here, the example of population $\geq$ 20,000 is based on the Summary of the Health Insurance Probability and Accountability Act (HIPAA) Privacy Rule (www.hhs.gov/hipaa/for-professionals/privacy/laws-regulations/index.html), and a minimum cancer count >15 is used by the State Cancer Profiles (https://statecancerprofiles.cancer.gov/).

Part III

Advanced Computational
Methods and Applications

10 System of Linear Equations and Application of the Garin–Lowry Model in Simulating Urban Population and Employment Patterns

This chapter introduces the method for solving a system of linear equations (SLE). The technique is used in many applications, including the popular input–output analysis (Appendix 10A). The method is fundamental in numerical analysis (NA) and is often used as a building block in other NA tasks, such as solving a system of nonlinear equations (Appendix 10B) and the eigenvalue problem.

Here, the SLE is illustrated in the Garin–Lowry model, a model widely used by urban planners and geographers for analyzing urban land use structure. An ArcGIS tool is developed to implement the Garin–Lowry model in a user-friendly interface. A case study using a hypothetical city shows how the distributions of population and employment interact with each other and how the patterns can be affected by various transportation networks.

10.1 SYSTEM OF LINEAR EQUATIONS

A system of n linear equations with n unknowns $x_1, x_2, \ldots, x_n$ is written as:

$$\begin{cases} a_{11}x_1 + a_{12}x_2 + \ldots + a_{1n}x_n = b_1 \\ a_{21}x_1 + a_{22}x_2 + \ldots + a_{2n}x_n = b_2 \\ \ldots\ldots \\ a_{n1}x_1 + a_{n2}x_2 + \ldots + a_{nn}x_n = b_n \end{cases}.$$

In the matrix form, it is:

$$\begin{bmatrix} a_{11} & a_{12} & \cdots & a_{1n} \\ a_{21} & a_{22} & \cdots & a_{2n} \\ \vdots & \vdots & \ddots & \vdots \\ a_{n1} & a_{n2} & \cdots & a_{nn} \end{bmatrix} \begin{bmatrix} x_1 \\ x_2 \\ \vdots \\ x_n \end{bmatrix} = \begin{bmatrix} b_1 \\ b_2 \\ \vdots \\ b_n \end{bmatrix},$$

DOI: 10.1201/9781003292302-13

Or simply:

$$Ax = b. \tag{10.1}$$

If the matrix A has a *diagonal* structure, system (10.1) becomes:

$$\begin{bmatrix} a_{11} & 0 & \cdots & 0 \\ 0 & a_{22} & \cdots & 0 \\ \vdots & \vdots & \ddots & \vdots \\ 0 & 0 & \cdots & a_{nn} \end{bmatrix} \begin{bmatrix} x_1 \\ x_2 \\ \vdots \\ x_n \end{bmatrix} = \begin{bmatrix} b_1 \\ b_2 \\ \vdots \\ b_n \end{bmatrix}.$$

The solution is simple, and it is:

$$x_i = b_i / a_{ii}.$$

If $a_{ii} = 0$ and $b_i = 0$, x_i can be any real number, and if $a_{ii} = 0$ and $b_i \neq 0$, there is no solution for the system.

There are two other simple systems with easy solutions. If the matrix A has a *lower triangular* structure (i.e., all elements above the main diagonal are 0), system 10.1 becomes:

$$\begin{bmatrix} a_{11} & 0 & \cdots & 0 \\ a_{21} & a_{22} & \cdots & 0 \\ \vdots & \vdots & \ddots & \vdots \\ a_{n1} & a_{n2} & \cdots & a_{nn} \end{bmatrix} \begin{bmatrix} x_1 \\ x_2 \\ \vdots \\ x_n \end{bmatrix} = \begin{bmatrix} b_1 \\ b_2 \\ \vdots \\ b_n \end{bmatrix}.$$

Assuming $a_{ii} \neq 0$ for all i, the *forward substitution* algorithm is used to solve the system by obtaining x_1 from the first equation, substituting x_1 in the second equation to obtain x_2, and so on.

Similarly, if the matrix A has an *upper triangular* structure (i.e., all elements below the main diagonal are 0), system 10.1 becomes:

$$\begin{bmatrix} a_{11} & a_{12} & \cdots & a_{1n} \\ 0 & a_{22} & \cdots & a_{2n} \\ \vdots & \vdots & \ddots & \vdots \\ 0 & 0 & \cdots & a_{nn} \end{bmatrix} \begin{bmatrix} x_1 \\ x_2 \\ \vdots \\ x_n \end{bmatrix} = \begin{bmatrix} b_1 \\ b_2 \\ \vdots \\ b_n \end{bmatrix}.$$

The *back substitution* algorithm is used to solve the system.

By converting system 10.1 to the simple systems as discussed earlier, one may obtain the solution for a general system of linear equations. Say, if the matrix A can be factored into the product of a lower triangular matrix L and an upper triangular matrix U, such as $A = LU$, system 10.1 can be solved in two stages:

1. $Lz = b$, solve for z.
2. $Ux = z$, solve for x.

The first one can be solved by the forward-substitution algorithm, and the second one by the back-substitution algorithm.

Among various algorithms for deriving the *LU-factorization* (or *LU-decomposition*) of A, one called *Gaussian elimination with scaled row pivoting* is used widely as an effective method. The algorithm consists of two steps: a factorization (or *forward-elimination*) phase and a solution (involving updating and back-substitution) phase (Kincaid and Cheney, 1991, p. 145). Computation routines for the algorithm of Gaussian elimination with scaled row pivoting can be found in various computer languages, such as FORTRAN (Press et al., 1992a; Wang, 2006, pp. 234–241), C (Press et al., 1992b), C++ (Press et al., 2002), or Python, as in the ArcGIS Pro tool used in Section 10.3. One may also use commercial software MATLAB (www.mathworks.com), Mathematica (www.wolfram.com), or R for the task of solving a system of linear equations.

10.2 THE GARIN–LOWRY MODEL

10.2.1 BASIC VS. NONBASIC ECONOMIC ACTIVITIES

There has been an interesting debate on the relation between population and employment distributions in a city. Does population follow employment (i.e., workers find residences near their workplaces to save commuting)? Or vice versa (i.e., businesses locate near residents for recruiting workforce or providing services)? The *Garin–Lowry model* (Lowry, 1964; Garin, 1966) argues that population and employment distributions interact with each other and are interdependent. However, different types of employment play different roles. The distribution of *basic employment* is independent of the population distribution pattern and may be considered exogenous. *Service (nonbasic) employment* follows population. On the other side, the population distribution is determined by the distribution patterns of both basic and service employment. See Figure 10.1 for illustration. The interactions between employment and population decline with distances, which are defined by a transportation network. Unlike the urban economic model built on the assumption of monocentric employment (see Section 6.2), the Garin–Lowry model has the flexibility of simulating a population distribution pattern corresponding to any given basic employment pattern. It can be used to examine the impact of basic employment distribution on population as well as the impact of transportation network.

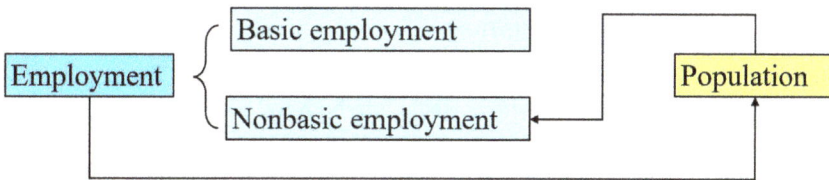

FIGURE 10.1 Interaction between population and employment distributions in a city.

The binary division of employment into basic and service employment is based on the *concept of basic and nonbasic activities*. A local economy (a city or a region) can be divided into two sectors: basic sector and nonbasic sector. The *basic sector* refers to goods or services that are produced within the area but sold outside of the area. It is the export or surplus that is independent of the local economy. The *nonbasic sector* refers to goods or services that are produced within the area and also sold within the area. It is local or dependent and serves the local economy. By extension, basic employment refers to workers in the basic sector, and service employment refers to those in the nonbasic sector.

This concept is closely related to the division of primary and secondary activities. According to Jacobs (1961, pp. 161–162), *primary activities* are "those which, in themselves, bring people to a specific place because they are anchorages," including manufactures and nodal activities at the metropolitan and regional levels, and *secondary activities* are "enterprises that grow in response to the presence of primary uses, to serve the people the primary uses draw." To some extent, primary and secondary activities correspond to basic and nonbasic activities, respectively. These two activities may also have distinctive location preference, as exemplified in correlations with various centrality measures introduced in Appendix 6B. The research on Barcelona, Spain, indicates that secondary economic activities have a higher correlation with street centrality than primary ones (Porta et al., 2012). Primary activities draw people as final destinations mainly because of their function, while secondary activities live solely on passers-by, therefore on their location.

The concept of basic and nonbasic activities is useful for several reasons (Wheeler et al., 1998, p. 140). It identifies basic activities as the most important to a city's viability. Expansion or recession of the basic sector leads to economic repercussions throughout the city and affects the nonbasic sector. City and regional planners forecast the overall economic growth based on anticipated or predicted changes in the basic activities. A common approach to determine employment in basic and nonbasic sectors is the *minimum requirements approach* by Ullman and Dacey (1962). The method examines many cities of approximately the same population size and computes the percentage of workers in a particular industry for each of the cities. If the lowest percentage represents the minimum requirements for that industry in a city of a given population size range, that portion of the employment is engaged in the nonbasic or city-serving activities. Any portion beyond the minimum requirements is then classified as basic activity. Classifications of basic and nonbasic sectors can also be made by analyzing export data (Stabler and St. Louis, 1990).

10.2.2 THE MODEL'S FORMULATION

In the Garin–Lowry model, an urban area is composed of n tracts. The population in any tract j is affected by employment (including both the basic and service employment) in all n tracts, and the service employment in any tract i is determined by population in all n tracts. The degree of interaction declines with distance measured by a gravity kernel. Given a basic employment pattern and a distance matrix, the model computes the population and service employment at various locations.

First, the service employment in any tract i, S_i, is generated by the population in all tracts j ($j = 1, 2, \ldots, n$), P_j, through a gravity kernel t_{ij}, with:

$$S_i = e\sum\nolimits_{j=1}^{n}\left(P_j t_{ij}\right) = e\sum\nolimits_{j=1}^{n}\left[P_j\left(d_{ij}^{-\alpha} \,/\, \sum\nolimits_{l=1}^{n} d_{lj}^{-\alpha}\right)\right], \tag{10.2}$$

where e is the service employment/population ratio (a simple scalar uniform across all tracts), d_{ij} (or d_{lj}) the distance between tracts i (or l) and j, and α the distance friction coefficient characterizing shopping (resident-to-service) behavior. The gravity kernel t_{ij} represents the proportion of service employment in tract i owing to the influence of population in tract j, out of its impacts on all tracts l. In other words, the service employment at i is a result of summed influences of population at all tracts j ($j = 1, 2, \ldots, n$), each of which is only a fraction of its influences on all tracts l ($l = 1, 2, \ldots, n$).

Second, the population in any tract j, P_j, is determined by the employment in all tracts i ($i = 1, 2, \ldots, n$), E_i, through a gravity kernel g_{ij}, with:

$$P_j = h\sum\nolimits_{i=1}^{n}\left(E_i g_{ij}\right) = h\sum\nolimits_{i=1}^{n}\left[\left(B_i + S_i\right)\left(d_{ij}^{-\beta} \,/\, \sum\nolimits_{k=1}^{n} d_{ik}^{-\beta}\right)\right], \tag{10.3}$$

where h is the population/employment ratio (also a scalar uniform across all tracts), and β the distance friction coefficient characterizing commuting (resident-to-workplace) behavior. Note that employment E_i includes both service employment S_i and basic employment B_i, that is, $E_i = S_i + B_i$. Similarly, the gravity kernel g_{ij} represents the proportion of population in tract j, owing to the influence of employment in tract I, out of its impacts on all tracts k ($k = 1, 2, \ldots, n$).

Let P, S, and B be the vectors defined by the elements P_j, S_i, and B_i, respectively, and G and T the matrices defined by g_{ij} (with the constant h) and t_{ij} (with the constant e), respectively. Equations 10.2 and 10.3 become:

$$S = TP, \tag{10.4}$$

$$P = GS + GB. \tag{10.5}$$

Combining 10.4 and 10.5 and rearranging, we have:

$$\left(I - GT\right)P = GB, \tag{10.6}$$

where I is the $n \times n$ identity matrix. Equation 10.6 in the matrix form is a system of linear equations with the population vector P unknown. Four parameters (the distance friction coefficients α and β, the population/employment ratio h, and service employment/population ratio e) are given; the distance matrix d is derived from a road network, and the basic employment B is predefined.

Plugging the solution P back to equation 10.4 yields the service employment vector S. For a more detailed discussion of the model, see Batty (1983).

The following subsection offers a simple example to illustrate the model.

10.2.3 AN ILLUSTRATIVE EXAMPLE

See Figure 10.2 for an urban area with five ($n = 5$) equal-area square tracts. The dashed lines are roads connecting them. Only two tracts (say, tract 1 and 2, shaded in the figure) need to be differentiated and carry different population and employment. Others (3, 4, and 5) are symmetric of tract 2. Assume all basic employment is concentrated in tract 1 and *normalized* as 1, that is, $B_1 = 1$, $B_2 = B_3 = B_4 = B_5 = 0$. This normalization implies that population and employment are relative, since we are only interested in their variation over space. The distance between tracts 1 and 2 is unit 1, and the distance *within a tract* is defined as 0.25 (i.e., $d_{11} = d_{22} = d_{33} = \ldots = 0.25$). Note that the distance is the travel distance through the transportation network (e.g., $d_{23} = d_{21} + d_{13} = 1 + 1 = 2$). For illustration, define constants $e = 0.3$, $h = 2.0$, $\alpha = 1.0$, and $\beta = 1.0$.

From equation 10.2, after taking advantage of the symmetric property (i.e., tracts 2, 3, 4, and 5 are equivalent in locations relative to tract 1), we have:

$$S_1 = 0.3 \left(\frac{d_{11}^{-1}}{d_{11}^{-1} + d_{21}^{-1} + d_{31}^{-1} + d_{41}^{-1} + d_{51}^{-1}} P_1 + \frac{d_{12}^{-1}}{d_{12}^{-1} + d_{22}^{-1} + d_{32}^{-1} + d_{42}^{-1} + d_{52}^{-1}} P_2 * 4 \right).$$

That is:

$$S_1 = 0.1500 P_1 + 0.1846 P_2, \tag{10.7}$$

where the distance terms have been substituted by their values.

Similarly:

$$S_2 = 0.3 \left(\begin{array}{c} \dfrac{d_{21}^{-1}}{d_{11}^{-1} + d_{21}^{-1} + d_{31}^{-1} + d_{41}^{-1} + d_{51}^{-1}} P_1 + \dfrac{d_{22}^{-1}}{d_{12}^{-1} + d_{22}^{-1} + d_{32}^{-1} + d_{42}^{-1} + d_{52}^{-1}} P_2 + \\[3ex] \dfrac{d_{23}^{-1}}{d_{13}^{-1} + d_{23}^{-1} + d_{33}^{-1} + d_{43}^{-1} + d_{53}^{-1}} P_3 * 2 + \dfrac{d_{24}^{-1}}{d_{14}^{-1} + d_{24}^{-1} + d_{34}^{-1} + d_{44}^{-1} + d_{54}^{-1}} P_4 \end{array} \right)$$

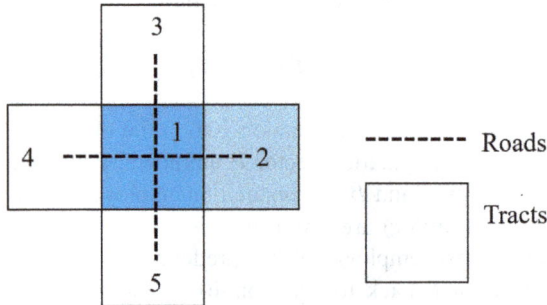

FIGURE 10.2 A simple city for illustration.

where tracts 3 and 5 are equivalent in locations relative to tract 2. Noting $P_4 = P_3 = P_2$, the preceding equation is simplified as:

$$S_2 = 0.0375P_1 + 0.2538P_2. \tag{10.8}$$

From equation 10.3, we have:

$$P_1 = S_1 + 1.2308S_2 + 1, \tag{10.9}$$

$$P_2 = 0.2500S_1 + 1.6923S_2 + 0.25. \tag{10.10}$$

Solving the system of linear equations (equations 10.7–10.10), we obtain: $P_1 = 1.7472$, $P_2 = 0.8136$; $S_1 = 0.4123$, $S_2 = 0.2720$. Both the population and service employment are higher in the central tract than others.

10.3   CASE STUDY 10: SIMULATING POPULATION AND SERVICE EMPLOYMENT DISTRIBUTIONS IN A HYPOTHETICAL CITY

A hypothetical city is here assumed to be partitioned by a transportation network made of 10 circular rings and 15 radial roads. See Figure 10.3. Areas around the CBD form a unique downtown tract, and thus the city has $1 + 9 \times 15 = 136$ tracts. The hypothetical city does not have any geographic coordinate system or unit for distance measurement, as we are only interested in the spatial structure of the city. The case study is built upon the work reported in Wang (1998) but revised extensively for clarity of illustration.

The project primarily uses the file geodatabase SimuCity.gdb under the data folder SimuCity, which contains:

1. A polygon feature tract (136 tracts)
2. A point feature trtpt (136 tract centroids)
3. A feature dataset road for the road network

Other data include two distance matrices, ODTime and ODTime1 (results from steps 1–3 in dBase format for those who want to skip these steps), and the toolkit file Garin-Lowry.tbx.

Features tract and trtpt contain similar attribute fields: area and perimeter for each tract; BEMP_CBD and BEMP_Unif for predefined basic employment in the basic case and the uniform distribution case (see steps 4 and 5); POP_CBD, POP_Unif, POP_A2B2, and POP_Belt for population to be calibrated in the four scenarios (basic case, uniform distribution case, $\alpha = \beta = 2.0$ [step 6], and the case with the beltway [step 7]); and SEMP_CBD, SEMP_Unif, SEMP_A2B2, and SEMP_Belt for service employment to be calibrated in the corresponding four scenarios.

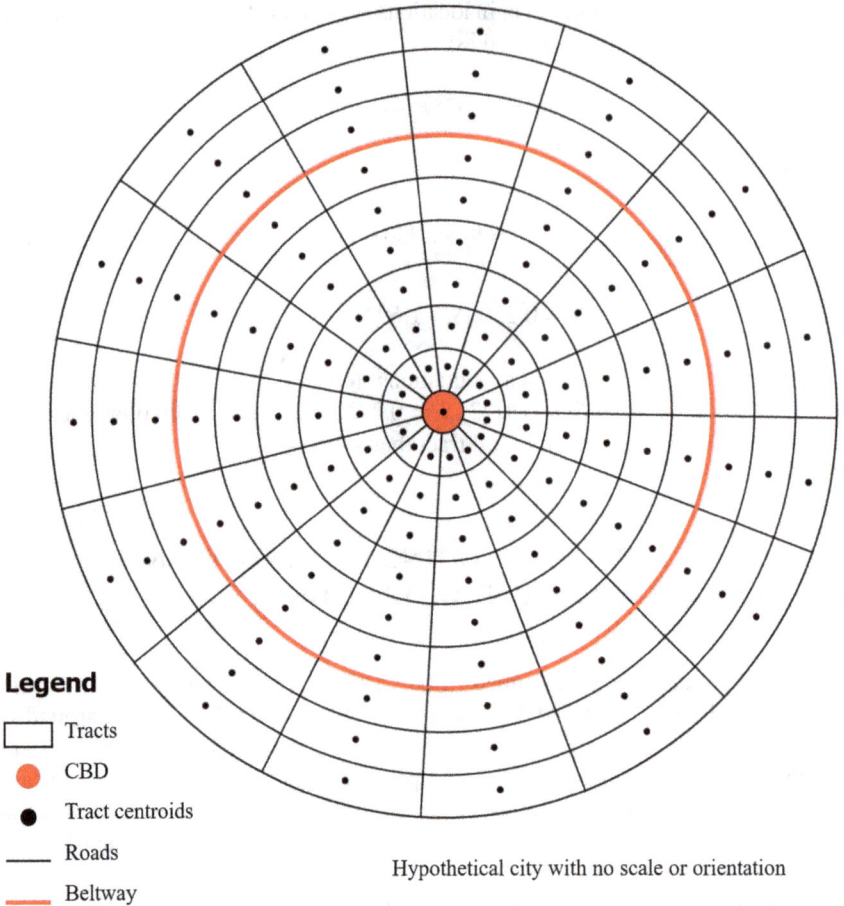

Legend

☐ Tracts

🔴 CBD

⬤ Tract centroids

── Roads

── Beltway

Hypothetical city with no scale or orientation

FIGURE 10.3 Spatial structure of a hypothetical city.

Prior to the construction of its corresponding network dataset, the road network feature dataset road contains a single feature road, which has two important attribute fields, LENGTH and LENGTH1. Field LENGTH is the length on each road segment and defines the standard impedance values in the network travel time computation. In this case, travel speed is assumed to be uniform on all roads, and travel distance is equivalent to travel time for measuring network impedance. Another field, LENGTH1, is defined as 1/2.5 of LENGTH for the seventh ring road and the same as LENGTH for other roads and will be used to define the new impedance values for examining the impact of a suburban beltway in step 7. That is to say, the travel speed on the beltway in the seventh ring is assumed to be 2.5 times (e.g., 75 mph) of the speed on others (e.g., 30 mph), and thus its travel time is 1/2.5 of others.

Step 1. Computing Network Travel Time Matrices in the Basic Case: If necessary, refer to detailed instructions in Section 2.2.2 for computing a network travel time matrix. The following provides a brief guideline.

In the Geoprocessing toolboxes pane, search and activate Create Network Dataset to generate a new Network Dataset `road_ND` based on Target Feature Dataset `road`. The newly created network dataset `road_ND` and feature class `road_ND_Junctions` are both placed under the feature dataset `road`. Remove the new dataset `road_ND` from the map. In the Catalog pane, right-click the feature dataset `road_ND` > Properties > in the window of Network Dataset Properties, click Travel Attributes > Costs > under Evaluators for road (Along), choose `Field Script` for Type, input `[LENGTH]` for Value[1] and for road (Against), keep `Same as Along` for Type, then click OK. Activate and apply Build Network on the network dataset `road_ND`. Refer to step 5 in Section 2.2.2 if needed.

In the Analysis tab, click Network Analysis and choose Origin–Destination Cost Matrix to add OD Cost Matrix layer to the Map pane > click OD Cost Matrix to activate the Network Analyst tab on the top ribbon. Under OD Cost Matrix tab, choose `trtpt` as both Origins and Destinations > click Run to compute the travel time matrix `ODTime` in the basic case. Note that the result is contained in the feature `Lines` with $136 \times 136 = 18{,}496$ records (its field `Total_Length` is the total network distance between two tracts). Refer to step 6 in Subsection 2.2.2 if needed.

Step 2. Amending Network Travel Time Matrices by Accounting for Intrazonal Travel Time in the Basic Case: One has to account for the intrazonal travel distance not only to avoid zero-distance terms in calibration of gravity model but also to capture more realistic travel impedance (Wang, 2003). In this case, the average within-tract travel distance is approximated as 1/4 of the tract perimeters.[2] This can also be considered as the segment one residing in a tract has to travel in order to reach the nearest junction on the road network. Therefore, the total network travel distance between two tracts is composed of the network distance segment derived from step 1 and intra-tract travel distances at both the origin and destination tracts.

In implementation, export the attribute table of `trtpt` as two identical tables `attr1` and `attr2`; join the table `attr1` (common key `OBJECTID`) to `Lines` (common key `OriginID`) and also the table `attr2` (common key `OBJECTID`) to `Lines` (`DestinationID`) > export the attribute table of `Lines` to a new table `ODTime`, and add a new field `NetwTime` to `ODTime` and calculate it as:

```
!Total_Length!+0.25*!PERIMETER!+0.25*!PERIMETER_1!.
```

Step 3. Computing Network Travel Time Matrices with a Suburban Beltway: Repeat steps 1–2 by using `LENGTH1` instead of `LENGTH` as the travel impedance values when creating a new network dataset (say, `road_1`), and output the travel times to a similar table `ODTime1`, where a different field name, `NetwkTime1`, is used to represent the time between tracts in the case with a suburban beltway. In this case, travel time is no longer equivalent to travel distance, because the speed on the seventh ring road is faster than others.

Step 4. Simulating Distributions of Population and Service Employment in the Basic Case: The basic case, as in the monocentric model, assumes that all basic employment (say, 100) is concentrated at the CBD. In addition, the basic case assumes that $\alpha = 1.0$ and $\beta = 1.0$ for the two distance friction coefficients in the gravity kernels. The values of h and e in the model are set equal to 2.0 and 0.3, respectively, based on data from the Statistical Abstract of the United States (Bureau of the Census, 1993). If P_T, B_T, and S_T are the total population and total basic and service employments, respectively, we have $S_T = eP_T$, and $P_T = hE_T = h(B_T + S_T)$, and thus $P_T = (h/(1 - he))B_T$. As B_T is normalized to 100, it follows that $P_T = 500$, and $S_T = 150$. Keeping h, e, and B_T constant throughout the analysis implies that the total population and employment (basic and service) remain constant. Our focus is on the effects of exogenous variations in the spatial distribution of basic employment and in the values of the travel friction parameters α and β and on the impact of building a suburban beltway.

Use the toolkit `Garin-Lowry.tbx` to implement the Garin–Lowry model. Since the distance (travel time) table is defined in steps 1–3, the tool "Garin–Lowry Model Pro" is used, in which `External Table` is set as Distance Type (and other steps throughout this case study).[3] Input the items in the interface as shown in Figure 10.4a.

Similar to other tools that utilize a distance matrix as input (e.g., the Huff model tool in Chapter 4, the 2SFCA method tool in Chapter 5), the distance matrix can be prepared externally or calibrated as a Euclidean or geodesic distance matrix internally by the tool. Figure 10.4b shows another option of calibrating the Euclidean or geodesic distance matrix by the tool itself.

Note that the basic employment pattern is defined in the field BEMP_CBD with its value = 100 in the CBD tract and 0 elsewhere, and it saves the results, that is, numbers of population and service employment, in the predefined fields POP_CBD and SEMP_CBD, respectively.

Since the values are identical for tracts on the same ring, ten tracts from different rings along the same direction are selected.[4] The results of simulated population and service employment are reported in the first column under Population and the first column under Service Employment, respectively, in Table 10.1. Figures 10.5 and 10.6 show the population and service employment patterns, respectively.

Step 5. Examining the Impact of Basic Employment Pattern: To examine the impact of basic employment pattern, this project simulates the distributions of population and service employment given a uniform distribution of basic employment. In this case, all tracts have the same amount of basic employment, that is, $100/136 = 0.7353$, which has been predefined in the field in BEMP_Unif. In the interface as shown in Figure 10.4a, change the inputs for "Basic Employment Field," "Service Employment Field (Output)," and "Population Field (Output)" to BEMP_Unif, SEMP_Unif, and POP_Unif, respectively. The results are reported under "Uniform Basic Employment" in Table 10.1, and similarly in Figures 10.5 and 10.6.

Note that both the population and service employment remain declining from the city center even when the basic employment is uniform across space. That is to say, the declining patterns with distances from the CBD are largely due to the location advantage (better accessibility) near the CBD shaped by the transportation network instead of job concentration in the CBD. The job concentration in CBD in the basic

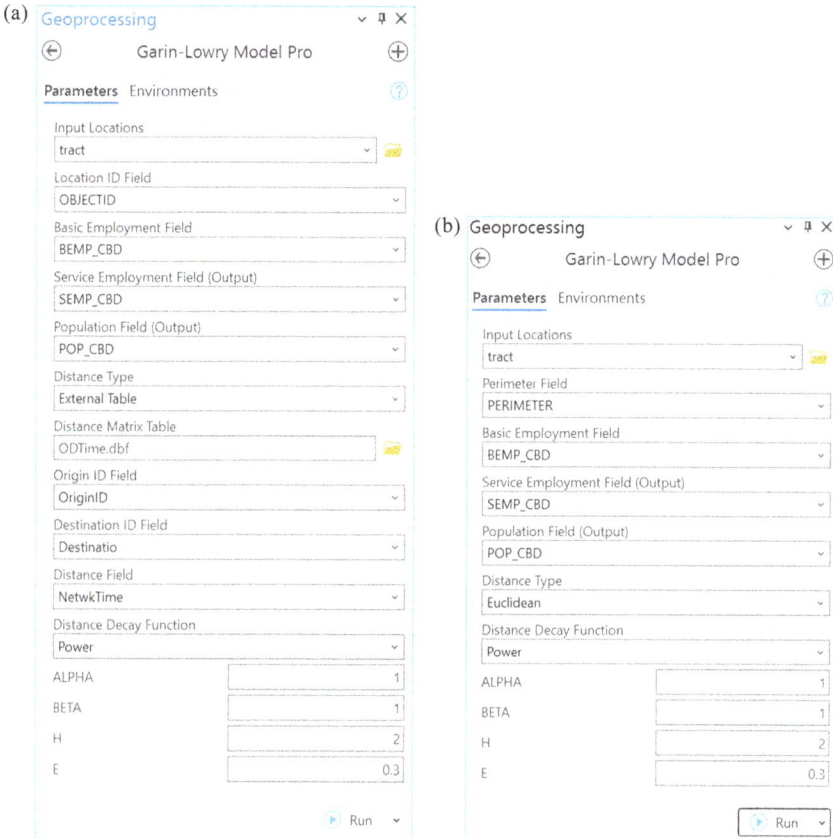

FIGURE 10.4 Interface of the Garin–Lowry model tool with (a) external table and (b) Euclidean distance.

case does enhance the effect, that is, both the population and service employment exhibit steeper slopes in the basic case (a monocentric pattern) than this case (uniform basic employment distribution). In general, service employment follows population, and their patterns are consistent with each other.

One may design other basic employment patterns (e.g., polycentric) and examine how the population and service employment respond to the changes. See Guldmann and Wang (1998) for more scenarios of different basic employment patterns.

Step 6. Examining the Impact of Travel Friction Coefficients: Keep all parameters in the basic case unchanged, except the two travel friction coefficients α and β. Compared to the basic case where $\alpha = 1$ and $\beta = 1$, this new case uses $\alpha = 2$ and $\beta = 2$. Input most items the same as shown in Figure 10.4a, but set "Service Employment Field (Output)" as SEMP _ A2B2, "Population Field (Output)" as POP _ A2B2, ALPHA as 2, and BETA as 2. The results are reported under "$\alpha, \beta = 2$" in Table 10.1, and similarly in Figures 10.5 and 10.6.

TABLE 10.1

Simulated Scenarios of Population and Service Employment Distributions

	Population				Service Employment			
Location	Basic Case[1]	Uniform Basic Employment	$\alpha, \beta = 2$	With a Suburban Beltway	Basic Case	Uniform Basic Employment	$\alpha, \beta = 2$	With a Suburban Beltway
1	9.2023	5.3349	22.4994	9.0211	1.7368	1.6368	3.2811	1.6700
2	6.3483	5.0806	9.7767	6.1889	1.6170	1.5540	2.5465	1.5536
3	5.1342	4.6355	6.1368	4.9930	1.4418	1.4091	1.8242	1.3839
4	3.9463	4.1184	3.6171	3.8529	1.2338	1.2384	1.2292	1.1946
5	3.8227	3.9960	3.3542	3.7318	1.1958	1.2008	1.1366	1.1576
6	3.2860	3.5643	2.3907	3.2494	1.0496	1.0632	0.8209	1.0345
7	2.8938	3.2303	1.8372	2.9197	0.9392	0.9578	0.6403	0.9515
8	2.5789	2.9374	1.4520	2.9394	0.8459	0.8666	0.5111	0.9831
9	2.3168	2.6737	1.1679	2.4016	0.7648	0.7858	0.4135	0.8037
10	2.0924	2.4315	0.9487	2.1532	0.6930	0.7130	0.3367	0.7209

[1] All basic employment is concentrated at the CBD; $\alpha, \beta = 1$; travel speed is uniform on all roads.

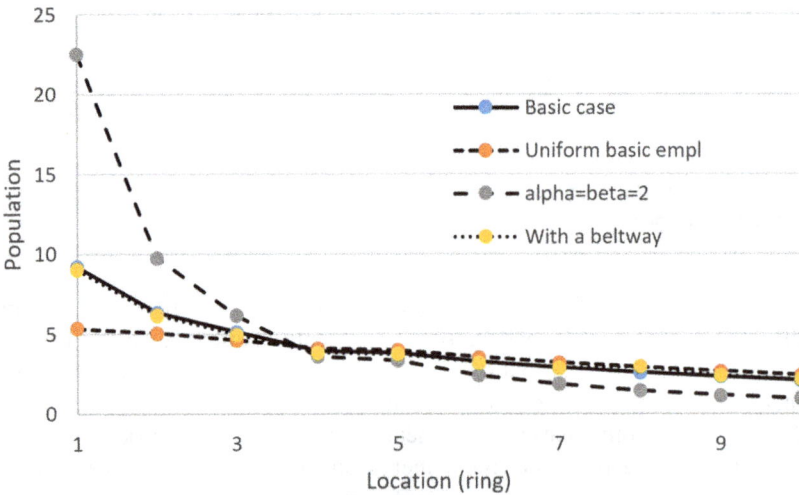

FIGURE 10.5 Population distributions in various scenarios.

Note the steeper slopes for both population and service employment in this case with larger α and β. The travel friction parameters indicate how much people's travel behavior (including both commuting to workplace and shopping) is affected by travel distance (time). As transportation technologies as well as road networks improve

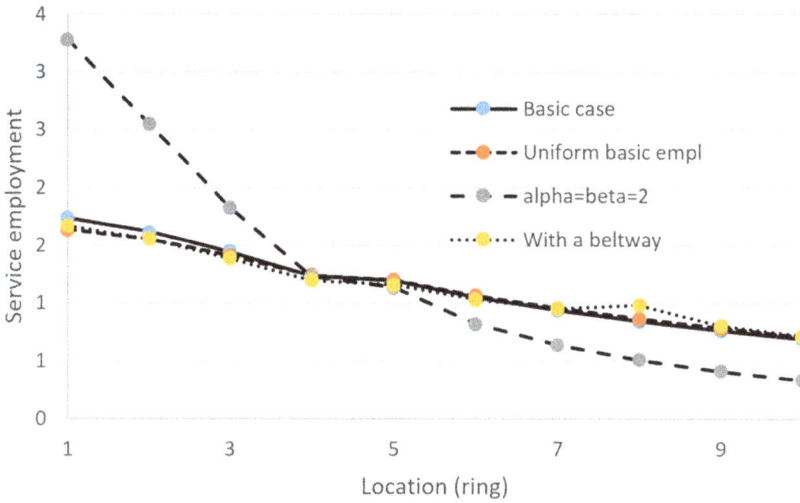

FIGURE 10.6 Service employment distributions in various scenarios.

over time, these two parameters have generally declined. In other words, the new case with $\alpha = 2$ and $\beta = 2$ may correspond to a city in earlier years. That explains the flattening population density gradient over time, an important observation in the study of population density patterns (see Chapter 6).

Step 7. Examining the Impact of Transportation Network: Finally, we examine the impact of transportation network, in this particular case, the building of a suburban beltway. Assume that the seventh ring road is a faster beltway: if the average travel speed on other urban road is 30 miles per hour (mph), the speed on the beltway is 75 miles per hour. Step 3 has already generated a different travel time dataset ODTime1 based on the new assumption. Input most items the same as shown in Figure 10.4a, but set "Service Employment Field (Output)" as SEMP _ Belt, "Population Field (Output)" as POP _ Belt, "Distance Matrix Table" as ODTime1, and "Distance Field" as NetwkTime1. The results are reported under "With a Suburban Beltway" in Table 10.1, and similarly in Figures 10.5 and 10.6.

The distribution patterns of population and service employment are similar to those in the basic case, but both are slightly flatter than those in the basic case (note the lower values near the CBD and higher values near the tenth ring). In other words, building the suburban beltway narrows the gap in location advantage between the CBD and suburbia and thus leads to flatter population and service employment patterns. More importantly, the presence of a suburban beltway leads to a local peak in population or service employment around there (note the higher values highlighted in *italics* in Table 10.1). That explains that the construction of suburban beltway in most large metropolitan areas in the United States helped accelerate the suburbanization process as well as the transformation of urban structure from monocentricity to polycentricity. See related discussion in Chapter 6 and Wang (1998).

10.4 DISCUSSION AND SUMMARY

The concept of basic and nonbasic activities emphasizes different roles in the economy played by basic and nonbasic sectors. The Garin–Lowry model uses the concept to characterize the interactions between employment and population distributions within a city. In the model, basic (export-oriented) employment serves as the exogenous factor, whereas service (locally oriented) employment depends on the population distribution pattern; on the other side, the distribution pattern of population is also determined by that of employment (including both basic and nonbasic employment). The interactions decline with travel distances or times as measured by gravity kernels. Based on this, the model is constructed as a system of linear equations. Given a basic employment pattern and a road network (the latter defines the matrix of travel distances or times), the model solves for the distributions of population and service employment.

Applying the Garin–Lowry model in analyzing real-world cities requires the division of employment into basic and nonbasic sectors. Various methods (e.g., the minimum requirements method) have been proposed to separate basic employment from total employment. However, the division is unclear in many cases, as most economic activities in a city serve both the city itself (nonbasic sector) and beyond (basic sector). The case study uses a hypothetical city to illustrate the impacts of basic employment patterns, travel friction coefficients, and transportation networks. The case study helps us understand the change of urban structure under various scenarios and explain many empirical observations in urban density studies.

For example, suburbanization of basic employment in an urban area leads to dispersion of population as well as service employment, but the dispersion does not change the general pattern of higher concentrations of both population and employment toward the CBD. As the improvements in transportation technologies and road networks enable people to travel farther in less time, the traditional accessibility gap between the central city and suburbs is reduced and leads to a more gradual decline in population toward the edge of an urban area. This explains the flattening density gradient over time as reported in many empirical studies on urban density functions. Suburban beltways "were originally intended primarily as a means of facilitating intercity travel by providing metropolitan bypass, it quickly became apparent that there were major unintended consequences for intracity traffic" (Taaffe et al., 1996, p. 178). The simulation in the case with a suburban beltway suggests a flatter population distribution pattern and, noticeably, a suburban density peak near the beltway even when all basic employment is assumed to be in the CBD.

The model may be used to examine more issues in the study of urban structure. For example, comparing population patterns across cities with different numbers of ring roads sheds light on the issue whether large cities exhibit flatter density gradients than smaller cities (McDonald, 1989, p. 380). Solving the model for a city with more radial or circular ring roads helps us understand the impact of road density. Simulating cities with different road networks (e.g., a grid system, a semicircular city) illustrates the impact of road network structure.

In addition to the challenges of decomposing basic and nonbasic employment and defining a realistic distance decay function for the gravity kernels, one major limitation of the Garin–Lowry model is its high level of abstraction in characterizing urban land uses. In the model, a city is essentially composed of three types of land uses: population reflecting residential land use, basic employment, and nonbasic employment, capturing a combination of industrial, commercial, and others where jobs are located. This simple conceptualization of urban land uses is also adopted in the case study of urban traffic simulation model in Section 12.4 of Chapter 12. However, urban land uses are far more diverse, and the interactions between them are much more dynamic than a gravity kernel.

APPENDIX 10A: THE INPUT–OUTPUT MODEL

The *input–output model* is widely used in economic planning at various levels of governments (Hewings, 1985). In the model, the output from any sector is also the input for all sectors (including the sector itself), and the inputs to one sector are provided by the outputs of all sectors (including itself). The key assumption in the model is that the *input–output coefficients* connecting all sectors characterize the technologies for a time period and remain unchanged over the period. The model is often used to examine how a change in production in one sector of the economy affects all other sectors, or how the production of all sectors needs to be adjusted in order to meet any changes in demands in the market.

We begin with a simple economy of two industrial sectors to illustrate the model. Consider an economy with two sectors and their production levels: X_1 for auto and X_2 for iron and steel. The total amount of auto has three components: (1) for each unit of output X_1, a_{11} is used as input (and thus a total amount of $a_{11}X_1$) in the auto industry itself; (2) for each unit of output X_2, a_{12} is used as input (and thus a total amount of $a_{12}X_2$) in the iron and steel industry; and (3) in addition to inputs that are consumed within industries, d_1 serves the final demand to consumers. Similarly, X_2 has three components: $a_{21}X_1$ as the total input of iron and steel for the auto industry, $a_{22}X_2$ as the total input for the iron and steel industry itself, and d_2 for the final demand for iron and steel in the market. It is summarized as:

$$\begin{cases} X_1 = a_{11}X_1 + a_{12}X_2 + d_1 \\ X_2 = a_{21}X_1 + a_{22}X_2 + d_2 \end{cases},$$

where the a_{ij} are the input–output coefficients.
  In matrix:

$$IX = AX + D,$$

where I is an identity matrix. Rearranging the equation yields:

$$(I - A)X = D,$$

which is a system of linear equations. Given any final demands in the future D and the input–output coefficients A, we can solve for the productions of all industrial sectors X.

APPENDIX 10B: SOLVING A SYSTEM OF NONLINEAR EQUATIONS

We begin with the solution of a single nonlinear equation by *Newton's method*. Say, f is a nonlinear function whose zeros are to be determined numerically. Let r be a real solution, and let x be an approximation to r. Keeping only the linear term in the *Taylor expansion*, we have:

$$0 = f(r) = f(x+h) \approx f(x) + hf'(x), \tag{A10.1}$$

where h is a small increment, such as $h = r - x$. Therefore:

$$h \approx -f(x) / f'(x)$$

If x is an approximation to r, $x - f(x)/f'(x)$ should be a better approximation to r. Newton's method begins with an initial value x_0 and uses iterations to gradually improve the estimate until the function reaches an error criterion. The iteration is defined as:

$$x_{n+1} = x_n - \frac{f(x_n)}{f'(x_n)}.$$

The initial value assigned (x_0) is critical for the success of using Newton's method. It must be "sufficiently close" to the real solution (r) (Kincaid and Cheney, 1991, p. 65). Also, it only applies to a function whose first-order derivative has a definite form. The method can be extended to solve a system of nonlinear equations.

Consider a system of two equations with two variables:

$$\begin{cases} f_1(x_1, x_2) = 0 \\ f_2(x_1, x_2) = 0 \end{cases}. \tag{A10.2}$$

Similar to equation A10.1, using the Taylor expansion, we have:

$$\begin{cases} 0 = f_1(x_1 + h_1, x_2 + h_2) \approx f_1(x_1, x_2) + h_1 \dfrac{\partial f_1}{\partial x_1} + h_2 \dfrac{\partial f_1}{\partial x_2} \\ 0 = f_2(x_1 + h_1, x_2 + h_2) \approx f_2(x_1, x_2) + h_1 \dfrac{\partial f_2}{\partial x_1} + h_2 \dfrac{\partial f_2}{\partial x_2} \end{cases}. \tag{A10.3}$$

The system of *linear* equations (A10.3) provides the basis for determining h_1 and h_2. The coefficient matrix is the *Jacobian matrix* of f_1 and f_2:

$$
J = \begin{bmatrix} \dfrac{\partial f_1}{\partial x_1} & \dfrac{\partial f_1}{\partial x_2} \\ \dfrac{\partial f_2}{\partial x_1} & \dfrac{\partial f_2}{\partial x_2} \end{bmatrix}.
$$

Therefore, Newton's method for system A10.2 is:

$$
x_{1,n+1} = x_{1,n} + h_{1,n}
$$

$$
x_{2,n+1} = x_{2,n} + h_{2,n},
$$

where the increments $h_{1,n}$ and $h_{2,n}$ are solutions to the rearranged system of linear equations A10.3:

$$
\begin{cases} h_{1,n} \dfrac{\partial f_1}{\partial x_1} + h_{2,n} \dfrac{\partial f_1}{\partial x_2} = -f_1 \left(x_{1,n}, x_{2,n} \right) \\ h_{1,n} \dfrac{\partial f_2}{\partial x_1} + h_{2,n} \dfrac{\partial f_2}{\partial x_2} = -f_2 \left(x_{1,n}, x_{2,n} \right) \end{cases},
$$

or

$$
J \begin{bmatrix} h_{1,n} \\ h_{2,n} \end{bmatrix} = - \begin{bmatrix} f_1 \left(x_{1,n}, x_{2,n} \right) \\ f_2 \left(x_{1,n}, x_{2,n} \right) \end{bmatrix}. \tag{A10.4}
$$

Solving the system of linear equations, A10.4 uses the method discussed in Section 10.1. Solution of a larger system of nonlinear equations follows the same strategy, and only the Jocobian matrix is expanded. For instance, for a system of three equations, the Jocobian matrix is:

$$
J = \begin{bmatrix} \dfrac{\partial f_1}{\partial x_1} & \dfrac{\partial f_1}{\partial x_2} & \dfrac{\partial f_1}{\partial x_3} \\ \dfrac{\partial f_2}{\partial x_1} & \dfrac{\partial f_2}{\partial x_2} & \dfrac{\partial f_2}{\partial x_3} \\ \dfrac{\partial f_3}{\partial x_1} & \dfrac{\partial f_3}{\partial x_2} & \dfrac{\partial f_3}{\partial x_3} \end{bmatrix}.
$$

NOTES

1 In step 3 for the case with a beltway, select LENGTH1 here to define travel impedance.
2 Another alternative proposed in the literature is to approximate the intrazonal distance as the radius of a circle with an equal area size (Frost et al., 1998).
3 Another option for Distance Type in "Garin–Lowry Model Pro" calibrates a Euclidean distance table internally and has fewer parameters to define. It requires the perimeter field for defining intrazonal travel distances.
4 For convenience, a field tract0_id is created on the feature tract (or trtpt) with default value 0, but 1–10 for ten tracts along one sector. One may extract the population and service employment patterns in this sector to prepare Table 10.1.

11 Linear and Quadratic Programming and Applications in Examining Wasteful Commuting and Allocating Healthcare Providers

This chapter introduces two popular methods in optimization, *linear programming* (LP) and *quadratic programming* (QP). QP is perhaps the simplest form of *nonlinear programming* (NLP). Both LP and QP are important techniques in socioeconomic analysis and planning. LP seeks to maximize or minimize an objective function subject to a set of constraints. Both the objective and the constraints are expressed in linear functions. QP has a quadratic objective function, but its constraints remain linear. It would certainly take more than one chapter to cover LP and QP in-depth, and many graduate programs in planning, engineering, or other fields use a whole course or more to teach the methods. This chapter discusses the basic concepts of LP and QP, their applications in location-allocation problems, and how these problems are solved in ArcGIS and R software.

Section 11.1 reviews the formulation of LP and the simplex method. The method is applied to examining the issue of wasteful commuting in Section 11.2. Commuting is an important research topic in urban studies for its theoretical linkage to urban structure and land use as well as its implications in public policy. Themes in the recent literature of commuting have moved beyond the issue of wasteful commuting and cover a diverse set of issues, such as relation between commuting and urban land use, explanation of intraurban commuting, and implications of commuting patterns in spatial mismatch and job access. However, strong research interests in commuting are, to some degree, attributable to the issue of wasteful commuting raised by Hamilton (1982) (see Appendix 11). A case study in Columbus, Ohio, is used to illustrate the method of measuring wasteful commuting by LP, and an R-ArcGIS tool is developed to solve the wasteful commuting LP.

Section 11.3 introduces the integer linear programming (ILP), in which some of the decision variables in a linear programming problem take integer values only. Some classic location-allocation problems, such as the *p*-median problem, the

DOI: 10.1201/9781003292302-14   **285** location set covering problem (LSCP), the maximum covering location problem (MCLP), and the minimax problem, are used to illustrate the formulation of ILP problems. Section 11.4 introduces QP, and then a new location-allocation problem termed "maximal accessibility equality problem (MAEP)," and how the problem is solved by QP. The MAEP seeks to minimize inequality in accessibility of facilities across geographic areas and, by extension, across population groups. Applications of these location-allocation problems can be widely seen in both private and public sectors. Section 11.5 uses a case study of allocating healthcare providers in a rural county in China to illustrate the implementation of these location-allocation problems in ArcGIS and R-ArcGIS tools. The chapter concludes with a brief summary in Section 11.6.

11.1 LINEAR PROGRAMMING AND THE SIMPLEX ALGORITHM

11.1.1 THE LP STANDARD FORM

The *linear programming (LP) problem* in the *standard form* can be described as follows:

Find the maximum of $\sum_{j=1}^{n} c_j x_j$ subject to the constraints $\sum_{j=1}^{n} a_{ij} x_j \leq b_i$ for all $i \in \{1,2,\ldots,m\}$ and $x_j \geq 0$ for all $j \in \{1,2,\ldots,n\}$.

The function $\sum_{j=1}^{n} c_j x_j$ is the *objective function*, and a solution $x_j \left(j \in \{1,2,\ldots,n\} \right)$ is also called the *optimal feasible point*.

In the matrix form, the problem is stated as:

Let $\boldsymbol{c} \in \boldsymbol{R}^n$, $\boldsymbol{b} \in \boldsymbol{R}^m$, and $\boldsymbol{A} \in R^{m \times n}$. Find the maximum of $\boldsymbol{c}^T \boldsymbol{x}$ subject to the constraints $\boldsymbol{A} \boldsymbol{x} \leq \boldsymbol{b}$ and $\boldsymbol{x} \geq 0$.

Since the problem is fully determined by the data A, b, c, it is referred to as problem (A, b, c).

Other problems not in the standard form can be converted to it by the following transformations (Kincaid and Cheney, 1991, p. 648):

1. Minimizing $\boldsymbol{c}^T \boldsymbol{x}$ is equivalent to maximizing $-\boldsymbol{c}^T \boldsymbol{x}$.

2. A constraint $\sum_{j=1}^{n} a_{ij} x_j \geq b_i$ is equivalent to $-\sum_{j=1}^{n} a_{ij} x_j \leq -b_i$.

3. A constraint $\sum_{j=1}^{n} a_{ij} x_j = b_i$ is equivalent to $\sum_{j=1}^{n} a_{ij} x_j \leq b_i, -\sum_{j=1}^{n} a_{ij} x_j \leq -b_i$.

4. A constraint $\sum_{j=1}^{n} \left| a_{ij} x_j \right| \leq b_i$ is equivalent to $\sum_{j=1}^{n} a_{ij} x_j \leq b_i, -\sum_{j=1}^{n} a_{ij} x_j \leq b_i$.

5. If a variable x_j can be negative, it is replaced by the difference of two variables, such as $x_j = u_j - v_j$.

11.1.2 THE SIMPLEX ALGORITHM

The *simplex algorithm* (Dantzig, 1948) is widely used for solving linear programming problems. By skipping the theorems and proofs, we move directly to illustrate the method in an example.

Consider a linear programming problem in the standard form:

Maximize: $\qquad\qquad z = 4x_1 + 5x_2$

Subject to: $\qquad\qquad 2x_1 + x_2 \leq 12$

$$-4x_1 + 5x_2 \leq 20$$

$$x_1 + 3x_2 \leq 15$$

$$x_1 \geq 0, x_2 \geq 0$$

The simplex method begins with introducing *slack variables* $u \geq 0$ so that the constraints $\boldsymbol{Ax} \leq \boldsymbol{b}$ can be converted to an equation form $\boldsymbol{Ax} + \boldsymbol{u} = \boldsymbol{b}$.

For the preceding problem, three slack variables $(x_3 \geq 0, x_4 \geq 0, x_5 \geq 0)$ are introduced. The problem is rewritten as:

Maximize: $\qquad\qquad z = 4x_1 + 5x_2 + 0x_3 + 0x_4 + 0x_5$

Subject to: $\qquad\qquad 2x_1 + x_2 + x_3 + 0x_4 + 0x_5 = 12$

$$-4x_1 + 5x_2 + 0x_3 + x_4 + 0x_5 = 20$$

$$x_1 + 3x_2 + 0x_3 + 0x_4 + x_5 = 15$$

$$x_1 \geq 0, x_2 \geq 0, x_3 \geq 0, x_4 \geq 0, x_5 \geq 0$$

The simplex method is often accomplished by exhibiting the data in a *tableau* form, such as:

$$
\begin{array}{ccccccc}
4 & 5 & 0 & 0 & 0 & \\
2 & 1 & 1 & 0 & 0 & 12 \\
-4 & 5 & 0 & 1 & 0 & 20 \\
1 & 3 & 0 & 0 & 1 & 15
\end{array}
$$

The top row contains coefficients in the objective function $\boldsymbol{c}^T \boldsymbol{x}$. The next m rows represent the constraints that are re-expressed as a system of linear equations. We leave the element at the top right corner blank as the solution to the problem z_{max} is yet to be determined. The tableau is of a general form:

$$
\begin{array}{ccc}
\boldsymbol{c}^T & 0 \\
\boldsymbol{A} & \boldsymbol{I} & \boldsymbol{b}
\end{array}
$$

If the problem has a solution, it is found at a finite stage in the algorithm. If the problem does not have a solution (i.e., an unbounded problem), it is discovered in the course of the algorithm. The tableau is modified in successive steps according to certain rules until the solution (or no solution for an unbounded problem) is found.

By assigning 0 to the original variables x_1 and x_2, the initial solution ($x_1 = 0$, $x_2 = 0$, $x_3 = 12$, $x_4 = 20$, $x_5 = 15$) certainly satisfies the constraints of equations. The variables x_j, whose values are zero, are designated *nonbasic variables*, and the remaining ones, usually nonzero, are designated *basic variables*. The tableau has n components of nonbasic variables and m components of basic variables, corresponding to the numbers of original and slack variables, respectively. In the example, x_1 and x_2 are nonbasic variables ($n = 2$), and x_3, x_4, and x_5 are basic variables ($m = 3$). In the matrix that defines the constraints, each basic variable occurs in only one row, and the objective function must be expressed only in terms of nonbasic variables.

In each step of the algorithm, we attempt to increase the objective function by converting a nonbasic variable to a basic variable. This is done through Gaussian elimination steps, since elementary row operations on the system of equations do not alter the set of solutions. The following summarizes the work on any given tableau:

1. Select the variable x_s whose coefficient in the objective function is the largest positive number, that is, $c_s = \{c_i > 0\}$. This variable becomes the new basic variable.
2. Divide each b_i by the coefficient of the new basic variable in that row, a_{ij}, and among those with $a_{is} > 0$ (for any i), select the minimum $\dfrac{b_i}{a_{ij}}$ as the pivot element and assign it to the new basic variable, that is, $x_s = \dfrac{b_k}{a_{kj}} = min\left\{\dfrac{b_i}{a_{ij}}\right\}$.

   If all a_{is} are ≤ 0, the problem has no solution.
3. Using the pivot element a_{ks}, create 0's in column s with Gaussian elimination steps (i.e., keeping all values in the k-th row with the pivot element and subtracting the k-th row from other rows by the values in the corresponding columns).
4. If all coefficients in the objective function (the top row) are ≤ 0, the current x is the solution.

We now apply the procedures to the example.

In step 1, x_2 becomes the new basic variable because 5 (i.e., its coefficient in the objective function) is the largest positive coefficient.

In step 2, a_{22} is identified as the pivot element (highlighted in underscore in the following tableau) because 20/5 is the minimum among {12/1, 20/5, 15/3}, and $x_2 = 20/5 = 4$. Then the original tableau is expressed as:

$$
\begin{array}{cccccc}
0.8 & 1 & 0 & 0 & 0 & \\
2 & 1 & 1 & 0 & 0 & 12 \\
-0.8 & 1 & 0 & 0.2 & 0 & 4 \\
0.3333 & 1 & 0 & 0 & 0 & 5
\end{array}
$$

In step 3, Gaussian eliminations yield a new tableau. That is, columns in each of the row 1, 2, and 4 minus the corresponding column in row 3. For example, the updated value in the fourth row and first column is equal to $0.3333 - (-0.8) = 1.1333$.

$$
\begin{array}{cccccc}
1.6 & 0 & 0 & -0.2 & 0 & \\
2.8 & 0 & 1 & -0.2 & 0 & 8 \\
-0.8 & 1 & 0 & 0.2 & 0 & 4 \\
1.1333 & 0 & 0 & -0.2 & 0.3333 & 1
\end{array}
$$

According to step 4, the process continues as $c_1 = 1.6 > 0$.

Similarly, x_1 is the new basic variable, a_{14} is the pivot element, $x_1 = 1/1.1333 = 0.8824$, and after step 2 and 3, the resulting new tableau is:

$$
\begin{array}{cccccc}
0 & 0 & 0 & 0.0515 & -0.2941 & \\
0 & 0 & 0.3571 & 0.1050 & -0.2941 & 1.9748 \\
0 & 1.25 & 0 & 0.0735 & 0.2941 & 5.8824 \\
1 & 0 & 0 & -0.1765 & 0.2941 & 0.8824
\end{array}
$$

The process continues and generates a new tableau, such as:

$$
\begin{array}{cccccc}
0 & 0 & -3.4 & 0 & -2.9413 & \\
0 & 0 & 3.4 & 1 & -2.8 & 18.8 \\
0 & 17 & -3.4 & 0 & 6.8 & 61.2 \\
5.6667 & 0 & 3.4 & 0 & -1.1333 & 23.8
\end{array}
$$

By now, all coefficients in the objective function (the top row) are ≤ 0, and the solution is $x_1 = 23.8/5.6667 = 4.2$ and $x_2 = 61.2/17 = 3.6$. The maximum value of the objective function is $z_{max} = 4 \times 4.2 + 5 \times 3.6 = 34.8$.

Many software packages (some free) are available for solving LP problems. In the following case study, we select the open-source R software to illustrate the LP implementation.

11.2 CASE STUDY 11A: MEASURING WASTEFUL COMMUTING IN COLUMBUS, OHIO

11.2.1 THE ISSUE OF WASTEFUL COMMUTING AND MODEL FORMULATION

The issue of wasteful commuting was first raised by Hamilton (1982). Assuming that residents can freely swap houses, the planning problem is to minimize total commuting given the locations of houses and jobs. Hamilton used the monocentric exponential function to capture the distribution patterns of both residents and employment. Since employment is more centralized than population (i.e., the employment density function has a steeper gradient than the population density function), the solution to the problem is that commuters always travel toward the CBD and stop at the nearest employment location. See Appendix 11 for details. Hamilton found an average of 87% wasteful commuting in 14 cities in the US. White (1988) proposed a simple LP model to measure wasteful commuting. White's study yielded very little wasteful commuting, likely attributable to a large area unit she used (Small and Song, 1992). Using a smaller unit, Small and Song (1992) applied White's LP approach to Los Angeles and found about 66% wasteful commuting, less than that in the Hamilton's model but still substantial. The impact of area units on the measurement of wasteful commuting is further elaborated in Horner and Murray (2002). The following formulation follows White's LP approach.

Given the number of resident workers P_i at i ($i = 1, 2, \ldots, n$) and the number of jobs at E_j at j ($j = 1, 2, \ldots, m$), the minimum commute is the solution to the following linear programming problem:

Minimize: $$\sum_{i=1}^{n}\sum_{j=1}^{m}c_{ij}x_{ij}$$

Subject to: $$\sum_{j=1}^{m}x_{ij} \leq P_i \text{ for all } i\,(= 1,2,\ldots,n)$$

$$\sum_{i=1}^{n}x_{ij} \leq E_j \text{ for all } j\,(= 1,2,\ldots,m)$$

$$x_{ij} > 0 \text{ for all } i\,(= 1,2,\ldots,n) \text{ and all } j\,(= 1,2,\ldots,m)$$

where c_{ij} is the commute distance (time) from residential location i to jobsite j, and x_{ij} is the number of commuters on that route.

The objective function is the total amount of commute in the city. The first constraint defines that the total commuters from each residential location to various job locations cannot exceed the number of resident workers there. The second constraint defines that the total commuters from various residential locations to each jobsite cannot exceed the number of jobs there. In the urbanized areas of most US metropolitan areas, it is most likely that the total number of jobs exceeds the total number of resident workers, that is, $\sum_{i=1}^{n}P_i \leq \sum_{j=1}^{m}E_j$.

11.2.2 DATA PREPARATION IN ARCGIS

The following datasets are prepared and provided in the geodatabase Columbus. gdb under the data folder Columbus:

1. An area feature class urbtaz and its corresponding point feature class urbtazpt with 991 TAZs (traffic analysis zones and their centroids, respectively)
2. A feature dataset roads containing a single feature class roads for the road network

The spatial data are extracted from the TIGER files. Specifically, the feature class urbtazpt is defined by combining all TAZs (traffic analysis zones) within the Columbus MSA, including seven counties (Franklin, Union, Delaware, Licking, Fairfield, Pickaway, and Madison) in 1990, and keeping only the urbanized portion. The same study area was used in Wang (2001b), which focused on explaining intraurban variation of commute (see Figure 11.1). The feature dataset road is defined similarly by combining all roads in the study area but covers a slightly larger area than the urbanized area for maintaining network connectivity. In the attribute table of urbtazpt, the field EMP is the number of employment (jobs) in a TAZ, and the field WORK is the number of resident workers. In addition, the field POPU is the population in a TAZ for reference. These attributes are extracted from the 1990 *CTPP (Census Transportation Planning Package)* (www.fhwa.dot.gov/planning/census_issues/ctpp/) Urban Element datasets for Columbus MSA. Specifically, the information for resident workers and population is based on Part 1 (by place of residence), and the information for jobs is based on Part 2 (by place of work).

The fields WORK and EMP in urbtazpt define the numbers of resident workers and employment in each TAZ, respectively, and thus the variables P_i $(i = 1, 2, \ldots, n)$ and E_j $(j = 1, 2, \ldots, m)$ in the LP problem stated in Subsection 11.2.1. In order to minimize the computation load, we need to restrict the origins and destinations to those TAZs with nonzero employment and nonzero resident worker counts, respectively,

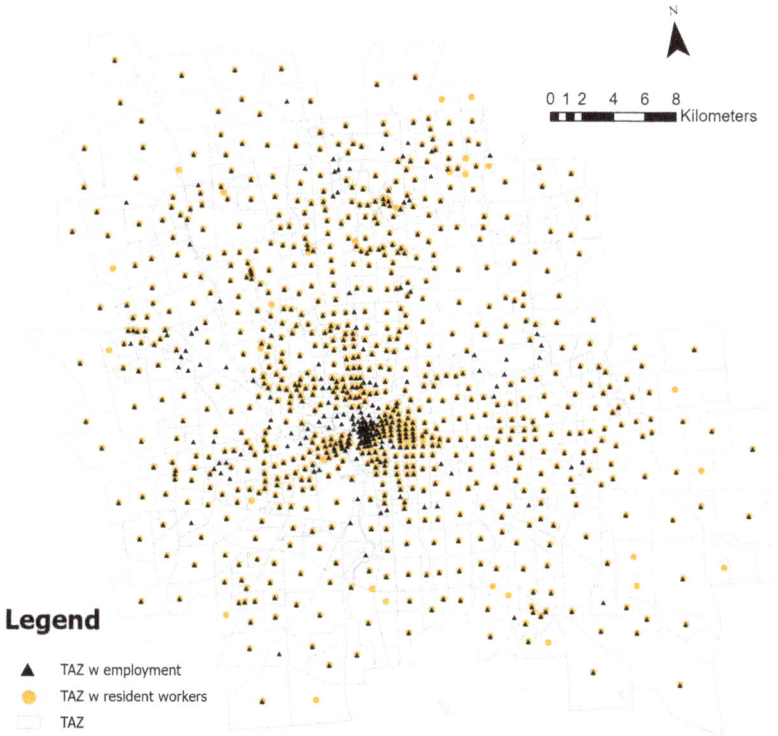

FIGURE 11.1 TAZs with employment and resident workers in Columbus.

and compute the OD travel time c_{ij} only between those TAZs. Subsection 2.2.2 in Chapter 2 has discussed the procedures for computing the OD travel time matrix, and this case study provides another opportunity to practice the technique. This study uses a simple approach for measuring travel time by assuming a uniform speed on the same level of roads. One may skip the steps and go directly to wasteful commuting analysis in Section 11.2.3, as the resulting files are also provided, as stated previously.

Step 1. Extracting Locations of Nonzero Employment and Nonzero Resident Workers: In ArcGIS Pro, open the layer urbtazpt, (1) select the TAZs with WORK > 0 and export the selected features to another new feature class restaz (812 resident worker locations), and (2) select the TAZs with EMP > 0 and export the selected features to a new feature class emptaz (931 employment locations). These two feature classes will be used as the origins and destinations in the OD time matrix in step 3 and subsequently in the wasteful commuting computation in Subsection 11.2.3.

Step 2. Defining Impedance for the Road Network: In ArcGIS Pro, open the attribute table of feature class roads, add a field Speed, and assign a default value of 25 (mph) for all road segments, then update the values according to the corresponding CFCC codes (Luo and Wang, 2003) by using a series of Select by Attributes and Field Calculator:

a. For CFCC >= 'A11' AND CFCC <= 'A18', Speed = 55
b. For CFCC >= 'A21' AND CFCC <= 'A28', Speed = 45
c. For CFCC >= 'A31' AND CFCC <= 'A38', Speed = 35

Another way is to use Calculate Field with Speed = Reclass(!CFCC!) and the following codes in the Code Block[1]:

```
def Reclass(CFCC):
n=int(CFCC.replace('A','').replace('P','9'))
if (n>=11 and n<=18): return 55
elif (n>=21 and n<=28): return 45
elif (n>=31 and n<=38): return 35
else: return 25
```

Add another field Minutes to the table and compute it as Minutes = (!Shape _ Leng!/ !Speed!)*60/1609 in minutes (note the unit for length is meter).

Step 3. Computing the OD Time Matrix between Resident Workers and Employment Locations: In the Geoprocessing pane, search and activate the tool Create Network Dataset. In the dialog, choose roads as Target Feature Dataset and type roads _ ND for Network Dataset Name, check roads under Source Feature Classes, keep other default settings, and click Run.

Remove roads _ ND in the Contents pane under Map.

In the Catalog pane, right-click the feature dataset roads _ ND > Properties > Travel Attributes > Costs tab. Under Properties, input Minutes for Name, and choose Minutes for Units. Under Evaluators, for roads (Along) under Edges, choose

Field Script for Type, and input [Minutes] for Value; for roads (Against), choose Same as Along. Keep other default settings, and click OK.

Search and activate Build Network in Geoprocessing pane, select road _ ND for Input Network Dataset, and click Run.

On the Analysis tab, click Network Analysis. In the drop-down menu, make sure road _ ND is the Network Data Source, and click Origin–Destination Matrix. Select the new OD Cost Matrix in the Map pane, click the top OD Cost Matrix Layer tab, use Import Origins for restaz and Import Destinations for emptaz without checking Append to Existing Locations and keeping other default settings in each dialog window, and click Run. Export the result table of Lines under OD Cost Matrix to a table ODTime under the geodatabase Columbus.gdb.

Similar to step 2 in Section 10.3, intrazonal time here is approximated as 1/4 of a TAZ's perimeter divided by a constant speed 670.56 meters/minute (i.e., 25 mph) in this study. The total travel time between two TAZs is composed of the aforementioned network time and the intrazonal time at both the origin and destination TAZs. Join the attribute table of feature class restaz to the table ODTime (based on the common fields OBJECTID and OriginID), and also join the attribute table of feature class emptaz to the table ODTime table (based on the common fields OBJECTID and DestinationID). By doing so, the perimeter values (in field Length) for both the origin and destination TAZs are attached to the table ODTime. Export it to a new table ODTimeNew.[2] Add a new field NetwTime (Double as Data Type) to the table ODTimeNew, and calculate it as !Total_Minutes! + 0.25*(!Length!+ !Length_1!) /670.56. This amends the network travel time by adding intrazonal times.

Export ODTimeNew to another table ODTimeAll. In the dialog of Export table, expand Fields; under Output Fields, keep only five fields, namely, OBJECTID (IDs for the origins or resident workers), WORK (resident workers), OBJECTID _ 1 (IDs for the destinations or employment), EMP _ 1(employment), and NetwTime; and under Properties, revise them to OBJECTID _ O, WORK, OBJECTID _ D, EMP, and NetwTime, respectively, and make sure each Field Name is identical to Alias. The new simplified OD travel time file is named ODTimeAll with 812×931 = 755,972 records (also provided in the data folder).

11.2.3 MEASURING WASTEFUL COMMUTING WITH THE WASTECOMMUTER TOOL

The LP problem for measuring wasteful commuting is defined by the spatial distributions of resident workers and employment and the travel time between them, which are prepared previously in table ODTimeAll with fields OBJECTID _ O, WORK, OBJECTID _ D, EMP, and NetwTime. Utilizing the lpSolve package in R, a tool WasteCommuteR is developed to solve the LP of wasteful commuting.

Under the Toolboxes, double-click R _ ArcGIS _ Tools.tbx to expand all tools. Click tool WasteCommuteR to activate it. In its dialog (Figure 11.2), select ODTimeAll for Input Table, OBJECTID _ O for Origin ID, WORK for Resident Worker, OBJECTID _ D for Destination ID, EMP for Employment, NetwTime for Commute Cost, and input Min _ Com for Output Table. Click Run.

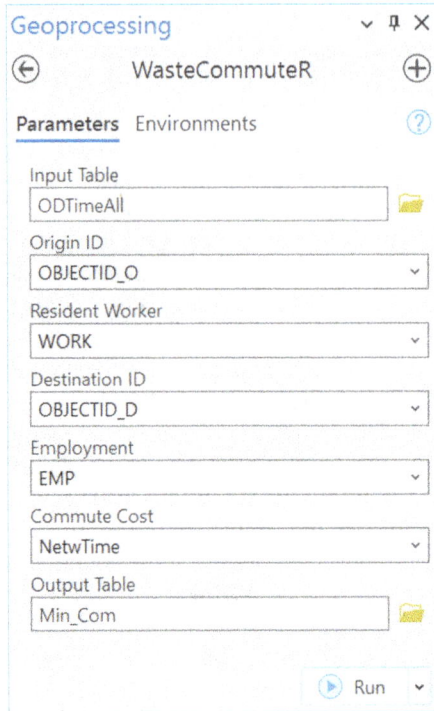

FIGURE 11.2 The interface of WasteCommuteR.

The output table `Min _ Com` contains the origin and destination IDs, the corresponding travel times, and the optimized numbers of commuters. The View Details window also shows that the objective function (total minimum commute time) is 3,805,721 minutes, that is, an average of 7.65 minutes per resident worker (with a total of 497,588 commuters). Given the actual mean commute time of 20.40 minutes in the study area in 1990 (Wang, 2001b, p. 173), there was 62.5% wasteful commuting, significantly below 87% reported by Hamilton (1982), and lower than 66% reported by Small and Song (1992).

Further examining the optimal commuting patterns in `Min _ Com` reveals that many of the trips have the same origin and destination TAZs, and therefore, accounting for intrazonal commute time is significant in reducing the resulting wasteful commuting. Studies on the 1990 CTPP Part 3 (journey-to-work) for the study area showed that among the 346 TAZs with nonzero intrazonal commuters, the average intrazonal drove-alone time was 16.3 minutes. Wang (2003) also revealed significant intrazonal commute time in Cleveland, Ohio (11.3 minutes). That is to say, the intrazonal time estimated in step 3 is much less than that reported in surveys. In addition, ArcGIS Pro also tends to underestimate travel time in comparison to time retrieved from Google Maps (Wang and Xu, 2011). In summary, a significant portion of so-called wasteful commuting may be attributable to the inaccurate travel time estimation. See Appendix 12 for a discussion of a more accurate measure of wasteful commuting by simulating individual commuters.

11.3   INTEGER PROGRAMMING AND LOCATION-ALLOCATION PROBLEMS

11.3.1   GENERAL FORMS AND SOLUTIONS FOR INTEGER PROGRAMMING

If some of the decision variables in a linear programming problem are restricted to only integer values, the problem is referred to as an *integer programming* problem.

If all decision variables are integers, it is an *integer linear programming (ILP) problem*. Similar to the LP standard form, it is written as:

Maximize: $\sum_{j=1}^{n} c_j x_j$

Subject to: $\sum_{j=1}^{n} a_{ij} x_j \leq b_i$ for all $i \in \{1, 2, \ldots, m\}$, and

$$\text{Integers } x_j \geq 0 \text{ for all } j \in \{1, 2, \ldots, n\}.$$

If some of the decision variables are integers and others are regular nonnegative numbers, it is a *mixed-integer linear programming (MILP) problem*. It is written as:

Maximize: $\sum_{j=1}^{n} c_j x_j + \sum_{k=1}^{p} d_k y_k$

Subject to: $\sum_{j=1}^{n} a_{ij} x_j + \sum_{k=1}^{p} g_{ik} y_k \leq b_i$ for all $i \in \{1, 2, \ldots, m\}$,

$$\text{Integers } x_j \geq 0 \text{ for all } j \in \{1, 2, \ldots, n\}, \text{and}$$

$$y_k \geq 0 \text{ for all } k \in \{1, 2, \ldots, p\}.$$

One may think that an easy approach to an ILP or MILP problem is to solve the problem as a regular LP problem and round the solution. In many situations, the rounded solution is not necessarily optimal (Wu and Coppins, 1981, p. 399). Solving the ILP or MILP requires special approaches, such as the *cutting planes method* or the *branch-and-bound method*, and the latter is more popular. The following summarizes a general branch-and-bound algorithm.

1. Find a feasible solution f_L as the lower bound on the maximum value of the objective function.
2. Select one of the remaining subsets, and separate it into two or more new subsets of solutions.
3. For each subset, compute an upper bound f_U on the maximum value of the objective function over all completions.
4. A subset is eliminated (fathomed) if (a) $f_U < f_L$, (b) its solution is not feasible, or (c) the best feasible solution in this subset has been found, in which case f_L is replaced with this new value.
5. Stop if there are no remaining subsets. Otherwise, go to step 2.

When the integer decision variables are restricted to value 0 or 1, the problem is said to be a *0–1 (binary) programming problem*. The 0–1 programming problem has wide applications in operations research and management science, particularly in location-allocation problems.

11.3.2 LOCATION-ALLOCATION PROBLEMS

We use four classic location-allocation problems to illustrate the formulation of ILP problems.

The first is the *p-median problem* (ReVelle and Swain, 1970). The objective is to locate a given number of facilities among a set of candidate facility sites so that the total travel distance or time to serve the demands assigned to the facilities is minimized. The *p*-median model formulation is:

Minimize: $Z = \sum\sum a_i d_{ij} x_{ij}$

Subject to: $x_{ij} \leq x_{jj} \ \forall i,j, i \neq j$ (each demand assignment is restricted to what has been located)

$\sum_{j=1}^{m} x_{ij} = 1 \forall i$ (each demand must be assigned to a facility)

$\sum_{j=1}^{m} x_{jj} = p \forall j$ (exactly p facilities are located)

$x_{ij} = 1,0 \forall i,j$ (a demand area is either assigned or not to a facility)

where i indexes demand areas ($i = 1, 2, \ldots, n$), j indexes candidate facility sites ($j = 1, 2, \ldots, m$), p is the number of facilities to be located, a_i is the amount of demand at area i, d_{ij} is the distance or time between demand i and facility j, and x_{ij} is 1 if demand i is assigned to facility j or 0 otherwise.

One may add a constraint that each demand site must be served by a facility within a critical distance or time $\left(d_{ij} \leq d_0\right)$. This formulation is known as the *p-median problem with a maximum distance constraint* (Khumawala, 1973; Hillsman and Rushton, 1975).

The second is the *location set covering problem (LSCP)* that minimizes the number of facilities needed to cover all demand (Toregas and ReVelle, 1972). The model formulation is:

Minimize: $Z = \sum_{j=1}^{m} x_j$

Subject to: $\sum_{j=1}^{N_i} x_j \geq 1 \forall i$ (a demand area must be within the critical distance or time of at least one open facility site)

$x_j = 1,0 \forall j$ (a candidate facility is either open or closed)

where N_i is the set of facilities, the distance or time between demand i and facility j is less than the critical distance or time d_0, that is, $d_{ij} \leq d_0$, x_j is 1 if a facility is open at candidate site j or 0 otherwise, and i, j, m, and n are the same as in the preceding *p*-median model formulation.

The third is the *maximum covering location problem (MCLP)* that maximizes the demand covered within a desired distance or time threshold by locating p facilities (Church and ReVelle, 1974). The model formulation is:

Minimize: $Z = \sum_{i=1}^{n} a_i y_i$

Subject to: $\sum_{j=1}^{N_i} x_j + y_i \geq 1 \forall i$ (a demand area must be within the critical distance or time of at least one open facility site or it is not covered)

$\sum_{j=1}^{m} x_j = p$ (exactly p facilities are located)

$x_j = 1,0 \, \forall j$ (a candidate facility is either open or closed)

$y_i = 1,0 \, \forall i$ (a demand area is either not covered or covered)

where i, j, m, n, and p are the same as in the *p-median* model formulation, and N_i and x_j are the same as in the preceding LSCP model formulation.

Note that y_i is 1 if a demand area i is not covered or 0 otherwise; thus, the objective function is structured to minimize the amount of demand not covered, equivalent to maximizing the amount covered. Similarly, one may add an additional constraint to the original MCLP that requires an uncovered demand point within a mandatory closeness constraint (the second and a larger distance threshold). The revised model is known as the *MCLP with mandatory closeness constraints* (Church and ReVelle, 1974).

The fourth is the *minimax problem*, also termed *p-center*. It minimizes the maximum distance between demand points and their nearest facilities by locating p facilities. The model is formulated as:

Minimize: Z

Subject to: $\sum_{j=1}^{m} x_{ij} = 1 \forall i$ (every demand area must have the only one nearest facility)

$\sum_{j=1}^{m} d_{ij} x_{ij} \leq Z \forall i$ (all the distances from the nearest facilities share a maximum value Z)

$x_{ij} \leq y_j \;\; \forall i, \forall j$ (every demand area must be connected to a chosen facility)

$\sum_{j=1}^{m} y_j = k$ (*k* facilities are chosen from *m* available)

(optional, $y_r = 1$, if the *r*-th facility is preselected)

$x_{ij} = 1,0 \, \forall i \, \forall j$ (a demand area is either assigned or not to a facility)

$y_i = 1,0 \, \forall i$ (a candidate facility is either open or closed)

where x_{ij}, i, j, m, n, and p are the same as in the *p-median* formulation.

The first three problems may be solved in the Network Analysis module in ArcGIS Pro. Specifically, under Location-Allocation > Location-Allocation Layer, the Type drop-down gallery in the Problem Type group allows users to specify different types of problems to be solved by the location-allocation solver. For example, "Minimize Weighted Impedance (P-Median)" solves the p-median problem, "Maximize Coverage and Minimize Facilities" solves the LSCP, and "Maximize Capacitated Coverage" solves the *MCLP*. The current version of ArcGIS Pro does not have a built-in tool to solve the p-center problem. Section 11.5 illustrates how to use the tools available in ArcGIS Pro to solve the first three problems, and an R-ArcGIS tool to solve the fourth problem. Table 11.1 summarizes the models and corresponding solution tools.

TABLE 11.1
Location-Allocation Models

Location-Allocation Model	Objective	Constraints	Solution Tool
1. *P*-median problem	Minimize total distance (time)	Locate p facilities; all demands are covered.	Minimize Impedance in ArcGIS Pro
1a. *P*-median with a max distance constraint	Minimize total distance (time)	In additional to the p-median constraints, demand must be within a specified distance of its assigned facility.	Minimize Impedance (by setting an impedance cutoff) in ArcGIS Pro
2. Location set covering problem (LSCP)	Minimize the number of facilities	All demands are covered.	Maximize Coverage and Minimize Facilities in ArcGIS Pro
3. Maximum covering location problem (MCLP)	Maximize coverage	Locate p facilities; demand is covered if it is within a specified distance (time) of a facility.	Maximize Capacitated Coverage (by setting an impedance cutoff) in ArcGIS Pro
3a. MCLP with mandatory closeness constraints	Maximize coverage	In additional to the MCLP constraints, demand not covered must be within a second (larger) distance (time) of a facility.	Maximize Capacitated Coverage (by setting two impedance cutoffs) in ArcGIS Pro
4. Minimax (*p*-center) problem	Minimize maximum distance (time)	Locate p facilities; all demands are covered.	MiniMaxR in R-ArcGIS
5. Maximal accessibility equality problem (MAEP)	Minimize inequality of accessibility	Locate p facilities, or redistribute a total capacity among p facilities.	MAEP in R-ArcGIS

11.4 QUADRATIC PROGRAMMING AND THE MAXIMAL ACCESSIBILITY EQUALITY PROBLEM (MAEP)

In quadratic programming (QP), the objective function becomes a quadratic function, but the constraints remain linear. This section uses a new location-allocation problem to illustrate the QP and its solution.

Planning often faces two competing goals, efficiency vs. equality. It has been long debated on which should be prioritized among academics and policy practitioners. Among the traditional location-allocation models listed in Table 11.1, the p-median problem minimizes total travel burden, the LSCP minimizes resource commitment, and the MCLP maximizes demand coverage. All three are designed for a goal related to the principle of efficiency, such as maximum payoff or minimum cost for the whole system. An exception is the minimax problem that seeks to minimize the largest travel burden. However, the minimax problem only attempts to reduce the inconvenience of the least accessible user without accounting for the distribution of demands across the whole spectrum and therefore is considered as marginally addressing the equality issue. These classic models and their variants have sustained extensive popularity, in part because of the convenience of solving them in ArcGIS.

Only modest progress has been made in research on location-allocation models addressing equality issues. One of such early efforts formulates the optimization objective as minimal inequality in accessibility of facilities across geographic areas (Wang and Tang, 2013). Specially, accessibility is measured by the 2SFCA method introduced in Chapter 5, and inequality can be defined as maximum deviation, mean absolute deviation, coefficient of variation, Gini coefficient, or variance of the accessibility index. When variance is used, the objective function is to minimize a quadratic function, and it becomes a QP problem.

Specifically, A_i is defined by the generalized 2SFCA method, such as:

$$A_i = \sum_{j=1}^{n} \frac{S_j f(d_{ij})}{\sum_{k=1}^{m} D_k f(d_{kj})}, \tag{11.1}$$

where D_k is demand at location k (= 1, 2, . . . , m), S_j is supply at location j (= 1, 2, . . . , n), d_{ij} or d_{kj} is the distance or travel time between demand (at i or k) and supply (at j), and f is a distance decay function that may take various forms, as discussed in Section 5.3.2 of Chapter 5.

Recall an important property of the 2SFCA accessibility index: the weighted average of accessibility (using the demand amount as weight) equals the ratio of total supply capacities $\left(\sum_{j=1}^{n} S_j = S \right)$ to total demand in the study area $\left(\left(\sum_{i=1}^{m} D_i = D \right) \right)$ (Appendix 5A). In other words, the mean of accessibility $\bar{A} = S / D$ is a constant.

In a reasonable scenario for planning new facilities (or adjusting the capacities of existing facilities), the number (n) and the sum of capacities of facilities are usually known, and the number (m) of demand locations and amount of individual demand (D_i) are also known. The objective function becomes:

$$min \sum_{i=1}^{m} \left(A_i - \bar{A} \right)^2 \ or \ min \sum_{i=1}^{m} \left(A_i - \bar{A} \right)^2 D_i, \tag{11.2}$$

which captures the total deviation of accessibility A_i across demand locations in the first formula, or weighted by their corresponding demand D_i there in the second formula. The latter, that is, minimum weighted variances, is adopted here. The decision variables are supply capacities for facilities (S_j) to be solved, and the optimization problem is subject to the total supply constraint.

Collecting the constant term in equation 11.1, such as $F_{ij} = \dfrac{f\left(d_{ij}\right)}{\sum_{k=1}^{m} D_k f\left(d_{kj}\right)}$, we rewrite the objective function in equation 11.2, such as:

$$min \sum_{i=1}^{m} \left(\sum_{j=1}^{n} S_j F_{ij} - \bar{A}\right)^2 D_i \qquad (11.3)$$

Subject to:

$$\sum_{j=1}^{n} S_j = S$$

In matrix notation, the objective function (11.2) is:

$$\left(FS - A\right)^T \left(DFS - DA\right) = S^T F^T DFS - A^T DFS - S^T F^T DA + A^T DA$$

$$= 2\left(\frac{1}{2} S^T F^T DFS - A^T DFS\right) + A^T DA$$

where:

$$\mathbf{S} = \begin{bmatrix} S_1 & S_2 & \cdots & S_p \end{bmatrix}^T$$

$$\mathbf{F} = \begin{bmatrix} F_{11} & \cdots & F_{1p} \\ \vdots & \ddots & \vdots \\ F_{m1} & \cdots & F_{1p} \end{bmatrix}$$

$$\mathbf{A} = \begin{bmatrix} \bar{A} & \bar{A} & \cdots & \bar{A} \end{bmatrix}^T, |A| = m$$

$$\mathbf{D} = \begin{bmatrix} D_1 & \cdots & 0 \\ \vdots & \ddots & \vdots \\ 0 & \cdots & D_m \end{bmatrix}$$

Dropping some constant terms, it is further transformed into a QP as follows:

$$\text{Min}\left(\frac{\mathbf{x}^T \mathbf{H} \mathbf{x}}{2}\right) + f^T \mathbf{x}$$

Subject to:

$$Cx \leq b$$

$$Ex = d$$

where:

$$\boldsymbol{x} = S$$

$$\mathbf{H} = \mathbf{F}^T\mathbf{DF}$$
$$\mathbf{f} = (-\mathbf{A}^T\mathbf{DF})^T = -\mathbf{F}^T\mathbf{DA}$$
$$\mathbf{C} = -\mathbf{1} \times \mathbf{I}, \text{ and } \mathbf{I} \text{ is an } n \times n \text{ identity matrix.}$$

$$\boldsymbol{b} = \begin{bmatrix} 0 & 0 & \ldots & 0 \end{bmatrix}^T, |b| = n$$

$$\boldsymbol{E} = \begin{bmatrix} 1 & 1 & \ldots & 1 \end{bmatrix}^T, |E| = n$$
$$\mathbf{d} = \mathbf{S}$$

There are various open-source programs for solving the QP problem (www.numerical.rl.ac.uk/qp/qp.html). The case study in Subsection 11.5.3 uses an R-based tool to implement the QP. See Wang and Tang (2013) and Li et al. (2017) for more discussion.

The preceding QP only solves the capacities for facilities. Many location-allocation problems also need to decide where to site the facilities (Luo et al., 2017). These spatial optimization models advocate a common objective of maximal equality (or minimal inequality) and emphasize that achieving such a goal begins with equal accessibility, not equal utilization or equal outcome, a principle consistent with the consensus reached by Culyer and Wagstaff (1993). They are grouped under the term "*maximal accessibility equality problem (MAEP)*" (Wang and Dai, 2020). Many variant models can be derived from this broad framework. For instance, beyond the variance definition in equation 11.2, inequality can be formulated as maximum deviation, mean absolute deviation, coefficient of variation, Gini coefficient, and others. Accessibility can also be measured in spatial proximity in distance, cumulative opportunities, or 2SFCA and its variants. The applications are seen in planning healthcare services (Zhang et al., 2019), schools (Dai et al., 2019), senior care facilities (Tao et al., 2014), and emergency medical services (Li et al., 2022; Luo et al., 2022).

11.5 CASE STUDY 11B: LOCATION-ALLOCATION ANALYSIS OF HOSPITALS IN RURAL CHINA

This case study is developed from the work reported in Luo et al. (2017). A rural county (Xiantao in Hubei Province) in China underwent its Regional Health Statistical Plan. The research team participated in data collection and expertise support for the plan. Part of the action plan called for building three new hospitals to reduce the travel time for residents in remote villages to seek basic medical care, especially for improving the response time for medical emergencies. Moreover, a survey of local residents indicated a strong desire for equal access to hospitals.

In response, the research team formulated a sequential decision-making approach, termed "*two-step optimization for spatial accessibility improvement (2SO4SAI).*" The first step is to find the best locations to site new hospitals by emphasizing accessibility as proximity to the nearest hospitals. The second step adjusts the capacities of facilities for minimal inequality in accessibility, measured by the 2SFCA method. The solution to the first step is to strike a balance among the solutions by three classic location-allocation models (*p*-median, MCLP, and minimax problems[3]), and the second step solves the QP problem as defined in equation 11.3.

Specifically, the case study is designed in three parts. Following data preparation in Subsection 11.5.1, Subsection 11.5.2 implements the first step by seeking a compromise from solutions (the *p*-median problem, the MCLP, and the minimax problem), and Subsection 11.5.3 carries out the second step by implementing the MAEP.

The following datasets for the study area are provided in the geodatabase XT.gdb under the data folder Xiantao:

1. Supply point feature class Hosp41 includes 41 existing hospitals with a field CHCI (comprehensive hospital capacity index) representing their capacities (Luo et al., 2017). Layer HospAll adds three newly sited hospitals to Hosp41 as a result from step 12.
2. Demand point feature class Village contains 647 villages with a field Popu indicating their population sizes.
3. Feature dataset Road contains three feature classes associated with the major road network: Road, Road_ND, and Road_ND_Junctions. Travel cost or impedance is defined in the field Length in Meters in the feature layer Road_ND.[4]
4. Base layers: Xiantao is the study area boundary, and Township contains the township administrative boundaries (each township includes multiple villages).
5. Intermediate results: layer LA _ chosen contains the four candidate facilities as solutions to the *p*-median and Maximize Coverage problem in steps 7 and 8, table ODnear is the distance matrix between extant 41 hospitals and all villages, and table ODmatrixCh is the distance matrix between the four candidate facilities and all villages.

11.5.1 DATA PREPARATION

Step 1. Building Network Dataset and Calculating OD Matrix: In ArcGIS Pro, create a new project Case11B and add all data into the map. Search and activate Build Network in the Geoprocessing pane, choose Road _ ND as Input Network Dataset, and run it.

In the top ribbon, select Analysis tab > Network Analysis, make sure that the Network Data Source is Road _ ND, and click Origin–Destination Cost Matrix.

Click the layer OD Cost Matrix in the Map pane, and click the top tab OD Matrix Layer. For Import Origins, select Village; for Import Destinations, choose Hosp41; uncheck Append to Existing Locations and leave other default settings in each dialog window; click the icon Σ ˅ for Cost Attribute for Accumulate Along Output Line; and check Length under Distance. Click Run. The result is saved in the layer Lines with a field Total _ Length in meters for $647 \times 41 = 26{,}527$ OD Pairs. Refer to Section 2.2.2 if needed.

Step 2. Accounting for Omitted Distances at the Trip Ends: Travel distances obtained from Step 1 are through the road network and omit the distances connecting the trip ends (i.e., villages and hospitals) and their nearest nodes. This step refines the estimate. Join the attribute tables of Origins and Destinations to Lines (all under OD Cost Matrix) based on the common fields of OriginID and DestinationID, respectively. Export the attribute table of Lines to a new table ODmatrix. Add a new field Dist (Double in Data Type) to table ODmatrix, and calculate it as "!Total _ Length!+!DistanceToNetworkInMeters! +!DistanceToNetworkInMeters _ 1!" Export the table ODmatrix to a new table ODdist while keeping the three key fields, OriginID, DestinationID, and Dist.

Step 3. Identifying the Distances between Demands and Their Nearest Facilities: On the table ODdist, right-click the field OriginID, and choose Summarize. In the dialog, note that Input Table is ODdist, then name Output Table as ODnear; under Statistics Fields, choose Dist for Field and Minimum for corresponding Statistics Type, ensure OriginID for Case Field, and click OK. The field MIN _ Dist in table ODnear contains the distances between villages and their nearest hospitals (647 records).

Step 4. Calculating the G2SFCA Accessibility Scores: For illustration, we assume the popular power function for distance decay, $f(d_{ij}) = dij{-}\beta$, and $\beta = 1$.

Activate the tool "Generalized 2SFCA/I2SFCA Pro" in the Catalog pane (refer to Subsection 5.6.2 if needed). In the dialog, choose Village as CustomerLayer, and its associated fields OBJECTID as Customer ID Field and Popu as Customer Size Field; choose Hosp41 as Facility Layer, and its associated fields OBJECTID as FacilityID Field, and CHCI as Facility Size Field; set 2SFCA for General Model; under Distance Type, choose External table for Distance Type, 0.001 for Distance Unit Conversion Factor; under Distance Decay, set Continuous for Distance Decay Function Category, Power for Continuous Function, 1 for Distance Decay Coefficient Beta; under External Table, choose ODdist as OD Matrix Table, and its associated fields OriginID as Customer ID Field, DestinationID as Facility ID Field, and Dist as Travel Cost; under Output setting, input 1 for Weight Scale Factor, DocpopR for Accessibility Field Name; keep No for Export ODmatrix Result; click Run.

The new field DocpopR in the attribute table of Village is the accessibility score.

Step 5. Generating the Accessibility Score Surface: This step uses an IDW-interpolated surface to visualize the spatial variability of accessibility.

In the Geoprocessing pane, search and activate the IDW tool. In the dialog, under Parameters tab, select Village for Input features, DocpopR for Z value field, V _ 2SFCA for Output Geostatistical layer, and V _ 2SFCA _ R for Output Raster; under Environments tab, input `Village` for Output Coordinate System, use `Village` for As Specified Below under Processing Extent, select `Xiantao` for Mask under Raster Analysis, and keep other default settings. Click Run. Figure 11.3 shows that areas around the hospitals (most noticeably, the area around the county seat with a cluster of hospitals) enjoy better accessibility, and remote countryside (especially the southeast corner) has the poorest accessibility scores.

11.5.2 Location Optimization for Site Selection

Step 6. Configuring Location-Allocation in Network Analysis: For all three location-allocation models in ArcGIS Pro, facilities are hospitals defined in the layer `Hosp41`, with their capacities defined in the field `CHCI`, and demands are villages in the layer `Village`, with their demand amount (or "weight") defined in the field `Popu`. We need to define three sets of locations: candidate facilities, required facilities (and their associated capacities), and demand locations (and their associated amounts).

In the Analysis tab, click Network Analysis, make sure that the Network Dataset Source remains `Road _ ND`, and choose Location-Allocation in the drop-down menu to activate the module. Click the new layer `Location-Allocation` in the Map pane to activate the Location-Allocation Layer tab at the top.

We first define Candidate Facilities.[5] Click Import Facilities (labeled "1" in Figure 11.4) to activate the dialog window of Add Locations. As shown in Figure 11.5, keep `Facilities` for Sub-Layer, choose `Village` for Input Locations, click FacilityType under Property, and choose `Candidate` for Default Value under Field. Uncheck Append to Existing Locations, and keep other default settings. Click Apply to add all village points as candidate locations. In theory, the number of candidate locations is countless. Since the villages are numerous and widely distributed across the study area, it is reasonable to assume that new hospitals will be sited at (or very close to) villages.

We then define Required Facilities. Still in the Add Locations dialog window of Import Facilities (similar to Figure 11.5), keep `Facilities` for Sub-Layer, choose `Hosp41` for Input Locations; under Property, click FacilityType, and under field, choose `Required` for Default Value; again, under Property, click Capacity, and under field, choose `CHCI` for Field Name; uncheck Append to Existing Locations; and keep other default settings. Click Apply to add all 41 existing hospitals as required facilities.

Finally, we define Demand Locations. Click Import Demand Points (labeled "2" in Figure 11.4) to activate the dialog window of Add Locations. In the dialog, choose `Demand Points` for Sub-Layer, select `Village` for Input Locations, click Weight under Property, choose `Popu` for Field Name under Field, uncheck Append to Existing Locations, and keep other default settings. Click Apply to add all villages as demand points.

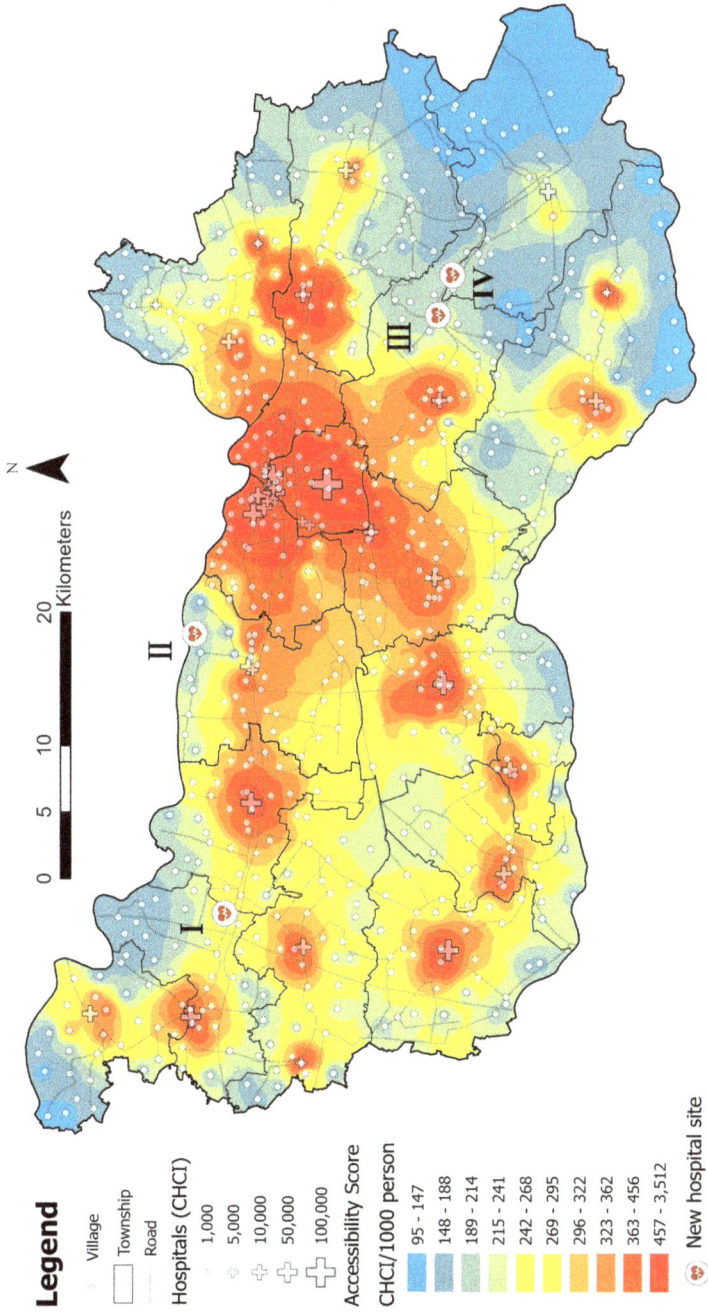

Legend

- Village
- Township
- Road

Hospitals (CHCI)

- 1,000
- 5,000
- 10,000
- 50,000
- 100,000

Accessibility Score

CHCI/1000 person

- 95 - 147
- 148 - 188
- 189 - 214
- 215 - 241
- 242 - 268
- 269 - 295
- 296 - 322
- 323 - 362
- 363 - 456
- 457 - 3,512

New hospital site

FIGURE 11.3 2SFCA-based accessibility surface in a rural county in China.

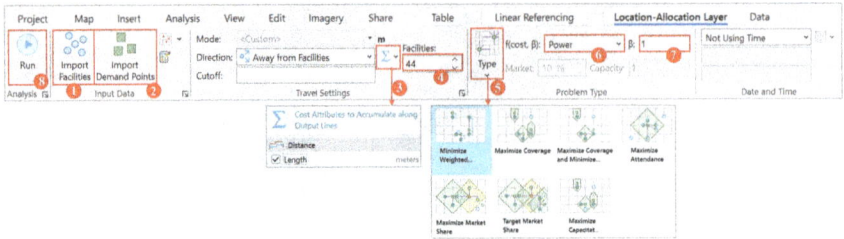

FIGURE 11.4 Interface and steps for solving the p-median problem.

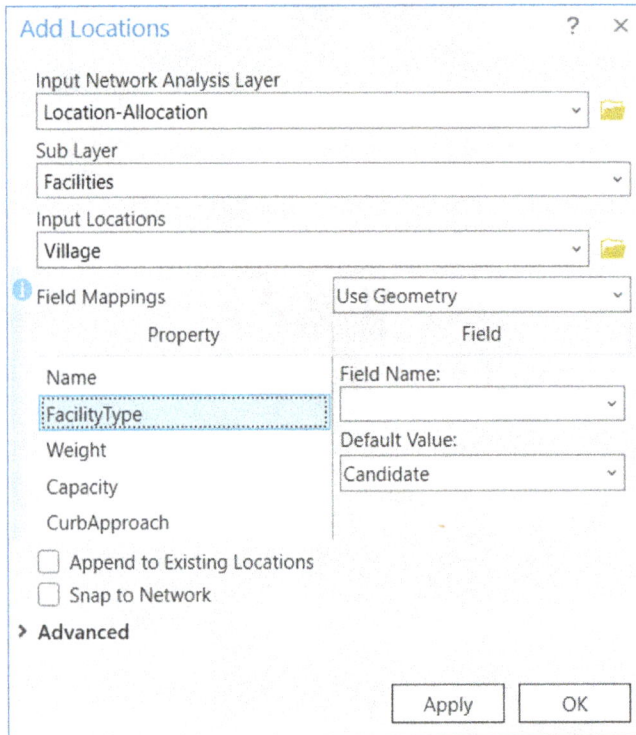

FIGURE 11.5 Interface defining candidate facilities in location-allocation analysis.

Step 7. Solving the p-Median Problem in ArcGIS: In the top Location-Allocation Layer tab (as shown in Figure 11.4, where each number labels the order of steps), (3) click the summation icon Σ ▾ for Cost Attribute for Accumulate along Output Line, and check Length under Distance; (4) input 44 for Facilities, which include 41 existing hospitals and 3 new hospitals to be sited; (5) click Type and choose Minimize Weighted Impedance (P-Median); (6) choose Power for f(cost, β) and (7) input 1 for β, which defines spatial impedance (cost) simply as distance[6]; and (8) click Run to solve the location-allocation problem.

The result is saved in the Facilities layer under Location-Allocation in the Map pane. Right-click the layer Facilities. Open the attribute table, click Select by Attributes, and input the expression "`FacilityType is equal to 3 - Chosen`." It exports the selected three points to a new feature class `LA _ pmedian`. As shown in Figure 11.3, the three new hospitals sited by the p-median model are labeled as I, II, and III.

Step 8. Solving the Maximize Coverage Problem in ArcGIS: Similar to step 7, in the same Location-Allocation layer, use the same settings, but input 10000 for Cutoff (i.e., 10 kilometers as a distance threshold), and choose Maximize Coverage for Type. Export the result as `LA _ Maxcov`. As shown in Figure 11.3, the three new hospitals sited by the maximize coverage problem are labeled as I, II, and III.

The two models yield very similar solutions. They choose the same two sites (I and II), and for the third one, location III by the p-median and location IV by the maximize coverage are within a short distance from each other.

The location-allocation module in ArcGIS Pro currently does not have a tool for the minimax model. Here we use an R-ArcGIS tool to implement its solution. Our experiment indicates that the computational time increases significantly as the number of candidate facilities increases. Here, for illustration, we use only four candidate sites. Specifically, the four sites I–IV suggested by the two previous models are considered. In preparing for the R-ArcGIS tool, a distance matrix between the demand locations and the candidate facilities needs to be calculated.

Step 9. Calculating the Distance Matrix between Demand Locations and Candidate Facilities: Note that the four candidate facilities correspond to four points from the layer `Village` with `OBJECTID` = 45, 436, 475, and 477. Use Select by Attributes to extract those points in `Village` and export them to a new feature class `LA _ chosen`.

Repeat steps 1–2 to calculate the OD matrix between `Village` and `LA _ chosen`, and save the table as `ODmatrixCh` with three fields, `OriginID`, `DestinationID`, and `DistChosen`. For the reader's convenience, table `ODmatrixCh` is provided under the geodatabase `XT.gdb`.

Step 10. Preparing Data for the MiniMaxR Tool: We now convert the OD matrix table to the format required by the MiniMaxR tool used in the next step. Search and activate the Pivot Table tool in the Geoprocessing pane. In the dialog, choose `ODmatrixCh` for Input Table, `OriginID` for Input Fields, `DestionationID` for Pivot Field, `DistChosen` for Value Field; input `ODchosenPv` for Output Table; and click Run. In the output table `ODchosenPv`, the first field `OriginID` is derived from the original table; the remaining fields are named, such as `Destionation1`, `Destionation2`, . . .; and their values are the corresponding distances between them.

Add a new field `DestinationID` to the table `ODnear` derived from step 3, and calculate it as `DestinationID` ="Extant." Apply the tool Pivot Table again on the updated table `ODnear` by using `OriginID` for Input Fields, `DestinationID` for Pivot Field,[7] and `MIN _ Didst` for Value Field, naming Output Table as `ODnearPv`.

Join the table `ODnearPv` to `ODchosenPv` based on the common filed `OriginID`, and then export the result to a new table `ODminimax` with six key columns: `OriginID`, `Destination1(2,3,4)` and `Extant`.

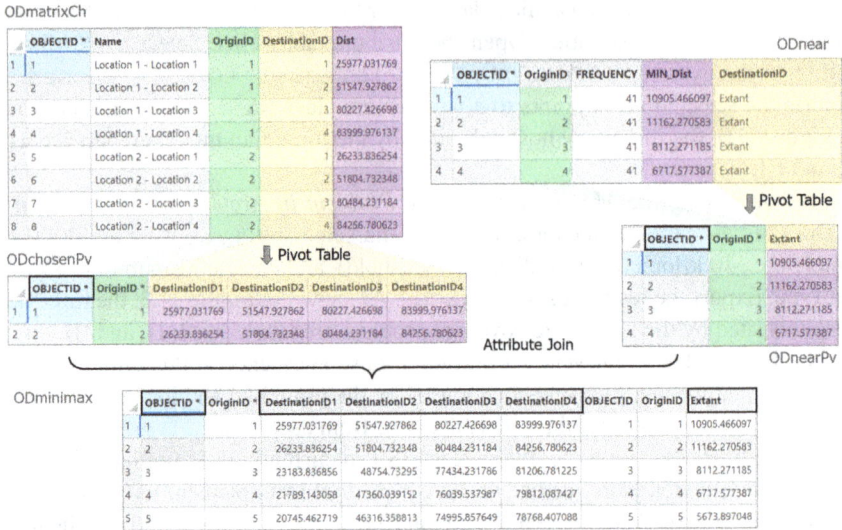

FIGURE 11.6 Data preparation for MiniMaxR tool.

See Figure 11.6 for illustrating the process. The Minimax problem is to choose the optimal combination of three distance columns from the candidates so that the maximum value among all the minimum distance values for each row (candidates and existing facilities) is minimal. In other words, each demand location (village) always chooses the nearest facility (hospital) from all 45 hospitals (here equivalent to choose among the nearest existing hospital and 4 candidate hospitals), and we look for the scenario of 3 candidate hospitals that yields the longest trip from various villages is minimal. Once again, solving the problem with only 4 candidate facilities is trivial, and it is designed for illustration.

Step 11. Solving the Minimax (p-Center) Problem with the MiniMaxR Tool: In the Catalog pane, activate the MiniMaxR tool under the folder R_ArcGIS_tools. In the dialog window (Figure 11.7a), select ODminimax for Input Table; choose OrigionID for OriginID; select DestinationID1, DestinationID2, DestinationID3, and DestinationID4 for Candidate Facilities; choose Yes for Use Required Facilities; select Extant for Distance to Nearest Required Facilities; input 4 for Number for *k* (i.e., adding 1 to 4 new candidate hospitals); input ODpcenter for Output Table; and click Run.

The table ODpcenter has the solutions for x_{ij}. Also, in the View Details window (Figure 11.7b), under Messages tab, solutions for y_j (= 1, 1, 1, 0, 1), reported under "Columns status as," correspond to DestionID1-4 and Extant, respectively, and the objective function value (minimax distance) is 16319.263 meters. That is to say, the first three candidate facilities (I, II, and III in Figure 11.3) are selected ($y_j = 1$), and site IV is not chosen ($y_j = 0$). The table ODpcenter contains several fields with their values shown in Figure 11.7c.

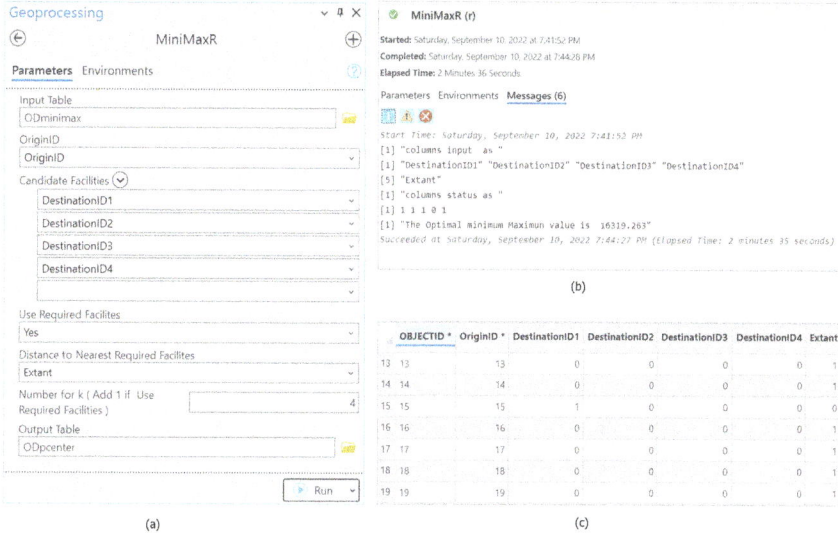

FIGURE 11.7 The MiniMaxR tool: (a) interface, (b) View Details, and (c) output table.

11.5.3 Capacity Optimization for Hospitals at Selected Sites

After the three new hospitals are sited by using the three classic location-allocation methods in Subsection 11.5.2, this subsection discusses the second task on how to plan for their capacities to max out their benefits, here by emphasizing maximizing equality in accessibility. Specifically, we use the 2SFCA method to measure accessibility to account for the intensity of competition for service (i.e., hospital capacity per capita). In other words, a hospital in an area with poorer access revealed by the 2SFCA method needs to be provided with more resources so that it would not be overcrowded to defeat its fundamental purpose of serving the neediest.

Here, the planning problem is to find the optimal values of S_j for the three newly sited facilities so that the disparity in accessibility across villages is minimized. As outlined in Section 11.4, the QP has an objective function in equation 11.2, where the key element F_{ij} (or matrix F) needs to be defined. An R-ArcGIS tool, "MAEP" (maximal accessibility equality problem), is developed to calibrate the matrix F and solve the QP problem.

Step 12. Preparing the Updated Facilities: To update facilities, we need to extract the three newly sited hospitals from feature class LA _ chosen and merge it with feature class Hosp41 that contains the existing 41 hospitals.

In ArcGIS Pro, add two new fields ID (Data Type: Long) and CHCI (Data Type: Double) into LA _ chosen, calculate ID as !OBJECTID! + 41, select the first three records (ID = 42, 43, 44), and export it as a new feature class LA _ new while keeping only two fields, ID and CHCI.

Search and activate Append tool in the Geoprocessing pane. In the dialog, input Hosp41 as Input Datasets and LA _ new as Target Dataset, select Input fields must match target fields for Field Matching Type, keep other default settings, and click Run. The feature layer LA _ new is updated by merging the two feature classes. Export it to a new feature class HospAll. Add a new field Status to the attribute table of HospAll and update the field values as "extant" for ID < 42 and "new" for ID > 41.

For your convenience, the layer HospAll with all 44 hospitals is provided in the geodatabase XT.gdb (OBJECTID = 42, 43, 44 index the three new hospitals).

Step 13. Preparing the OD Distance Matrix between Demands and Facilities: Recall that the table ODdist in step 1 stores the distance matrix between all 647 villages (OriginID = 1–647) and 41 hospitals (DestinationID = 1–41), and the table ODmatrixCh stores the distance matrix between the same 647 villages (OriginID = 1–647) and 4 candidate hospitals (DestinationID = 1–4).

Similar to step 12, on the table ODmatrixCh, extract the records with "DestinationID is less than 4" and export to a new table ODdistCh. On the new table, calculate its field DestinationID as !DestinationID! + 41. Use the Append tool to append the input table ODdist to the target dataset ODdistCh.

Step 14. Solving the Maximal Accessibility Equality Problem with the MAEP Tool: In the Catalog pane, activate the MAEP tool under the folder R _ ArcGIS _ tools. In the dialog as shown in Figure 11.8a, select ODdistCh for Input ODmatrix Table, Village for Input Demand Table, Hosp41 for Input Fixed Facilities Table, and their corresponding fields and parameters, name ODdistCh _ MAEP for Output Capacity Table and ODdistCh _ ACC for Output Accessibility Table, and click Run.

The View Details in Figure 11.8b reports the general information of this tool and the standard deviation of accessibility score in four scenarios.

The table ODdistCh _ MAEP, shown in Figure 11.8c, has four columns for the four different scenarios: Original for the initial status with the original capacities of the 41 hospitals, Average for distributing only the total new capacities to the 3 new hospitals equally, All for using the MAEP tool to optimally distribute the entire hospital capacities across all hospitals, and New for using the MAEP tool to optimally distribute only the total new capacities among the 3 new hospitals while keeping the capacities of the original hospitals unchanged. The last three rows with OBJECTID = 42, 43, and 44 in Figure 11.8c are the capacities of the 3 new hospitals in various scenarios.

The other table, ODdistCh _ ACC, shown in Figure 11.8d, reports the corresponding accessibility scores of demand locations (villages) in each scenario.

Our planning problem is to redistribute the resources saved from the closure of three existing hospitals to the three new hospitals. The capacity measure CHCI not only is more comprehensive than a singular measure, such as number of doctors or bed size, but also offers some flexibilities in the solution for hospital administrators to reach the best combination of personnel, equipment, and facility parameters. One may interpret it as a total financial commitment. The optimal allocation of resource (via capacity optimization) yields a slightly lower standard deviation, 0.1893 (thus less disparity), for accessibility than an arbitrary scenario, here evenly dividing the resource among the three hospitals (with a standard deviation of accessibility of 0.1895).

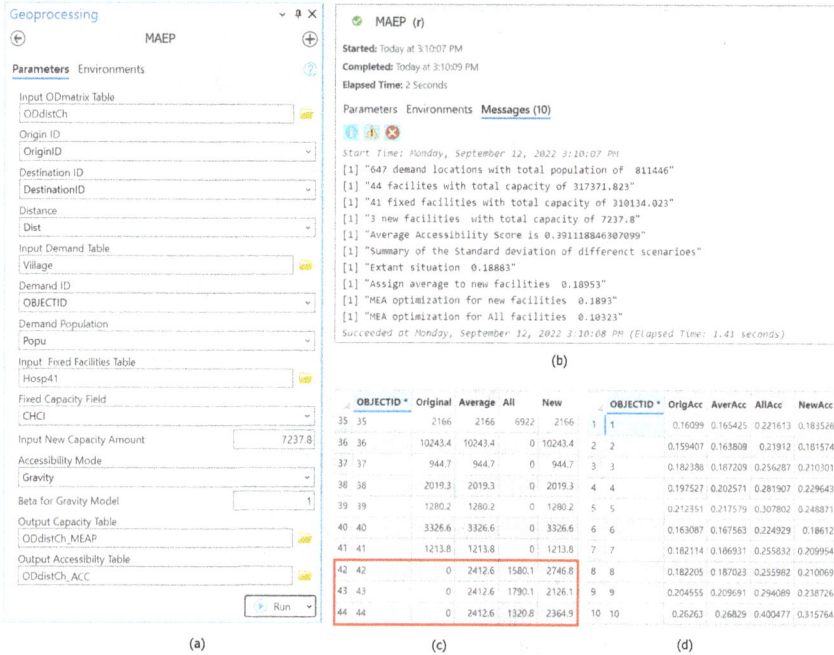

FIGURE 11.8 The MAEP tool: (a) interface, (b) "View Details," (c) "Output Capacity Table," and (d) "Output Accessibility Table."

11.6 SUMMARY

Applications of linear programming (LP) are widely seen in operational research, engineering, socioeconomic planning, location analysis, and others. This chapter first discusses the simplex algorithm for solving LP. Case study 11A uses a customized "WasteCommuteR" tool to solve a linear programming problem, such as the wasteful commuting problem. The solution to minimizing total commute in a city is termed "optimal (required or minimum) commute." In the optimal city, many journey-to-work trips are intrazonal (i.e., residence and workplace are in the same zone) and yield very little required commute and thus lead to a high percentage of wasteful commuting when compared to actual commuting. Computation of intrazonal travel distance or time is particularly problematic, as the lengths of intrazonal trips vary. Unless the journey-to-work data are down to street addresses, it is difficult to measure commute distances or times accurately. The interest in examining wasteful commuting, to some degree, stimulates the studies on commuting and issues related to it. One area with important implications in public policy is the issue of spatial mismatch and social justice (see Kain, 2004 for a review).

The location-allocation models are a set of applications of integer linear programming (ILP) in location analysis. Popular location-allocation models include the p-median problem, location set covering problem (LSCP), maximum covering location problem (MCLP), and minimax (p-center) problem. The p-median and MCLP problems can be solved by built-in Location Allocation Analysis tools in ArcGIS Pro as demonstrated in Case Study 11B. However, the solution to the minimax problem takes more effort in both data preparation and computational design and usually consumes significant computational time. Case study 11B uses an R tool "MinimaxR" to implement solving the minimax problem. We are not able to conceptualize a scenario for LSCP in the case study, and readers may use the "Maximize Coverage and Minimize Facilities" tool in ArcGIS Pro to implement it.

This chapter also proposes a new location-allocation problem, the maximal accessibility equality problem (MAEP). The optimization objective of MAEP is to minimize inequality in accessibility of facilities and is currently formulated as minimal variance across geographic areas. Specifically, it becomes a nonlinear programming (NLP) or quadratic programming (QP) problem. A MAEP tool in R is developed to solve the QP problem in Case Study 11B. Many variant models can be derived from this broad framework. For instance, beyond the variance definition in equation 11.2, inequality can be formulated as maximum deviation, mean absolute deviation, coefficient of variation, Gini coefficient, and others. Inequality or disparity can be assessed not only across geographic areas (where we are) but also between demographic groups (who we are). Beyond the 2SFCA method, accessibility can also be measured in spatial proximity in distance, cumulative opportunities, gravity-based potential model, and others as discussed in Chapter 5. The applications are seen in planning healthcare services, schools, senior care facilities, and emergency medical services (EMS).

APPENDIX 11: HAMILTON'S MODEL ON WASTEFUL COMMUTING

Economists often make assumptions in order to simplify a model with manageable complexity while capturing the most important essence of real-world issues. Similar to the monocentric urban economic model, Hamilton (1982) made some assumptions for urban structure. One also needs to note two limitations to Hamilton in the early 1980s: the lack of intraurban employment distribution data at a fine geographic resolution and the GIS technology at its developmental stage.

First, consider the commuting pattern in a monocentric city where all employment is concentrated at the CBD (or the city center). Assume that population is distributed according to a density function $P(x)$, where x is the distance from the city center. The concentric ring at distance x has an area size $2\pi x dx$, and thus population $2\pi x P(x) dx$ who travels a distance x to the CBD. Therefore, the total distance D traveled by a total population N in the city is the aggregation over the whole urban circle with a radius R:

$$D = \int_o^R x \left(2\pi x P(x) \right) dx = 2\pi \int_o^R x^2 P(x) dx.$$

Therefore, the average commute distance per person A is:

$$A = \frac{D}{N} = \frac{2\pi}{N} \int_o^R x^2 P(x) dx. \tag{A11.1}$$

Now, assume that the employment distribution is decentralized across the whole city according to a function $E(x)$. Hamilton believed that this decentralized employment pattern was more realistic than the monocentric one. Similar to equation A11.1, the average distance of employment from the CBD is:

$$B = \frac{2\pi}{J}\int_o^R x^2 E(x)\,dx, \tag{A11.2}$$

where J is the total number of employments in the city.

Assuming that residents can freely swap houses in order to minimize commute, the planning problem here is to minimize total commuting, given the locations of houses and jobs. Note that employment is usually more centralized than population. The solution to the problem is that commuters always travel toward the CBD and stop at the nearest employer. Compared to a monocentric city, "displacement of a job from the CBD can save the worker a commute equal to the distance between the job and the CBD" (Hamilton, 1982, p. 1040). Therefore, "optimal" commute or *required commute* or *minimum commute* per person is the difference between the average distance of population from the CBD (A) and the average distance of employment from the CBD (B):

$$C = A - B = \frac{2\pi P_0}{N}\int_o^R x^2 e^{-tx}\,dx - \frac{2\pi E_0}{J}\int_o^R x^2 e^{-rx}\,dx. \tag{A11.3}$$

where both the population and employment density functions are assumed to be exponential, that is, $P(x) = P_0 e^{-tx}$ and $E(x) = E_0 e^{-rx}$, respectively.

Solving equation A11.3 yields:

$$C = -\frac{2\pi P_0}{tN}R^{-2}e^{-tR} + \frac{2}{t} + \frac{2\pi E_0}{rJ}R^{-2}e^{-rR} - \frac{2}{r}.$$

Hamilton studied 14 American cities of various sizes and found that the required commute only accounts for 13% of the actual commute, and the remaining 87% is wasteful. He further calibrated a model in which households choose their homes and jobsites at random, and found that the random commute distances are only 25% over actual commuting distances, much closer than the optimal commute!

There are many reasons that people commute more than the required commute predicted by Hamilton's model. Some are recognized by Hamilton himself, such as bias in the model's estimations (residential and employment density functions, radial road network)[8] and the assumptions made. For example, residents do not necessarily move close to their workplaces when they change their jobs because of relocation costs and other concerns. Relocation costs are likely to be higher for homeowners than renters, and thus, home ownership may affect commute. There are also families with more than one worker. Unless the jobs of all workers in the family are at the same location, it is impossible to optimize commute trips for each income earner. More importantly, residents choose their homes for factors not related to their jobsites, such as accessibility to other activities (shopping, services, recreation, entertainment, etc.), quality of schools and public services, neighborhood safety, and others. Some of these factors are considered in research on explaining intraurban variation of actual commuting (e.g., Shen, 2000; Wang, 2001b).

NOTES

1 Since CFCC is in a string format, line 2 extracts A-series codes as two-digit numbers and transforms P-series codes to three-digit numbers starting with 9. The subsequent lines assign the speeds of 55, 45, and 35 according to their corresponding two-digit road numbers, and a speed of 25 to all three-digit roads.

2 Another option is to use the Join Field tool in the Geoprocessing pane twice to join the necessary fields in `restaz` and `emptaz` to the table `ODTime`. There would be no need to export it to a new table with too many fields.

3 The LSCP minimizes the number of facilities and thus does not apply here as the number of facilities is preset at 3.

4 The study assumes a uniform travel speed on all roads as the dominant transportation mode is electric motorcycles with an average speed of 25 km/h in the countryside.

5 The Location IDs in Facilities under Location-Allocation in Map pane are related to the order of importing data. We import `Village` as candidate facilities first (before we import it again as Demand Points) so that the Location IDs of Candidates will be identical to the `OBJECTID` of `Village`. This helps us identify them easily.

6 Similar to how distance decay is conceptualized and defined in Section 2.3, a user may choose a specific function and its associated parameters to adjust the weighting scheme for impedance. In the case of a power function, a smaller β (e.g., 0.5 instead of 1) implies that the impedance does not increase as fast as increasing distance.

7 For step 11, only the demand location IDs and the distances from their nearest existing facilities matter. The value "Extant" is assigned to the field `DestinationID` to identify those facilities from candidate facilities.

8 Hamilton did not differentiate residents (population in general, including dependents) and resident workers (those actually in the labor force who commute). By doing so, it was assumed that the labor participation ratio was 100% and uniform across an urban area.

12 Monte Carlo Method and Applications in Urban Population and Traffic Simulations

Monte Carlo simulation is a numerical analysis technique that uses repeated random sampling to obtain the distribution of an unknown probabilistic entity. It provides a powerful computational framework for spatial analysis and has become increasingly popular with rising computing power. Some applications include data disaggregation, designing a statistical significance test, and detecting spatial patterns. This chapter demonstrates the value of Monte Carlo technique in spatial analysis. One case study simulates individual residents consistent with census data and land use pattern, and another simulates individual commuting trips and then aggregates the trips to form traffic flows between areas.

The chapter is organized as follows. Section 12.1 provides a brief introduction to the Monte Carlo simulation technique and its applications in spatial analysis. Section 12.2 uses a case study to disaggregate area-based population data from the census to individual residents, aggregates back to areas of uniform size, and examines the zonal effect on urban population density patterns. Section 12.3 summarizes the four-step travel demand forecast model. Section 12.4 uses another case study to demonstrate how the Monte Carlo method is used to simulate individual commuting trips from home to work and then aggregates the trips to form traffic flows between areas. Section 12.5 concludes the chapter with a brief summary.

12.1 MONTE CARLO SIMULATION METHOD

12.1.1 Introduction to Monte Carlo Simulation

Basically, a Monte Carlo method generates suitable random numbers of parameters or inputs to explore the behavior of a complex system or process. The random numbers generated follow a certain *probability distribution function* (PDF) that describes the occurrence probability of an event. Some common PDFs include:

1. A *normal* distribution is defined by a mean and a standard deviation, similar to the Gaussian function introduced in Section 2.3. Values in the middle near the mean are most likely to occur, and the probability declines symmetrically from the mean.

DOI: 10.1201/9781003292302-15

2. If the logarithm of a random variable is normally distributed, the variable's distribution is *lognormal*. For a lognormal distribution, the variable takes only positive real values and may be considered as the multiplicative product of several independent random variables that are positive. The left tail of a lognormal distribution is short and steeper and approaches toward 0, and its right tail is long and flatter and approaches toward infinity.
3. In a *uniform* distribution, all values have an equal chance of occurring.
4. A *discrete* distribution is composed of specific values, each of which occurs with a corresponding likelihood. For example, there are several turning choices at an intersection, and a field survey suggests a distribution of 20% turning left, 30% turning right, and 50% going straight.

In a Monte Carlo simulation, a set of random values is generated according to a defined probability distribution function (PDF). Each set of samples is called an iteration and recorded. This process is repeated a large number of times. The larger the number of iteration times are simulated, the better the simulated samples conform to the predefined PDF. Therefore, the power of a Monte Carlo method relies on a large number of simulations. By doing so, Monte Carlo simulation provides a comprehensive view of what may happen and the probability associated with each scenario.

Monte Carlo simulation is a means of statistical evaluation of mathematical functions using random samples. Software such as MATLAB, Python, and R package provides several random number generators corresponding to different PDFs (e.g., a normal distribution). Others use pseudo-random number sampling. For example, the *inverse transformation* generates sample numbers at random from a probability distribution defined by its cumulative distribution function (CDF) and is therefore also called inverse CDF transformation. Specifically, one can generate a continuous random variable X by (1) generating a uniform random variable U within (0, 1) and (2) setting $X = F^{-1}(U)$ for transformation to solve X in terms of U. Similarly, random numbers following any other probability distributions could be obtained.

12.1.2 Monte Carlo Applications in Spatial Analysis

The Monte Carlo simulation technique is widely used in spatial analysis. Here, we briefly discuss its applications in spatial data disaggregation and statistical testing.

Spatial data often come as aggregated data in various areal units (sometimes large areas) for various reasons. One likely cause is the concern of geo-privacy, as discussed in Section 9.1. Others include administrative convenience, integration of various data sources, limited data storage space, etc. Several problems are associated with analysis of aggregated data, such as:

1. Modifiable areal unit problem (MAUP), that is, instability of research results when data of different areal units are used, as raised in Section 1.3.2.
2. *Ecological fallacy*, when one attempts to infer individual behavior from data of aggregate areal units (Robinson, 1950).
3. Loss of spatial accuracy when representing areas by their centroids in distance measures.

Therefore, it is desirable to disaggregate data in area units to individual points in some studies. For example, Watanatada and Ben-Akiva (1979) used the Monte Carlo technique to simulate representative individuals distributed in an urban area in order to estimate travel demand for policy analysis. Wegener (1985) designed a Monte Carlo–based housing market model to analyze location decisions of industry and households, corresponding migration and travel patterns, and related public programs and policies. Poulter (1998) employed the method to assess uncertainty in environmental risk assessment and discussed some policy questions related to this sophisticated technique. Luo et al. (2010) used it to randomly disaggregate cancer cases from the zip code level to census blocks in proportion to the age–race composition of block population and examined implications of spatial aggregation error in public health research. Gao et al. (2013) used it to simulate trips proportionally to mobile phone Erlang values and to predict traffic flow distributions by accounting for the distance decay rule.

Another popular application of Monte Carlo simulation in spatial analysis is to test statistical hypotheses using randomization tests. The tradition can be traced back to the seminal work by Fisher (1935). In essence, it returns test statistics by comparing observed data to random samples that are generated under a hypothesis being studied, and the size of the random samples depends on the significance level chosen for the test. A major advantage of Monte Carlo testing is that investigators could use flexible informative statistics rather than a fixed, known distribution theory. Besag and Diggle (1977) described some simple Monte Carlo significance tests in analysis of spatial data, including point patterns, pattern similarity, and space–time interaction. Clifford et al. (1989) used a Monte Carlo simulation technique to assess statistical tests for the correlation coefficient or the covariance between two spatial processes in spatial autocorrelation. Anselin (1995) used Monte Carlo randomization in the design of statistical significance tests for global and local spatial autocorrelation indices. Shi (2009) proposed a Monte Carlo–based approach to test whether the spatial clustering pattern of actual cancer incidences is statistically significant, by computing p-values based on hundreds of randomized cancer distribution patterns. Hu et al. (2018) also used a similar approach to simulate randomized crime patterns and designed a statistical significance test to identify spatiotemporal hotspots.

Case Study 12A in Section 12.2 utilizes Monte Carlo simulation for disaggregating area-based population data randomly to individual points so that the density pattern of simulated points reflects the pattern at the zonal level. Case Study 12B introduces another data disaggregation approach from area-based flow data to individual OD trips by randomly connecting the points between zones to ensure that the volume of resulting connections is consistent with a predefined (observed or projected) inter-zonal flow pattern. Both have potential for broad applications. Data disaggregation from areas to points enables subsequent aggregation back to areas by design (e.g., of uniform area size) to mitigate zonal effect. Disaggregation from inter-zonal flow data to flows between points can be used in analysis beyond traffic simulation (e.g., population or wildlife migration, disease spread, journey to crime).

12.2 CASE STUDY 12A: DERIVING URBAN POPULATION DENSITY FUNCTIONS IN UNIFORM AREA UNIT IN CHICAGO BY MONTE CARLO SIMULATION

The study area is composed of seven counties in northeast Illinois (Cook, DuPage, Kane, Kendall, Lake, McHenry, and Will) as the core of Chicago CMSA (Consolidated Metropolitan Statistical Area), served by the Chicago Metropolitan Agency for Planning (CMAP). Main data sources include the 2010 Census data and the 2010 Land Use Inventory via the CMAP Data Hub (https://datahub.cmap.illinois. gov/group/land-use-inventories). As the measure of population density derived from the census is residents-based, we limit the simulation of population to the residential land use category from the CMAP data.

As stated in Section 6.3 of Chapter 6, it is desirable to have analysis areas of identical or similar area size in fitting urban density functions. Wang F et al. (2019b) developed six uniform units representing three distinctive shapes (square, triangle, and hexagon) and two scales to facilitate examining both zonal and scale effects. This case study uses only one shape (hexagon) in one size (1 km^2) for illustration. These areas are uniform in area size except for those on the edge of the study area.

The following datasets are prepared and provided for this case study in the geodatabase `ZoneEffect.gdb` under the data folder `Chicago`:

1. Feature `residential` represents residential land use.
2. Feature `block` represents census blocks with field `POP100` for population in 2010.
3. Features `tract` and `blockgroup` represent census tracts and block groups, respectively.
4. Feature `citycenter` is the city center (commonly recognized at the intersection of Michigan Avenue and Congress Parkway).

Step 1. Allocating Population to Residential Portion of Census Blocks: We utilize the land use information to eliminate the error of distributing population to non-residential land use types and thus improve the accuracy of population distribution simulation based on the census block level population data alone.

In ArcGIS Pro, search and activate Pairwise Intersect[1] in the Geoprocessing pane. In the dialog, choose `residential` and `block` as Input Features, input `ResBlk` for Output Feature Class, choose `All attributes` for Join Attributes, select `Same as input` for Output Type, and click Run. The resulting feature `ResBlk` only contains residential portions of the census blocks, referred to as "residential blocks."

On the attribute table of `ResBlk`, apply Summarize on the field `BLOCKID10` to sum up the field of `Shape _ Area`, and save the result to a new table `ResSum`. Join the table `ResSum` back to `ResBlk` based on the common field `BLOCKID10` to attach the field `Sum _ Shape _ Area`. Add a new field `Popu` (DataType: Long) on `ResBlk`, and calculate it as `!ResBlk.Shape _ Area! *!ResBlk.POP10!`

/!ResSum. SUM _ Shape _ Area! This distributes the population in a census block to residential block(s) proportionally to their area sizes.[2]

To ensure that all residential blocks have nonzero population, on the layer ResBlk, select the features with "Popu is greater than 0," and export the selected records to a new feature ResBlkPop.

Step 2. Simulating Random Points to Represent Population in Residential Blocks: In the Geoprocessing pane, search and activate the tool Create Random Points. In the dialog, keep the default geodatabase for Output Location (e.g., Case12a.gdb), input ResSim for Output Point Feature, choose ResBlkPop for Constraining Feature Class, choose Field for Number of Points and Popu underneath it as the field value, and keep other default settings. Click Run. The process takes several minutes on a desktop computer.

Step 3. Designing a Uniform Analysis Area Unit: In the Geoprocessing pane, search and activate the tool Generate Tessellation. In the dialog, input Hexagon for Output Feature Class; for Extent, choose As Specified Below and select tract. Choose Hexagon for Shape Type, set Size as 1 and Square Kilometers, and for Spatial reference, choose ResSim. Click Run. The resulting layer Hexagon covers the envelope of tract. Use Select by Location to choose the hexagons intersected with tract and export to a new feature HexUnit.

Step 4. Aggregating Population as Individual Points to the Designed Area Unit: Right-click the layer HexUnit, and activate the tool Add Spatial Join. In the dialog, keep HexUnit as Target Features, select ResSim as Join Features, check Keep All Target Features, select Interset for Match Operation, and click Run. In the updated layer HexUnit, the field Join _ Count is the total population in each hexagon, which is also population density in each hexagon, since its area size is 1 square kilometer.

Repeat steps 3–4 to try different Shape Type in Generate Tessellation, and use Census Block Group or Census Tract as aggregate units. For census tracts, users need to add a new field PopDen and calculate it as !Join _ Count! / (!Shape _ Area!/1000000). Figure 12.1a–b shows the population density pattern across census tracts and hexagons, respectively. The general patterns between the two are largely consistent.

Step 5. Measuring Distances from the City Center and Classifying Distance Ranges: In the Geoprocessing pane, search and activate the tool Feature to Point. Use the tool to convert the polygon feature HexUnit to a point feature HexUnitPt.

Search and activate the tool Near. In the dialog, choose HexUnitPt as Input Feature and citycenter as Near Feature, and run it. In the updated attribute table of HexUnitPt, the added field NearDist contains the distance from each hexagon's centroid to the city center.

Add a new field DistClass (Data Type: Long) to the attribute table of HexUnitPt, and calculate it as math.ceil(!NearDist!/10000). This classifies the distances with equal intervals of 10 kilometers.

Repeat the same process on the census tract feature tract to (1) generate a tract centroid feature tractPt, (2) calculate and save its distances from the city center in a field NearDist, and (3) classify the distance intervals in a field DistClass.

(a)   (b)

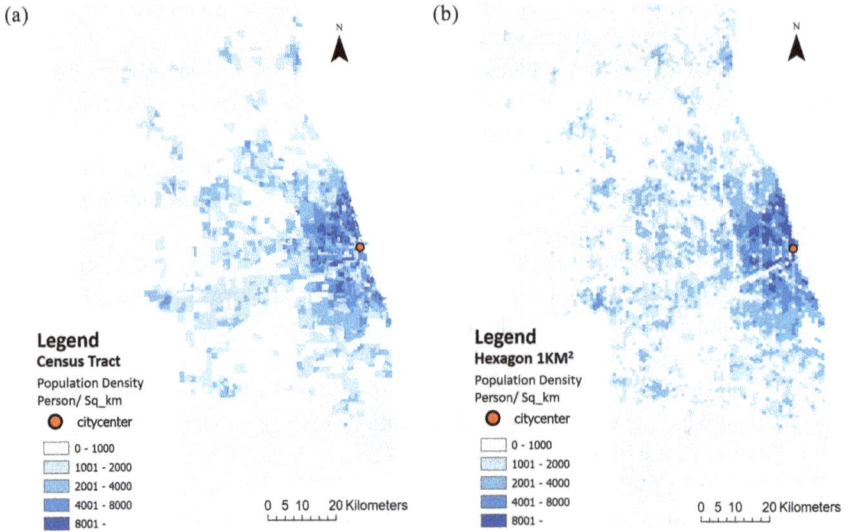

FIGURE 12.1 Population density pattern in Chicago: (a) census tracts, and (b) uniform hexagons,

Step 6. Examining Density Frequency Distribution Patterns: Summarize both attribute tables of `HexUnitPt` and `tractPt` based on the field `DistClass` and set Statistic Type as `Count`, and save the two summarized tables as `HexSumCount` and `trtSumCount`. Join tables `HexSumCount` and `trtSumCount` based on their common field `DistClass`, and rename their count fields (`COUNT_DistClass` for both tables) as `HexagonDist` and `TractDist`, respectively.

Activate the Bar Chart tool to graph the distribution pattern by setting `DistClass` as Category or Data, `<none>` for Aggregation, and `HexagonDist` and `TractDist` for Numeric Field(s).

Figure 12.2 shows how the numbers of observations vary by distance ranges. Indeed, there are many census tracts in short distances from the CBD (e.g., 357 tracts in 0–10 km and 499 tracts in 10–20 km), where tracts are small in area size, and the numbers decline toward longer distance ranges as tracts get larger. On the other side, the largest number of hexagons is observed in the middle-distance range (1,453 hexagons in 60–70 km), and the numbers decline both toward the CBD (less space to be divided by fewer hexagons) and toward the urban edge (less area due to the rectangular shape of the study area). The number of census tracts generally declines with increasing distance from the city center and thus suggests oversampling in near distance ranges, an issue raised in Section 6.3 of Chapter 6. In contrast, the distribution of hexagon count largely conforms to a normal distribution, which is more desirable when applying OLS regression analysis.

Step 7. Modeling Urban Density Functions: Apply Select by Attributes on `HexUnitPt` to select the records which population is larger than 0 and export it as a new feature `HexUnitPt1`. Add two new fields `LnDist` and `LnPopuDen` onto the attribute table of `HexUnitPt1`, and calculate them as `math.log(!NEAR _ DIST!)`

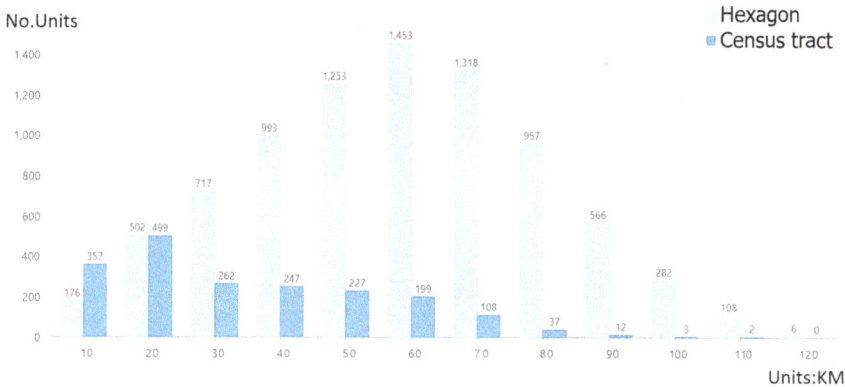

FIGURE 12.2 Numbers of census tracts and uniform hexagons by distance ranges.

and `math.log(!Join _ Count!)`, respectively. Repeat the same step on the feature `tractPt`, and update `LnDist` with the same formula and `LnPopuDen` as `math.log(!PopDen!)`.

A monocentric density function captures how population density in an area D_r varies with its distance r from the city center.[3] Users can explore various regression models between the distance and density terms, such as `LnDist`, `LnPopuDen`, `NearDist`, and `PopuDen`. Refer to Section 6.2 of Chapter 6 for detail. The results are summarized in Table 12.1.

For census tracts, the best-fitting function is exponential, as shown in Figure 12.3a, that is, $lnD_r = A + br$ (in its log transform). For uniform hexagons, the best-fitting function is logarithmic, as shown in Figure 12.3b, that is, $D_r = A + blnr$. In other words, the commonly observed negative exponential function for the urban population density pattern could be an artifact of data reporting based on the census unit. When the analysis unit is area of uniform size, the logarithmic function emerges as the best fit. This difference may be attributable to both zonal effect (uniform hexagons vs. census tracts of various sizes) and scale effect (far more hexagons than census tracts). One may isolate the zonal effect by generating hexagons with an area size similar to the average area size of census tracts.

Figure 12.4 shows the main workflow of this case study when using uniform hexagons.

12.3 TRAVEL DEMAND MODELING

The history of travel demand modeling, especially in the United States, has been dominated by the four-step model (McNally, 2008), composed of trip generation, trip distribution, mode choice, and trip assignment. The first comprehensive application of the four-step model was in the Chicago Area Transportation Study (Weiner, 1999). This section provides a brief overview of each step.

Traffic originated from areas or going to the same units is collectively referred to as *trip generation* (Black, 2003, p. 151). The former is sometimes called *trip*

TABLE 12.1

Regression Results of Population Density Functions by OLS

Area Unit	Coefficients and R^2	$D_r = a + br$	$\ln D_r = A + br$	$D_r = a + b\ln r$	$\ln D_r = A + b\ln r$
Census tract	a (or A)	5497.712	8.936	20917.255	17.522
($n = 1,953$)	b	−0.019	−0.00004	−1598.968	−0.983
	R^2	0.0004	**0.5192**	0.0005	0.4978
Uniform	a (or A)	3178.403	8.773	21724.899	29.927
hexagon	b	−0.041	0.00005	−1928.722	−2.278
($n = 8,331$)	R^2	0.3442	0.412	**0.4899**	0.3526

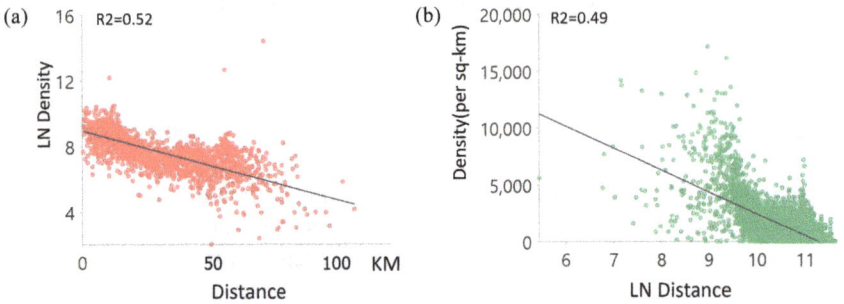

FIGURE 12.3 Population density vs. distance: (a) census tracts, (b) uniform hexagons.

production, and the latter *trip attraction*. Trip production is usually estimated by trip purposes (e.g., work, school, shopping, etc.). Take work trips produced by residential areas as an example. A multivariate regression model can be used to estimate the number of trips by several independent variables, such as population, labor participation ratio, vehicle ownership, income, and others at a zonal (e.g., subdivision, census tract, or traffic analysis zone) level. Trip attraction is often modeled by different types of land use (e.g., port, factory, hotel, hospital, school, office, store, service, park), and similarly with various regression models. In either trip production or attraction, there are hundreds of equations with coefficients tabulated for various sizes of metropolitan areas and different regions based on data collected over the years (ITS, 2012).

Based on the results from trip generation, *trip distribution* distributes the total number of trips from an origin (or to a destination) to specific OD pairs. In other words, trip generation estimates O_i and D_j, and trip distribution breaks them down to T_{ij} so that $O_i = \sum_j T_{ij}$ and $D_j = \sum_i T_{ij}$. Among various models, the gravity model discussed in Section 2.3 of Chapter 2 remains the most popular approach to estimating inter-zonal traffic:

$$T_{ij} = aO_i D_j d_{ij}^{-\beta},$$

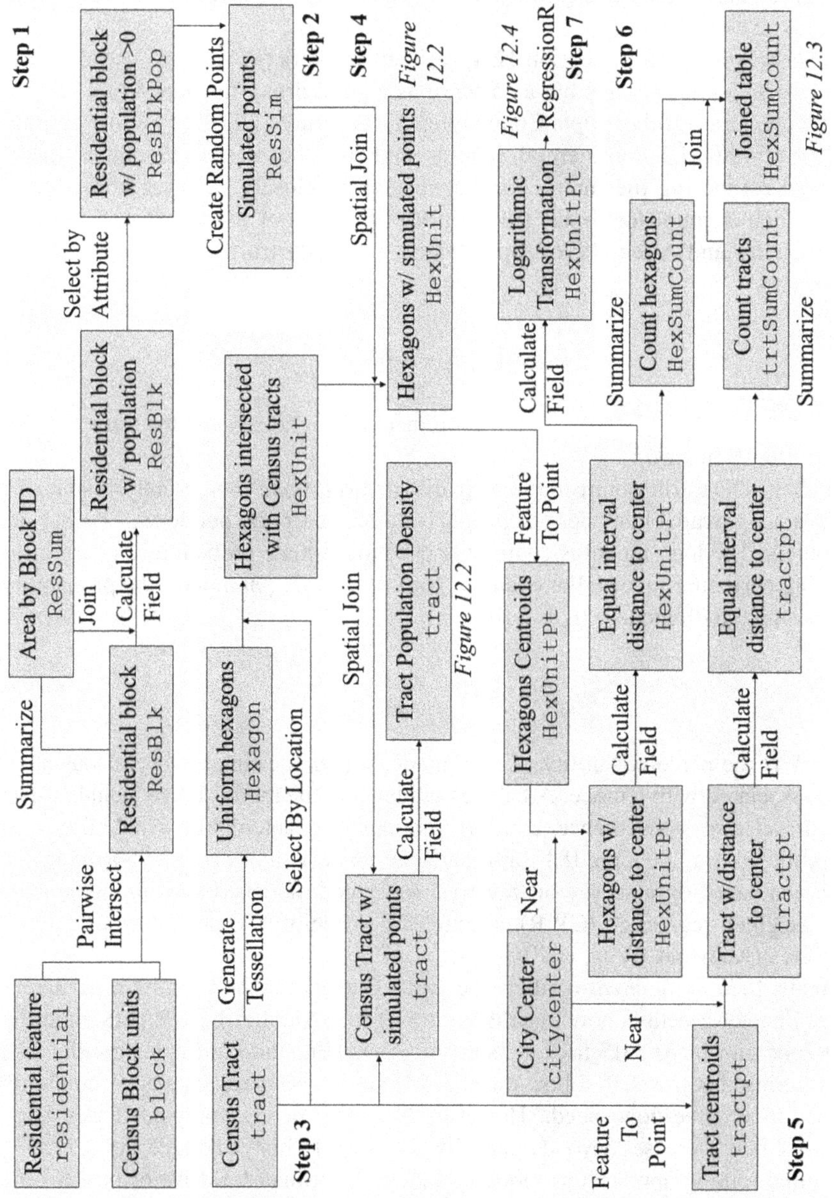

FIGURE 12.4 Case Study 12A workflow.

where T_{ij} is the number of trips between zones (e.g., tracts) i and j, O_i is the size (e.g., number of resident workers W_i) of an origin i, D_j is the size (e.g., number of employment E_j) of a destination j, a is a scalar factor, d_{ij} is the distance or travel time between them, and β is the distance friction coefficient. Here, the conventional power function is chosen as the distance decay function, and other functions discussed in Section 2.3 may also be considered.

In Case Study 12B in Section 12.4, resident workers (W) do not just represent those producing work trips but also serve as a good proxy for population in general that generates other trips. Employment (E) captures all other non-residential land uses (industrial, commercial, schools, and other businesses) that attract trips. A model estimating the interaction between them closely captures home-based trips, which account for by far the majority (70–75%) of urban trips by individuals (Schultz and Allen, 1996). Specifically, the daily traffic from zone i to j is modeled as:

$$T_{ij} = a\left(W_i E_j + E_i W_j\right)d_{ij}^{-\beta}, \tag{12.1}$$

where the term $W_i E_j$ accounts for trips originated from homes and the term $E_i W_j$ for trips ending at homes.

With the OD traffic obtained from trip distribution, *mode choice* analysis splits the traffic among available modes, such as drive-alone, carpool, public transit, bicycle, and others. The logit model is commonly used to estimate probabilities of travelers choosing particular modes. For example, for $i = 1, 2, \ldots, m$ modes, the probability of choosing the ith mode, $P(i)$, is written as:

$$P(i) = e^{V(i)} / \sum_{j=1}^{m} e^{V(j)},$$

where V is the perceived utility of each mode, which is estimated by various attributes associated with a mode. Attributes influencing the mode choice include travel time, travel cost, convenience, comfort, trip purpose, automobile availability, and reliability (Black, 2003, pp. 188–189). The logit model can be estimated by a logistic regression based on a survey of travelers. Advanced models, including geographically weighted regression (GWR), account for variability of mode choice across a study area (Kuai and Wang, 2020).

Finally, *trip assignment* predicts the paths (routes) for the trips by a particular mode. Chapter 2 details how to calibrate travel time via the shortest path either by driving or transit. ArcGIS uses its Network Analyst module to implement the task by assuming that travelers follow the speed limit on each road segment, commonly referred to as "free-flow speed." However, the travel speed usually decreases when the traffic flow increases beyond a capacity constraint (Khisty and Lall, 2003, p. 533). An online routing app such as Google Maps API estimates a drive or transit time matrix that accounts for real traffic conditions (Wang and Xu, 2011).

Transportation modeling is conducted at different levels of detail, from microscopic models for individual vehicles to mesoscopic models for aggregated behavior of vehicles and to macroscopic models of traffic flow between areas (Kotusevski and Hawick, 2009). Among a large number of transportation modeling packages for completing some or all the four steps of travel demand forecasting, there are *SUMO* (namely, Simulation of Urban Mobility) (www.eclipse.org/sumo/), *CORSIM* (known as Corridor Simulation) (https://mctrans.ce.ufl.edu/tsis-corsim/), *TRANSIMS* (Transportation Analysis and Simulation System) (www.fhwa.dot.gov/planning/ tmip/resources/transims/), and *TransModeler* (www.caliper.com/transmodeler/).

Traffic simulation is a major task in transportation planning and analysis. The aforementioned programs require extensive data preparation and input (detailed land use data, road network data with turning lanes, and traffic signals), have built-in algorithms unknown to users, and incur a significant learning curve for users. The previous version of the book (Wang, 2015) introduced a prototype model with four *traffic simulation modules for education* purposes (termed "TSME") that was coded in C#. In the next section, the case study implements similar tasks in ArcGIS so that readers can develop a better understanding of specific ArcGIS tools in accomplishing related Monte Carlo–based spatial simulations. This is enabled by a much-expanded set of ArcGIS tools available in the recent ArcGIS Pro version. With minimal data requirement (using publicly accessible data of road network and land use), the model illustrates the major steps of travel demand forecasting in distinctive modules.

12.4 CASE STUDY 12B: MONTE CARLO–BASED TRAFFIC SIMULATION IN BATON ROUGE, LOUISIANA

This case study is developed to illustrate the application of Monte Carlo simulation in traffic simulation. It is composed of four modules, roughly in correspondence to the four-step travel demand forecast model, except for the absence of "mode choice" step and the addition of "inter-zonal trip estimation" module (considered an extended task of data preparation). The study focuses on East Baton Rouge Parish (EBRP) of Louisiana, the urban core of Baton Rouge MSA. We also include eight neighboring parishes in traffic simulation in order to account for internal–external, external–internal, and through traffic. Validation of simulation results is confined to EBRP, where the annual average daily traffic (AADT) data are recorded at the traffic monitoring stations (Figure 12.5). Hereafter, the study area is simply referred to as Baton Rouge.

The spatial data for census tracts and road network were based on the 2010 TIGER files (www.census.gov/geographies/mapping-files/time-series/geo/tiger-line-file.html). Information on numbers of resident workers and jobs in census tracts was extracted from the 2006–2010 five-year Census Transportation Planning Package (CTPP) for the study area (https://ctpp.transportation.org/ctpp-data-set-informa-tion/). Similar to the datasets used in Case Study 11A in Section 11.2.2, the information for resident workers and population is based on Part 1, and the information for jobs is based on Part 2. The traffic count data in 816 monitoring stations in the central parish (EBRP) in 2005 was downloaded from the Louisiana Department of Transportation and Development (DOTD) website (https://ladotd.public.ms2soft.com/tcds/tsearch.asp?loc=ladotd).

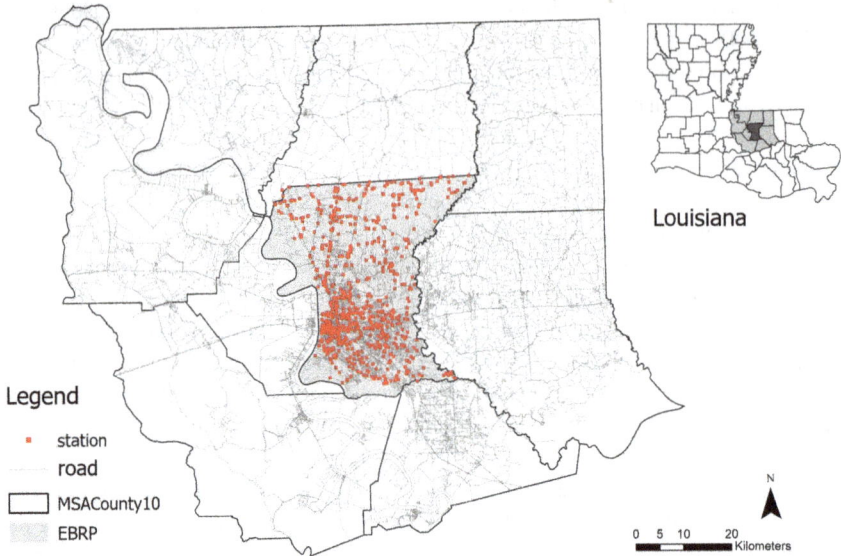

FIGURE 12.5 Traffic monitoring stations and adjacent areas in Baton Rouge.

The following datasets are prepared and provided for this case study under the data folder BRMSA:

1. Geodatabase BRMSAmc.gdb includes (a) an area feature trt and its corresponding point feature trtpt for 151 census tracts and their centroids, respectively (each containing fields res and emp for numbers of resident workers and jobs, respectively), (b) a feature dataset rd containing all feature classes associated with the road network dataset, including the edges and junctions, and (c) a point feature station for 816 traffic count stations (its field ADT1 representing the annual average daily traffic or AADT in 2005).
2. For convenience, geodatabase BRMSAmc.gdb also includes other features that are to be produced in the project. For example, ODtable is the OD Cost Matrix generated from steps 1–2, and SimTrip includes simulated 150,000 points of origins and destinations from step 6.

Two Python-based tools, ProbSampling and RepeatRow, are in the MCSimulation.tbx under the subfolder AcrPy-tools in the folder CMGIS-V3-Toolbox.

Figure 12.6 outlines the workflow. Based on the numbers of resident workers and employment (jobs) in census tracts, module 1 uses a gravity model to estimate the zone-level traffic between tracts (steps 1–2). Module 2 uses the Monte Carlo method to randomly simulate individual trip origins and destinations, proportional to the distributions of resident workers and employment, respectively (steps 3–4). Module 3

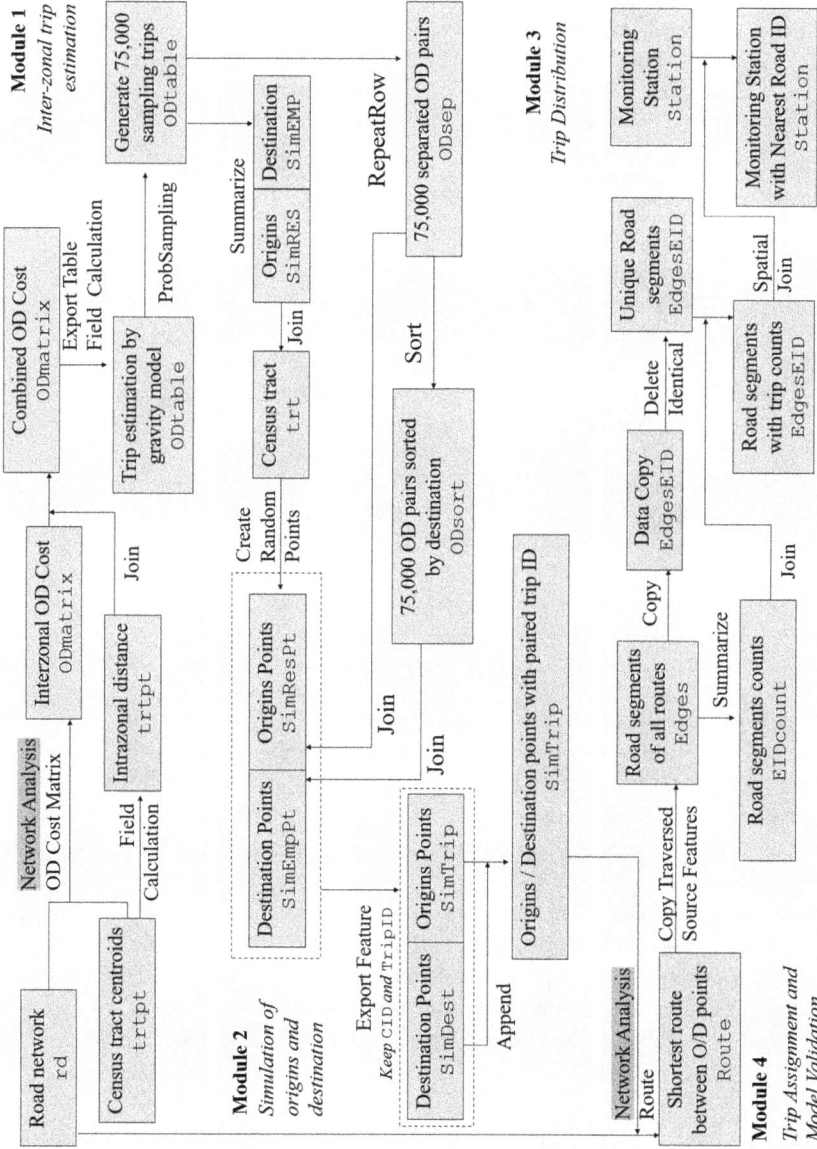

FIGURE 12.6 Workflow for Monte Carlo–based traffic simulation.

TABLE 12.2
Major Tasks and Estimated Computation Time in Traffic Simulation

Step Index	Task	Data Size	Computation Time (Minutes)*
1	Computing OD time matrix between census tracts	$151 \times 151 = 22{,}801$ records	<1.0
2	Sampling 75,000 trips according to estimated trips with ProbSampling	75,000 OD trips	<1.0
4	Simulating origins and destinations with Create Random Points	150,000 Origins/ Destinations	<1.0
	Computing shortest paths: loading stops	$2 \times 75{,}000 = 150{,}000$ locations	8.0
7	Computing shortest paths: solving routes	75,000 routes	65.0
	Executing "Copy Traversed Source Features"	75,000 routes	16.5
8	Applying "Delete Identical" on the copy of Edges	8,349,981 lines	14.0
	Applying "Summarize" on Edges	50,052 lines	2.3

Note: On a laptop described in the Preface.

connects the origins and destinations randomly and caps the volume of OD trips proportionally to the estimated inter-zonal traffic from module 1 (steps 5–6). Module 4 calibrates the shortest routes for all OD trips, measures the simulated traffic through each monitoring station, and compares it to the observed traffic for validation (steps 7–9). Table 12.2 lists estimated computing time for major tasks (excluding any steps with negligible time).

Step 1. Computing OD Time Matrix between Census Tracts: In ArcGIS Pro, create a new project and add trtpt, trt, and rd _ ND to the Map pane. Follow the network analysis steps (e.g., steps 4–6 in Subsection 2.2.2 of Chapter 2) to compute the OD time between tract centroids. Note that feature trtpt is loaded as both origins and destinations. Set Search Tolerance as 7,500 meters. In the attribute table of Lines in OD Cost Matrix, the field Total _ Minutes records the travel time in minutes.

Similar to step 2 in Section 10.3 of Chapter 10, intrazonal time is approximated as 1/4 of a tract's perimeter divided by a constant speed (670.56 m/min, that is, 25 mph, in this study). The total travel time between two tracts is composed of the aforementioned network time and the intra-tract times at both the origin and destination tracts. On the table of trtpt, add a new field Minutes as the intrazonal

time, and calculate it as 0.25*!Length! / 670.56. Copy trtpt to new feature trtpt1. Join the two attribute tables of trtpt and trtpt1 to Lines in the
result OD Cost Matrix based on their common fields OBJECTID, OriginID, and
DestinationID, and export the updated Lines as a new table ODmatrix. Add
a new field Dist to ODmatrix and calculate Dist as !Total _ Minutes! +
!Minutes! + !Minutes _ 1! Export ODmatrix to a new table ODtable,
which contains $151 \times 151 = 22{,}801$ records.

Step 2. *Estimating Inter-Zonal Traffic Volumes*: As an example, this case
study sets the parameter values in equation 12.1 as $\beta = 2$ and $a = 0.001$. On the
table ODtable, add a new field Trip (Data Type: Long) and calculate it as
0.001*(!emp!*!res1!+!res! *!emp _ 1!) *!Dist!**(-2). The sum on the
field Trip is 362,859.

Due to computation limitations, our simulation is downscaled to 75,000 trips.
Activate the Python tool ProbSampling in the toolbox MCSimulation.tbx
to implement a Monte Carlo simulation of OD traffic volumes between census
tracts. The OD traffic volumes total 75,000 and are distributed proportionally to
the numbers of estimated trips, 362,859 in total. In the dialog of ProbSampling
(Figure 12.7a), input ODtable as Input Table, choose OBJECTID for ID, choose
Trip for Probability Field, input 75,000 in the box of Simulation Count, and click
Run to update the input table ODtable. The field SimN indicates the simulated
number of trips. As shown in Figure 12.7b, the values of SimN and Trip are highly
correlated.

This completes module 1 for estimating inter-zonal OD trip volumes.

Step 3. *Calibrating Numbers of Simulated Origin and Destination Points in
Census Tracts*: Apply Summarize to sum up the field SimN on the table ODtable
based on the field OriginID (and DestinationID), and save the resulting table as
SimRES (and SimEMP), whose field SUM _ SimN represents the number of resident
workers (and jobs) in each census tract to be simulated. Join these two tables to the
Census Tract feature trt based on their common fields OriginID, DestinationID,
and OBJECTID.

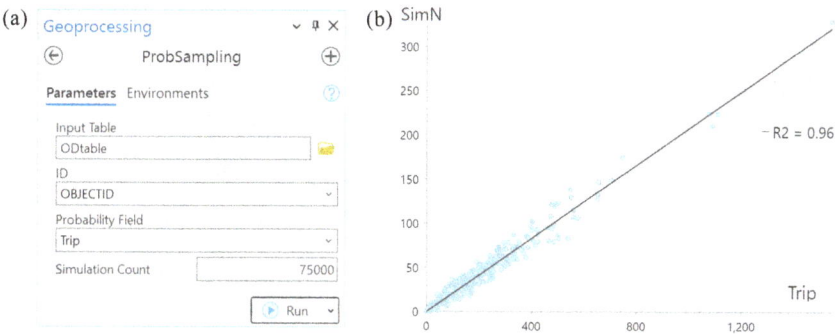

FIGURE 12.7 (a) Interface of Python-based tool ProbSampling, and (b) correlation of
estimated trips vs. simulated trips

Step 4. Simulating Origin and Destination Points in Census Tracts: Similar to step 2 in Section 12.2, activate the tool Create Random Points. In the dialog, keep the default geodatabase for Output Location (e.g., Case12b.gdb), input SimResPt for Output Point Feature, choose trt for Constraining Feature Class, choose Field for Number of Points and SimRES.SUM _ SimN underneath it as the field value, keep other default settings, and click Run. Repeat this step with SimEmpPt as Output Point Feature, SimEMP.SUM _ SimN as Field Value, and others the same. Note that field CID in point features SimResPt and SimEmpPt and field OBJECTID in polygon feature trt share the same values.

This completes module 2 for randomly simulating individual trip origins and destinations.

Step 5. Disaggregating Inter-Zonal OD Traffic Flows to Individual OD Pairs: Recall that the OD table ODtable contains three key fields, OriginID, DestinationID, and SimN. To generate individual trips, we repeat each row in the table up as many times as inter-zonal trip counts (i.e., value of field Sum). This is accomplished by using the Python tool RepeatRow in the toolbox MCSimulation. tbx. In the dialog of RepeatRow, choose ODtable as Input Table and set OriginID and DestinationID for Remain Field, SimN for Count Field, ODsep for Output Table, and click Run. The output table ODsep has 75,000 rows for trips to be simulated.

Step 6. Appending Simulated Origins and Destinations to Individual OD Pairs: The number of points sharing a certain CID in SimResPt (or SimEmpPt) is identical to the number of trips with a corresponding OriginID (or DestinationID) in the table ODsep. Such a relationship is utilized to randomly assign the points of SimResPt and SimEmpPt to trips in ODsep.

First, join ODsep to SimResPt based on their common field OBJECTID. Note that both are automatically sorted by OriginID (or CID) and identical to OBJECTID of feature trt. Then use the tool Sort to (1) sort ODsep by field DestinationID (option Ascending) and save the Output Dataset as a new table ODsort, and (2) join ODsort to SimEmpPt based on their field OBJECTID directly since both have been sorted by DestinationID.

Now we are ready to combine the points of SimResPt and SimEmpPt to form unique OD pairs. Export SimResPt to a new feature SimTrip while keeping two fields CID and OBJECTID and changing the field name OBJECTID to TripID. Similarly, export SimEmpPt to a new feature SimDest while keeping fields CID and ORIG _ FID and changing the field name ORIG _ FID to TripID.

Activate the tool Append. In the dialog, define SimDest as Input Datasets, SimTrip for Target Dataset, and Input fields must match target fields for Field Matching Type. Click Run to update SimTrip, and its field TripID identifies a unique OD pair.

This completes module 3 for connecting the origins and destinations randomly to form unique OD pairs.

Step 7. Computing Shortest Paths between Origins and Destinations: In ArcGIS Pro, make sure that the road network dataset rd _ ND is in the active project. Under

Analysis tab > Network Analysis, activate the Route tool. Similar to the OD Cost Matrix tool used in step 1, we need to define the Stops by loading the origins and destinations for all trips. Click the layer `Route` in the Map pane, and Route Layer in the top bar, and then Import Stops. In the dialog, select Stops for Sub-Layer and `SimTrip` for Input Locations, choose Use Geometry for Field Mapping, and under Property, specify the field for RouteName as `TripID`. Under Advance, set the Search Tolerance to 7,500 meters, and click OK to load points.

After loading the $2 \times 75,000 = 150,000$ locations, click Solve to compute the shortest paths for 75,000 OD trips.[4] The result is saved in the layer `Route`.

Step 8. Recovering Route Connections for Each Shortest Path: The previous step does not save the specific route connections for each path, which is needed to detect simulated traffic flows. Search and activate the tool Copy Traversed Source Features. In the dialog, choose `Route` for Input Network Analysis Layer, select `BRMSAmc.gdb` as Output Location, keep other default settings such as names for `Edges` and `Junctions` feature classes and `Turns` table, and click OK to run the tool. The resulting edge feature `Edges` will be used in the subsequent analysis.

Step 9. Counting Simulated Traffic: In the attribute table of `Edges`, values in the fields `SourceOID` and `EID` of the same road segment in different trips are identical. Apply Summarize on the field `EID` in the attribute table of `Edges` by setting `EID` as Case Field, `Count` for Statistic Type, and `EIDcount` for Output Table. The values in the field `COUNT _ EID` indicate the traffic volumes (i.e., number of times of all simulated trips on each road segment).

It may be desirable to get unique road segments for mapping and other purposes. Copy the feature `Edges` to `EdgesEID`. Apply the tool Delete Identical on `EdgesEID` to remove duplicated edges identified by the field `EID`. In this case, it takes several minutes to complete the process and reduces the number of segments to 50,052. Join the table `EIDcount` to the updated feature `EdgesEID` based on the common field `EID`. Figure 12.8 shows the variation of traffic volumes in Baton Rouge.

This completes module 4 of calibrating the shortest path of each origin–destination trip and counting simulated traffic volumes through unique road segments.

Step 10. Comparing the Simulated and Observed Traffic Patterns: Use a spatial join to identify the nearest road segments (feature `EdgesEID`) from traffic monitoring stations (feature `station`) by setting Closest as Match Option and 5,000 Meters as search Radius. Fields `STATION` and `ADT1` are the traffic monitoring station ID and its annual average daily traffic, and `OBJECTID _ 1` and `COUNT _ EID` are the closest road IDs and its simulated traffic. Apply a Scatter Plot to the two traffic fields. As shown in Figure 12.9, the two are largely consistent with $R^2 = 0.38$ (i.e., correlation coefficient = 0.62). Considering that the model uses limited data with a much-simplified structure, the result is encouraging.

One may repeat the analysis by experimenting with different parameters (e.g., β value and scalar factor in the gravity model) in module 1, using a different number of OD trips in module 2 and 3, and examine whether the result could be improved.

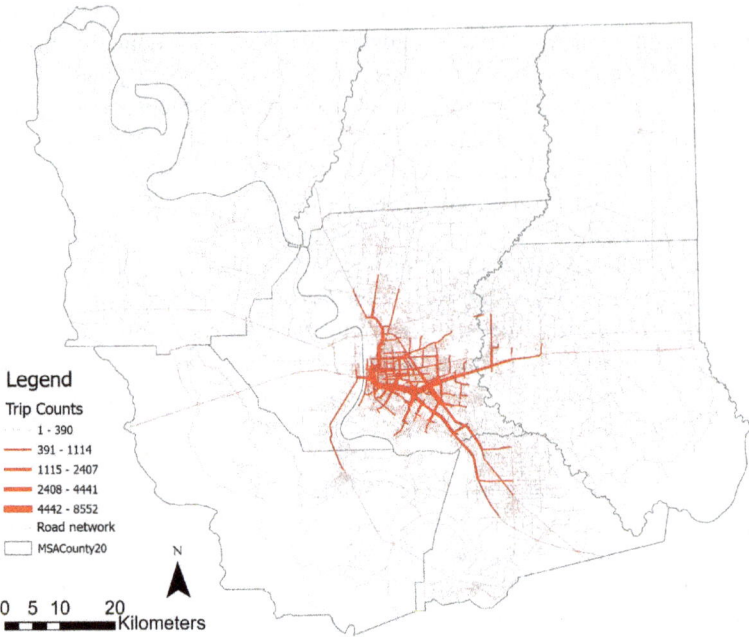

FIGURE 12.8 Simulated trip counts of road segments.

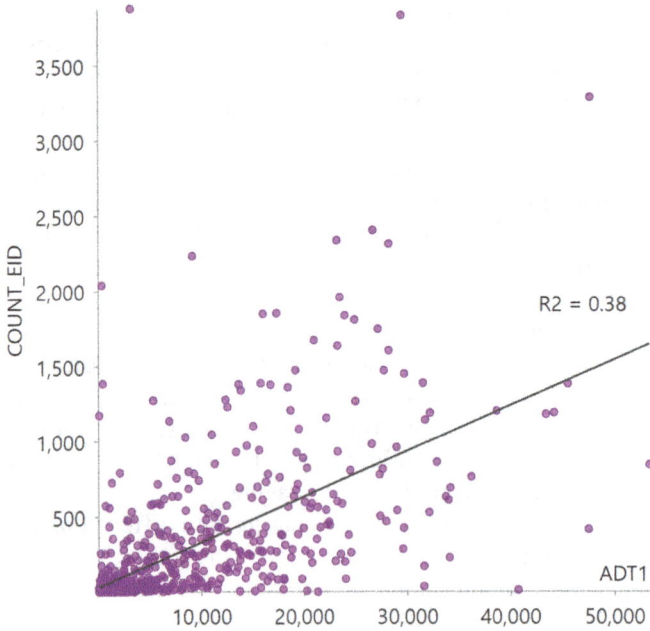

FIGURE 12.9 Observed vs. simulated traffic in Baton Rouge.

12.5 SUMMARY

This chapter introduces the Monte Carlo simulation technique, a numerical analysis technique with increasing applications for spatial analysis. The Monte Carlo method generates random numbers by following a certain probability distribution function. The two case studies illustrate the applications in spatial data disaggregation: one disaggregating area data to individual points within the areas, and another disaggregating inter-zonal flow data to individual trips between points. Simulation of individual OD trips (e.g., journey-to-work trips) is the foundation for modeling spatiotemporal trajectories of various agents and predicting where and when they converge in an agent-based model, as discussed in Chapter 13.

Many spatial analysis tasks have traditionally relied on data aggregated in various area units, such as census or administrative units, and are subject to criticisms, including MAUP, unfairness of sampling, and also inaccuracy or uncertainty in distance measure by using areal centroids. Some of the concerns are raised in empirical studies of urban population density function fittings in Section 6.3 of Chapter 6. The Monte Carlo simulation of individual resident locations utilizes the census data and other accessory data, such as land use inventory data, and randomly generates individuals that are consistent with known patterns of population distribution. The simulated data enables us to aggregate population back to area units of any scale in any shape so that the scale effect and the zonal effect can be examined explicitly. When an area is represented as the average location of individuals within the area, it also yields a more accurate centroid to facilitate the distance calibration.

Much of the interest in intraurban commuting in literature may be traced back to the issue of wasteful commuting raised by Hamilton (1982), as discussed in Appendix 11 in Chapter 11. In the formulation of wasteful commuting, Hamilton made several assumptions in order to simplify the urban structure (both employment and population) as concentric and the underlying commuting pattern as a singular direction toward the CBD so that the problem could be solved elegantly by calculus. One may imagine that commuters in an optimal (ideal) city conceptualized by Hamilton abide the rule of minimizing total citywide commuting, and their behaviors are perhaps the opposite of randomness. It is the drive to identify the difference between the optimal and actual commuting that motivates researchers to model and understand intraurban commuting. Case Study 12B demonstrates the value of applying the Monte Carlo technique in simulating urban traffic flows. It utilizes publicly available data sources, adopts a simple gravity model to estimate inter-zonal traffic flows, and decomposes those estimated flows to individual trips between points that route through the road network. Most of the researchers attribute the discrepancy in measuring the extent of wasteful commuting to scale effect and believe that a sharper resolution (smaller areal unit) helps mitigate the problem. The smallest resolution is individual, and only simulation can lead us there when commonly available data are area-based (Appendix 12).

In addition to several routine GIS tasks that have been introduced in previous chapters, the case studies in this chapter introduce several new tools, such as Create Random Points, Copy Traversed Source Features, Sort, Sampling by Probability, Repeat Rows, etc. These tools are useful for other purposes beyond urban population and traffic simulations.

APPENDIX 12: IMPROVING MEASUREMENT OF WASTEFUL COMMUTING BY MONTE CARLO SIMULATION

The wasteful commuting problem was raised by Hamilton (1982), as discussed in Section 11.2 of Chapter 11. In short, given the locations of houses and jobs, if resident workers are free to swap houses in order to minimize (optimize) the overall commuting in a city, the resulting commuting is considered optimal (required) commuting. The difference between the actual and optimal commuting lengths, measured either in distance or time, is wasteful.

White (1988) formulated the wasteful commuting problem in LP. Given the number of resident workers P_i in zone (e.g., census tract) i ($i = 1, 2, \ldots, n$) and the number of jobs E_j in zone j ($j = 1, 2, \ldots, m$), one may define the LP as:

Minimize:

$$\sum_{i=1}^{n}\sum_{j=1}^{m}c_{ij}x_{ij}$$

Subject to:

$$\sum_{j=1}^{m}x_{ij} \leq P_i \text{ for all } i\left(= 1, 2, \ldots, n\right)$$

$$\sum_{i=1}^{n}x_{ij} \leq E_j \text{ for all } j\left(= 1, 2, \ldots, m\right)$$

where c_{ij} is the commute distance (time) from residential location i to jobsite j, and x_{ij} is the number of commuters on that route and non-negative.

By following steps 3–9 in Section 12.4, one can simulate individual resident workers that amount to P_i in zone i, individual jobs that amount to E_j in zone j, and also OD flows that cap at x_{ij} from i to j. In practice, simulating the same total number of individual commuters as the actual number in a study area may be computationally cost-prohibitive, especially for a large city. One can simulate a fraction of the total workers and standardize the measure of wasteful commuting on a per capita basis.

Similar to the LP at the zonal level, the formulation of optimal commute at the individual level utilizes the simulated resident workers and jobs by the Monte Carlo method. Index the locations for simulated individual resident workers and jobs as k and l, respectively. The total number of simulated workers is the same as that of simulated jobs, denoted by N. The optimization problem is:

Minimize:

$$\sum_{i=1}^{N}\sum_{j=1}^{N}c_{kl}f_{kl} \tag{A12.1}$$

Subject to:

$$\sum_{l=1}^{N} f_{kl} \leq 1 \text{ for all } k \left(= 1, 2, \ldots, N\right) \tag{A12.2}$$

$$\sum_{k=1}^{N} f_{kl} \leq 1 \text{ for all } l \left(= 1, 2, \ldots, N\right), \tag{A12.3}$$

where c_{kl} is the estimated travel distance (time) from resident worker location k to job location l, and f_{kl} indicates whether a journey-to-work flow is chosen by the optimization algorithm (= 1 when chosen, and 0 otherwise). The objective function (A12.1) is to minimize the average commute time of N simulated commuters. Equations A12.2 and A12.3 ensure that each worker can be assigned to one unique job and vice versa. Since the variable f is a binary integer, it is an integer linear program (ILP) problem. Since the origins and destinations are individual points, both optimal commute and existing commute are estimated from the point-to-point OD trips. The approach is thus independent from the zonal effect or MAUP.

Based on the 2006–2010 CTPP data in Baton Rouge, Hu and Wang (2015) first measured wasteful commuting at the census tract level, and then at the simulated individual level. At the zonal (census tract) level, average estimated commuting time based on the observed commuting pattern was 12.76 minutes, and the average minimum commuting time from the LP was 6.61 minutes, and therefore the excess (wasteful) commuting was 48.20%. At the simulated individual level, average estimated commuting time was 12.65 minutes and average minimum commuting time from the ILP was 4.75 minutes, and thus the excess (wasteful) commuting was 62.45%. In short, formulating the wasteful commuting problem at the zonal level tends to underestimate the extent of wasteful commuting significantly.

NOTES

1 Different from the Intersect tool that intersects all (multiple) input layers and calculates the intersection of all input features, the Pairwise Intersect tool takes features from the first input layer and intersects them with each feature in the second input layer, one at a time.

2 For a block is composed of multiple residential blocks, its total population is allocated proportionally to the area sizes of residential blocks. For a block is composed of one single residential block, which is either the whole block or part of the block, it inherits the full population of the block.

3 As the points are randomly generated across the study area, a user may have a slightly different regression result on the hexagons from Table 12.1.

4 Due to random simulation, limitation of search radius in Adding location, or connectedness of road network, there may be a small number of unsolvable locations. Network Analysis skips those points and completes the calibration of valid route pairs.

13 Agent-Based Model and Application in Crime Simulation[1]

The pursuit of understanding individual behavior is a constant theme in social and behavioral sciences, and any significant advancement in related theory is marked by our ability to model human behavior more accurately. Spatial analysis may begin with describing and characterizing spatial patterns at aggregate levels (e.g., aggregated in an area unit or for a demographic group). For example, spatial smoothing and interpolation in Chapter 3 intend to visualize the spatial variability of a measure more accurately, spatial accessibility in Chapter 5 strives to capture such a variability or disparity in access, analysis of urban density in Chapter 6 attempts to characterize the pattern by a single function with reference to distance from a center or centers, and much of spatial statistical analysis in Chapter 8 focuses on statistical patterns of spatial clustering of one feature and relationship between features. Chapters 4 and 9 construct areas (functional and homogeneous regions) so that the patterns in those areas can be more reliable. However, a pattern is observed, and understanding or predicting such a pattern requires probing the process that shapes and eventually forms that pattern. The best social science strives to model individual behaviors across space and over time and thus reveals the process behind patterns. While public policy seeks desirable outcomes in a society often measured at a community level for a duration of time, its effectiveness relies on influencing individual behaviors that interact with the environment and change in space and time.

Spatial analysis also strives for a sharper spatial resolution. The finest resolution comes down to individuals, as the Monte Carlo simulation in Chapter 12 is designed to imitate. The simulation is limited to static points (e.g., individual residents in Case Study 12A in Section 12.2) or connection between those points (e.g., commuting trips from residence to workplace in Case Study 12B in Section 12.4). The focus of Chapter 12 remains on modeling spatial patterns. This chapter furthers the endeavor of simulation to model individual spatiotemporal trajectories. Simulated agents (various groups of individuals) guided by defined rules (routines) move in space (a study area) over time, and their convergence in both space and time predicts the likelihood for an event to take place. In a way, an agent-based model mimics human movements in real life. When the functions and parameters used in the model are well informed from increasingly detailed real-world empirical data (e.g., big data discussed in Chapter 14), the model has great potentials for capturing the dynamics of human–environment and human-to-human interactions and thus explaining and predicting spatial patterns and underneath processes more accurately. Until then, assessment of a policy scenario would not be feasible in a lab setting before conducting costly field experiments.

DOI: 10.1201/9781003292302-16

The chapter is structured as follows. Section 13.1 provides an overview of the agent-based model (ABM), including its evolution and major functionalities. Section 13.2 discusses theoretical or behavioral foundation for and the overall design of crime modeling in an ABM setting. Section 13.3 explains GIS implementation of our ABM-based crime simulation. A case study of simulating the spatiotemporal trajectory of robbery in Baton Rouge is presented in Section 13.4 with detailed step-by-step illustrations for various scenarios. Section 13.5 concludes the chapter with a summary.

13.1   FROM CELLULAR AUTOMATA TO AGENT-BASED MODEL

Analysis of aggregate data in areas or groups assumes that relationships or behaviors derived can be reasonably used as proxies for component (individual) behaviors. Such an assumption does not hold for a complex system, which is composed of many autonomous, interrelated, and interacting elements and features heterogeneity, nonlinear dynamics, path dependence, feedback loop, emergence, self-organization, and adaptation (Benenson and Torrens, 2004). Heterogeneity is inherent in any social system with human actions at the center and organizations of all sizes, and the simplistic approach of aggregate data analysis often leads to "*ecological fallacy*" (Robinson, 1950). In a complex system, past behaviors can lead to events with lingering effects on the system and make its future outcome dependent on not only current behaviors of system elements but also the historical trajectory of system. The best example is the so-called "*butterfly effect*" in a deterministic chaos system (Lorenz, 1995). Relatively simple behaviors for basic elements and local interactions among them can scale up to surprising global patterns from collective behaviors not directly observable at the individual level.

These properties of a complex system have inspired GIS simulation models to grow complex spatial patterns, such as the structure and form of urban settlements, from the bottom up, with relatively simple rules governing low-level movements and interactions. One school representing this bottom-up approach is the *automata system*, with great potential for modeling the complexity of an urban area (Dijkstra et al., 2001; O'Sullivan and Perry, 2013). The system is composed of a set of automata, and the actions of each automaton over time are based on its internal characteristics, behavioral rules, and environmental information from its surroundings. Through behavioral rules, an automaton also can modify its internal characteristics after processing information gathered on itself and its surroundings. Time is explicitly integrated into an automata system in either discrete or continuous manner. With discrete time, the timeline is marked with a series of evenly spaced ticks based on an appropriate temporal resolution. At every tick, each automaton carries out its intended actions defined by behavioral rules. With continuous time, there is no tick on the timeline and time flows forward continuously. Events of interest can occur at any moment and are dispatched to automata that need to decide whether to respond to these events or not. Interactions between automata are naturally reflected in the information each automaton collects from its surrounding automata and subsequently acts on. An automaton can also actively alter various environmental qualities within its vicinity. In general, automata systems used for modeling spatially explicit social dynamics fall into two broad categories based on how automata are situated in simulation space and the way their internal characteristics are encoded—cellular automata (CA) and agent-base model (ABM). The key difference between CA and ABM in geo-simulation closely mirrors that between raster and vector in GIS data model.

In *cellular automata (CA)*, the simulation space is tessellated into a grid of regularly spaced cells. Each cell is potentially an automaton. All cells share the same finite set of possible states used to represent internal characteristics of automata. At any time, each cell can only be in one state. At each time step, based on current states of itself and its neighboring cells, each cell computes a new state according to the state transition rules. Cell neighborhood is defined by a local template, such as a 3×3 block centered at each cell. A transition rule specifies the mapping from one state to another when certain conditions are met. In CA, system dynamics is equivalent to the evolving cell states over time.

Formally, a cellular automaton can be expressed as:

$$T_S : \left(S_t(L), N_L \right) \rightarrow S_{t+1}(L)$$
$$N_L = [s_t(L_0), S_t(L_1), ..., S_t(L_{n-1})]^\cdot$$

where T_S is the transition rule for state S of cellular automaton situated on a grid; L is the grid location of cellular automaton as a pair of row and column indices; $S_t(L)$ and $S_{t+1}(L)$ are the current and updated states of cellular automaton at location L, respectively, which can be either scalar for a single attribute or vector for multiple attributes; and N_L is the collection of current states S_t of all cellular automata within the neighborhood of location L. The simplicity of the CA model lies in the fact that the same transition rule is applied to all cellular automata in a simulation.

The biggest challenge in applying CA to geographic problems is that locations of automata are fixed. Although the illusion of moving can be created by propagating information continuously through neighboring cells, CA is never designed with the wide variety of truly mobile automata in mind. Its strict grid-based tessellation and restricting neighborhood rules are unable to reflect sophisticated network topologies, hierarchies, and varying spatial scales underlying many human activities.

The *agent-based model (ABM)* eliminates the grid structure imposed by CA on simulation space and defines *automaton* as an autonomously behaving agent (Bonabeau, 2002; Turner and Penn, 2002). As with CA, ABM encodes the internal characteristics of an agent as state variables and defines a set of transition rules representing agent behaviors. However, state variables are not attached to the space each agent occupies, but to each agent itself. At every time step, each agent evaluates current conditions of itself and its neighborhood before applying transition rules to update its states. Unlike CA, agent neighborhood is loosely defined in ABM and can take any arbitrary form deemed appropriate by specific domain knowledge. For a human agent in communicable disease models, its neighborhood can be the street block it resides in or its social network, depending on what type of interaction and proximity is considered vital to disease transmission.

ABM was originally developed for generic agents and did not explicitly include space and spatial behaviors as model components. Spatial extensions to ABM, such as geographic automata system (GAS), addressed this oversight in order to accommodate mobile agents in geo-simulations (Torrens, 2006). In geo-simulation, the location of each agent can be geo-referenced in a Cartesian space and is often part of agent states. Spatial relationships between agents can be inferred from their locations and function as the basis of agent neighborhood. Besides state transition rules, agents in geo-simulation can also move by updating their locations directly at each time step. Formally, a geo-simulation agent can be expresses as:

$$T_S : (S_t, L_t, N_t) \rightarrow S_{t+1}$$
$$M_L : (S_t, L_t, N_t) \rightarrow L_{t+1}$$
$$R_N : (S_t, L_t, N_t) \rightarrow N_{t+1}$$

where T_S, M_L, and R_N are state transition rules, movement rules, and neighborhood transition rules that take current agent states (S_t), location (L_t), and neighborhood (N_t) as inputs and update agent states (S_{t+1}), location (L_{t+1}), and neighborhood (N_{t+1}), respectively.

13.2 SIMULATING CRIMES IN THE AGENT-BASED MODEL

Since the early 2000s, there has been steady adoption of ABM in crime geography ranging from theorization of driving factors and processes behind crime places to the more practical analysis of spatiotemporal patterns in crime rates, prediction of hotspots, and evaluation of how effective different intervention measures might be. When built with the right amount of detail for the problem at hand through proper abstraction, ABM can function as a virtual sandbox for researchers to test the validity of alternative urban theories linking actions taken by autonomous entities at all levels of an urban system, including individual residents, businesses, and government agencies, to the spatial structures and functionalities of urban sub-systems and their interconnections. Under this holistic view, while crimes are committed by individuals or groups with motivation, the availability of suitable crime opportunities in space and time is conditioned by the backdrop of local demographic, socioeconomic, and environmental factors at much larger geographic scales. ABM has also proven itself as a promising policy-making tool for urban planning practitioners to explore what-if scenarios in situations where controlled experiments are deemed infeasible or impractical. It has been used to assess the pros and cons of competing crime reduction measures without incurring massive disruption or extra financial burden to the public.

The relatively mature state of ABM modeling in crime geography research and the availability of related data at reasonable spatiotemporal scales enable us to use crime simulation as the application domain for demonstrating the general workflow of a spatial ABM framework and the value of such an exercise. The remainder of this section (1) begins with the theoretical background of ABM crime simulation in the field of environmental criminology, then (2) outlines the general architecture and necessary components of an ABM crime simulation framework that can support the spatiotemporal colocation events from the routine activity theory, and (3) highlights how to extend ABM with various Python-based GIS modules and create agents that are cognitive of geographic space, capable of making spatial decisions, and able to act on these decisions.

The ABM crime simulation model and related case study discussed in this chapter are based on a recent study reported in Zhu and Wang (2021).

13.2.1 Theoretical Foundation in Environmental Criminology

Traditional criminology theories, such as strain theory or social disorganization theory, are usually centered on the development of criminality (Bonger, 1916; Gibbons, 1977; Glaser, 1971). While these theories are instrumental to uncovering psychological, social, political, or economic factors behind crime motivations, they offer very little insight into when and where crimes may take place. They may aid long-term crime reduction strategies such as job creation initiatives or education reform but are hardly applicable to day-to-day police tactics for curbing crimes. *Environmental criminology* marks a significant departure from these theories by taking a closer look at the geographic settings under which crime act happens. The three most important theories proposed in environmental criminology are the routine activity theory, the rational choice theory, and the spatial pattern theory. They all emphasize the space–time convergence of a capable offender, a vulnerable victim, and lack of place guardianship in any crime requiring physical contact, and differ largely in the perspectives, from which this space–time convergence is studied.

In the *routine activity theory*, both potential offenders and victims are assumed to frequent certain places at certain times as part of their daily routines. Such places include residence, workplace, school, grocery store, gymnasium, and places for other recreational activities. Places where a person engages in mandatory and optional activities along with routes to reach these places form the skeleton of their daily activity space. Under this assumption, crime opportunities can arise from overlapped activity spaces of a motivated offender and a potential victim. Activity spaces of different individuals are considered overlapped when these individuals are in close proximity both spatially and temporally.

While the routine activity theory outlines the necessary conditions for crime opportunities to occur, not every crime opportunity eventually turns into crime. *Rational choice theory* asserts that even potential criminals are rational beings. When facing crime opportunities, a rationale offender bases their decisions on potential risks and rewards. Limited by offenders' information gathering and processing capability, these decisions are often stochastic due to uncertainties associated with partial knowledge and information bias.

The *spatial pattern theory* focuses on the role of environmental features in shaping spatial and temporal crime patterns. By linking the well-known crime hotspot phenomenon to places acting as crime attractors or generators, it acknowledges the importance of place in facilitating or inhibiting crime incidences, by increasing or reducing the chances potential offenders can interact with victims. These places either attract more potential targets or fail to provide sufficient guardianship due to design flaws. The theory also recognizes the influence of road network topology and land use pattern on the spatial distribution of crime. The concentration of economic activities at nodes and pathways of a transportation network and within certain land use types makes these locations better candidates for opportunity-based crimes.

Among the three theories, elements of routine activity theory are best suited to be translated into ABM with spatial extension. The necessary conditions for a crime incident outlined in this theory are concerned with individual activities in space, which fit well for a model focusing on autonomous agents and their behaviors. The other two theories are more about offender choice behaviors and connections between environmental qualities and crime opportunities. Based on the routine activity theory, we design an ABM framework capable of simulating crime patterns from the daily routines of individual residents.

13.2.2 COMPONENTS OF ABM FRAMEWORK FOR CRIME SIMULATION

The design of ABM for crime simulation consists of model execution flow, typology of agents, states and transition rules for each agent type, road network, and neighborhood environment.

The execution flow is composed of model initialization, iterative model updates, and termination condition check (Figure 13.1). The simulation model starts with initialization of model world and populating it among different types of agents. It then proceeds to update all agents in cycles. At the end of each update cycle, conditions for simulation termination are checked. Simulation terminates if all conditions are met.

Agents in a crime simulation framework can be broadly categorized into reactive and proactive agents (Ben-Ari and Mondada, 2018). A *reactive agent* does not possess any decision-making ability and has a set of states whose transitions over time simply follow some predefined patterns. In crime simulation, a reactive agent includes

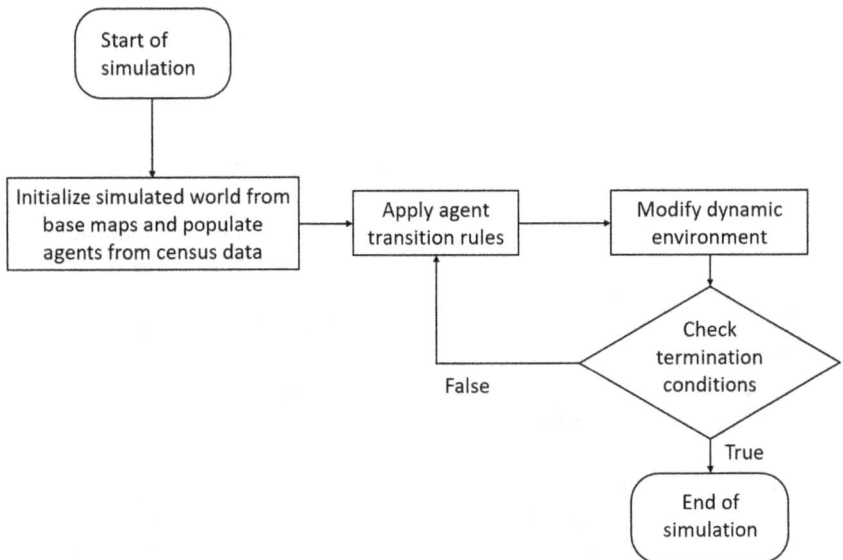

FIGURE 13.1 Execution flow of agent-based crime simulation model.

activity locations that human agents can visit in simulation for flexible activities, such as shopping or recreation. They have states for their locations and opening statuses. While the former is fixed, the latter changes according to facility operation hours. *Proactive agents* are most suitable for modeling intelligent behaviors of individuals or organizations, such as information-gathering, planning, and deliberation. Here, all residents in simulation are modeled as proactive agents. They are capable of keeping track of daily schedules, current locations, and next destinations. More importantly, they are aware of the simulation world, including activity locations, road network, neighborhood conditions, and the existence of other residents within a radius of their current locations. Using global information gathered on the simulation world and local information from its surrounding, a resident can proactively plan its next destination, shortest path to reach it through road network, and the way to interact with environment and possibly other agents. The update of agent states is managed by transition rules. Two common sets of transition rules for all residents are responsible for regulating their daily schedules and locations. The first set of rules dynamically removes items from residents' schedules as soon as they are accomplished and generates new items under the constraint of available time budget. The second set of transition rules updates their locations based on their current travel paths and speeds. Whenever necessary, these rules can also compute new travel paths.

Three types of resident agents can be identified for most crime simulations: motivated offender, potential victim, and police agents. Besides states and transition rules common to all residents, each of these three types of agents also has unique states and transition rules. A *motivated offender agent* has states for motivation level, distance to its home, number of police and number of targets within its awareness zone, and tolerance for number of witnesses within this zone. It also has transition rules to determine whether there is suitable target within its awareness zone, risk of available opportunity, and how fast its motivation goes back to normal after successfully committing a crime. A *potential victim agent* has an additional state for self-defense and corresponding transition rules on change of self-defense after victimization. This setup allows us to control how often re-victimization occurs based on empirical data. A *police agent* has a state for patrol strategy and transition rules to determine locations and lengths of time to patrol. The presence of police within an offender's awareness zone deters offenders from committing any crime. The offender–police and offender–victim interactions are encapsulated in these type-specific transition rules.

For crime simulation situated in any real-world location, a simulated world needs to be initialized from base maps for this location and act as the container for all resident agents and crime-related events. In order to use advanced network algorithms for planning shortest paths, vector-based street network and activity locations are used to build the foundation of simulated world. A graph-based data structure composed of directed links, nodes, and topology can be derived from road network and activity locations. The network graph provides information on traversing cost of each street segment and connectivity between street segments. To more efficiently handle fluctuating social environment that a resident agent can sense and change through its own actions, we can use raster-like surfaces to create a set of environmental grids to represent qualities like street lighting, ambient traffic at different hours of day, and neighborhood reputation. In addition, demographic and commuting flow data are used to seed a sample population with matching statistical distributions for residential and employment locations, time to leave home for work, and average daily work hours.

13.2.3 INTEGRATION OF GIS FUNCTIONALITIES WITH SIMULATION

To extend generic ABM for simulations situated in geographic space, various GIS functionalities must be integrated into this framework. At the highest level, it is composed of three layers built on top of each other with distinct functionalities (Figure 13.2). It mirrors the common three-tier software architecture used by most customized GIS toolkits that keep data access, application logic, and visualization in three separate layers.

The *data storage and access layer* forms the infrastructure of the entire simulation model. Its main function is to provide other model layers with easy access to GIS and other related data. These data include tabular data and vector or raster-based GIS data. The *simulation logic layer* relies on high-quality GIS data to build a realistic urban environment with enough details where simulated residents live, work, move around, and interact with each other. This layer is also responsible for managing updates to dynamic data and efficient queries on spatial and non-spatial data. One major feature that distinguishes the ABM from other dynamic GIS modeling methods is its iterative update process controlled by agent and its neighborhood transition rules. In every simulation cycle, these rules use the current states of agents and their neighborhoods and generate a set of new states for the next time-step according to predefined logics. At the end of each cycle, current states in the storage layer are updated with these new states and stay up-to-date before the next simulation cycle starts. Historical states also need to be created and saved in every cycle if the trajectory of an entire simulation run is to be analyzed.

The core simulation layer makes extensive use of spatial query and shortest-path routing algorithm (Dijkstra, 1959). These functions are frequently called by resident agents for pathfinding and, in the case of offender agent, for detecting the existence of other agents within its awareness zone. These operations are very costly in terms of computation time and do not lend themselves well to real-time simulations with

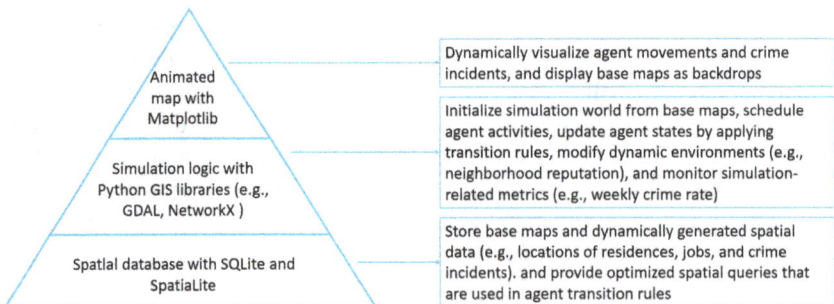

FIGURE 13.2 Agent-based crime simulation model architecture.

a large number of cycles at high temporal resolution and potentially filled with tens of thousands of agents. In addition, because resident agents are mobile, distances between them constantly change, which makes traditional buffer-based spatial queries highly inefficient. To cope with these two computational efficiency–related issues, we can use a grid-based spatial index to significantly speed up these calculations. Network-based distances between centers of spatial index grids are cached in lookup tables for quick searches by resident agents when they need estimation on travel distances to new destinations. A resident agent also dynamically registers itself to the correct spatial index grid where it is currently located. This allows buffer-based spatial queries on moving agents to use spatial index grids overlapping with a buffer zone's bounding box for quick narrowing-down of potential candidates that require exact distance test.

The *visualization layer* provides real-time animation for simulation output. One challenge coming from spatial data with timestamps is to visualize and analyze temporal patterns and the evolution of spatial patterns as simulation progresses. A straightforward extension to traditional mapping and spatial analysis techniques is to create a series of maps and summary statistics from snapshot data and plot them side-by-side in the order of timestamps. The visualization module of this model is based on a similar concept but takes a step further by creating maps and computer summary statistics in real time and displaying them as computer animation that is synchronized with model update. Compared to offline rendering, real-time animation is more efficient for users to inspect ongoing simulations for errors and patterns without having to wait for the entire simulation to terminate and resorting to third-party mapping tools.

13.3 IMPLEMENTING AN ABM CRIME SIMULATOR WITH IMPROVED DAILY ROUTINES

13.3.1 RESIDENT AGENT MOVEMENT AND RISK AVOIDANCE BEHAVIORS

An important aspect of this crime simulation is the modeling of agents' daily routine activities and movement between activity locations. These routine activities include fixed and flexible activities. Our simulation defines *fixed activity* as mandatory in terms of participation, location, starting time, and duration. Currently, only two types of fixed activities are identified, namely, work and rest. Their locations, starting time, and duration are generated during model initialization. Any other activity an agent chooses to participate in, if it still has disposable time to spend, is considered a *flexible activity*. Flexible activity is scheduled during the model update phase. This process is based on time geography and starts with the calculation of *potential path area (PPA)* on the road network with the amount of disposable time left until the start of the next fixed activity. This network-based PPA is more accurate than the ellipse-based PPA often used in time geography studies. In order to improve its performance in real-time simulation, an optimized heuristic PPA is developed with bounding boxes of ellipse-based PPA and pre-calculated network distances between scalable grids (Figure 13.3). An agent can travel to any location within this PPA without being late to its next scheduled fixed activity. A probability-based selection

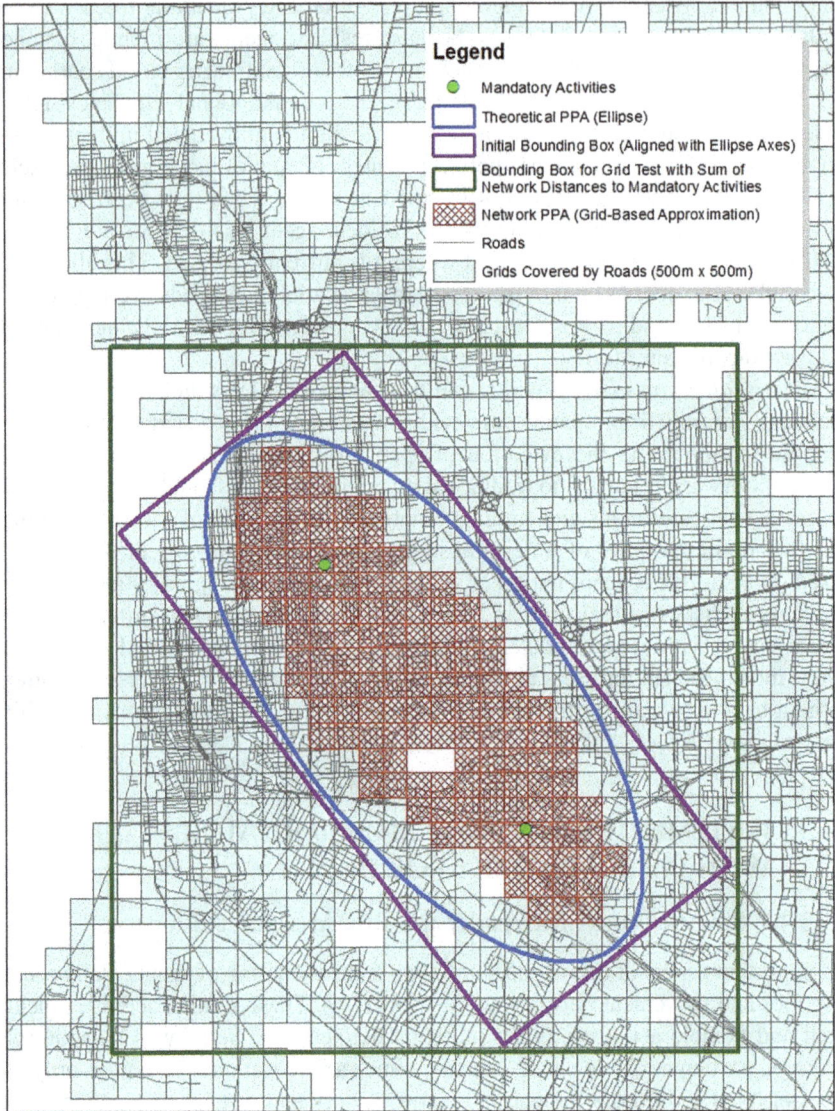

FIGURE 13.3 Illustration of network-based potential path area (PPA) calculation.

scheme is implemented to imitate discrete choice behavior and randomly choose a location inside the PPA as the new destination for a flexible activity. To differentiate an agent's preferences of eligible locations, an attractiveness score is assigned to every candidate location. In current implementation, the attractiveness score for a candidate place is inverse to its deviation distance $d_{deviation}$ under the assumption that an agent prefers less deviation from its original path in order to minimize extra travel and save disposable time for more flexible activities. $d_{deviation}$ is calculated as:

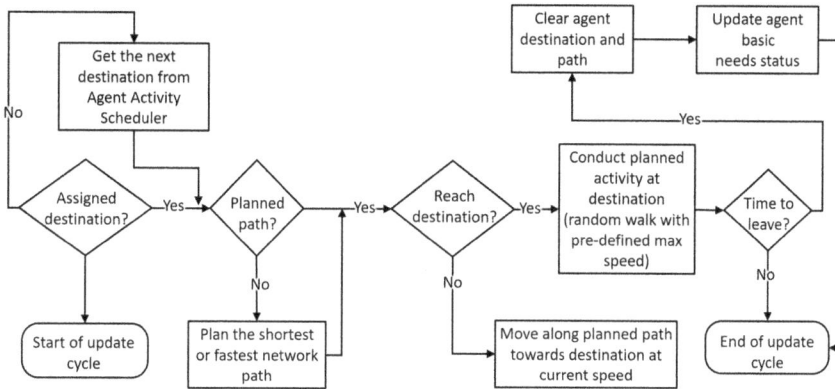

FIGURE 13.4 Update cycle for a resident agent.

$$d_{deviation} = d_{p,f1} + d_{p,f2} - d_{f1,f2},$$

where $d_{p,f1}$ and $d_{p,f2}$ are the network distances from a candidate place p to fixed activity locations $f1$ and $f2$, and $d_{f1,f2}$ is the network distance between fixed activity locations $f1$ and $f2$.

With the next destination determined for an agent (either a fixed or flexible activity location), the movement of this agent is simulated by updating its location along the shortest network path between previous and current destinations. In addition to movement, a resident agent's update cycle is also responsible for spending time at destination, planning next destination, routing, and updating other agent states (Figure 13.4).

While an agent moves through the road network, its risk of being victimized is affected by criminogenic features in its immediate environment and its own decision processes for risk avoidance. Such a decision is guided by past studies on vulnerability and exposure to crime with risk terrain modeling, repeat victimization and near-repeat, fear of crime associated with neighborhood reputation, crime reduction through the presence of police patrol, and natural surveillance provided by ambient traffic.

13.3.2 OFFENDER MOTIVATION, AWARENESS SPACE, AND RISK ASSESSMENT

A given percentage of resident agents in simulation is initialized with random crime motivations. This motivation is generated from a normal distribution and controls how often a potential offender checks its awareness space for suitable targets. Police agents are introduced into simulation runs as a form of place guardianship. In current implementation, police agents can either patrol random destinations or actively move toward crime hotspots within their assigned police districts.

During each simulation cycle, a potential offender, defined as a resident agent with a positive crime motivation, evaluates the following series of conditions in its awareness zone sequentially: (1) randomly initiating a search for any suitable target,

(2) finding no presence of police agent, (3) identifying at least one suitable target, and (4) not deterred by a large-enough crowd. If any condition fails, a potential offender aborts the evaluation and carries on with its ongoing legitimate activity. Currently, there is no arrest in simulation, as this requires removing and reintroducing offender agents and thus adds another layer of complexity.

With all four conditions met, a potential offender attempts a robbery by randomly selecting one vulnerable target from a list of candidates within its awareness zone. Since all resident agents are initialized with self-defense capabilities, this robbery attempt may fail when offender faces resistance from its target. A successful robbery is committed when:

$$random(0,1) > self\_defence_{target},$$

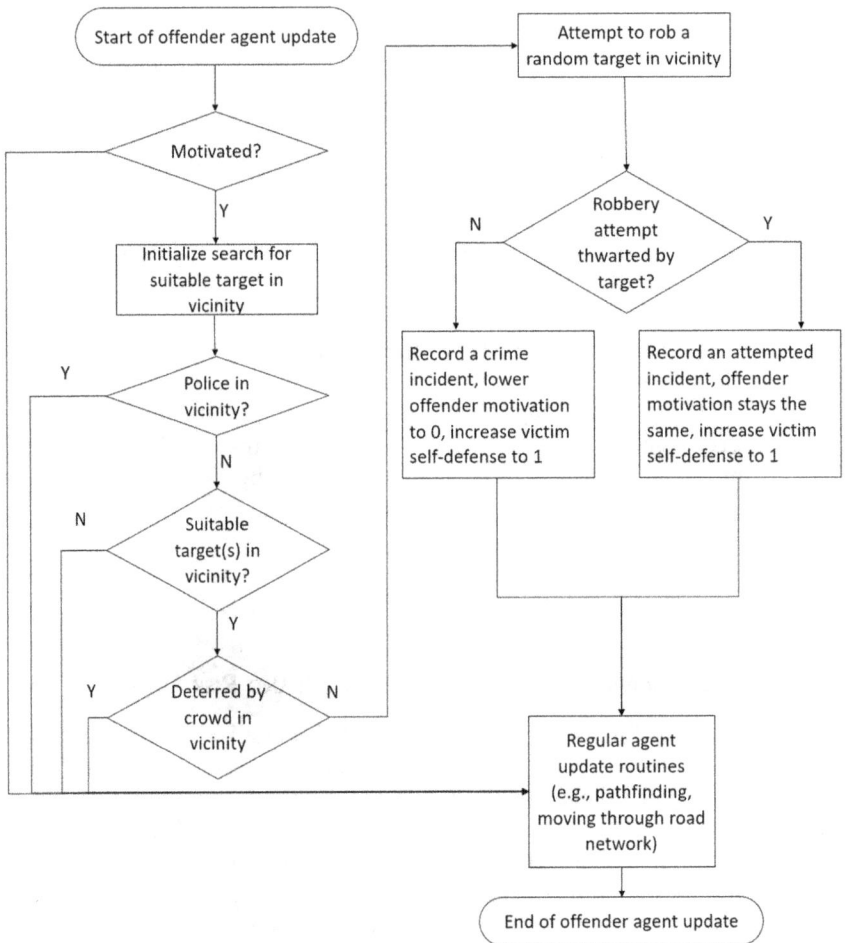

FIGURE 13.5 Update cycle for an offender agent.

where *random(0,1)* returns a random number between 0 and 1, and *self_defense-*~target~ is the self-defense capability of resident agent being targeted. This condition is more likely to fail with higher self-defense capability. A robbery attempt is always recorded as a simulated crime incident whether it succeeds or not. However, its outcome affects an offender's crime motivation. A successful robbery will keep an offender's crime motivation at zero for some time. A failed robbery will leave an offender's crime motivation unchanged. Figure 13.5 summarizes the cycle for an offender agent. See Table 13.1 for model parameters governing the behaviors of victim, offender, and police agents.

TABLE 13.1
Model and Behavioral Parameters for Victim, Offender, and Police Agents

Parameter	Effects on Simulation
Total simulation cycles and unit time	Affecting model running time and stability of simulation output (fewer simulation cycles lead to more variabilities between runs with different initializations; a larger unit of time lowers temporal resolution and underestimates the chances of agents in proximity).
Total number of agents and composition of agent types	Computing time increases slightly worse than linearly with number of agents (offender agents must examine more candidates as potential targets with increasing agent population; percent of potential offenders can be lowered with a large total number of agents).
Agent moving speed	Faster moving speed inflates agents' daily activity space and creates more potential crime opportunities (setting minimum speed too high can severely under-count close-proximity encounters between agents; time agents spend on road segments is less dependent on segment lengths with more variability in moving speed).
Flexible activity time and police patrol time	More crime opportunities are available in flexible activity sites if potential targets and offenders linger there longer; longer police patrol time helps deter crime activities there but leaves elsewhere susceptible.
Crime motivation value and recovery time	Higher average crime motivation leads to more robberies; shorter offender motivation recovery is likely to cause a disproportionately large number of robberies within a short period, followed by another period with very few (our model assigns offender agents a random period of motivation recovery time at the beginning of simulation).
Self-defense normalization time	Affecting the frequency of repeated victimization (after being victimized, an agent is invulnerable during the time before its self-defense gradually decreases to normal).
Target search radius	Very sensitive to the number of agents in simulation (use a large search radius for locating suitable targets when agents are sparsely distributed, but a smaller one with a large agent population to avoid inflating crime opportunities and reduce computing cost).
Ambient population thresholds	Using two ambient populations (broad-phase risk calculation counts targets within all grids overlapping with offender's search circle; narrow-phase risk calculation counts targets within search circle with precise point-to-point distances).

13.4 CASE STUDY 13: AGENT-BASED CRIME SIMULATION IN BATON ROUGE, LOUISIANA

This case study applies the ABM in simulating the spatiotemporal distribution of street robberies in East Baton Rouge Parish, Louisiana (hereafter referred to as Baton Rouge). It highlights the role of ABM as an artificial laboratory for social science experiments by simulating scenarios to explore crime theory and alternative policing strategies.

Simulation runs are compared by their abilities to predict crime hotspots from actual data. For each simulation run, a series of kernel density estimate (KDE) maps is created from simulated robbery events with different bandwidths. On a KDE map, areas with crime density above two standard deviations are treated as predicted crime hotspots from the corresponding simulation run. The predictive ability of crime hotspots from simulation is evaluated with three metrics: hit rate (H), prediction accuracy index (PAI), and forecast accuracy index (FAI). The first two metrics, H and PAI, are discussed in step 7 in Section 3.7 of Chapter 3. The third metric, forecast accuracy index (FAI), measures the difference between reported and simulated crimes falling inside predicted hotspots adjusted by total crime counts:

$$FAI = \frac{n_r / n_s}{N_r / N_s},$$

where n_r and n_s are the numbers of reported and simulated robberies inside hotspots, and N_r and N_s are the total numbers of reported and simulated robberies, respectively. A simulation run with FAI closer to 1.0 is considered to more accurately replicate the relative sizes of spatial clusters in reported crimes.

The datasets are prepared and provided in the subfolder Data under the folder ABMSIM. While previous chapters use GIS data in Esri's file geodatabase format, this chapter uses GIS and tabular data in traditional Esri shapefile (.shp), dBASE (.dbf), and comma-separated values (.csv) formats. These formats were chosen for their better compatibility with the open-source Python GIS libraries used to build the core of this simulation model, including GDAL, Shapely, and NetworkX. The toolbox ABM Crime Simulator for ArcGIS Pro.tbx and its associated folder ABMscripts are in the root directory.[2]

The major files in the subfolder Data are organized as follows:

1. Under the subfolder ebr _ grid, (1) ebr _ grid _ 500m _ with _ roads.shp is a spatial grid layer covering the entire study area with a 500-meter resolution, and (2) np _ odmatrix.npy is an origin–destination matrix, created for network-based travel distances between grid centers. Since resident agents can only travel on roads, only those grids with roads inside are kept for the spatial index layer. They are used by agents in heuristic estimates on travel costs.

2. In the root directory, (1) ebr _ boundary.shp represents the boundary of East Baton Rouge Parish, (2) ctpp _ ebr _ taz _ utm _ fixed. shp provides the commuting flows between TAZs extracted from the CTPP (Census Transportation Planning Products) data, (3) ebr _ taz _ flow _ total _ workers _ sparse.dbf records the worker flows

for TAZs, (4) tl _ 2015 _ 22033 _ road _ segments _ utm.shp is the line layer for the road network, (4) ebr _ random _ points _ grid _ 500m _ taz.shp is a point layer with all randomly generated points on road network, and (5) ebr _ acs _ 2015 _ 5yr _ bg.shp is the census block group layer.

Refer to Table 13.2 for more detailed information on the model input data in simulation mode. Note that the spatial data are in shapefiles, not in geodatabase. The model uses open-source Python GIS libraries that do not read files in geodatabase.

TABLE 13.2
Model Input Datasets in Simulation Mode

Name and Description	Key Fields
Boundary of study area ebr_boundary.shp A polygon layer to display the boundary of study area on animated map. It is also used to clip other GIS datasets, such as roads, to the study area, if necessary.	GEOID10: Unique FIPS (Federal Information Processing Standards) identifier for county or parish from 2010 US census data. NAME10: Name for county or parish.
CTPP commuting flows for TAZs ctpp_ebr_taz_utm_fixed.shp CTPP (Census Transportation Planning Products) commuting data for traffic analysis zones (TAZs) in study area. It contains number of employed residents, numbers of people leaving for work at certain hour in several categories, numbers of people working for a certain number of hours every day in several categories, and number of people arrested in 2010 at TAZ level.	GEOID10: Unique FIPS (Federal Information Processing Standards) identifier for TAZ. ID: Unique sequential identifier created for worker flow matrix and origin–destination matrix with network-based travel costs between TAZs. TOT_WORKER: Total number of workers in this TAZ. PCT_WORKER: Percent of workers relative to the total number of workers in study area. T1P_12A_5A, T2P_5A_8A, T3P_8A_11A, T4P_11A_4P, T5P_4P_8P, T6P_8P_12A: Percent of total number of workers in this TAZ leaving for work in the following 6 groups: midnight to 5:00 a.m., 5:00 a.m. to 8:00 a.m., 8:00 a.m. to 11:00 a.m., 11:00 a.m. to 4:00 p.m., 4:00 p.m. to 8:00 p.m., and 8:00 p.m. to midnight. HOUR_0_2, HOUR_2_3, HOUR_3_5, HOUR_5_6, HOUR_6_8, HOUR_8: Percent of total number of workers in this TAZ with daily working hours in the following 6 groups: 0 to 2, 2 to 3, 3 to 5, 5 to 6, 6 to 8, and above 8. T_ARREST10: Total number of people arrested in 2010 in this TAZ. P_ARREST10: Percent of people arrested in 2010 relative to the total number of people arrested in study area. PCT_A2W_10: Percent of people arrested in 2010 relative to the total number of workers in this TAZ.

(Continued)

TABLE 13.2 (*Continued*)
Model Input Datasets in Simulation Mode

Name and Description	Key Fields

Worker flows for TAZs
`ebr_taz_flow_total_workers_`
`sparse.dbf`
A stand-alone table for number of employed residents traveling from one TAZ (home) to another TAZ (job). Because such flow may not exist between some pairs of TAZs due to total lack of resident or job in some TAZs, a full N × N matrix (*N* is the total number of TAZs) contains a lot of wasteful zero flows. Instead, a sparse matrix for only pairs of TAZs with nonzero flows is used here.
	`ID_FROM`: Identifier for the origin TAZ of worker flow.
	`ID_TO`: Identifier for the destination TAZ of worker flow.
	Both `ID_FROM` and `ID_TO` are based on the field ID from the "CTPP commuting flows for TAZs" layer.
	`TOT_WORKER`: Total number of workers who live in origin TAZ (`ID_FROM`) and commute to destination TAZ (`ID_TO`) for work.

Road network
`tl_2015_22033_road_segments_`
`utm.shp`
A line layer for the road network. It is used in both animated map as backdrop and simulation as graph for calculating shortest or quickest path between two locations. All activities in simulation are located on this network.
	`SegmentId`: Unique identifier for each road segment in the network. A road polyline is divided into multiple segments whenever it intersects with other roads.
	`LENGTH`: The length of this segment in projected unit, which is meter in this case, with UTM coordinates.

Spatial grid `ebr_grid_500m_with_roads.shp` A 500m × 500m grid layer covering the entire study area. Because all activities in simulation are located on road networks, grids not covered by any road are removed from this layer. These grids are used to keep track of moving agents during simulation so that an agent can quickly retrieve a list of agents within a buffer zone from its neighboring grids.	`Row`: The row number of this grid from top (starting from 0). `Col`: The column number of this grid from left (starting from 0). `Row_Col`: Concatenation of row and column numbers of this grid with "_" between them. `ID`: Unique sequential number created for each grid. `UTILITY`: Contains a constant value of 1 that is used to represent the equal utilities of all grids in this layer. `CRIME2010`: The total number of reported robberies in 2020 that were located inside this grid. `EBRSO`: Sector of East Baton Rouge Sheriff's Office that this grid belongs to. Sector number matches the field `EBRSO_ID` in the "East Baton Rouge Sheriff's Office Sectors" layer (`ebrso_sectors.shp`). `BRPD`: District of Baton Rouge Police Department that this grid belongs to. District number matches the field `BRPD_ID` in the "Baton Rouge Police Department Districts" layer (`brpd_districts.shp`). A value of 0 means this grid is outside the jurisdiction of Baton Rouge Police Department, which is within the city limit of Baton Rouge.

TABLE 13.2 (*Continued*)
Model Input Datasets in Simulation Mode

Name and Description	Key Fields
OD matrix for spatial grid `np_odmatrix.npy` A matrix of network-based travel distance or time between all pairs of spatial grids. This matrix is pre-calculated using road network and grid centers. The value in this matrix is used as a heuristic measure of network-based travel cost between two arbitrary locations. It should be provided in. npy format, which can be created by NumPy package in Python. This file was converted from the stand-alone table: `ebr_grid_500m_odmatrix.dbf` in the folder `Data\ebr_grid`.	`FROM_ID`: Identifier for the origin grid. `TO_ID`: Identifier for the destination grid. Both `FROM_ID` and `TO_ID` are based on the field `ID` from the "Spatial grid" layer. `LENGTH`: Network-based travel distance from the center of origin grid (`FROM_ID`) to the center of destination grid (`TO_ID`).
Activity locations `ebr_random_points_ grid_500m_taz.shp` A point layer with all randomly generated points on road network. These points are 10 meters apart on average. They are used as candidate locations for agents' home, job, and flexible activities. Because of the random nature, point density in an area is determined solely by the road density. A more realistic spatial distribution of activity locations requires the use of high-resolution point-of-interest data for study area.	`FID`: Unique feature identifier for each activity location. `X`: X-coordinate in UTM projection. `Y`: Y-coordinate in UTM projection. `SegmentId`: Identifier for the road segment that this location falls on. `SegmentId` is based on the field `SegmentId` from the "Road network" layer. `Ratio`: This location's ratio (a fractal number between 0 and 1) on the road segment measured from its starting vertex. This value is used by routing algorithm. `ID_TAZ`: Identifier for the TAZ that this location falls inside of. It is used to generate a sample population that matches the spatial distribution of population at the TAZ level. `ID_TAZ` is based on the field `ID` from the "CTPP commuting flows for TAZs" layer.
Census block groups `ebr_acs_2015_5yr_bg.shp` A polygon layer for the census block groups in study area. While currently only used as backdrop in animated map, it can also be used for statistical analysis of census demographic variables and simulated crime rates at census block group level.	`GEOID`: Unique FIPS (Federal Information Processing Standards) identifier for census block groups from 2010 US Census data. `B01003e1`: Estimate of total population for census block groups.

(Continued)

TABLE 13.2 (*Continued*)
Model Input Datasets in Simulation Mode

Name and Description	Key Fields

Reported robberies
`ebr_2010_robbery.shp`
`ebr_2011_robbery.shp`
Point layers representing reported robbery incidents for East Baton Rouge Parish, Louisiana, in 2010 and 2011.
	`ReportNumb`: Police report number.
	`OffenseDat`: Offense date.
	`OffenseTim`: Offense time of day.
	`OffenseYea`: Offense year.
	`ViolentCri`: UCR violent crime classification code (robbery is 3).
	`X_UTM`: X-coordinate in UTM projection.
	`Y_UTM`: Y-coordinate in UTM projection.

Baton Rouge Police Department districts and East Baton Rouge Sheriff's Office sectors
`brpd_districts.shp`
`ebrso_sectors.shp`
Polygon layers that are used to assign values to BRPD and EBRSO fields of each grid in "Spatial Grid" layer through spatial overlay operation during data pre-processing. If a grid is outside BRPD's jurisdiction, a 0 is assigned to its BRPD field.
	`BRPD_ID`: Unique identifier for BRPD district.
	`EBRSO_ID`: Unique identifier for EBRSO sector.

13.4.1 SIMULATING CRIMES IN THE BASE SCENARIO

The case study begins with the base scenario. This sets up the baseline for testing some hypotheses on offender behaviors, such as the journey-to-crime behavior and the neighborhood reputation effect in Subsection 13.4.2.

Step 1. Preparing Running Environment for the Python-Based Simulation Tool: The tool "Agent-Based Crime Simulation Model" requires extra Python packages. Before running the tool, go to Settings > Package Manager > Installed Packages. Make sure all required packages are among the list of installed packages. Refer to Appendix 4B in Chapter 4 for procedures to install a third-party package with the Python Package Manager in ArcGIS Pro.

Under Options > Geoprocessing, uncheck "Add output datasets to an open map." This enables simulation results to be overwritten in new simulation runs so that users can run the model repeatedly.

In ArcGIS Pro, create a new project. Go to the Catalog pane, browse to the folder ABMSIM, and add all files under Data to the Map.

Step 2. Running the Tool with Default Setting for the Base Scenario: Still under the Catalog pane, expand the toolbox "ABM Crime Simulator for ArcGIS Pro.tbx" and double-click on the script tool "Agent-Based Crime Simulation Model." As shown in Figure 13.6a (or Figure 13.7a), the interface has simulation options and model parameters organized in six groups. Expand each category and examine the default values defined for various options and parameters (see Figure 13.6b-d and Figure 13.7b-d).

Geoprocessing   ∨ ⁴ ✕

← Agent-Based Crime Simulation Model ⊕

Parameters Environments   ?

> Input data
> Visualization, export, and animation playback
> Model parameters
> Police agent behaviors
> Offender agent behaviors
> Victim agent behaviors

▶ Run ∨

(a)

∨ **Input data**

Base data folder

Data

Boundary of study area

ebr_boundary.shp

CTPP commuting flows for TAZs

ctpp_ebr_taz_utm_fixed.shp

Worker flows for TAZs

ebr_taz_flow_total_workers_sparse.dbf

Road network

tl_2015_22033_road_segments_utm.shp

Spatial grid

ebr_grid_500m_with_roads.shp

OD matrix for spatial grid

E:\CMS_GISV3\Case13\ABMSIM\Data\ebr_grid\np.

Activity locations

ebr_random_points_grid_500m_taz.shp

Census blockgroups

ebr_acs_2015_5yr_bg.shp

(b)

∨ **Visualization, export, and animation playback**

Running mode

Simulation (Normal)   ∨

☑ Real-time visualization (simulation mode only)

☐ Show agent home & job locations in real-time
   visualization (simulation mode only)

☑ Export simulated crime incidents (simulation
   mode only)

Output file for crime incidents (simulation mode
only)

sim_crime_incidents.csv

☑ Export agent location history from simulation
   (simulation mode only)

Output file for agent location history (simulation
mode only)

sim_agent_location_history.csv

Output file for agent ID, type, and locations of home
& job (simulation mode only)

sim_agent_type_home_work.csv

Animation file for agent location history (animation
mode only)

anim_agent_location_history.csv

Animation file for agent ID, type, and locations of
home & job (animation mode only)

anim_agent_type_home_work.csv

Animation file for crime incidents (animation mode
only)

anim_crime_incidents.csv

☐ Display "Activity locations" layer from "Input
   data" in animation (animation mode only)

(c)

∨ **Model parameters**

Simulation cycle time (minutes)	5
Total simulation cycles	288
No. of agents	500
Percent of offender agents	0.25
Percent of police agents	0.1
Min. recreate cycles	1
Max. recreate cycles	6
Min. moving speed (meters per cycle)	400
Max. moving speed (meters per cycle)	4000
Spatial index resolution for agents (meters)	500
Spatial index resolution for environment (meters)	500

(d)

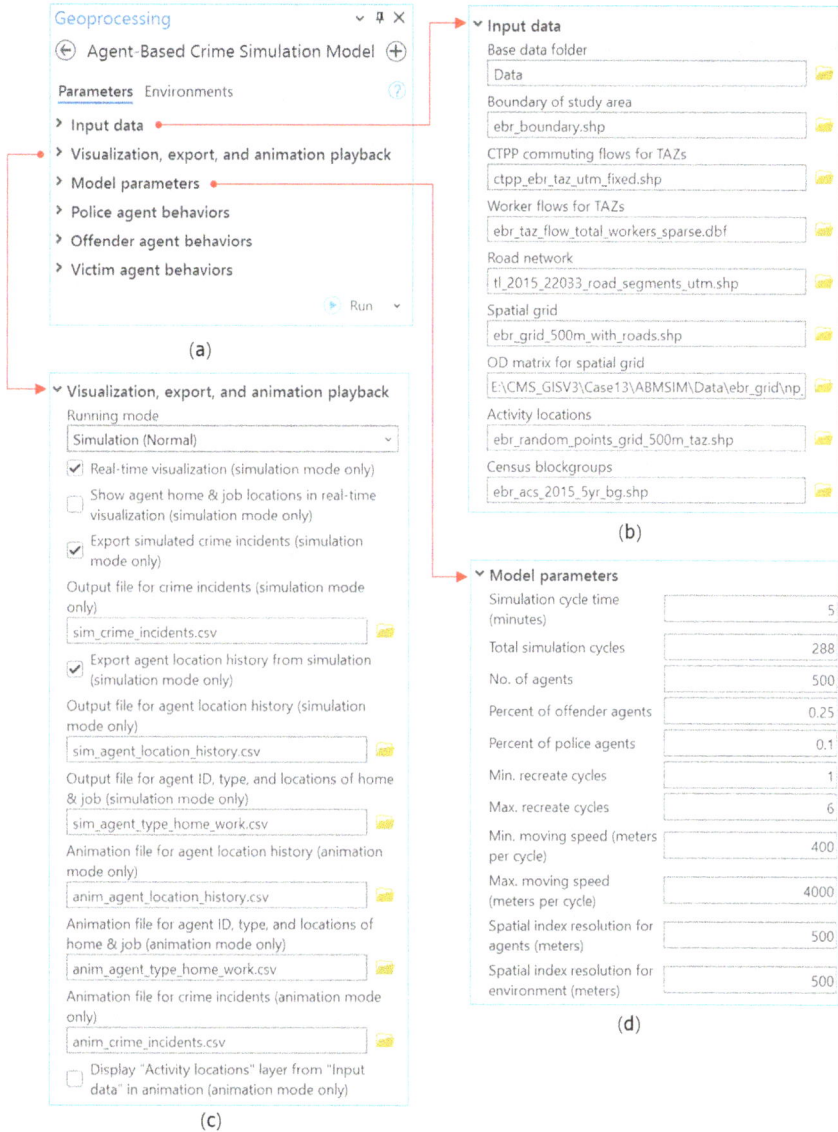

FIGURE 13.6 Interface of Agent-Based Crime Simulation Model: (a) main screen; (b) input data; (c) visualization, export, and animation playback; and (d) model parameters.

For the base scenario, only a few global parameters need to be adjusted before running the simulation. Expand the "Model parameters" group (Figure 13.6d) and change the values of "Total simulation cycles" and "No. of agents" to 288 and 500, respectively. Given the size of our study area for about 470 square miles, a simulation run needs to have a large-enough number of agents so that offender agents have enough opportunities to be in close proximity with potential victim agents.

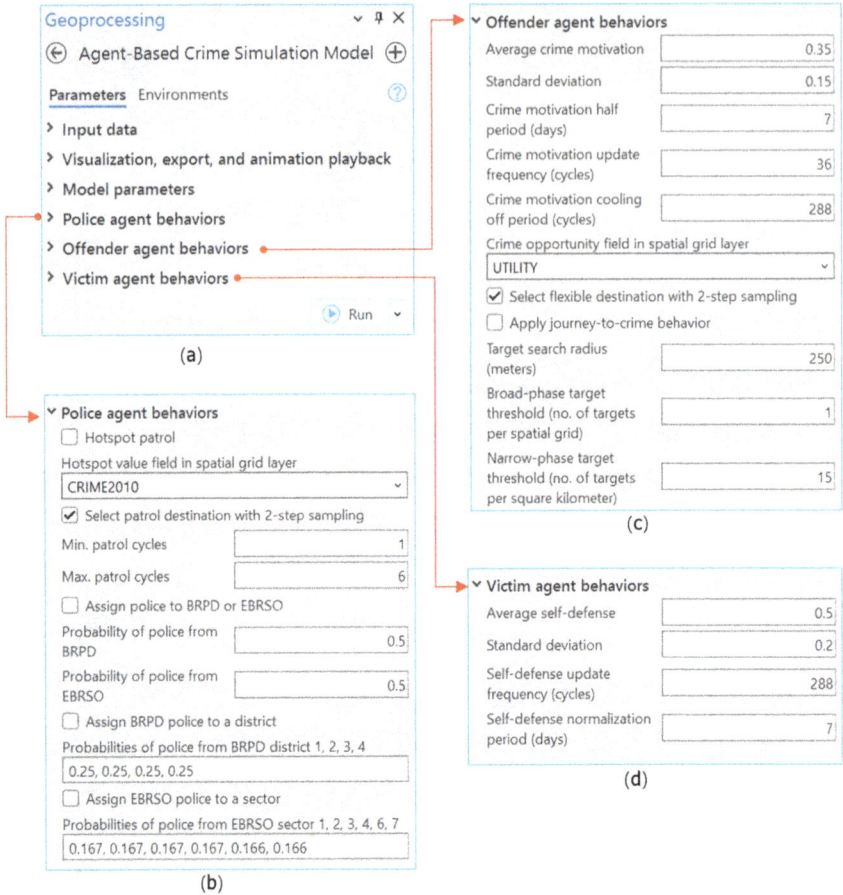

FIGURE 13.7 Interface of Agent-Based Crime Simulation Model (continued): (a) main screen, (b) police agent behaviors, (c) offender agent behavior, and (d) victim agent behaviors.

In addition, it is necessary to run enough simulation cycles in order to stabilize simulated crime patterns.[3] Click Run.

By default, an animated map opens and shows movement of agents against the backdrop of boundary, road network, and census block groups in our study area (Figure 13.8). When a simulation run concludes, the animated map closes itself.

Hereafter, the subsequent analyses are based on the results generated from the default parameter setting, that is, simulated crime incidents in the file sim _ crime _ incidents.csv. This CSV file, along with the other two resulting files, such as sim _ agent _ location _ history.csv and sim _ agent _ type _ home _ work.csv, can be used for animation. See Table 13.3.

Step 3. Applying the KDE Analysis on the Simulated Result: Add the result sim _ crime _ incidents.csv in the default folder to the Contents pane.

Animated Map — □

Agent-Based Crime Simulator (Mode: Simulation)

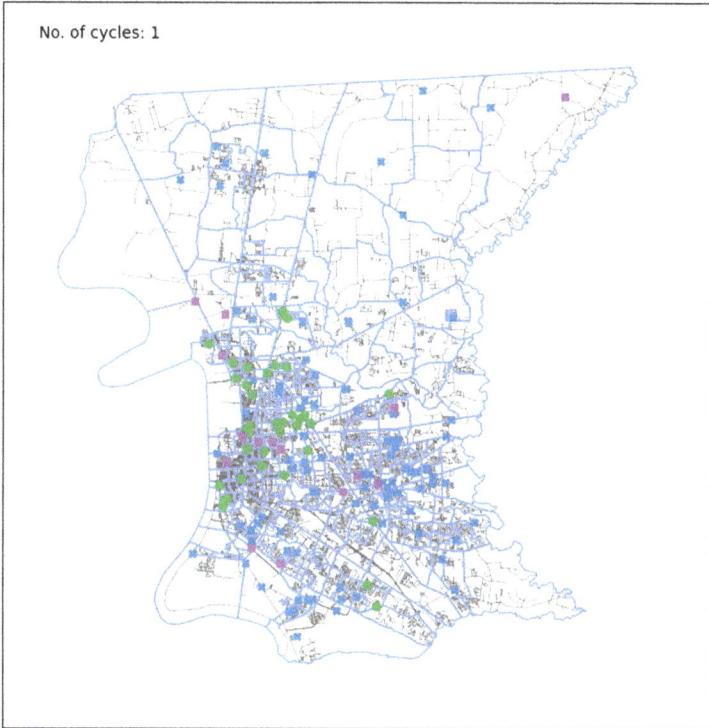

No. of cycles: 1

FIGURE 13.8 Animated map in simulation mode.

Right-click on it and select Display XY Data to activate the dialog box. In the dialog, input sim _ crime _ incidents _ base _ scenario for Output Feature Class, X for X Field, and Y for Y Field; set NAD _ 1983 _ UTM _ Zone _ 15N for Coordinate System; and click OK. Refer to step 1 in Subsection 2.2.1 of Chapter 2 if needed. Simulated crime incidents are added to the map.

Search and active the tool Kernel Density in the Geoprocessing pane. In the dialog, set sim_crime_incidents_base_scenario for Input point or polyline features, KernelD_base_scenario for Output raster, and 200 for Output cell size; under the Environments tab, select sim_crime_incidents_base_scenario

TABLE 13.3
Model Output Datasets in Simulation Mode

Crime incidents file `sim_crime_incidents.csv` A text file in comma-separated format (.csv) that records all simulated crime incidents. Each row contains simulation cycle number, agent identifies for offender and victim, and UTM coordinates of this incident.	`Cycle`: Simulation cycle number. `AgentID_Offender`: Agent ID for offender. `AgentID_Target`: Agent ID for target. `X`: X-coordinate in UTM projection. `Y`: Y-coordinate in UTM projection.
Agent location history file `sim_agent_location_history.csv` A text file in comma-separated format (.csv) that records the locations of all agents for the entire simulation run. Each row contains simulation cycle number, agent identifier, and UTM coordinates of agent location for this cycle. These rows are exported in the order of simulation cycles. Within the same simulation cycle, rows are ordered by agent identifiers.	`AgentID`: Agent ID. `Cycle`: Simulation cycle number. `X`: X-coordinate in UTM projection. `Y`: Y-coordinate in UTM projection.
Agent home and job locations file `sim_agent_type_home_work.csv` A text file in comma-separated format (.csv) that records the type, home, and job locations of all agents. Each row contains agent identifier, agent type, location identifiers, and UTM coordinates for this agent's home and job. Location identifier is based on those in activity locations dataset.	`AgentID`: Agent ID. `Type`: Agent type. `Home_FID`: Identifier for agent home location. It is based on the field `FID` from the "Activity locations" layer. `Home_X`: X-coordinate of home location in UTM projection. `Home_Y`: Y-coordinate of home location in UTM projection. `Work_FID`: Identifier for agent job location. It is based on the field `FID` from "the Activity locations" layer. `Work_X`: X-coordinate of job location in UTM projection. `Work_Y`: Y-coordinate of job location in UTM projection.

for Output Coordinate System, and in the drop-down menus for Extent and Mask, select `ebr_boundary.shp` for both. Click Run. Refer to step 9 in Subsection 3.3.2 of Chapter 3 if needed.

The resulting raster `KernelD_base_scenario` is added to the map. Right-click on it in the Contents pane and select Properties. In the dialog, select "Source" in the left pane and expand "Statistics" in the right pane, and record the "Mean" and "Std. Deviation" values, 0.062704 (μ) and 0.085265 (σ), which are the mean and standard deviation of simulated crime density in raster cells (no. of crimes per square kilometer). Due to the nature of randomization in simulation, your values are likely to be different.

Here, we classify *hotspot* as any raster cell with two standard deviations above the average. In our case, the threshold for hotspot is $\mu + 2\sigma = 0.062704 + 2 \times 0.085265 = 0.233236$.

Step 4. Reclassifying the Crime Simulation Raster to Generate Crime Hotspots: Search and activate the tool Reclassify in the Geoprocessing pane. On the Parameters tab, set `KernelD _ base _ scenario` for Input raster and `Reclass _ KernelD _ base _ scenario` for Output raster, and click Classify to bring out the Classify dialog. In the dialog, set 2 for classes, and use the hotspot threshold from step 3 to reclassify any values below the threshold to 1 and any values above threshold to 2. Click OK to confirm and turn off the dialog of Classify, and click Run to execute it.

Add the resulting raster `Reclass _ KernelD _ base _ scenario` to the map.

Step 5. Calculating Simulation Evaluation Indices Based on Crime Hotspots: In order to calibrate indices *H*, *FPI*, and *FAI*, we need to convert the hotspot raster to a polygon feature class to facilitate point-in-polygon selections.

Search and activate the tool "Raster to Polygon" in the Geoprocessing pane. In the dialog, set `Reclass _ KernelD _ base _ scenario` for Input raster, Value for Field, and `polygon _ base _ scenario` for Output polygon features; uncheck Simplify Polygons and check Create Multipart Features. Click Run. Add and move the new polygon layer `polygon _ base _ scenario` to the top of the map and apply a "Unique Values" symbology to map the field `gridcode`. Due to the nature of randomization, simulated crime patterns differ in various simulation rounds (Figure 13.9).

Now we can calculate *H* (hit rate). Turn off all layers except `ebr _ 2011 _ robbery.shp` (reported robberies for 2011), `polygon _ base _ scenario` (polygon layer converted from reclassified KDE), `ebr _ boundary.shp` (boundary of East Baton Rouge Parish), world topographic map, and world hillshade. Move `ebr _ 2011 _ robbery.shp` on top of `polygon _ base _ scenario`.

Now we use Definition Query to filter the data displayed in the Map pane. Right-click on the layer `ebr _ 2011 _ robbery.shp` > Properties > Definition Query > + New Definition Query, and set `OffenseMon is equal to 01` in the query expression box for robberies from January 2011. It identifies 76 robberies for this month ($N_r = 76$). Similarly, apply Definition Query on `polygon _ base _ scenario`, and set `gridcode is equal to 2` in the query expression box to select the polygons as hotspots. Click the data `ebr _ 2011 _ robbery.shp` in the map pane, and use Select by Location to select the points inside the filtered layer `polygon _ base _ scenario` as the reported robberies falling in predicted crime hotspots. It identifies 15 robberies ($n_r = 15$). Therefore, $H = \dfrac{15}{76} (= 19.74\%)$.

To calculate *FPI* (forecast precision index), we need to calculate the ratio of predicted hotspot size to that of study area (a/A). Open the attribute table of `Reclass _ KernelD _ base _ scenario`. It shows the total number of cells is 30,463, with 1,589 classified as hotspots. Therefore, $a/A = 1589/30463$, and $FPI = \dfrac{15/76}{1589/30463} = 3.784$.

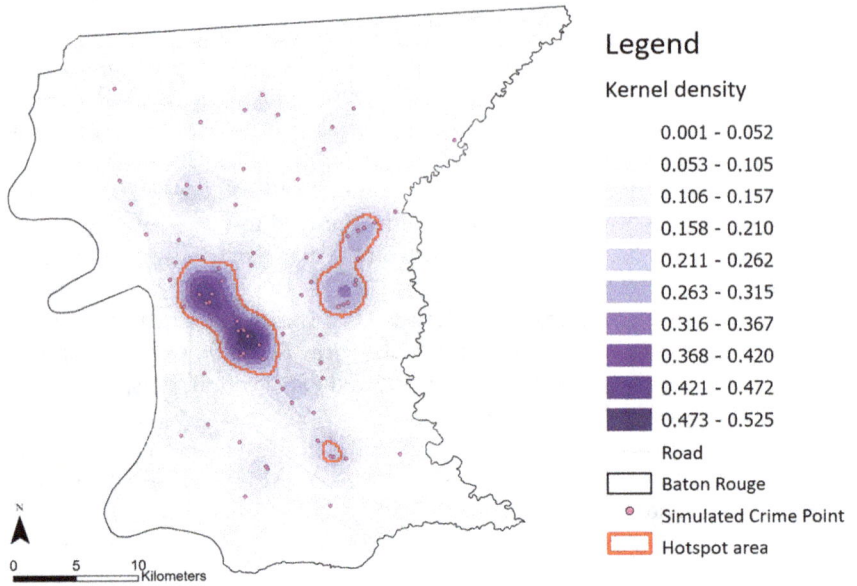

(a)

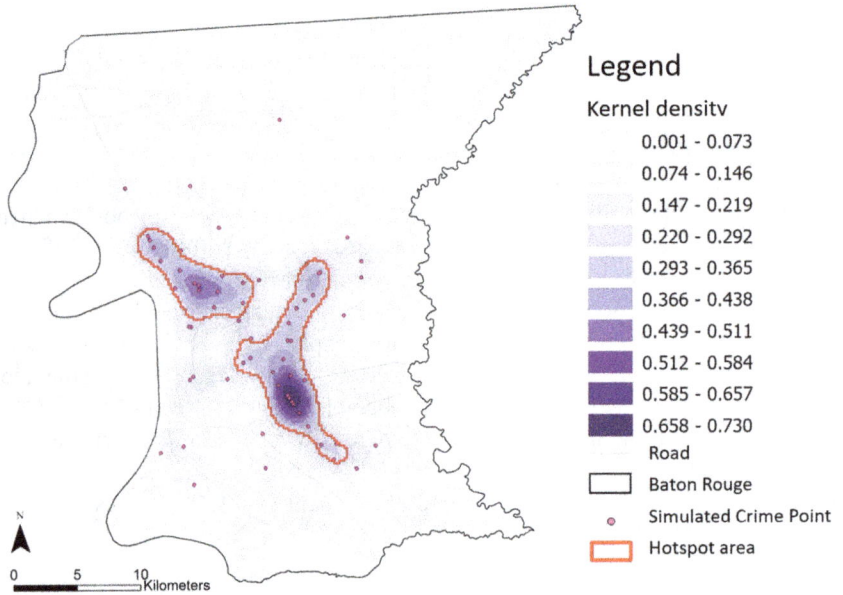

(b)

FIGURE 13.9 Simulated crime incidents and hotspots at two different times in the base scenario.

To calculate *FAI* (forecast accuracy index), click "Select by Location" on the Map ribbon again and select simulated crime incidents in predicted crime hotspots. Notice that out of 77 simulated crime incidents ($N_s = 77$), 29 of them are inside the predicted hotspots ($n_s = 29$). Therefore, $FAI = \dfrac{15/29}{76/77}(= 0.524)$.

This concludes simulating the base scenario and calculating its goodness-of-fit measures. For every alternative scenario in the following, repeat steps 3 to 5, and compare it against the base scenario based on the three goodness-of-fit measures.

13.4.2 TESTING THE HYPOTHESES ON OFFENDER BEHAVIORS

We first test the journey-to-crime hypothesis that, with all other things being equal, a motivated offender is more likely to commit crime near its familiar locations because of better knowledge on target availability and avoiding apprehension. It assumes that the distance between an offender's home and current location is negatively associated with the probability for the offender to commit a robbery at its current location. Note that we use pre-calculated distance between grid centers for offender's home and current location in order to avoid costly network-based distance calculation in real time. The result is compared to the base scenario with hotspot metrics.

Step 6. Simulating the Scenario with Offender Journey-to-Crime Behavior: Open the Agent-Based Crime Simulation Model tool again. Refer to step 2 to adjust "Total simulation cycles" and "No. of agents" to match with the settings in the base scenario. Expand the group "Offender agent behaviors" (Figure 13.7c) and check the option "Apply journey-to-crime behavior." Expand the group "Visualization, export, and animation playback" (Figure 13.6c), and input sim_crime_incidents_j2c.csv for "Crime incidents file (simulation mode only)." Click Run to start the simulation.

After the simulation finishes, repeat steps 3–5 to visualize simulated crime incidents, generate the hotspot polygons, calculate the three goodness-of-fit measures with the new simulated crime incidents file, and compare the goodness-of-fit measures to those in the base scenario.

Another alternative scenario is designed to test the effect of neighborhood reputation on offender behavior. The base scenario assumes that offender agents have no information on the reputation of neighborhoods, and only use travel costs to select locations for flexible activities. Here, offender agents possess some global knowledge on how easy it is to find suitable targets and get away from committing crime in each neighborhood. In this case, the total number of robberies inside each spatial grid from the previous year 2010 is used as a proxy measure to this knowledge. In general, more reported crimes within an area in recent months indicate the abundance of crime opportunities and lack of sufficient guardianship. Motivated offenders use both neighborhood reputation and travel cost to determine the probability of choosing a spatial grid for candidate locations. Within the chosen grid, a random location on a road network becomes the destination of potential victim's next activity. This two-stage random destination selection process is designed to optimize simulation runs with a large number of agents or network locations. Again, pre-calculated distances between grid centers are used in the first stage as fast heuristics for travel costs to different neighborhoods.

Step 7. Simulating the Scenario with Predefined Crime Opportunities: Open the attribute table of ebr _ grid _ 500m _ with _ roads.shp in the Contents pane and examine the attribute values for the fields UTILITY and CRIME2010. The UTILITY field has a constant value of 1 and assigns equal reputation to every grid in the base scenario. The CRIME2010 field, on the other hand, records the total number of robberies that took place in each grid in 2010. This field can be used by offender agents to differentiate the possibility to successfully commit a robbery in each grid.

Open the tool Agent-Based Crime Simulation Model again. Refer to step 2 for matching "Total simulation cycles" and "No. of agents" with the settings for base scenario. Expand the group "Offender agent behaviors" (Figure 13.7c) and choose CRIME2010 for "Crime opportunity field in spatial grid layer." Expand the group "Visualization, export, and animation playback" (Figure 13.6c), and input sim _ crime _ incidents _ crime2010.csv for "Crime incidents file (simulation mode only)." Click Run to start the simulation.

Repeat steps 3 to 5. The evaluation metrics indicate whether the inclusion of neighborhood reputation in offender decision-making process improves simulation performance in predicting crime hotspots.

13.4.3 TESTING THE EFFECTS OF POLICE PATROL STRATEGIES

The previous subsection tests competing theories on crime-related behaviors. This subsection evaluates what-if scenarios for police patrol strategies.

A patrol strategy is regarded more effective in curbing crimes when it can disrupt more crime opportunities with less manpower. In the base scenario, a police agent randomly travels to a patrol destination within its designated police district. After reaching this location, it stays there for a random period of up to 30 minutes. In the alternative scenario, a police agent uses the crime counts from the previous year to determine the probability of choosing a spatial grid for candidate patrol destinations. Grids with higher crime counts are favored by the hotspot patrol strategy. Within the chosen grid, a random location on the road network becomes the patrol destination. This two-stage destination selection process is used again for optimizing simulation runs.

Step 8. Exploring the Scenario with Hotspot Police Patrol: Open the tool Agent-Based Crime Simulation Model. Again, match "Total simulation cycles" and "No. of agents" with the settings in the base scenario. In the Contents pane, open the attribute table of ebr _ grid _ 500m _ with _ roads.shp layer and note the fields BRPD and EBRSO representing Baton Rouge Police Department district and East Baton Rouge Sheriff's Office sector, respectively.

A police agent uses a field in the spatial index grid layer to determine their preferences for grids within their designated police district as possible patrol destinations. While the base scenario randomly assigns a police agent a grid to patrol with an equal probability for any grid within its designated patrol zone, the hotspot patrol scenario uses reported robberies in each grid for the year 2010 to estimate its probability of being a crime hotspot. To switch the patrol strategy, expand the "Police agent behaviors" group (Figure 13.7b) and check "Hotspot patrol." Make sure that CRIME2010 is selected in the drop-down menu of "Hotspot value field in spatial grid layer."[4]

Expand the group "Visualization, export, and animation playback" (Figure 13.6c) and input `sim _ crime _ incidents _ hotspot _ patrol.csv` for "Crime incidents file (simulation mode only)." Click Run to start the simulation.

Again, repeat steps 3 to 5.

While the model can animate agent movement and crime incident locations during simulation in real time, the resulting animation is often not smooth enough for pattern detection. The model includes a function to animate agent movement and crime incident locations from three required CSV files for agent location history, home and job locations, and crime incidents, as summarized in Table 13.4. These CSV files can be created by running simulations with their corresponding export options turned on. At every animation cycle, the current location of each agent and its identifier are retrieved from a row in the agent location history file. The location and initial animation cycle of every crime are retrieved from a row in the crime incidents file. Agent home and job locations are displayed as a static layer. Users can also choose whether to display the `Activity locations` layer from "Input data" in animation for reference purpose. However, too many activity locations may clutter the animated map and make it more difficult to detect dynamic spatial patterns.

TABLE 13.4
Model Input Datasets in Animation Mode

Animation file for agent location history `anim_agent_location_history.csv` A text file in comma-separated format (.csv) that provides information on the locations of all agents for the entire animation. Its contents and data organization are the same as agent location history file that can be exported from simulation runs.	`AgentID`: Agent ID. `Cycle`: Animation cycle number. `X`: X-coordinate in UTM projection. `Y`: Y-coordinate in UTM projection.
Animation file for agent home and job locations `anim_agent_type_home_work.csv` A text file in comma-separated format (.csv) that provides the type, home, and job locations of all agents. Its contents and data organization are the same as agent home and job locations file that can be exported from simulation runs. Different point symbols are used to display three types of agents. Unique symbols are also applied to distinguish home, job, and optionally flexible activity locations.	`AgentID`: Agent ID. `Type`: Agent type. `Home_FID`: Identifier for agent home location. It is based on the field `FID` from the "Activity locations" layer. `Home_X`: X-coordinate of home location in UTM projection. `Home_Y`: Y-coordinate of home location in UTM projection. `Work_FID`: Identifier for agent job location. It is based on the field `FID` from "the Activity locations" layer. `Work_X`: X-coordinate of job location in UTM projection. `Work_Y`: Y-coordinate of job location in UTM projection.

(Continued)

TABLE 13.4 (*Continued*)
Model Input Datasets in Animation Mode

Animation files for crime incidents	Cycle: Animation cycle number.
anim_crime_incidents.csv	AgentID_Offender: Agent ID for offender.
A text file in comma-separated format (.csv) that	AgentID_Target: Agent ID for target.
provides information on all crime incidents for	X: X-coordinate in UTM projection.
animation purpose. Its contents are the same as	Y: Y-coordinate in UTM projection.
crime incidents file that can be exported from	
simulation runs.	

Step 9. Optional: Animating Agent Movement and Crime Incident Locations:
In Notepad, open anim _ agent _ location _ history.csv, anim _
agent _ type _ home _ work.csv, and anim _ crime _ incidents.csv,
which by default are located under the folder Data\animation, and examine
their contents. In addition to a header for field names, they contain information on
an agent's current location, locations of its home and job, and a crime incident for
each line.

Open the tool Agent-Based Crime Simulation Model again. Expand the group
"Visualization, export, and animation playback" and choose Animation (With
home and job locations) for "Running mode." The other option "Animation
(Normal)" mode is similar but does not display home and job locations. Click Run
to start the animation.

13.5 SUMMARY

No longer constrained by area-aggregated data and snapshots of system states, the
agent-based model (ABM) is capable of representing discrete places with unique
geographic features, heterogeneous population of autonomous agents, and dynamic
interactions between an agent and places and among agents themselves. The ABM
crime simulation model introduced in this chapter has largely met our goal of modeling individual behaviors across space and over time in order to reveal the process behind patterns. Our choice of application in simulating crimes is related to a
high-quality dataset of crimes available to the research team and the advancement
of crime theory that accounts for the role of environment and interactions of various
agents (e.g., motivated offenders, potential victims, and police) in time and space.

The case study of robberies in Baton Rouge, Louisiana, examines how criminogenic features of local places, daily routines of vulnerable targets, and motivated
offenders affect the spatiotemporal crime pattern. One objective is to test some
hypotheses suggested by crime theories, such as the routine activity theory and the
environmental criminology theory. How well simulated robberies from this model
predict hotspots in reported crimes sets a benchmark on the theory's usefulness in
applied crime research. Another objective is to test the impact or effectiveness of various police patrol strategies, such as hotspot policing. The simulated crime pattern
by the model largely replicates prominent robbery hotspots in the study area. It still

lacks enough realistic elements to constrain the spatial distribution of crime-related attraction and risk. One direction for improving the work is to expand modeling daytime (work) and nighttime (rest at home) activities to evening-time activity (dining out, shopping, or entertainment) and model them more accurately. Such an effort needs to be informed by analysis of quality data of individuals with spatial and temporal precisions, which may come from big data, as discussed in Chapter 14.

ABM can also benefit studies of spatial convergence in other domains. For example, transmission of communicable diseases also requires multiple parties, such as infected and susceptible populations or disease vectors, to be within close proximity spatially and temporally. However, communicable diseases do not always require synchronous interaction and can transmit further down from infected subjects. These properties require extension of our ABM with more relaxed colocation conditions and more sophisticated contact structures within and between groups (Bian, 2004). Spatialized epidemic models, such as epidemic tree (Li et al., 2019), can be readily used as a starting point to include spatially heterogeneous epidemiological processes in ABM.

For illustration, our case study in this chapter implements each scenario with a single simulation run. Due to the extensive use of randomization in the ABM model, it produces different metrics from multiple simulation runs under the same parameter setting. When using hotspot metrics to calibrate model parameters against empirical data and compare simulation scenarios with different parameter sets, one needs to run a scenario multiple times and use the statistical distribution of these metrics, instead of values from a single simulation run, to assess model stability and identify the best combination of parameters.

NOTES

1 This chapter was drafted by Haojie Zhu, Department of Geography and Anthropology, Louisiana State University, Baton Rouge, LA 70803, USA.

2 Unlike the tools in other chapters stored under CMGIS-V3_Toolbox, this tool highly depends on the relative path of data and Python scripts.

3 On a laptop as described in the Preface, it took 27 minutes to run this setting, wherein 288 simulation cycles are equivalent to 1 day, with 5 minutes per cycle. The default parameters are set at 8,928 and 1,000, respectively, and took more than 1 day on the laptop, where in 8,928 simulation cycles are equivalent to 31 days (one month).

4 Optionally, users can also assign police agents to different jurisdictions with separate configurations of patrol zone.

14 Spatiotemporal Big Data Analytics and Applications in Urban Studies[1]

One major challenge in urban studies is to examine the spatiotemporal dynamics of human mobility in cities, especially in fine spatial and temporal resolutions, as we are increasingly interested in questions, such as where people are, at what time, and in what activity they are engaged. The availability of big data and rising computational power provides us an unprecedented opportunity to answer these questions in both spatial and temporal precisions quickly and effectively. The term "big data" is not necessarily defined by its size but rather emphasizes the sources other than surveys. It is "found" or "observed" "passively as the digital exhausts of personal and commercial activities" (US Census Bureau, 2022). This chapter uses taxi trajectory data as an example to illustrate how to process and analyze spatiotemporal big data in urban studies. XSTAR, a software product for analyzing taxi trajectory data (Li et al., 2021), is used to implement the study.

Section 14.1 offers an overview of big data, with a focus on taxi trajectory data, and explains how urban studies can benefit from such a data source. Section 14.2 uses a scaled-down taxi trajectory dataset in Shanghai to explore the process of big data analysis in ArcGIS. Section 14.3 introduces approaches to processing taxi trajectory data with XSTAR. Section 14.4 uses the same data source but a larger dataset in Shanghai to demonstrate the functionality of XSTAR in analyzing origin–destination (OD) traffic flows with big data. The chapter is concluded with a summary in Section 14.5.

14.1 USING TAXI TRAJECTORY BIG DATA IN URBAN STUDIES

14.1.1 WHY TAXI TRAJECTORY BIG DATA?

Big data are data of large volumes incessantly collected through various online services, such as shopping records, web exploring footprints, travel itineraries, etc. The advances of the Internet of Things (IoT) link more and more devices and users to the Internet and thus further increase the velocity of collecting data and the scope and volume of collected data. Meanwhile, the breakthroughs of data storage and computing technologies make big data analysis more accessible and feasible. Big data are spatiotemporal when each record contains the location and time generated by a device. Among most popular data sources, taxi trajectory data are the most applicable and accessible due to the highly intensive and nearly full-time driving activities of taxis.

DOI: 10.1201/9781003292302-17

Most importantly, there is a relatively low risk of violating the privacy of taxi passengers compared to other data. For example, smart card (those for public transit like bus or subway) data is usually linked to a specific individual. It is easy to reveal this individual's itinerary through reading his or her smart card data, and the consequence is that the person is identified easily. It is also the case for many other big data, such as mobile phone data, etc. In some situations, we may use those data for urban studies or other purposes. To protect the privacy of data providers, their spatiotemporal granularity and data accuracy have to be degraded. As a result, analysis based on these data might be unreliable. On the other hand, taxi passengers cannot be identified at most time, and thus it is hard to match a specific trip to an individual based on taxi trajectory data. Therefore, there is less need to compromise data quality in order to protect passengers' privacy.

Taxi trajectory data have been used in many aspects of urban studies. Some scholars focus on developing trajectory data mining methods (Dodge et al., 2009; Izakian et al., 2016; Liu and Karimi, 2006; Pfoser and Theodoridis, 2003; Zhou et al., 2017), while others may be interested in discovering knowledges behind data, such as residents' travel characteristics and patterns (Hu et al., 2014; Kang et al., 2015; Liu et al., 2012; Torrens et al., 2012), spatiotemporal features of traffic flows (Ge et al., 2010; J Wang et al., 2017; Wei et al., 2012; Zhang Z et al., 2011), or travel time prediction (Jiang and Li, 2013; Xu et al., 2018). It is understandable that large data volume is necessary to figure out patterns in a large city, but meanwhile, it might be a challenge to deal with such big data. Efficient data storage and accessing techniques are needed. For example, in case study 14B, the size of data is 10 GB, consisting of about 100 million records from about 13,000 taxis traveling on a road network across 2,500 square kilometers for 24 hours. It is not an easy task to handle the data for an analyst who has few experiences in computer programming or data processing.

14.1.2 Taxi Trajectory Data Format

Usually, raw taxi trajectory data consists of a huge number of records. Each record includes at least four fields, namely, identification, time, location, and status, where identification is used to label individual taxi, time and location store the instantaneous spatiotemporal information of the taxi, and status is usually a binary variable (e.g., "loaded" and "unloaded") representing whether the taxi is carrying passengers or not. The interval between each two records may be fixed, such as 10 seconds, or flexible.

It is possible to tell a taxi's driving mode by retrieving the instant when its status is switched between "loaded" and "unloaded" along the timeline. If sorting all records of one taxi by its sampling time, then any two successive records having different loading statuses mean that the beginning or ending of a passenger's trip occurs between the two records. By this means, it is possible to extract both complete trips of passengers and cruising trajectories.

Part of sample records are given in Table 14.1, where location is expressed as a pair of coordinates, and status assumes a value "loaded" or "unloaded." Obviously, records in this table are not sorted by either identifier or time. If we group records by taxi identifiers and sort records in each group by time, then we can draw and link their

TABLE 14.1

Sample Records in Raw Taxi Trajectory Data

Identifier	Time	Location	Status
. . . . . .	. . . . . .	. . . . . .	. . . . . .
T23323	2022/5/1 19:38	89.221123, 43.654654	Loaded
T33424	2022/5/1 2:55	89.222653, 43.654432	Loaded
T53234	2022/5/1 23:23	89.221267, 43.655322	Unloaded
T64543	2022/5/1 4:49	89.221324, 43.654675	Loaded
T98787	2022/5/1 1:12	89.221232, 43.654162	Unloaded
T16374	2022/5/1 17:37	89.221975, 43.654023	Loaded
T77473	2022/5/1 2:28	89.221732, 43.654643	Unloaded
T97643	2022/5/1 0:24	89.221775, 43.654912	Loaded
T35577	2022/5/1 12:05	89.221115, 43.654780	Unloaded
T18654	2022/5/1 3:46	89.221438, 43.654520	Loaded
. . . . . .	. . . . . .	. . . . . .	. . . . . .

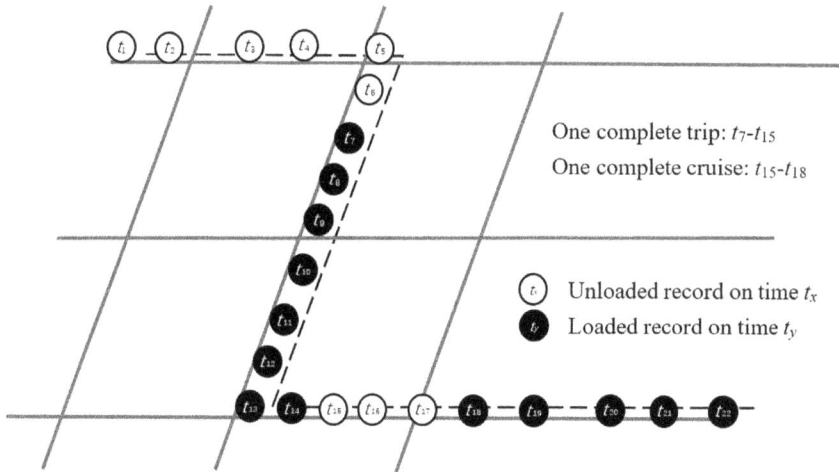

FIGURE 14.1 Temporally continuous positioning records of one taxi.

locations on road network. The resultant path is the trajectory of this taxi. For instance, as shown in Figure 14.1, where a taxi's locations are represented with white or black markers reflecting the taxi's status on this location, and dashed lines are used to link successive markers. The trajectory of one taxi consists of *trips* of taxi passengers and *cruises* for searching passengers alternately. The taxi cruises from t_1 to t_6, picks up passengers at t_7, drops off passengers at t_{15}, and picks up other passengers at t_{18}.

Based on the example, we define related variables as: $D_k = \left[\{t_i, l_i, s_i\} \mid 1 \leq i \leq n\right]$ denotes a set of n positioning records of taxi k, where t_i, l_i, and s_i are the i-th sampling time, location, and loading status, respectively, and s_i equals *loaded* or *unloaded.* D_k is ordered by time, that is, $t_i < t_{i+1}$. We define a trip or a cruise as a set of $\left[\{t_i, l_i\} \mid p \le i \le q \right]$, where $p, q \in [1, n]$. The set is a trip if $s_{p-1} = unloaded$, $s_{j \mid p \le j \le q-1} = loaded$, and $s_q = unloaded$, and otherwise, the set is a cruise. For example, in Figure 14.1, it is a complete trip from t_7 to t_{15}, and a complete cruise from t_{15} to t_{18}. It is noted that the last record in a trip (or a cruise) is always the first record belonging to a cruise (or a trip) alternately. That means the trajectory of a taxi could also be a list of $\{Trip, Cruise, Trip, Cruise, \ldots\ldots\}$ or $\{Cruise, Trip, Cruise, Trip, \ldots\ldots\}$.

14.1.3 APPLICATION-ORIENTED TAXI TRAJECTORY DATA QUERIES

Applications including the case studies in this chapter utilize taxi trajectory data in the past, not present (e.g., real-time traffic) or future (e.g., travel time estimation). Urban studies usually rely on aggregate measures meeting conditions without taxi identifiers. Each aggregate query has explicit spatial and temporal query conditions. For example, how many trips have happened from district A to district B during rush hours yesterday? In the query, "from district A to district B" and "rush hours yesterday" are spatial and temporal conditions, respectively. A temporal condition can either be an instant (e.g., 10:30 am) or an interval (e.g., 8 am-2 pm) according to the temporal length it refers to, while a spatial condition can be the whole study area, one or a group of spatial analysis units, paths constrained by underlying road networks, points on road segments, etc. The combination of temporal and spatial conditions may be used to suit various query objectives. Table 14.2 lists some typical query

TABLE 14.2
Typical Queries on Taxi Trajectory Data

Type	Example
Instant origin–destination analysis	Calculate the number of ongoing trips at 11:0 a.m. yesterday for each pair of origin and destination
Interval origin–destination analysis	Calculate the number of trips starting and/or ending between 12:00 a.m. and 3:00 p.m. yesterday for each pair of origin and destination
Instant distribution analysis	Calculate the number of loaded/unloaded taxis for each of given spatial analysis units at 11:00 a.m. yesterday
Interval distribution analysis	Calculate the number of locations where passengers got on/off taxis in each of given spatial analysis units between 12:00 a.m. and 3:00 p.m. yesterday
Location-based network-constrained analysis	Calculate the volume and average speed of taxi traffic flow passing through the third checkpoint of Road A between 9:00 a.m. and 11:00 a.m. yesterday
Component-based network-constrained analysis	Calculate the volume and average speed of taxi traffic flow traveling on Road Segment B at 8:00 a.m. or between 8:00 a.m. and 10:00 a.m. yesterday
Path-based network-constrained analysis	Calculate the volume and average speed of taxi traffic flow traveling from Points C to D along Road Segments E, F, and G at 9:00 a.m. or between 9:00 a.m. and 10:00 a.m. yesterday

examples. A combination of these queries may satisfy most application purposes. Most of these queries can be evaluated with XSTAR.

The following outlines some popular tasks based on taxi trajectory data queries.

Origin–destination (OD) analysis targets trips of taxi passengers. Each trip has an origin and a destination, and aggregation of trips reveals the intensity of connectivity for unique origin–destination (OD) pairs. The result is often a matrix with origins and destinations enumerated as labels of rows and columns, respectively, and each element of the matrix as the number of trips for an OD pair. In space, origins or destinations are usually represented as small polygons, such as traffic analysis zones, etc. Spatial query condition of an OD analysis may include one/many origins and one/many destinations. A one-to-many OD analysis (i.e., one origin and many destinations) discovers the distribution of destinations of trips from the same origin. A many-to-one OD analysis indicates the spatial sources of travelers targeting the same destination. Temporal query conditions could be either instants or intervals. As shown in Table 14.2, an instant query is usually applied to ongoing trips, while an interval query is to completed trips.

Distribution analysis describes spatial distributions of taxis based on their loading statuses across a study area measured in an area unit. The spatial query condition of a distribution analysis may be all or part of the study area. If the objective is to show the number of loaded or unloaded taxis in each unit, the temporal condition is usually an instant. If its objective is to summarize the number of trips starting or ending in each unit, an interval query condition is adopted. Based on the result, a density or clustering analysis can be conducted, for example, finding hotspots of passengers getting on taxis during specific periods.

Since the movement of taxis is mostly constrained by road networks, *network-constrained analysis* aligns taxi trajectories with road segments through map-matching algorithms. Spatial query conditions in network-constrained analysis are usually composed of network locations, network components, or network paths. Network locations refer to locations associated with road networks, such as street addresses. Location-based network-constrained analysis is often used to analyze profiles of traffic flows passing through a location toward different directions. A network component can be a road segment or an intersection, while a network path consists of a series of network components. Component-based or path-based network-constrained analysis focuses on summarized or average measures of traffic flow related to one component or one path. In applications, location-based network-constrained analysis is usually accompanied by an interval query condition, while the other two types of network-constrained analysis could be applied with either instant or interval temporal conditions.

14.2 CASE STUDY 14A: EXPLORING TAXI TRAJECTORY IN ARCGIS

The case study area uses a sample dataset in Shanghai, China, to illustrate some basic properties of taxi trajectory data. Shanghai is a large city with a population of over 25 million and an area size of about 6,400 square kilometers. The sample dataset contains 900,862 records for taxis with a vacant rate ranging 24–25%, that is, about 0.8% of a larger taxi trajectory big data with more than 100 million locational records collected from more than 13,000 taxis running 24 hours in Shanghai. Each data entry contains 13 fields: taxi ID, alarm status, vacant (passenger loading status), roof light, elevated road, brake, GPS receive time, GPS sent time, longitude, latitude, speed, heading, and number of GPS satellites.

The following datasets are prepared and provided in the geodatabase `Shanghai.gdb` under the data folder `Shanghai`:

1. Feature `Shanghai` represents the administration boundary, and feature `SHroad` is road network.
2. File `SHtaxi.csv` includes sample taxi trajectory records.

Step 1. Constructing Taxi Trajectories from Spatiotemporal GPS Tracking Points: This step utilizes the tool Points to Line in ArcGIS Pro to map taxi trajectories from the GPS tracking points based on the group field and sort field.

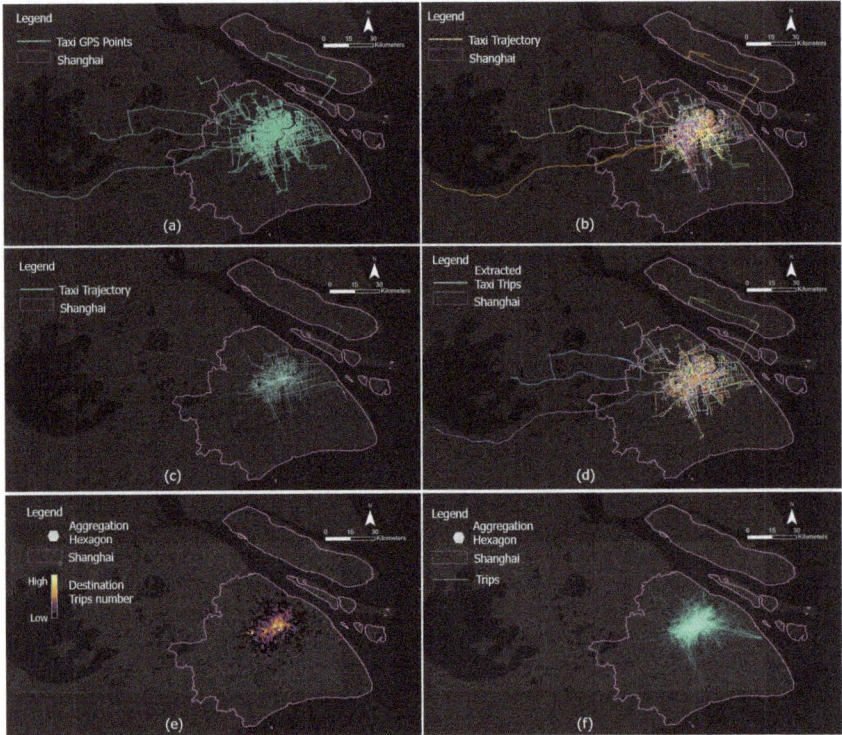

FIGURE 14.2 Rebuild taxi trajectories: (a) taxi GPS tracking points, (b) taxi trajectories with unique ID visualization, (c) taxi trajectories with single symbol visualization, (d) extracted trips with loading passengers, (e) trips number in destination hexagon units, and (f) aggregated trips with hexagon unit.

In ArcGIS Pro, add the file SHtaxi.csv and save it as a feature class SHtaxi in the geodatabase.[2] Apply the tool XY Table to Point on the feature SHtaxi by defining its fields Longitude and Latitude as X and Y Fields, GCS _ WGS _ 1984 as Coordinate System, and save the result as SHtaxiPt. Figure 14.2a shows the presence of taxis on the road network of Shanghai with ArcGIS built-in base map.

Search and activate the tool Points to Line. In the dialog, input SHtaxiPt as Input Features, SHtaxiL as Output Feature Class, TaxiID for Line Field, and GPSreceive for Sort Field. The resulting feature SHtaxiL indicates the trajectories of individual taxis based on their IDs (Figure 14.2b). The subset of taxi trajectory data reveals remarkable details of traffic intensity in Shanghai by visualization with transparency (Figure 14.2c).

Users can add a new field Length on SHtaxiL and apply Calculate Geometry on the field to check the statistical property of those trajectories, wherein the projected boundary feature Shanghai can be used for Coordinate System. The taxis have a mean drive length of 382 kilometers, a minimum of 48 kilometers, and a maximum of 702 kilometers.

Step 2. Building Tags for Trip and Cruise Based on the Field Vacant and GPS Time Sorting: The previous step maps the entire movements of taxis, and this step segments them into trips with origins and destinations, which captures the spatial connection between places. The type of trajectory segments, trip vs. cruise, is differentiated by the field Vacant in the feature SHtaxiPt.

Search and activate the tool Sort. In the dialog, choose SHtaxiPt as Input Dataset, use TaxiID and GPSreceive as Sort Field, Acending as Sort Method, and save the result as SHtaxiSort.

On the table SHtaxiSort, add a new field TOC (Datatype: Text), use Calculate Field to update it as TOC = accumulate(!TaxiID!,!Vacant!) with the following scripts in the Code Block:

```
total = 0

def accumulate(ID,Vacant):

  global total

  if total:

    total += Vacant

  else:

    total = Vacant

  TOC=str(ID)+"_"+str(total)

  return TOC
```

The records sharing the same value in the updated field of TOC are the component points of an entire taxi trip, such as all the records with a TOC label of 10461 _ 0. The records with ascending TOC labels (e.g., 10490 _ 5113, 10490 _ 5114, . . .) are the points of an entire taxi cruise.

Step 3. Constructing the Trips from Points by the TOC Label: This step extracts the trips records that share the same TOC values. In contrast, the cruise records use ascending and unique TOC labels. Apply Summarize on the field TOC of SHtaxiSort, choose TaxiID for Fields and Count for Statistics Type under Statistics Fields, then save the output as SHtaxiToc. Apply Select by Attribute on the new feature SHtaxiToc with the expression FREQUENCY is greater than 1 to extract the points of trips, and export the records as a new table SHtaxiTrip. All the values in the field TOC of SHtaxiTrip are the labels for taxi trip.

Join the table SHtaxiTrip to the point feature SHtaxiSort with the common field TOC, and make sure to uncheck Keep All Target Features to filter out non-trip records. Export the joined feature SHtaxiSort to a new feature SHtaxiTP (*n* = 683,695) without the joined fields coming from SHtaxiTrip to avoid duplicated fields. The output feature SHtaxiTP contains all the points of trips with their unique trip IDs in the field TOC.

Rerun the tool Points to Line by choosing SHtaxiTP as Input Features, SHtaxiTPLine as Output Feature Class, TOC for Line Field, and GPSreceive for Sort Field. The resulting feature SHtaxiTPLine shows the trips (*n* = 4,122).

An additional step is needed to filter out those very short trips (e.g., < 0.2 kilometers) caused by data errors. Add a new field Length on the table SHtaxiTPLine and apply Calculate Geometry to update it as Length (geodesic) in kilometer. Select the records with the expression Length is larger than 0.2, and export it as a new table SHtaxiTPL (*n* = 3,645). As shown in Figure 14.2d, most of the trips are in the inner city, and a large portion of the trajectories are in cruise mode.

As discussed in Section 12.4 of Chapter 12, researchers are often interested in examining spatial interaction between places, exemplified in traffic volumes. The following two steps aggregate taxi trips between designed areas of uniform size, such as hexagon of 1 square kilometer, and visualize the OD traffic volumes across the study area.

Step 4. Aggregating Numbers of Trips by Origins and Destinations: Search and activate the tool Calculate Geometry Attribute. In the dialog, choose SHtaxiTPL as Input Feature; under Geometry Attributes, input ox, oy, dx, and dy for Field (Existing or New), and for Property, choose Line start (and end) x(and y)-coordinates, respectively. Choose Same as Input for Coordinate Format and GCS_WGS_1984 for Coordinate System, and click Run. This adds the start and end points of the line feature of taxi trips.

Search and activate the tool Generate Tessellation. In the dialog, input Hexagon as Output Feature Class; for Extent, choose As Specified Below and the feature SHtaxiTPL; for Size, input 1 Square Kilometers and GCS-WGS_1984 for Spatial Reference; and click Run. This generates a hexagon feature Hexagon within the extent of SHtaxiTPL.

Use the tool Calculate Geometry Attribute again. In the dialog, set Hexagon as Input Features, hx and hy for Centroid x-coordinates and Centroid y-coordinates,

Same as Input for Coordinate Format, and GCS_WGS_1984 for Coordinate System. The fields hx and hy are the centroid coordinates of hexagons.

Use the tool XY Table to Point. In the dialog, set SHtaxiTPL as Input feature, generate origin and destination points from coordinates ox and oy, dx and dy, respectively, and save them as two new point features SHtaxiOr and SHtaxiDe. Apply Add Spatial Join on SHtaxiOr and SHtaxiDe to attach the IDs of Hexagon that contains the origin and destination points, and only keep the three fields of Hexagon such as GRID _ ID, hx, and hy.

Apply Summarize on the updated tables of SHtaxiOr and SHtaxiDe by setting TOC for Field, Count for Statistic Type, and GRID _ ID for Case Field, and save the Output Tables as SHtaxiHexO and SHtaxiHexD, respectively. The field FREQUENCY indicates the total number of origins or destinations in the hexagons with corresponding GRID _ ID. Join SHtaxiHexO and SHtaxiHexD to the polygon layer Hexagon based on the common key GRID _ ID to visualize the field FREQUENCY.[3] Figure 14.2e visualizes the total counts for the hexagons as destinations, with higher concentrations toward the central area.

Step 5. Calibrating and Visualizing Interaction Volumes between Hexagons: Export the attribute tables of SHtaxiOr and SHtaxiDe to two new tables, SHtaxiOt and SHtaxiDt, while keeping only fields TOC, hx, hy, and GRID _ ID, and renaming GRID _ ID as GRID _ OID (or GRID _ DID), hx as hxo (or hxd), hy as hyo (or hyd). As SHtaxiOt and SHtaxiDt share the same field TOC, join SHtaxiDt to SHtaxiOt based on the common field TOC. In the updated table SHtaxiOt, the origin and destination points are now associated with GRID _ OID and GRID _ DID identifying the hexagons, within which they reside.

To exclude trips within a same hexagon, apply Select by Attribute on the updated table SHtaxiOt with the expression GRID _ OID is not equal to GRID _ DID, and save the selected records as a new table SHtaxiHexOD while keeping key fields such as TOC, GRID _ OID, hxo, hyo, GRID _ DID, hxd, and hyd. Apply Summarize on SHtaxiHexOD by using GRID _ OID and GRID _ DID as Case Field to calculate the trips between different hexagons, setting TOC with Count for Statistic Type, hxo, hyo, hxd, and hyd with First for Statistic Type, and saving the Output table as SHtaxiHexSum (its field FREQUENCY or COUNT _ TOC represents total trips between GRID _ OID and GRID _ DID).

To visualize the OD flows, activate the tool XY to Line. In the dialog, set SHtaxiHexSum as Input Table; ShtaxiHexTrip for Output Feature Class; hxo, hyo, hxd, and hyd for Start (End) and X (Y) Field; Geodesic as Line Type; GCS_ WGS_1984 as Spatial Reference; and check Preserve Attributes. Figure 14.2f shows the taxi OD volumes in central Shanghai.

The workflow from steps 1 to 5 is illustrated in Figure 14.3.

14.3 PROCESSING TAXI TRAJECTORY BIG DATA WITH XSTAR

Taxi trajectory big data for urban studies often come in a large volume and capture the spatiotemporal variability at a fine scale. Some analysts may employ popular database management systems (DBMS), for example, MySQL, to handle the big data. However, these DBMSs are rarely equipped with indexing technologies to

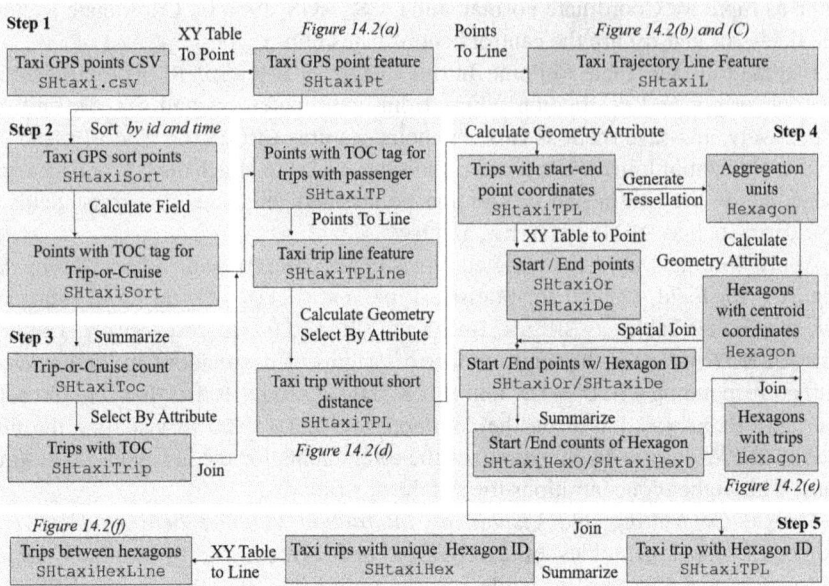

FIGURE 14.3 Workflow for Case Study 14A.

support queries, as proposed in Section 14.1.3. Users may have to code programs based on indexing methods embedded in those DBMS, which can be time-consuming, with no guarantee in performance. Another alternative is to use popular distributed or parallel big data computational frameworks, such as Hadoop (Dittrich et al., 2012), Spark (Zaharia et al., 2016), or Storm (Toshniwal et al., 2014), to process data, which also require advanced programming skills and costly computational resources.

Behind these analytical platforms, data index structures are key to the acceleration of accessing large volumes of data. Most existing works on spatiotemporal data indexing are variants of R-tree (Guttman, 1984). Initially, R-tree was only used to index two-dimensional spatial data. Extensions based on R-tree have been proposed to also account for the temporal dimension. See Li et al. (2021) for a brief review. When applied to taxi trajectory big data distributed within a fixed spatial and temporal range, the tree-based index structures become inefficient. It is expensive to maintain index structure, and the overlapping area among entries increases quickly. After evaluating all these challenges, a new index structure is designed in XSTAR, built upon some previous works (Li and Lin, 2006; Xu et al., 2018).

14.3.1 Taxi Trajectory Data Index Structure in XSTAR

XSTAR adopts a four-level data index structure as shown in Figure 14.4. The four levels are identification level, trajectory level, trip/cruise level, and location level,

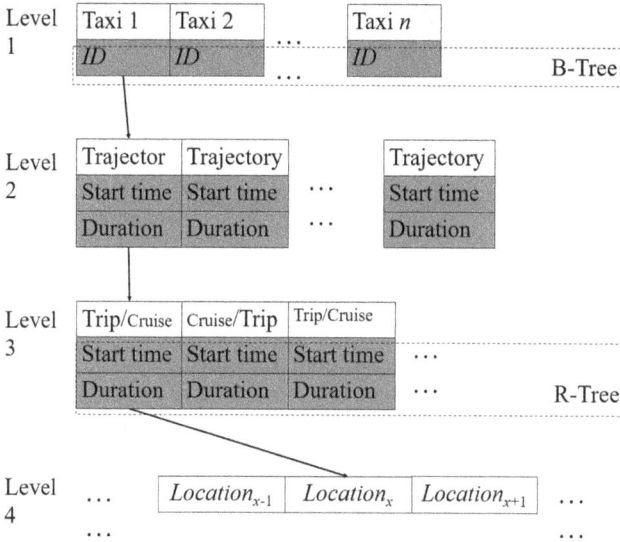

FIGURE 14.4 A conceptual multi-level model of data storage and index structures in XSTAR.

respectively. The top level consists of identifications of taxis. A pointer attached to each identification is linked to a set of trajectories at the second level. These trajectories belong to the same taxi, and each of them includes a group of trips and cruises in the third level. Each trip or cruise consists of a set of locations at the fourth level.

Each record in the first three levels has a pointer to records at the next level. Besides the pointer, the identification level has a property of unique ID (i.e., identifier) to distinguish individual taxis; the trajectory and trip/cruise levels have properties of start time and duration to record associated temporal information. The location level is the simplest. At most cases, it only includes a list of coordinates. However, they are not direct copies of raw trajectory data but are resampled or interpolated from raw data based on a fixed temporal granularity given by users. Since locational records of each trip or cruise are physically saved in an ordered list sequentially, it is possible to quickly point to the position of a locational record in the list at any temporal instant without fetching each record from the beginning.

Besides the basic structure, a B-tree is created to index records in the identification level. An R-tree is used to index two-tuple temporal properties based on Li and Lin (2006) to facilitate evaluating queries with temporal conditions.

The number of locational records in the software might be smaller than the number of raw sampling points if the given temporal granularity is larger than the original average sampling interval, and vice versa. Even when the amount is large, with the preceding structure, it is still fast to locate a taxi at any time. Furthermore, since the properties of taxi identifiers and sampling time are not physically saved in locational records, we employ other measures to reduce the length of bytes for saving coordinates in the software (e.g., replacing 64-bit double numbers with 32-bit decimal numbers), and thus, storage space is saved to a great extent. One may note a lack of spatial indexing in this structure as each locational record can only be accessed from taxi to its trajectory and trip or cruise. However, each query on a taxi trajectory must be accompanied with temporal conditions to narrow the search. It is efficient to enumerate each taxi to meet spatial query conditions.

The four-level structure is implemented as follows:

1. Find taxi identifier from raw data.
2. Sort all raw records by taxi identifier (primary) and sampling time (secondary).
3. Split each taxi record into one or more trajectories (if the difference of sampling time between continuous records is less than a given threshold, the two records belong to one trajectory; otherwise, two trajectories).
4. Regroup each trajectory to trip or cruise by its loading status.
5. Generate resampled locational records for each trip or cruise with an identical interval according to a given temporal granularity.
6. Build auxiliary B-tree and R-tree index structures (Figure 14.4).

The preceding design saves computational time and storage space significantly.[4] XSTAR also supports parallel computations on a computer with multiple-core processors for better performance.

14.3.2 Data Processing Steps in XSTAR

XSTAR implements data processing in four steps. One advantage of the stepwise approach is that users can save intermediate results from each step and experiment with different parameters without the need to complete the whole process, which may be time-consuming.

Step 1 detects the required fields in the raw data (e.g., a text file), such as taxi identifier, time, location, loading status, etc., and saves the result in a raw-data-structure (RDS) file. Step 2 reads the RDS file to identify the number of taxis and saves it in an object-identifier (OID) file. Step 3 sorts locational records by taxi identifiers as the primary key and sampling time as the secondary key and saves the result in a sorted-data (in short, SDT) file. Users can split the large file into several smaller SDT files with different periods. Step 4 reads a user-defined temporal granularity to determine the interval between interpolated locations, builds index structures, and saves the result in an index-data (IDT) file.

In these steps, users can define other parameters, such as the number of parallel threads, thresholds of recognizing and removing wrong locational records, type of coordinate conversion, etc. From the case studies in Section 14.4, XSTAR is computationally efficient in processing large tax trajectory data with reasonable time. The SDT file size is only one-tenth of the raw data file size, and the IDT file size varies with the given temporal granularity denoted with tg.

14.3.3 ANALYZING TAXI TRAJECTORY BIG DATA WITH XSTAR

This subsection overviews the main functions of XSTAR. Readers are referred to the user's manual for details. The program groups all analytical functions into three modules.

The objective of module A is to visualize trajectories after grouping by taxi identifiers and sorting by time. As shown in Figure 14.5a, when users open an SDF file, and the software loads a list of taxi identifiers. Users can select one or several identifiers and customize the time interval. The display screen updates the selected trajectories represented by different colors. Sampling points and trajectories are labelled, and the labels can be customized. Results can be saved as shapefiles or images.

Functions based on instantaneous analysis are grouped in module B when the temporal condition in a data query is an instant. IDF files are input of this module. After opening one IDF file, users can visualize the location of each taxi at the currently selected time instant, as shown in Figure 14.5b. Each taxi is labelled with its identifier, and distinctive symbols are used to differentiate its loading status. When set to the animation mode, the current instant changes automatically, and the map panel shows the movement of taxis in a movie fashion. Similar to module A, results can be exported as shapefiles or images.

Module C implements interval analysis of trajectories when the temporal condition is a given time period. This module possesses the most functions compared to other modules. It also reads an IDF file. Figure 14.5c shows the trip origins (in red dots) meeting defined temporal conditions. Similarly, destinations of trips or cruises can be extracted and displayed. By sliding the two buttons in the top panel, users can easily define the time interval range.

In addition to temporal conditions, modules B and C support queries based on the area of interests (AOI). For instance, when one creates an origin AOI, all trips or cruises departing from this AOI are selected (Figure 14.6a). If the user changes it to a destination AOI, the origins of all trips or cruises that end in this AOI are displayed (Figure 14.6b). One can also define multiple origin or destination AOIs. For example, Figure 14.6c shows all trips that begin with one origin AOI, end at one destination AOI, but take different network paths.

To summarize the distribution of discrete records, a user can define an aggregated analysis zone by a polygon shapefile or let XSTAR generate analysis zones of a regular shape. For example, in Figure 14.6d, each analysis zone is a hexagon, and the number in each hexagon is an aggregated analytical result (e.g., total number of origins). Analysis zones are also needed in OD analysis. As shown in Figure 14.6e, a line connects any OD pairs between two analysis zones, and each is labeled by the number of trips. In profile analysis, as shown in Figure 14.6f, when lines intersect with road segments to form analysis zones, the system records the total numbers and average speeds of the trips or cruises passing through these lines.

XSTAR supports combining temporal conditions and spatial conditions for more complicated analysis, such as meeting analysis to find overlapping trajectories in space and time, stay point analysis to extract locations where taxis keep static for a while, etc. XSTAR can also be applied to analysis of trajectories of other moving objects (e.g., human beings, private vehicles, ships, airplanes, etc.).

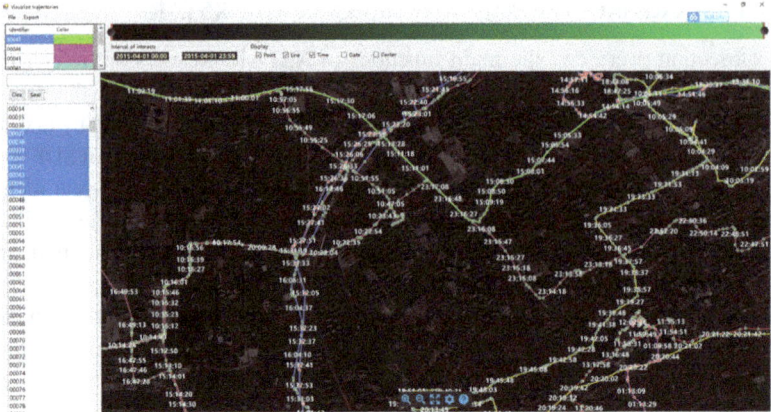

(a)

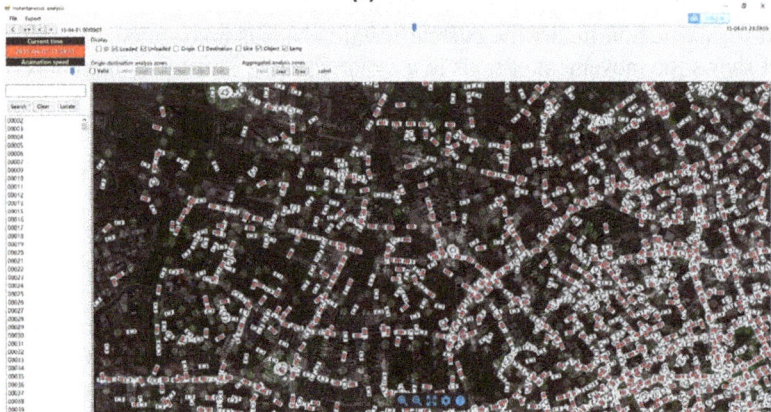

(b)

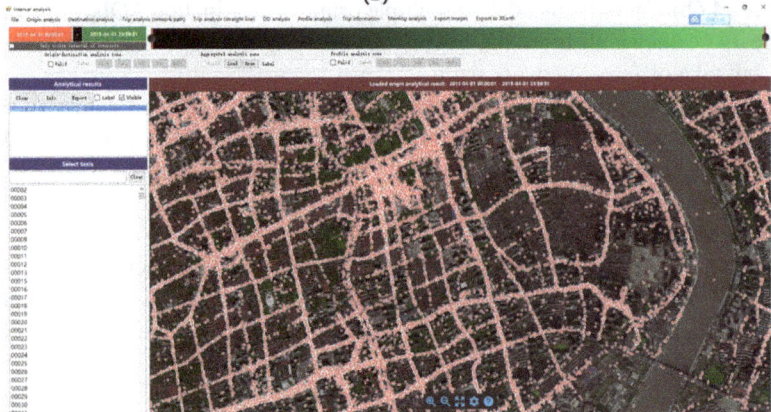

(c)

FIGURE 14.5 XSTAR modules: (a) visualizing selected trajectories, (b) distribution of taxis at a given time instant, and (c) origin distribution of trips within a given time interval.

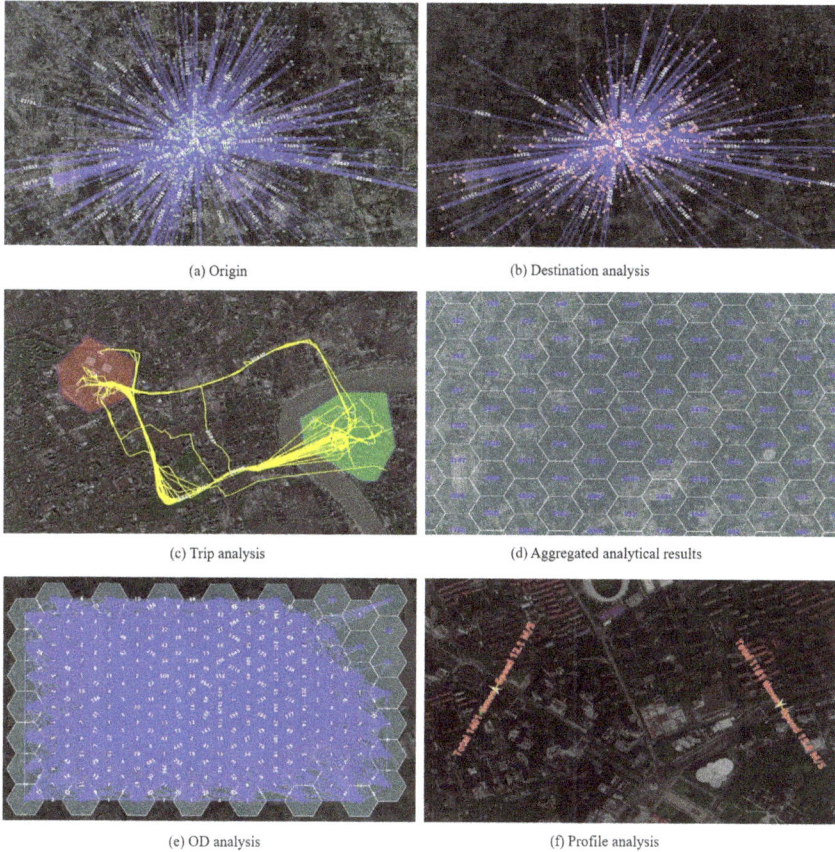

(a) Origin   (b) Destination analysis

(c) Trip analysis   (d) Aggregated analytical results

(e) OD analysis   (f) Profile analysis

FIGURE 14.6 Data analysis examples with XSTAR.

14.4   CASE STUDY 14B: IDENTIFYING HIGH TRAFFIC CORRIDORS AND DESTINATIONS IN SHANGHAI

In urban planning and management, it is important to examine connections between places. For example, identifying major corridors with high traffic volumes helps facilitate land use plans and adjust public transit routes. This case study uses XSTAR to reveal the intensity of connectivity between locations and thus identifies the most connected places in downtown Shanghai during morning rush hours. A 10 GB taxi trajectory data file consists of more than 100 million locational records from more than 13,000 taxis running 24 hours. While the constitution of traffic includes other modes, taxi accounts for a large proportion of traffic in Shanghai, and their trajectory data are the most accessible.

The origin–destination (OD) analysis usually uses area units such as traffic analysis zones (TAZ), census units, or any areas of interest (AOI). XSTAR can use a predefined analysis area unit, one drawn by users, or create regular analytical zones. This study uses the third approach, as in Case Study 14A in Section 14.2.

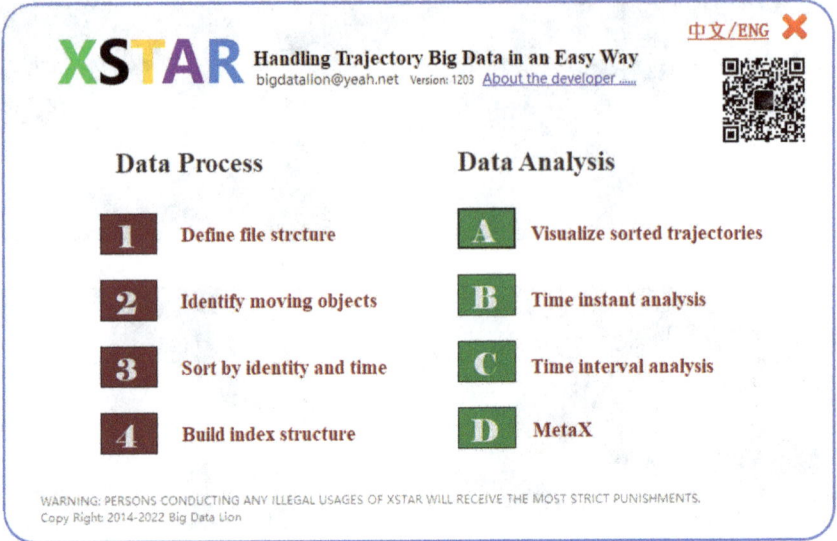

FIGURE 14.7 The main portal of XSTAR.

Figure 14.7 shows the main portal of XSATR. The first part uses the data process module in XSTAR to process the raw data (Subsection 14.4.1), and the second part uses the data analysis to implement the OD analysis (Subsection 14.4.2) and destination analysis (Subsection 14.4.3).

14.4.1 TAXI TRAJECTORY RAW DATA PROCESSING IN XSTAR

Step 1. Defining File Structure: In the XSTAR main portal, click Define File Structure to activate the dialog, as shown in Figure 14.8a. For Data file, click Browse to select the CSV file raw. The input file could be other structured text files. Part of the sample data is displayed, and each line is a record. The file shares the same column names as SHtaxi.csv used in Case Study 14A. The group box "Parameters" includes a checkbox to determine whether the first line of the file is field names, the separator between values, and the prefix and suffix of values. Click Update after defining these parameters.

XSTAR automatically extracts field names (or indexes) and fills in drop boxes in the group box "Fields." Users need to adjust and verify the matches[5] and make sure to identify the unit of coordinates (degree) and the format of time value. Some values are compulsory, such as object ID, X, Y, and Time, while others are optional. In the group box "Trip," XSTAR provides several options to define a trip. When selecting Loading Status, indicate the value standing for loaded (in this case, the field value is 0).

Click Verify to confirm all settings. If there are no errors, click Save to save the settings in an RDS file sh.rds. Users can retrieve the file and modify any settings as needed.

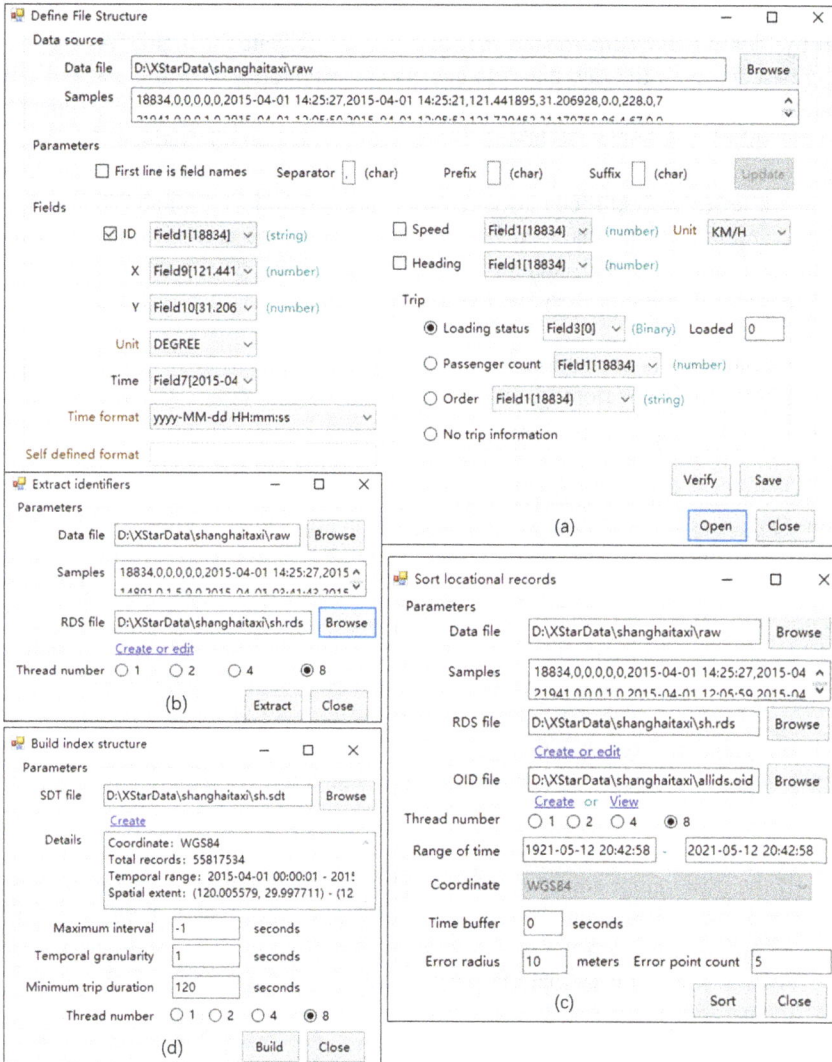

FIGURE 14.8 Interface for preprocessing data in XSTAR: (a) define file structure, (b) identify moving objects, (c) sort by identity and time, and (d) build index structure.

Step 2. Identifying Moving Objects: In the XSTAR main portal, click Identify Moving Objects to activate the dialog, as shown in Figure 14.8b. Open the raw data file again and its corresponding RDS file. Then, select the number of parallel threads to apply parallel computing. If the computer is equipped with a multi-core CPU, a larger number of threads accelerate the process. Click Extract to save the result into an OID file, which records a list of unique values representing each moving object (taxi in this case).

Step 3. Sorting by Identity and Time: Back to the XSTAR main portal, click Sort by Identity and Time to start the function and activate the dialog. As shown in Figure 14.8c, open the raw data file and its corresponding RDS file and OID file, then choose the number of threads for parallel computing. A complete trajectory may be split by defining the boundary of a temporal range to reduce the file size and processing time. If it is desirable to keep the whole trajectory, users can define a time buffer (e.g., 300 seconds) and the actual range of time to add the buffer at its head and tail. Recall that step 1 selects `Degree` as the unit of locations in geographical coordinates. This step defines the datum of the geographical coordinates in order to align the trajectory data with other spatial data, thus selects `WGS84` for Coordinate.

Locational errors are common in trajectory data. When taxis or other moving objects stop, their positioning function remains on and collects some drift locational records. These records usually carry an instant speed of zero, while their locations vary within an area. XSTAR uses an algorithm to remove these points. It needs two parameters, namely, error radius (e.g., $r = 10$ meters) and error point count (e.g., $c = 5$). That is to say, if there are small values within a circle with a radius of 10 meters or five temporally continuous points with a zero speed, the algorithm treats them as drift points, removes them, and keeps their centroid as the only record.

Click `Sort` and save the resulting SDT file locally. The SDT file is a restructured trajectory data file sorted by identity of moving objects and sampling time.

Step 4. Building Index Structure: Go back to the XSTAR main portal again and click Build Index Structure, that is, the last step of data processing. As shown in Figure 14.8d, click Browse to open an SDT file, and define three temporal parameters as follows.

1. If the positioning device in a taxi experiences a glitch for a long time, the time span between two continuous locational records may be large without recording a trajectory. *Maximum interval* is used to define the maximum span. If one believes that the positioning device does not malfunction, set the parameter as –1 to ignore it.
2. *Temporal granularity* defines the minimum temporal unit for analysis (e.g., 1 second).
3. Users can apply the parameter *Minimum Trip Duration* (e.g., 120 seconds) to eliminate very short trips (<120 seconds), which are most likely caused by errors.

Finally, choose the number of parallel threads and click Build. The result is an IDT file (e.g., `sh.idt`) saved locally. It will be used in the following OD analysis.

Table 14.3 summarizes the computational time and file size associated with each step for processing raw taxi trajectory data consisting of 100 million locational records and about 13,000 moving objects.

TABLE 14.3

Computational Time and File Size for Steps 1–4

	Output	Computational Time*	File Size
Step 1: define file structure	RDS file	Up to users	<1 KB
Step 2: extract identifiers	OID file	312 seconds	121 KB
Step 3: sort records	SDT file	507 seconds	1,034 MB
Step 4: build index structure	IDT file (tg = 10 seconds)	253 seconds	1,245 MB
	IDT file (tg = 20 seconds)	122 seconds	640 MB
	IDT file (tg = 30 seconds)	98 seconds	432 MB

* On a laptop as described in the Preface.

14.4.2   OD ANALYSIS IN XSTAR

In the main portal of XSTAR, under Data Analysis, click module C Time Interval Analysis. Explore the background map and make sure the study area (Downtown Shanghai) is completely visible in the map window, as shown in Figure 14.9a. Click File > Open IDT file to open `sh.idt` for implementing the following steps.

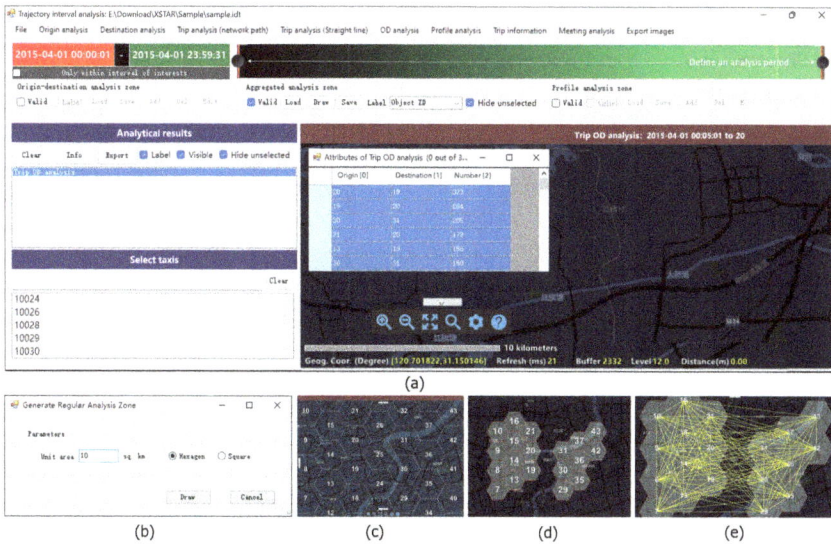

FIGURE 14.9 OD analysis in XSTAR: (a) interface, (b) define parameters for Generate Regular Analysis Zone, (c) regularly shaped analysis zones with labels, (d) selected analysis zones, (e) OD analysis results.

Step 5. Defining Analysis Zones: OD analysis begins with defining analysis zones for trip origins and destinations. Users can click the button Draw under the toolbar Aggregated Analysis Zone to define analysis zones. If you already have a polygon layer for analysis zone, click Load to load the layer. Here, we create analysis zones of regular shape, like step 4 in Section 14.2. Click Draw to open the tool Generate Regular Analysis Zones (Figure 14.9b). Define Unit Area as 10 square kilometers, check Hexagon, and click Draw to generate a polygon layer covering the study area like a fishnet. Users can label each hexagon by choosing `Object ID` as label field (Figure 14.9c).

If users want to examine the interaction between any two analysis zones in the study areas, select a polygon in the map by left-clicking while holding the key Alt, then select another polygon by left-clicking while holding the key combination Ctrl + Alt.[6] If desirable, check "Hide unselected" in the toolbar to make unselected polygons less visible (Figure 14.9d).

Step 6. Defining Analysis Period: Once an IDT file is opened, its time span is displayed in the top-left corner in red and green, that is, from 00:00 to 23:59 on April 1, 2015, as shown in Figure 14.9a. In the top panel "Define an analysis period," slide the two buttons to select the morning rush hour period from 7:00 a.m. to 10:00 a.m.

If the box "Only within interval of interests" is checked, only trips starting and ending within the defined analysis period will be included in the following analysis. Otherwise, all trips with their duration intersecting with the period will be included.

Step 7. Conducting OD Analysis: In the menu strip, click OD analysis > Trip OD analysis.

The result is shown in Figure 14.9e, where the number of trips between two zones is labeled. More details can be revealed in the resulting table. In the box "Analytical results," select "Trip OD analysis" and click Info. The table includes Origin, Destination, and Number (of trips between them). Users can click the header of one column to sort. For example, check Hide Unselected in the box Analytical Results (as shown in Figure 14.9a), and sort by number of trips in a descending order. Users can select the most connected OD pairs in the top rows, and the corresponding lines will be highlighted in the map window.

14.4.3 DESTINATION ANALYSIS IN XSTAR

The final part of the case study demonstrates the functionality of destination analysis in XSTAR. We use a practical problem of locating shared canteens for taxi drivers to showcase some valuable big data analysis techniques. In central Shanghai, taxi drivers look for free parking places to have a quick lunchtime (often in their vehicles) and then move on with their busy day. One solution is to organize some canteens in a large parking lot for serving taxi drivers during lunchtime (11:00–13:00). It can be formulated as a location-allocation problem, that is, where to locate the canteens and how much demand to be met. The solution relies on destination analysis based on the taxi trajectory data. Specifically, a taxi driver ready to have lunch in designated canteens usually makes such a decision in an instant, when the taxi changes its status from loaded to unloaded (e.g., t_{14} and t_{15} in Figure 14.1).

Step 8. Implementing Destination Analysis: In the main portal of XSTAR, under Data Analysis, open the application Time Interval Analysis, and then click File > Open IDT file to open sh.idt. Define an analysis period from 11:00 to 13:00. Click Destination analysis > Trip destination analysis. This loads all taxis with the instantaneous status changed from loaded to unloaded, as shown in Figure 14.10a.

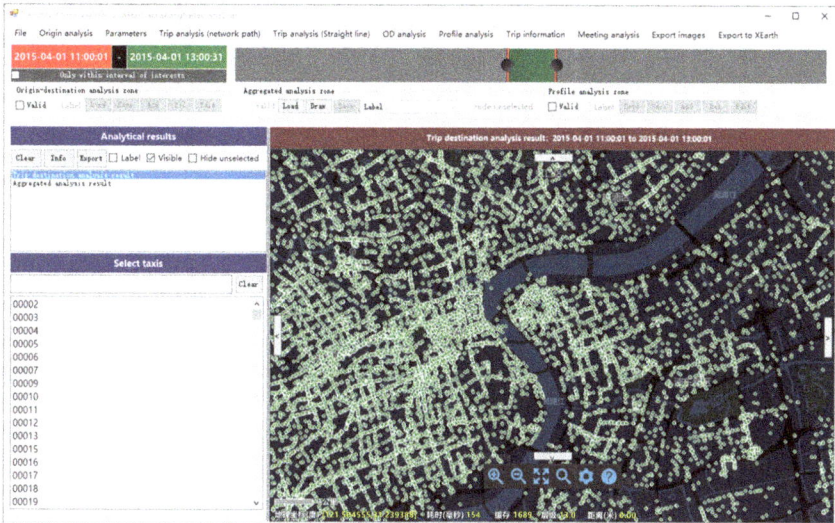

(a)

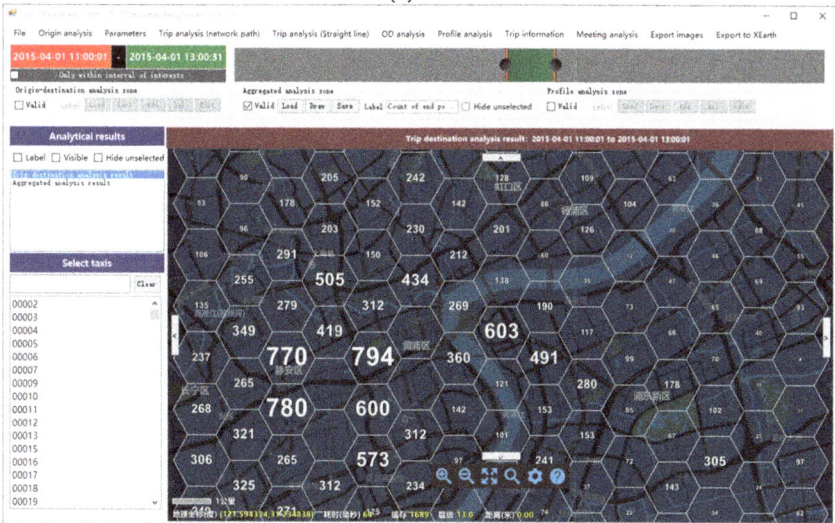

(b)

FIGURE 14.10 Destination analysis in XSTAR: (a) individual taxis represented by green dots, and (b) aggregated taxis in hexagons.

Step 9. Aggregating Trip Destinations to Areas: Similar to step 5, use the tool Generate Regular Analysis Zones to draw hexagons with an area size of 2 square kilometers, as shown in Figure 14.10b. Repeat the "Trip destination analysis" function again. It implements a zone-based destination analysis.

In the box Analytical Results, select the item "trip destination analysis result," then uncheck the box Label and Visible. In the toolbar Aggregated Analysis Zone, choose "Count of end point" as the label option. It visualizes the distribution in an aggregated form. As shown in Figure 14.10b, the label of each zone is the count of taxis covered by this zone as the font size is proportional to its taxi volume.

In the box Analytical Results, select the item "trip destination analysis result" or "aggregated analysis result," and then click Export to save the result as a point shapefile or a polygon shapefile, respectively. The result, as a record of demands, can be used in the decision-making of location-allocation of shared canteens.

14.5 SUMMARY

From urban structure in terms of population density patterns in Chapter 6 to connection between land uses via commuting in Chapter 11, urban studies strive for spatial analysis with high resolution and precision. Scale effect or zonal effect leads to modifiable areal unit problem (MAUP), which shakes our confidence in findings often derived from area-based data. Chapter 12 uses Monte Carlo simulation to generate random points or individual trips that are consistent with observed area-based spatial patterns. Chapter 13 simulates agents of various behaviors and models their spatiotemporal trajectories in a realistic urban environment. Simulated individuals enable an analyst to aggregate data back to area units of any size or shape, identify spatial patterns, and test impacts of policy scenarios. However, simulated individuals or agents are not real-world residents, workers, or commuters.

Big data of real individuals in real time provide us an opportunity to examine the spatiotemporal patterns of human mobility in a fine resolution. This chapter uses taxi trajectory data as an example to showcase the potential of big data. When the file size is moderate, ArcGIS can be used to process data, examine activity patterns, and derive OD trips between areas as illustrated in case study 14A. When the dataset is large, data processing and analysis require advanced programming or special software, which becomes a major technical barrier. XSTAR is a free package for analysis of large volumes of taxi trajectory data. Its core design is a scalable index and storage structure and facilitates efficient data retrieval and processing. It fulfills, as demonstrated in case study 14B, challenging spatial analysis tasks with a user-friendly interface on a desktop computer.

NOTES

1 This chapter was drafted by Xiang Li, School of Geographic Sciences, East China Normal University, Shanghai 200241, China.

2 The conversion from CSV to feature helps accelerate data processing.

3 If the process of Attribute Join generates NULL values, right-click the table > Properties > Indexes, and delete the Attribute Index for GRID_ID. Another option is to export

the attribute tables as CSV files or refer to https://support.esri.com/en/technical-article/000008715.

4 On a laptop as described in the Preface, the computational time was 15 minutes for building index structure for 100 million records (about 10GB), and the size of the resultant data file was about 1 GB.

5 Check the metadata or data dictionary provided by the data source for such information.

6 Alt + left click selects a single feature and deselects all other feature simultaneously. Ctrl + Alt + left click adds selected features to existing selection.

References

Abu-Lughod, J. 1969. Testing the theory of social area analysis: The ecology of Cairo, Egypt. *American Sociological Review* 34, 198–212.

Agency for Healthcare Research and Quality (AHRQ). 2011. *Healthcare Cost and Utilization Project (HCUP) State Inpatient Databases (SID)—Florida*. Rockville, MD. Available www.hcup-us.ahrq.gov/sidoverview.jsp.

Agnew, R. 1985. A revised strain theory of delinquency. *Social Forces* 64, 151–167.

Alford-Teaster, J., F. Wang, A. N. A. Tosteson, and T. Onega. 2021. Incorporating broadband durability in measuring geographic access to healthcare in the era of telehealth: A case example of the two-step virtual catchment area (2SVCA). *Journal of the American Medical Informatics Association* 28, 2526–2530.

Alonso, W. 1964. *Location and Land Use*. Cambridge, MA: Harvard University.

Anderson, J. E. 1985. The changing structure of a city: Temporal changes in cubic spline urban density patterns. *Journal of Regional Science* 25, 413–425.

Anselin, L. 1995. Local indicators of spatial association—LISA. *Geographical Analysis* 27, 93–115.

Anselin, L. 2018. *Maps for Rates or Proportions*. Available https://geodacenter.github.io/workbook/3b_rates/lab3b.html.

Anselin, L. 2020a. *Spatial Clustering (1): Spatializing Classic Clustering Methods*. Available https://geodacenter.github.io/workbook/9a_spatial1/lab9a.html.

Anselin, L. 2020b. *Spatial Clustering (2): Spatially Constrained Clustering—Hierarchical Methods*. Available https://geodacenter.github.io/workbook/9c_spatial3/lab9c.html#spatially-constrained-hierarchical-clustering-schc.

Anselin, L. 2020c. *Spatial Clustering (3): Spatially Constrained Clustering—Partitioning Methods*. Available https://geodacenter.github.io/workbook/9d_spatial4/lab9d.html#introduction.

Anselin, L., and A. Bera. 1998. Spatial dependence in linear regression models with an introduction to spatial econometrics. In A. Ullah and D. E. Giles (eds.), *Handbook of Applied Economic Statistics*. New York: Marcel Dekker, 237–289.

Antrim, A., and S. J. Barbeau. 2013. The many uses of GTFS data–opening the door to transit and multimodal applications. *Location-Aware Information Systems Laboratory at the University of South Florida*, 4.

Assunção, R. M., M. C. Neves, G. Câmara, and C. D. C. Freitas. 2006. Efficient regionalization techniques for socio-economic geographical units using minimum spanning trees. *International Journal of Geographical Information Science* 20, 797–811.

Bailey, T. C., and A. C. Gatrell. 1995. *Interactive Spatial Data Analysis*. Harlow, England: Longman Scientific & Technical.

Baller, R., L. Anselin, S. Messner, G. Deane, and D. Hawkins. 2001. Structural covariates of U.S. County homicide rates: Incorporating spatial effects. *Criminology* 39, 561–590.

Barkley, D. L., M. S. Henry, and S. Bao. 1996. Identifying "spread" versus "backwash" effects in regional economic areas: A density functions approach. *Land Economics* 72, 336–357.

Bartholdi, J. J. III, and L. K. Platzman. 1988. Heuristics based on spacefilling curves for combinatorial problems in Euclidean space. *Management Science* 34, 291–305.

Batty, M. 1983. Linear urban models. *Papers of the Regional Science Association* 52, 141–158.

Bauer, J., and D. A. Groneberg. 2016. Measuring spatial accessibility of health care providers—introduction of a variable distance decay function within the Floating Catchment Area (FCA) method. *PLoS One* 11, e0159148.

Becker, G. S. 1968. Crime and punishment: An economic approach. *Journal of Political Economy* 76, 169–217.

Beckmann, M. J. 1971. On Thünen revisited: A neoclassical land use model. *Swedish Journal of Economics* 74, 1–7.

Bellair, P. E., and V. J. Roscigno. 2000. Local labor-market opportunity and adolescent delinquency. *Social Forces* 78, 1509–1538.

Ben-Ari, M., and F. Mondada. 2018. *Elements of Robotics*. New York: Springer.

Benenson, I., and P. M. Torrens. 2004 *Geosimulation: Automata-Based Modeling of Urban Phenomena*. West Sussex, UK: John Wiley.

Berry, B. J. L. 1967. *The Geography of Market Centers and Retail Distribution*. Englewood Cliffs, NJ: Prentice-Hall.

Berry, B. J. L. 1972. *City Classification Handbook, Methods, and Applications*. New York: Wiley-Interscience.

Berry, B. J. L., and H. Kim. 1993. Challenges to the monocentric model. *Geographical Analysis* 25, 1–4.

Berry, B. J. L., and R. Lamb. 1974. The delineation of urban spheres of influence: Evaluation of an interaction model. *Regional Studies* 8, 185–190.

Berry, B. J. L., and P. H. Rees. 1969. The factorial ecology of Calcutta. *American Journal of Sociology* 74, 445–491.

Besag, J., and P. J. Diggle. 1977. Simple Monte Carlo tests for spatial pattern. *Applied Statistics* 26, 327–333.

Besag, J., and J. Newell. 1991. The detection of clusters in rare diseases. *Journal of the Royal Statistical Society Series A* 15, 4143–4155.

Beyer, K. M., and G. Rushton. 2009. Mapping cancer for community engagement. *Preventing Chronic Disease* 6, A03.

Bian, L. 2004. A conceptual framework for an individual-based spatially explicit epidemiological model. *Environment and Planning B: Planning and Design* 31, 381–395.

Black, R. J., L. Sharp, and J. D. Urquhart. 1996. Analysing the spatial distribution of disease using a method of constructing geographical areas of approximately equal population size. In P. E. Alexander and P. Boyle (eds.), *Methods for Investigating Localized Clustering of Disease*. Lyon, France: International Agency for Research on Cancer, 28–39.

Black, W. 2003. *Transportation: A Geographical Analysis*. New York: Guilford.

Block, C. R., R. L. Block, and the Illinois Criminal Justice Information Authority (ICJIA). 1998. *Homicides in Chicago, 1965–1995 [Computer File]*. 4th ICPSR version. Chicago, IL: ICJIA [producer]. Ann Arbor, MI: Inter-university Consortium for Political and Social Research [distributor].

Block, R., and C. R. Block. 1995. Space, place and crime: Hot spot areas and hot places of liquor-related crime. In John E. Eck and David Weisburd (eds.), *Crime Places in Crime Theory*. Newark: Criminal Justice Press.

Blondel, V. D., J. L. Guillaume, R. Lambiotte, and E. Lefebvre. 2008. Fast unfolding of communities in large networks. *Journal of Statistical Mechanics: Theory and Experiment* 10, 1–12.

Bonabeau, E. 2002. Agent-based modeling: Methods and techniques for simulating human systems. *Proceedings of the National Academy of Sciences* 99003, 7280–7287.

Bonger, W. 1916. *Criminality and Economic Conditions*. Boston: Little, Brown and Co.

Brown, B. B. 1980. Perspectives on social stress. In H. Selye (ed.), *Selye's Guide to Stress Research*, vol. 1. Reinhold, New York: Van Nostrand, 21–45.

Brunsdon, C., J. Corcoran, and G. Higgs. 2007. Visualising space and time in crime patterns: A comparison of methods. *Computers, Environment and Urban Systems* 31, 52–75.

Bureau of the Census. 1993. *Statistical Abstract of the United States* (113rd ed.). Washington, DC: US Department of Commerce.

Burgess, E. 1925. The growth of the city. In R. Park, E. Burgess and R. Mackenzie (eds.), *The City*. Chicago: University of Chicago Press, 47–62.

Cadwallader, M. 1975. A behavioral model of consumer spatial decision making. *Economic Geography* 51, 339–349.

Cadwallader, M. 1981. Towards a cognitive gravity model: The case of consumer spatial behavior. *Regional Studies* 15, 275–284.

Cadwallader, M. 1996. *Urban Geography: An Analytical Approach.* Upper Saddle River, NJ: Prentice Hall.

Casetti, E. 1993. Spatial analysis: Perspectives and prospects. *Urban Geography* 14, 526–537.

Cervero, R. 1989. Jobs-housing balance and regional mobility. *Journal of the American Planning Association* 55, 136–150.

Chang, K-T. 2004. *Introduction to Geographic Information Systems* (2nd ed.). New York: McGraw-Hill.

Chavent, M., V. Kuentz-Simonet, A. Labenne, and J. Saracco. 2018. ClustGeo: An R package for hierarchical clustering with spatial constraints. *Computational Statistics* 33, 1799–1822.

Chiricos, T. G. 1987. Rates of crime and unemployment: An analysis of aggregate research evidence. *Social Problems* 34, 187–211.

Chu, R., C. Rivera, and C. Loftin. 2000. Herding and homicide: An examination of the Nisbett-Reaves hypothesis. *Social Forces* 78, 971–987.

Church, R. L., and C. S. ReVelle. 1974. The maximum covering location problem. *Papers of the Regional Science Association* 32, 101–118.

Clark, C. 1951. Urban population densities. *Journal of the Royal Statistical Society* 114, 490–494.

Clayton, D., and J. Kaldor. 1987. Empirical bayes estimates of age-standardized relative risks for use in disease mapping. *Biometrics* 43, 671–681.

Cliff, A., P. Haggett, J. Ord, K. Bassett, and R. Davis. 1975. *Elements of Spatial Structure.* Cambridge, UK: Cambridge University.

Cliff, A. D., and J. K. Ord. 1973. *Spatial Autocorrelation.* London: Pion.

Clifford, P., S. Richardson, and D. Hémon. 1989. Assessing the significance of the correlation between two spatial processes. *Biometrics* 45, 123–134.

Cockings, S., and D. Martin. 2005. Zone design for environment and health studies using pre-aggregated data. *Social Science & Medicine* 60, 2729–2742.

Colwell, P. F. 1982. Central place theory and the simple economic foundations of the gravity model. *Journal of Regional Science* 22, 541–546.

Cornish, D. B., and R. V. Clarke (eds.). 1986. *The Reasoning Criminal: Rational Choice Perspectives on Offending.* New York: Springer-Verlag.

Cressie, N. A. C. 1991. *Statistics for Spatial Data.* New York: Wiley.

Cressie, N. A. C. 1992. Smoothing regional maps using empirical bayes predictors. *Geographical Analysis* 24, 75–95.

Cromley, E., and S. McLafferty. 2002. *GIS and Public Health.* New York: Guilford Press.

Cromley, R. G., D. M. Hanink, and G. C. Bentley. 2014. Geographically weighted colocation quotients: Specification and application. *The Professional Geographer* 66, 138–148.

Culyer, A. J., and A. Wagstaff. 1993. Equity and equality in health and health care. *Journal of Health Economics* 12, 431–457.

Cuzick, J., and R. Edwards. 1990. Spatial clustering for inhomogeneous populations. *Journal of the Royal Statistical Society Series B* 52, 73–104.

Dai, D. 2010. Black residential segregation, disparities in spatial access to health care facilities, and late-stage breast cancer diagnosis in metropolitan Detroit. *Health and Place* 16, 1038–1052.

Dai, T., C. Liao, and S. Zhao. 2019. Optimizing the spatial assignment of schools through a random mechanism towards equal educational opportunity: A resemblance approach. *Computers, Environment and Urban Systems* 76, 24–30.

Dantzig, G. B. 1948. *Programming in a Linear Structure*. Washington, DC: U.S. Air Force, Comptroller's Office.

Davies, W., and D. Herbert. 1993. *Communities Within Cities: An Urban Geography*. London: Belhaven.

De Vries, J. J., P. Nijkamp, and P. Rietveld. 2009. Exponential or power distance-decay for commuting? An alternative specification. *Environment and Planning A* 41, 461–480.

Delamater, P. L., J. P. Messina, S. C. Grady, V. WinklerPrins, and A. M. Shortridge. 2013. Do more hospital beds lead to higher hospitalization rates? a spatial examination of Roemer's Law. *PLoS One* 8, e54900.

Diggle, P. J., and A. D. Chetwynd. 1991. Second-order analysis of spatial clustering for inhomogeneous populations. *Biometrics* 47, 1155–1163.

Dijkstra, E. W. 1959. A note on two problems in connection with graphs. *Numerische Mathematik* 1, 269–271.

Dijkstra, J., J. Jessurun, and H. Timmermans. 2001. A multi-agent cellular automata model of pedestrian movement. In M. Schreckenberg and S. D. Sharma (eds.), *Pedestrian and Evacuation Dynamics*. Berlin: Springer-Verlag, 173–181.

Dittrich, J., and J. A. Quiané-Ruiz. 2012. Efficient big data processing in Hadoop MapReduce. *Proceedings of the VLDB Endowment* 5(12), 2014–2015.

Dobrin, A., and B. Wiersema. 2000. Measuring line-of-duty homicides by law enforcement officers in the United States, 1976–1996. Paper presented at the 52nd Annual Meeting of the American Society of Criminology, November 15–18, San Francisco, CA.

Dodge, S., R. Weibel, and E. Forootan. 2009. Revealing the physics of movement: Comparing the similarity of movement characteristics of different types of moving objects. *Computers, Environment and Urban Systems* 33(6), 419–434.

Duque, J. C., L. Anselin, and S. J. Rey. 2012. The Max-P-regions problem. *Journal of Regional Science* 52, 397–419.

Everitt, B. S., S. Landau, and M. Leese. 2001. *Cluster Analysis* (4th ed.). London: Arnold.

Fisch, O. 1991. A structural approach to the form of the population density function. *Geographical Analysis* 23, 261–275.

Fisher, R. A. 1935. *The Design of Experiments*. Edinburgh: Oliver and Boyd.

Flowerdew, R., and M. Green. 1992. Development in areal interpolation methods and GIS. *The Annals of Regional Science* 26, 67–78.

Forstall, R. L., and R. P. Greene. 1998. Defining job concentrations: The Los Angeles case. *Urban Geography* 18, 705–739.

Fotheringham, A. S., C. Brunsdon, and M. Charlton. 2000. *Quantitative Geography: Perspectives on Spatial Data Analysis*. London: Sage.

Fotheringham, A. S., C. Brunsdon, and M. Charlton. 2002. *Geographically Weighted Regression: The Analysis of Spatially Varying Relationships*. West Sussex: Wiley.

Fotheringham, A. S., and M. E. O'Kelly. 1989. *Spatial Interaction Models: Formulations and Applications*. London: Kluwer Academic.

Fotheringham, A. S., and D. W. S. Wong. 1991. The modifiable areal unit problem in multivariate statistical analysis. *Environment and Planning A* 23, 1025–1044.

Fotheringham, A. S., and B. Zhan. 1996. A comparison of three exploratory methods for cluster detection in spatial point patterns. *Geographical Analysis* 28, 200–218.

Fox, J. A. 2000. *Uniform Crime Reports [United States]: Supplementary Homicide Reports, 1976–1998 [Computer File]*. ICPSR version. Boston, MA: Northeastern University, College of Criminal Justice [producer]. Ann Arbor, MI: Inter-university Consortium for Political and Social Research [distributor].

Franke, R. 1982. Smooth interpolation of scattered data by local thin plate splines. *Computers and Mathematics with Applications* 8, 273–281.

Frankena, M. W. 1978. A bias in estimating urban population density functions. *Journal of Urban Economics* 5, 35–45.

Freeman, L. 1979. Centrality in social networks: Conceptual clarification. *Social Networks* 1, 215–239.

Frost, M., B. Linneker, and N. Spence. 1998. Excess or wasteful commuting in a selection of British cities. *Transportation Research Part A: Policy and Practice* 32, 529–538.

Gaile, G. L. 1980. The spread-backwash concept. *Regional Studies* 14, 15–25.

GAO. 1995. *Health Care Shortage Areas: Designation Not a Useful Tool for Directing Resources to the Underserved.* Washington, DC: GAO/HEHS-95-2000, General Accounting Office.

Gao, S., Y. Wang, Y. Gao, and Y. Liu. 2013. Understanding urban traffic-flow characteristics: A rethinking of betweenness centrality. *Environment and Planning B* 40, 135–153.

Garin, R. A. 1966. A matrix formulation of the Lowry model for intrametropolitan activity allocation. *Journal of the American Institute of Planners* 32, 361–364.

Ge, Y., H. Xiong, A. Tuzhilin, K. Xiao, M. Gruteser, and M. Pazzani. 2010, July. An energy-efficient mobile recommender system. In proceedings of the 16th ACM SIGKDD International Conference on Knowledge Discovery and Data Mining, pp. 899–908.

Geary, R. 1954. The contiguity ratio and statistical mapping. *The Incorporated Statistician* 5, 115–145.

Getis, A., and D. A. Griffith. 2002. Comparative spatial filtering in regression analysis. *Geographical Analysis* 34, 130–140.

Getis, A., and J. K. Ord. 1992. The analysis of spatial association by use of distance statistics. *Geographical Analysis* 24, 189–206.

Ghosh, A., and S. McLafferty. 1987. *Location Strategies for Retail and Service Firms.* Lexington, MA: D. C. Heath.

Gibbons, D. C. 1977. *Society, Crime and Criminal Careers.* Englewood Cliffs, NJ: Prentice-Hall.

Giuliano, G., and K. A. Small. 1991. Subcenters in the Los Angeles region. *Regional Science and Urban Economics* 21, 163–182.

Giuliano, G., and K. A. Small. 1993. Is the journey to work explained by urban structure. *Urban Studies* 30, 1485–1500.

Glaser, D. 1971. *Social Deviance.* Chicago: Markham.

Goodchild, M. F., L. Anselin, and U. Deichmann. 1993. A framework for the interpolation of socioeconomic data. *Environment and Planning A* 25, 383–397.

Goodchild, M. F., and N. S.-N. Lam. 1980. Areal interpolation: A variant of the traditional spatial problem. *Geoprocessing* 1, 297–331.

Goodman, D. C., S. S. Mick, D. Bott, T. Stukel, C.-H. Chang, N. Marth, J. Poage, and H. J. Carretta. 2003. Primary care service areas: A new tool for the evaluation of primary care services. *Health Services Research* 38, 287–309.

Gordon, P., H. Richardson, and H. Wong. 1986. The distribution of population and employment in a polycentric city: The case of Los Angeles. *Environment and Planning A* 18, 161–173.

Grady, S. C., and H. Enander. 2009. Geographic analysis of low birthweight and infant mortality in Michigan using automated zoning methodology. *International Journal of Health Geographics* 8(10).

Greene, D. L., and J. Barnbrock. 1978. A note on problems in estimating exponential urban density models. *Journal of Urban Economics* 5, 285–290.

Griffith, D. A. 1981. Modelling urban population density in a multi-centered city. *Journal of Urban Economics* 9, 298–310.

Griffith, D. A., and C. G. Amrhein. 1997. *Multivariate Statistical Analysis for Geographers.* Upper Saddle River, NJ: Prentice Hall.

Grimson, R. C., and R. D. Rose. 1991. A versatile test for clustering and a proximity analysis of neurons. *Methods of Information in Medicine* 30, 299–303.

Gu, C., F. Wang, and G. Liu. 2005. The structure of social space in Beijing in 1998: A socialist city in transition. *Urban Geography* 26, 167–192.

Guagliardo, M. F. 2004. Spatial accessibility of primary care: Concepts, methods and challenges. *International Journal of Health Geography* 3, 3.

Guldmann, J. M., and F. Wang. 1998. Population and employment density functions revisited: A spatial interaction approach. *Papers in Regional Science* 77, 189–211.

Guo, D. 2008. Regionalization with dynamically constrained agglomerative clustering and partitioning (REDCAP). *International Journal of Geographical Information Science* 22, 801–823.

Guttman, A. 1984. R-trees: A dynamic index structure for spatial searching. Proceedings of the 1984 ACM SIGMOD International Conference on Management of Data, June, pp. 47–57.

Haining, R. F., S. Wises, and M. Blake. 1994. Constructing regions for small area analysis: Material deprivation and colorectal cancer. *Journal of Public Health Medicine* 16, 429–438.

Haining, R. F., S. Wise, and J. Ma. 2000. Designing and implementing software for spatial statistical analysis in a GIS environment. *Journal of Geographical Systems* 2, 257–286.

Halás, M., P. Klapka, and P. Kladivo. 2014. Distance-decay functions for daily travel-to-work flows. *Journal of Transport Geography* 35, 107–119.

Hamilton, B. 1982. Wasteful commuting. *Journal of Political Economy* 90, 1035–1053.

Hamilton, L. C. 1992. *Regression with Graphics*. Belmont, CA: Duxbury.

Hansen, W. G. 1959. How accessibility shapes land use. *Journal of the American Institute of Planners* 25, 73–76.

Harrell, A., and C. Gouvis. 1994. *Predicting Neighborhood Risk of Crime: Report to the National Institute of Justice*. Washington, DC: The Urban Institute.

Harris, C. D., and E. L. Ullman. 1945. The nature of cities. *The Annals of the American Academy of Political and Social Science* 242, 7–17.

Hartschorn, T. A. 1992. *Interpreting the City: An Urban Geography*. New York: Wiley.

Haynes, A. G., M. M. Wertli, and D. Aujesky. 2020. Automated delineation of hospital service areas as a new tool for health care planning. *Health Services Research* 55, 469–475.

Heikkila, E. P., P. Gordon, J. Kim, R. Peiser, H. Richardson, and D. Dale-Johnson. 1989. What happened to the CBD-distance gradient? Land values in a polycentric city. *Environment and Planning A* 21, 221–232.

Hewings, G. 1985. *Regional Input-Output Analysis*. Beverly Hills, CA: Sage.

Hillier, B. 1996. *Space Is the Machine: A Configurational Theory of Architecture*. Cambridge: Cambridge University Press.

Hillier, B., and J. Hanson. 1984. *The Social Logic of Space*. Cambridge: Cambridge University Press.

Hillier, B., A. Penn, and J. Hanson, et al. 1993. Natural movement: Or, configuration and attraction in urban pedestrian movement. *Environment and Planning B* 20, 29–66.

Hillsman, E., and G. Rushton. 1975. The *p*-median problem with maximum distance constraints. *Geographical Analysis* 7, 85–89.

Hirschi, T. 1969. *Causes of Delinquency*. Berkeley, CA: University of California Press.

Horner, M. W., and A. T. Murray. 2002. Excess commuting and the modifiable areal unit problem. *Urban Studies* 39, 131–139.

Hoyt, H. 1939. *The Structure and Growth of Residential Neighborhoods in American Cities*. Washington, DC: USGPO.

Hu, Y., H. J. Miller, and X. Li. 2014. Detecting and analyzing mobility hotspots using surface networks. *Transactions in GIS* 18(6), 911–935.

Hu, Y., and F. Wang. 2015. Decomposing excess commuting: A Monte Carlo simulation approach. *Journal of Transport Geography* 44, 43–52.

Hu, Y., F. Wang, C. Guin, and H. Zhu. 2018. A spatio-temporal kernel density estimation framework for predictive crime hotspot mapping and evaluation. *Applied Geography* 99, 89–97.

Huff, D. L. 1963. A probabilistic analysis of shopping center trade areas. *Land Economics* 39, 81–90.

Huff, D. L. 2000. Don't misuse the Huff model in GIS. *Business Geographies* 8, 12.

Huff, D. L. 2003. Parameter estimation in the Huff model. *ArcUser* 2003 (October–December), 34–36.

Immergluck, D. 1998. Job proximity and the urban employment problem: Do suitable nearby jobs improve neighborhood employment rates? *Urban Studies* 35, 7–23.

Institute of Transportation Engineers (ITS). 2012. *Trip Generation Manual* (9th ed.). Washington, DC: ITS.

Izakian, Z., M. S. Mesgari, and A. Abraham. 2016. Automated clustering of trajectory data using a particle swarm optimization. *Computers, Environment and Urban Systems* 55, 55–65.

Jacobs, J. 1961. *The Death and Life of Great American Cities*. New York: Random House.

Jacquez, G. M. 1998. GIS as an enabling technology. In A. C. Gatrell and M. Loytonen (eds.), *GIS and Health*. London: Taylor & Francis, 17–28.

Jia, P., F. Wang, and I. M. Xierali. 2017. Using a huff-based model to delineate hospital service areas. *Professional Geographer* 69, 522–530.

Jia, P., F. Wang, and I. M. Xierali. 2019. Differential effects of distance decay on hospital inpatient visits among subpopulations in Florida, USA. *Environmental Monitoring and Assessment* 191(2), 1–16.

Jiang, B., and C. Claramunt. 2004. Topological analysis of urban street networks. *Environment and Planning B* 31, 151–162.

Jiang, Y., and X. Li. 2013. Travel time prediction based on historical trajectory data. *Annals of GIS* 19(1), 27–35.

Jin, F., F. Wang, and Y. Liu. 2004. Geographic patterns of air passenger transport in China 1980–98: Imprints of economic growth, regional inequality and network development. *Professional Geographer* 56, 471–487.

Joseph, A. E., and P. R. Bantock. 1982. Measuring potential physical accessibility to general practitioners in rural areas: A method and case study. *Social Science and Medicine* 16, 85–90.

Joseph, A. E., and D. R. Phillips. 1984. *Accessibility and Utilization—Geographical Perspectives on Health Care Delivery*. New York: Happer & Row.

Joseph, M., L. Wang, and F. Wang. 2012. Using Landsat imagery and census data for urban population density modeling in Port-au-Prince, Haiti. *GIScience & Remote Sensing* 49, 228–250.

Kain, J. F. 2004. A pioneer's perspective on the spatial mismatch literature. *Urban Studies* 41, 7–32.

Kang, C., Y. Liu, and L. Wu. 2015. Delineating intra-urban spatial connectivity patterns by travel-activities: A case study of Beijing, China. 23rd International Conference on Geoinformatics, June, pp. 1–7.

Khan, A. A. 1992. An integrated approach to measuring potential spatial access to health care services. *Socio-Economic Planning Science* 26, 275–287.

Khisty, C. J., and B. K. Lall. 2003. *Transportation Engineering: An Introduction* (3rd ed.). Upper Saddle River, NJ: Prentice Hall.

Khumawala, B. M. 1973. An efficient algorithm for the p-median problem with maximum distance constraints. *Geographical Analysis* 5, 309–321.

Kincaid, D., and W. Cheney. 1991. *Numerical Analysis: Mathematics of Scientific Computing*. Belmont, CA: Brooks/Cole Publishing Co.

Knorr-Held, L. 2000. Bayesian modelling of inseparable space-time variation in disease risk. *Statistics in Medicine* 19, 2555–2567.

Knorr-Held, L., and G. Rasser. 2000. Bayesian detection of clusters and discontinuities in disease maps. *Biometrics* 56, 13–21.

Knox, P. 1987. *Urban Social Geography: An Introduction* (2nd ed.). New York: Longman.

Koenderink, J. J. 1984. The structure of images. *Biological Cybernetics* 50, 363–370.

Kotusevski, G., and K. A. Hawick. 2009. A review of traffic simulation software. In *Computational Science Technical Note CSTN-095*. Auckland, New Zealand: Massey University.

Krige, D. 1966. Two-dimensional weighted moving average surfaces for ore evaluation. *Journal of South African Institute of Mining and Metallurgy* 66, 13–38.

Kuai, X., and F. Wang. 2020. Global and localized neighborhood effects on public transit ridership in Baton Rouge, Louisiana. *Applied Geography* 124, 102338.

Kuby, M., S. Tierney, T. Roberts, and C. Upchurch. 2005. *A Comparison of Geographic Information Systems, Complex Networks, and Other Models for Analyzing Transportation Network Topologies*. NASA/CR-2005–213522.

Kulldorff, M. 1997. A spatial scan statistic. *Communications in Statistics: Theory and Methods* 26, 1481–1496.

Kulldorff, M. 1998. Statistical methods for spatial epidemiology: Tests for randomness. In A. C. Gatrell and M. Loytonen (eds.), *GIS and Health*. London: Taylor & Francis, 49–62.

Kwan, M-P. 2012. The uncertain geographic context problem. *Annals of the Association of American Geographers* 102, 958–968.

Ladd, H. F., and W. Wheaton. 1991. Causes and consequences of the changing urban form: Introduction. *Regional Science and Urban Economics* 21, 157–162.

Lam, N. S.-N., and K. Liu. 1996. Use of space-filling curves in generating a national rural sampling frame for HIV-AIDS research. *Professional Geographer* 48, 321–332.

Land, K. C., P. L. McCall, and L. E. Cohen. 1990. Structural covariates of homicide rates: Are there any in variances across time and social space? *American Journal of Sociology* 95, 922–963.

Land, K. C., P. L. McCall, and D. S. Nagin. 1996. A comparison of poisson, negative binomial, and semiparametric mixed poisson regression models: With empirical applications to criminal careers data. *Sociological Methods and Research* 24, 387–442.

Langford, I. H. 1994. Using empirical Bayes estimates in the geographical analysis of disease risk. *Area* 26, 142–149.

Lee, R. C. 1991. Current approaches to shortage area designation. *Journal of Rural Health* 7, 437–450.

Leslie, T. F., and B. J. Kronenfeld. 2011. The colocation quotient: A new measure of spatial association between categorical subsets of points. *Geographical Analysis* 43, 306–326.

Leung, Y., J. S. Zhang, and Z. B. Xu. 2000. Clustering by scale-space filtering. *IEEE Transactions on Pattern Analysis and Machine Intelligence* 22, 1396–1410.

Levitt, S. D. 2001. Alternative strategies for identifying the link between unemployment and crime. *Journal of Quantitative Criminology* 17, 377–390.

Li, M., X. Shi, X. Li, W. Ma, J. He, and T. Liu. 2019. Epidemic forest: A spatiotemporal model for communicable diseases. *Annals of the American Association of Geographers* 109(3) 812–836.

Li, M., F. Wang, M-P. Kwan, J. Chen, and J. Wang. 2022. Equalizing the spatial accessibility of emergency medical services in Shanghai: A trade-off perspective. *Computers, Environment and Urban Systems* 92, 101745.

Li, X., and H. Lin. 2006. Indexing network-constrained trajectories for connectivity-based queries. *International Journal of Geographical Information Science* 20(3), 303–328.

Li, X., J. Mango, J. Song, et al. 2021. XStar: A software system for handling taxi trajectory big data. *Computational Urban Science* 1, 17.

Li, X., F. Wang, and H. Yi. 2017. A two-step approach to planning new facilities towards equal accessibility. *Environment and Planning B: Urban Analytics and City Science* 44, 994–1011.

Liberman, N., Y. Trope, and E. Stephan. 2007. Psychological distance. In A. W. Kruglanski and E. T. Higgins (eds.), *Social Psychology: Handbook of Basic Principles* (2nd ed.). New York: Guilford Publications, 353–384.

Liu, L., J. Alford-Teaster, T. Onega, and F. Wang. 2023. Refining 2SVCA method for measuring telehealth accessibility of primary care physicians in Baton Rouge, Louisiana. *Cities* 138, 104364.

Liu, X., and H. A. Karimi. 2006. Location awareness through trajectory prediction. *Computers, Environment and Urban Systems* 30(6), 741–756.

Liu, Y., F. Wang, C. Kang, Y. Gao, and Y. Lu. 2014. Analyzing relatedness by toponym co-occurrences on web pages. *Transactions in GIS* 18, 89–107.

Liu, Y., F. Wang, Y. Xiao, and S. Gao. 2012. Urban land uses and traffic 'source-sink areas': Evidence from GPS-enabled taxi data in shanghai. *Landscape and Urban Planning* 106, 73–87.

Lorenz, E. N. 1995. *The Essence of Chaos*. Washington, DC: Taylor & Francis.

Louisiana Hospital Association. 2021. *Louisiana Hospital Association Membership Directory*. Available https://cdn.ymaws.com/www.lhaonline.org/resource/resmgr/about/2020_21LHADirectory.pdf.

Lowry, I. S. 1964. *A Model of Metropolis*. Santa Monica, CA: Rand Corporation.

Luo, J., L. Tian, L. Luo, H. Yi, and F. Wang. 2017. Two-step optimization for spatial accessibility improvement: A case study of health care planning in rural China. *BioMed Research International* 2017, 2094654.

Luo, L., S. McLafferty, and F. Wang. 2010. Analyzing spatial aggregation error in statistical models of late-stage cancer risk: A Monte Carlo simulation approach. *International Journal of Health Geographics* 9, 51.

Luo, W., and Y. Qi. 2009. An enhanced two-step floating catchment area (E2SFCA) method for measuring spatial accessibility to primary care physicians. *Health and Place* 15, 1100–1107.

Luo, W., and F. Wang. 2003. Measures of spatial accessibility to healthcare in a GIS environment: Synthesis and a case study in Chicago region. *Environment and Planning B: Planning and Design* 30, 865–884.

Luo, W., J. Yao, R. Mitchell, X. Zhang, and W. Li. 2022. Locating emergency medical services to reduce urban-rural inequalities. *Socio-Economic Planning Sciences* 56, 101416.

Mandloi, D. 2009. *Partitioning Tools*. Redlands, CA: ESRI.

Marshall, R. J. 1991. Mapping disease and mortality rates using empirical Bayes estimators. *Applied Statistics* 40, 283–294.

McDonald, J. F. 1989. Econometric studies of urban population density: A survey. *Journal of Urban Economics* 26, 361–385.

McDonald, J. F., and P. Prather. 1994. Suburban employment centers: The case of Chicago. *Urban Studies* 31, 201–218.

McGrail, M. R., and J. S. Humphreys. 2009a. Measuring spatial accessibility to primary care in rural areas: Improving the effectiveness of the two-step floating catchment area method. *Applied Geography* 29, 533–541.

McGrail, M. R., and J. S. Humphreys. 2009b. A new index of access to primary care services in rural areas. *Australian and New Zealand Journal of Public Health* 33, 418–423.

McNally, M. G. 2008. The four step model. In D. A. Hensher and K. J. Button (eds.), *Transport Modelling* (2nd ed.). Pergamon: Elsevier Science.

Messner, S. F., L. Anselin, R. D. Baller, D. F. Hawkins, G. Deane, and S. E. Tolnay. 1999. The spatial patterning of county homicide rates: An application of exploratory spatial data analysis. *Journal of Quantitative Criminology* 15, 423–450.

Mills, E. S. 1972. *Studies in the Structure of the Urban Economy*. Baltimore: Johns Hopkins University.

Mills, E. S., and J. P. Tan. 1980. A comparison of urban population density functions in developed and developing countries. *Urban Studies* 17, 313–321.

Mitchell, A. 2005. *The ESRI Guide to GIS Analysis Volume 2: Spatial Measurements & Statistics*. Redlands, CA: ESRI Press.

Moran, P. A. P. 1950. Notes on continuous stochastic phenomena. *Biometrika* 37, 17–23.

Morenoff, J. D., and R. J. Sampson. 1997. Violent crime and the spatial dynamics of neighborhood transition: Chicago, 1970–1990. *Social Forces* 76, 31–64.

Mu, L., and F. Wang. 2008. A scale-space clustering method: Mitigating the effect of scale in the analysis of zone-based data. *Annals of the Association of American Geographers* 98, 85–101.

Mu, L., F. Wang, V. W. Chen, and X. Wu. 2015. A place-oriented, mixed-level regionalization method for constructing geographic areas in health data dissemination and analysis. *Annals of the Association of American Geographers* 105, 48–66.

Muth, R. 1969. *Cities and Housing*. Chicago: University of Chicago.

Nakanishi, M., and L. G. Cooper. 1974. Parameter estimates for multiplicative competitive interaction models—Least square approach. *Journal of Marketing Research* 11, 303–311.

Newling, B. 1969. The spatial variation of urban population densities. *Geographical Review* 59, 242–252.

Niedercorn, J. H., and B. V. Bechdolt Jr. 1969. An economic derivation of the "gravity law" of spatial interaction. *Journal of Regional Science* 9, 273–282.

Oden, N., G. Jacquez, and R. Grimson. 1996. Realistic power simulations compare point- and area-based disease cluster tests. *Statistics in Medicine* 15, 783–806.

O'Kelly, M. E., W. Song, and G. Shen. 1995. New estimates of gravitational attraction by linear programming. *Geographical Analysis* 27, 271–285.

Onega, T., E. J. Duell, X. Shi, D. Wang, E. Demidenko, and D. Goodman. 2008. Geographic access to cancer care in the U. S. *Cancer* 112, 909–918.

Openshaw, S. 1977. A geographical solution to scale and aggregation problems in region-building, partitioning, and spatial modelling. *Transactions of the Institute of British Geographers NS* 2, 459–472.

Openshaw, S. 1984. *Concepts and Techniques in Modern Geography, Number 38. The Modifiable Areal Unit Problem*. Norwich: Geo Books.

Openshaw, S., M. Charlton, C. Mymer, and A. W. Craft. 1987. A mark 1 geographical analysis machine for the automated analysis of point data sets. *International Journal of Geographical Information Systems* 1, 335–358.

Openshaw, S., and L. Rao. 1995. Algorithms for reengineering 1991 census geography. *Environment & Planning A* 27, 425–446.

Osgood, D. W. 2000. Poisson-based regression analysis of aggregate crime rates. *Journal of Quantitative Criminology* 16, 21–43.

Osgood, D. W., and J. M. Chambers. 2000. Social disorganization outside the metropolis: An analysis of rural youth violence. *Criminology* 38, 81–115.

O'Sullivan, D., and G. Perry. 2013. *Spatial Simulation: Exploring Pattern and Process*. West Sussex, UK: John Wiley.

Parr, J. B. 1985. A population density approach to regional spatial structure. *Urban Studies* 22, 289–303.

Parr, J. B., and G. J. O'Neill. 1989. Aspects of the lognormal function in the analysis of regional population distribution. *Environment and Planning A* 21, 961–973.

Parr, J. B., G. J. O'Neill, and A. G. M. Nairn. 1988. Metropolitan density functions: A further exploration. *Regional Science and Urban Economics* 18, 463–478.

Paskett, E. D., and R. A. Hiatt. 2018. Catchment areas and community outreach and engagement: The new mandate for NCI-designated cancer centers. *Cancer Epidemiology, Biomarkers & Prevention* 27, 517–519.

Peano, G. 1890. Sur une courbe qui remplit toute en aire plaine [On a curve which completely fills a planar surface]. *Mathematische Annalen* 36, 157–160.

Peng, Z. 1997. The jobs-housing balance and urban commuting. *Urban Studies* 34, 1215–1235.

Pfoser, D., and Y. Theodoridis. 2003. Generating semantics-based trajectories of moving objects. *Computers, Environment and Urban Systems* 27(3), 243–263.

Press, W. H., et al. 1992a. *Numerical Recipes in FORTRAN: The Art of Scientific Computing* (2nd ed.). Cambridge: Cambridge University Press.

Press, W. H., et al. 1992b. *Numerical Recipes in C: The Art of Scientific Computing* (2nd ed.). Cambridge: Cambridge University Press.

Press, W. H., et al. 2002. *Numerical Recipes in C++: The Art of Scientific Computing* (2nd ed.). Cambridge: Cambridge University Press.

Price, M. H. 2020. *Mastering ArcGIS Pro.* New York: McGraw-Hill.

Porta, S., V. Latora, F. Wang, et al. 2012. Street centrality and location of economic activities in Barcelona. *Urban Studies* 49, 1471–1488.

Poulter, S. R. 1998. Monte Carlo simulation in environmental risk assessment: Science, policy and legal issues. *Risk: Health, Safety and Environment* 9, 7–26.

Rees, P. 1970. Concepts of social space: Toward an urban social geography. In B. Berry and F. Horton (eds.), *Geographic Perspectives on Urban System.* Englewood Cliffs, NJ: Prentice Hall, 306–394.

Reichardt, J., and S. Bornholdt. 2006. Statistical mechanics of community detection. *Physical Review E* 74, 1–14.

Reilly, W. J. 1931. *The Law of Retail Gravitation.* New York: Knickerbocker.

Rengert, G. F., A. R. Piquero, and P. R. Jones. 1999. Distance decay reexamined. *Criminology* 37, 427–445.

ReVelle, C. S., and R. Swain. 1970. Central facilities location. *Geographical Analysis* 2, 30–34.

Revelle, W. 2022. *How To: Use the Psych Package for Factor Analysis and Data Reduction.* Available https://cran.r-project.org/web/packages/psychTools/vignettes/factor.pdf.

Ripley, B. D. 1977. Modelling spatial patterns. *Journal of the Royal Statistical Society. Series B (Methodological)* 39, 172–212.

Robinson, W. S. 1950. Ecological correlations and the behavior of individuals. *American Sociological Review* 15, 351–357.

Roos, N. P. 1993. Linking patients to hospitals: Defining urban hospital service populations. *Medical Care* 31(5), YS6–YS15.

Rose, H. M., and P. D. McClain. 1990. *Race, Place, and Risk: Black Homicide in Urban America.* Albany, NY: SUNY Press.

Rushton, G., and P. Lolonis. 1996. Exploratory spatial analysis of birth defect rates in an urban population. *Statistics in Medicine* 7, 717–726.

Schroeder, J. P. 2007. Target-density weighting interpolation and uncertainty evaluation for temporal analysis of census data. *Geographical Analysis* 39, 311–335.

Schultz, G. W., and W. G. Allen, Jr. 1996. Improved modeling of non-home-based trips. *Transportation Research Record* 1556, 22–26.

Scott, D., and M. Horner. 2008. Examining the role of urban form in shaping people's accessibility to opportunities: An exploratory spatial data analysis. *Journal of Transport and Land Use* 1, 89–119.

Sevtsuk, A., M. Mekonnen, and R. Kalvo. 2013. *Urban Network Analysis: A Toolbox v1.01 for ArcGIS.* Available http://cityform.mit.edu/projects/urban-network-analysis.html.

Shao, Y., and W. Luo. 2022. Supply-demand adjusted two-steps floating catchment area (SDA-2SFCA) model for measuring spatial access to health care. *Social Science & Medicine* 296, 114727.

Shao, Y., and W. Luo. 2023. Enhanced Two-Step Virtual Catchment Area (E2SVCA) model to measure telehealth accessibility. *Computational Urban Science* 3, 16.

Shen, Q. 1994. An application of GIS to the measurement of spatial autocorrelation. *Computer, Environment and Urban Systems* 18, 167–191.

Shen, Q. 1998. Location characteristics of inner-city neighborhoods and employment accessibility of low-income workers. *Environment and Planning B: Planning and Design* 25, 345–365.

Shen, Q. 2000. Spatial and social dimensions of commuting. *Journal of the American Planning Association* 66, 68–82.

Sheps Center for Health Services Research. 2021. *Rural Hospital Closures: More Information.* Available www.shepscenter.unc.edu/programs-projects/rural-health/rural-hospital-closures/ (last accessed 6–24–2021).

Sherratt, G. 1960. A model for general urban growth. In C. W. Churchman and M. Verhulst (eds.), *Management Sciences: Models and Techniques.* Oxford: Pergamon Press.

Shevky, E., and W. Bell. 1955. *Social Area Analysis.* Stanford, CA: Stanford University.

Shevky, E., and M. Williams. 1949. *The Social Areas of Los Angeles.* Los Angeles: University of California.

Shi, X. 2009. A geocomputational process for characterizing the spatial pattern of lung cancer incidence in New Hampshire. *Annals of the Association of American Geographers* 99, 521–533.

Shi, X., J. Alford-Teaster, T. Onega, and D. Wang. 2012. Spatial access and local demand for major cancer care facilities in the United States. *Annals of the Association of American Geographers* 102, 1125–1134.

Shi, X., E. Duell, E. Demidenko, T. Onega, B. Wilson, and D. Hoftiezer. 2007. A polygon-based locally-weighted-average method for smoothing disease rates of small units. *Epidemiology* 18, 523–528.

Silverman, B. W. 1986. *Density Estimation for Statistics and Data Analysis.* London: Chapman and Hall.

Small, K. A., and S. Song. 1992. "Wasteful" commuting: A resolution. *Journal of Political Economy* 100, 888–898.

Small, K. A., and S. Song. 1994. Population and employment densities: Structure and change. *Journal of Urban Economics* 36, 292–313.

Sorensen, M. J., S. Bessen, J. Danford, C. Fleischer, and S. L. Wong. 2020. Telemedicine for surgical consultations—pandemic response or here to stay? A report of public perceptions. *Annals of Surgery* 272, e174–e180.

Stabler, J., and L. St. Louis. 1990. Embodies inputs and the classification of basic and nonbasic activity: Implications for economic base and regional growth analysis. *Environment and Planning A* 22, 1667–1675.

Taaffe, E. J., H. L. Gauthier, and M. E. O'Kelly. 1996. *Geography of Transportation* (2nd ed.). Upper Saddle River, NJ: Prentice Hall.

Talbot, T. O., M. Kulldorff, S. P. Forand, and V. B. Haley. 2000. Evaluation of spatial filters to create smoothed maps of health data. *Statistics in Medicine* 19, 2399–2408.

Talen, E. 2001. School, community, and spatial equity: An empirical investigation of access to elementary schools in West Virginia. *Annals of the Association of American Geographers* 91, 465–486.

Talen, E., and L. Anselin. 1998. Assessing spatial equity: An evaluation of measures of accessibility to public playgrounds. *Environment and Planning A* 30, 595–613.

Tanner, J. 1961. *Factors Affecting the Amount of Travel*, Road Research Technical Paper No. 51. London: HMSO.

Tao, Z., Y. Cheng, and T. Dai. 2014. Spatial optimization of residential care facility locations in Beijing, China: Maximum equity in accessibility. *International Journal of Health Geographics* 13, 1–11.

Taylor, P. J. 1983. Distance decay in spatial interactions. In The Study Group in Quantitative Methods, of the Institute of British Geographers (eds.), *Concepts and Techniques in Modern Geography.* Norwich, England: Geo Books.

Tiwari, C., and G. Rushton. 2004. Using spatially adaptive filters to map late stage colorectal cancer incidence in Iowa. In P. Fisher (ed.), *Developments in Spatial Data Handling.* New York: Springer-Verlag US, 665–676.

Tobler, W. R. 1970. A computer movie simulating urban growth in the Detroit region. *Economic Geography* 46, 234–240.

Toregas, C., and C. S. ReVelle. 1972. Optimal location under time or distance constraints. *Papers of the Regional Science Association* 28, 133–143.

Torrens, P. M. 2006. Simulating sprawl. *Annals of the Association of American Geographers* 96, 248–275.

Torrens, P. M., A. Nara, X. Li, H. Zhu, W. A. Griffin, and S. B. Brown. 2012. An extensible simulation environment and movement metrics for testing walking behavior in agent-based models. *Computers, Environment and Urban Systems* 36(1), 1–17.

Toshniwal, A., S. Taneja, A. Shukla, K. Ramasamy, J. M. Patel, S. Kulkarni, . . . D. Ryaboy. 2014. Storm@ twitter. Proceedings of the 2014 ACM SIGMOD International Conference on Management of Data, June, pp. 147–156.

Traag, V. A., L. Waltman, and N. J. Van Eck. 2019. From Louvain to Leiden: Guaranteeing well-connected communities. *Scientific Reports* 9, 1–12.

Turner, A., and A. Penn. 2002. Encoding natural movement as an agent-based system: An investigation into human pedestrian behaviour in the built environment. *Environment and Planning B* 29, 473–490.

Ullman, E. L., and M. Dacey. 1962. The minimum requirements approach to the urban economic base. In *Proceedings of the IGU Symposium in Urban Geography*. Lund: Lund Studies in Geography, 121–143.

U.S. Census Bureau. 2022. *Big Data*. Available www.census.gov/topics/research/big-data.html#:~:text=Big%20data%20is%20a%20term,from%20sources%20other%20than%20 surveys [last accessed 1–6–2023].

Von Thünen, J. H. 1966. In C. M. Wartenberg (trans.) and P. Hall (Ed.), *Von Thünen's Isolated State*. Oxford: Pergamon.

Wang, C., F. Wang, and T. Onega. 2021. Network optimization approach to delineating health care service areas: Spatially constrained Louvain and Leiden algorithms. *Transactions in GIS* 25, 1065–1081.

Wang, F. 1998. Urban population distribution with various road networks: A simulation approach. *Environment and Planning B: Planning and Design* 25, 265–278.

Wang, F. 2000. Modeling commuting patterns in Chicago in a GIS environment: A job accessibility perspective. *Professional Geographer* 52, 120–133.

Wang, F. 2001a. Regional density functions and growth patterns in major plains of China 1982–90. *Papers in Regional Science* 80, 231–240.

Wang, F. 2001b. Explaining intraurban variations of commuting by job accessibility and workers' characteristics. *Environment and Planning B: Planning and Design* 28, 169–182.

Wang, F. 2003. Job proximity and accessibility for workers of various wage groups. *Urban Geography* 24, 253–271.

Wang, F. 2005. Job access and homicide patterns in Chicago: An analysis at multiple geographic levels based on scale-space theory. *Journal of Quantitative Criminology* 21, 195–217.

Wang, F. 2006. *Quantitative Methods and Applications in GIS*. Boca Raton, FL: CRC Press.

Wang, F. 2012. Measurement, optimization, and impact of healthcare accessibility: A methodological review. *Annals of the Association of American Geographers* 102, 1104–1112.

Wang, F. 2015. *Quantitative Methods and Socioeconomic Applications in GIS* (2nd ed.). Boca Raton, FL: CRC Press.

Wang, F. 2018. Inverted two-step floating catchment area method for measuring facility crowdedness. *Professional Geographer* 70, 251–260.

Wang, F. 2021. From 2SFCA to i2SFCA: Integration, derivation and validation. *International Journal of Geographical Information Science* 35, 628–638.

Wang, F. 2022. Four methodological themes in computational spatial social science. In B. Li, X. Shi, A. Zhu, C. Wang and Lin (eds.), *New Thinking in GIScience*. Singapore: Springer, 275–282.

Wang, F., A. Antipova, and S. Porta. 2011. Street centrality and land use intensity in Baton Rouge, Louisiana. *Journal of Transport Geography* 19, 285–293.

Wang, F., and T. Dai. 2020. Spatial optimization and planning practice towards equal access of public services. *Journal of Urban and Regional Planning* 12, 28–40 (in Chinese).

Wang, F., and J. M. Guldmann. 1996. Simulating urban population density with a gravity-based model. *Socio-Economic Planning Sciences* 30, 245–256.

Wang, F., and J. M. Guldmann. 1997. A spatial equilibrium model for region size, urbanization ratio, and rural structure. *Environment and Planning A* 29, 929–941.

Wang, F., D. Guo, and S. McLafferty. 2012. Constructing geographic areas for cancer data analysis: A case study on late-stage breast cancer risk in Illinois. *Applied Geography* 35, 1–11.

Wang, F., Y. Hu, S. Wang, and X. Li. 2017. Local indicator of colocation quotient with a statistical significance test: Examining spatial association of crime and facilities. *Professional Geographer* 69, 22–31.

Wang, F., C. Liu, and Y. Xu. 2019. Mitigating the zonal effect in modeling urban population density functions by Monte Carlo simulation. *Environment and Planning B: Urban Analytics and City Science* 46, 1061–1078.

Wang, F., and W. Luo. 2005. Assessing spatial and nonspatial factors in healthcare access in Illinois: Towards an integrated approach to defining health professional shortage areas. *Health and Place* 11, 131–146.

Wang, F., S. McLafferty, V. Escamilla, and L. Luo. 2008. Late-stage breast cancer diagnosis and healthcare access in Illinois. *Professional Geographer* 60, 54–69.

Wang, F., and W. W. Minor. 2002. Where the jobs are: Employment access and crime patterns in Cleveland. *Annals of the Association of American Geographers* 92, 435–450.

Wang, F., and V. O'Brien. 2005. Constructing geographic areas for analysis of homicide in small populations: Testing the herding-culture-of-honor proposition. In F. Wang (ed.), *GIS and Crime Analysis*. Hershey, PA: Idea Group Publishing, 83–100.

Wang, F., and Q. Tang. 2013. Planning toward equal accessibility to services: A quadratic programming approach. *Environment and Planning B* 40, 195–212.

Wang, F., M. Vingiello, and I. Xierali. 2020a. Serving a segregated metropolitan area: Disparities in spatial accessibility of primary care in Baton Rouge, Louisiana. In Y. Lu and E. Delmelle (eds.), *Geospatial Technologies for Urban Health*. Cham: Springer, 75–94.

Wang, F., and C. Wang. 2022. *GIS-Automated Delineation of Hospital Service Areas*. Boca Raton, FL: CRC Press.

Wang, F., C. Wang, Y. Hu, J. Weiss, J. Alford-Teaster, and T. Onega. 2020b. Automated delineation of cancer service areas in northeast region of the United States: A network optimization approach. *Spatial and Spatio-Temporal Epidemiology* 33, 100338.

Wang, F., and Y. Xu. 2011. Estimating O-D matrix of travel time by google maps API: Implementation, advantages and implications. *Annals of GIS* 17, 199–209.

Wang, F., Y. Zeng, L. Liu, and T. Onega. 2023. Disparities in spatial accessibility of primary care in Louisiana: From physical to virtual accessibility. *Frontiers in Public Health* 11, 1154574.

Wang, F., L. Zhang, G. Zhang, and H. Zhang. 2014. Mapping and spatial analysis of multi-ethnic toponyms in Yunnan, China. *Cartography and Geographic Information Science* 41, 86–99.

Wang, F., and Y. Zhou. 1999. Modeling urban population densities in Beijing 1982–90: Suburbanisation and its causes. *Urban Studies* 36, 271–287.

Wang, J., C. Wang, X. Song, and V. Raghavan. 2017. Automatic intersection and traffic rule detection by mining motor-vehicle GPS trajectories. *Computers, Environment and Urban Systems* 64, 19–29.

Wang, Y., F. Wang, Y. Zhang, and Y. Liu. 2019. Delineating urbanization "source-sink" regions in China: Evidence from mobile app data. *Cities* 86, 167–177.

Watanatada, T., and M. Ben-Akiva. 1979. Forecasting urban travel demand for quick policy analysis with disaggregate choice models: A Monte Carlo simulation approach. *Transportation Research Part A* 13, 241–248.

Webber, M. J. 1973. Equilibrium of location in an isolated state. *Environment and Planning A* 5, 751–759.

Webster, R., and P. Burrough. 1972. Computer-based soil mapping of small areas from sample data II. Classification smoothing. *Journal of Soil Science* 23, 222–234.

Wegener, M. 1985. The Dortmund housing market model: A Monte Carlo simulation of a regional housing market. *Lecture Notes in Economics and Mathematical Systems* 239, 144–191.

Wei, L. Y., Y. Zheng, and W. C. Peng. 2012. Constructing popular routes from uncertain trajectories. Proceedings of the 18th ACM SIGKDD International Conference on Knowledge Discovery and Data Mining, August, pp. 195–203.

Weibull, J. W. 1976. An axiomatic approach to the measurement of accessibility. *Regional Science and Urban Economics* 6, 357–379.

Weisbrod, G. E., R. J. Parcells, and C. Kern. 1984. A disaggregate model for predicting shopping area market attraction. *Journal of Marketing* 60, 65–83.

Weisburd, D., and L. Green. 1995. Policing drug hot spots: The Jersey City drug market analysis experiment. *Justice Quarterly* 12, 711–735.

Weiner, E. 1999. *Urban Transportation Planning in the United States: An Historical Overview*. Westport, CT: Praeger.

Wheeler, J. O., et al. 1998. *Economic Geography* (3rd ed.). New York: John Wiley & Sons.

White, M. J. 1988. Urban commuting journeys are not "wasteful". *Journal of Political Economy* 96, 1097–1110.

Whittemore, A. S., N. Friend, B. W. Brown, and E. A. Holly. 1987. A test to detect clusters of disease. *Biometrika* 74, 631–635.

Williams, S., and F. Wang. 2014. Disparities in accessibility of public high schools in metropolitan Baton Rouge, Louisiana 1990–2010. Forthcoming in *Urban Geography*.

Wilson, A. G. 1967. Statistical theory of spatial trip distribution models. *Transportation Research* 1, 253–269.

Wilson, A. G. 1969. The use of entropy maximizing models in the theory of trip distribution, mode split and route split. *Journal of Transport Economics and Policy* 3, 108–126.

Wilson, A. G. 1974. *Urban and Regional Models in Geography and Planning*. London: Wiley.

Wilson, A. G. 1975. Some new forms of spatial interaction models: A review. *Transportation Research* 9, 167–179.

Wong, D. W. S., and J. Lee. 2005. *Statistical Analysis and Modeling of Geographic Information*. Hoboken, NJ: John Wiley & Sons.

Wong, D. W. S., and F. Wang. 2018. Spatial analysis methods. In B. Huang (ed.), *Comprehensive Geographic Information Systems*, vol. 1. Oxford: Elsevier, 125–147.

Wu, N., and R. Coppins. 1981. *Linear Programming and Extensions*. New York: McGraw-Hill.

Xiao, Y., F. Wang, Y. Liu, and J. Wang. 2013. Reconstructing city attractions from air passenger flow data in China 2001–2008: A PSO approach. *Professional Geographer* 65, 265–282.

Xie, Y. 1995. The overlaid network algorithms for areal interpolation problem. *Computer, Environment and Urban Systems* 19, 287–306.

Xu, T., X. Li, and C. Claramunt. 2018. Trip-oriented travel time prediction (TOTTP) with historical vehicle trajectories. *Frontiers of Earth Science* 12, 253–263.

Xu, T., X. Zhang, C. Claramunt, and X. Li. 2018. TripCube: A trip-oriented vehicle trajectory data indexing structure. *Computers, Environment and Urban Systems* 67, 21–28.

Yang, D., R. Goerge, and R. Mullner. 2006. Comparing GIS based methods of measuring spatial accessibility to health services. *Journal of Medical Systems* 30, 23–32.

Zaharia, M., R. S. Xin, P. Wendell, T. Das, M. Armbrust, A. Dave, . . . I. Stoica. 2016. Apache spark: A unified engine for big data processing. *Communications of the ACM* 59, 56–65.

Zhang, X., H. Lu, and J. B. Holt. 2011. Modeling spatial accessibility to parks: A national study. *International Journal of Health Geographics* 10, 31.

Zhang, Y., H. Yang, and J. Pan. 2019. Maximum equity in access of top-tier general hospital resources in China: A spatial optimisation model. *The Lancet* 394, S90.

Zhang, Z., D. Chen, W. Liu, J. S. Racine, S. Ong, Y. Chen, et al. 2011. Nonparametric evaluation of dynamic disease risk: A spatio-temporal kernel approach. *PLoS One* 6, e17381.

Zheng, X.-P. 1991. Metropolitan spatial structure and its determinants: A case study of Tokyo. *Urban Studies* 28, 87–104.

Zhou, Y., Y. Zhang, Y. Ge, Z. Xue, Y. Fu, D. Guo, J. Shao, T. Zhu, X. Wang, and J. Li. 2017. An efficient data processing framework for mining the massive trajectory of moving objects. *Computers, Environment and Urban Systems* 61, 129–140.

Zhu, H., and F. Wang. 2021. An agent-based model for simulating urban crime with improved daily routines. *Computers, Environment and Urban Systems* 89, 101680.

Zipf, G. K. 1949. *Human Behavior and the Principle of Least Effort.* Cambridge, MA: Addison-Welsey.

Index

For Product Safety Concerns and Information please contact our EU
representative GPSR@taylorandfrancis.com
Taylor & Francis Verlag GmbH, Kaufingerstraße 24, 80331 München, Germany

www.ingramcontent.com/pod-product-compliance
Lightning Source LLC
Chambersburg PA
CBHW060745220326
41598CB00022B/2337